Cosmetic Creams

Cosmetic Creams

Development, Manufacture and Marketing of Effective Skin Care Products

Wilfried Rähse

WILEY-VCH

Author

Dr. Wilfried Rähse
Dahlenstr. 168
40589 Düsseldorf
Germany

Cover Images: © keng88/Shutterstock,
© Arthur S. Aubry/Getty Images

Library of Congress Card No.:
applied for

British Library Cataloguing-in-Publication Data
A catalogue record for this book is available from the British Library.

Bibliographic information published by the Deutsche Nationalbibliothek
The Deutsche Nationalbibliothek lists this publication in the Deutsche Nationalbibliografie; detailed bibliographic data are available on the Internet at <http://dnb.d-nb.de>.

Print ISBN: 978-3-527-34398-0
ePDF ISBN: 978-3-527-81243-1
ePub ISBN: 978-3-527-81245-5
oBook ISBN: 978-3-527-81221-9

Typesetting SPi Global, Chennai, India
Printing and Binding

10 9 8 7 6 5 4 3 2 1

Contents

Preface *xi*
Prior Publications *xiii*

1 General and Legal Aspects of Cosmetics *1*
1.1 Short Look at the History of Cosmetics *1*
1.2 Definition of Cosmetics *2*
1.3 Typical Cosmetic Products *3*
1.4 Legal Regulations of Cosmetics in Europe *4*
1.5 Label Lettering and Trademark *7*
1.6 Mandatory Registration of Cosmetic Products *11*
1.7 Databases for Ingredients *11*
1.8 Regulations in the United States *15*
1.9 Regulations of the Cosmetics Markets in Asia *16*
1.10 Delimitation of Cosmetic Products *18*
1.11 Learnings *23*
References *23*

2 Economic Importance of Cosmetics *27*
2.1 Cosmetics Market and Distribution Channels in Germany *27*
2.2 Shopping on the Internet in Germany *30*
2.3 European Cosmetics Markets *31*
2.4 Cosmetics Market in USA *35*
2.5 Cosmetics Market and Distribution Channels in Japan *35*
2.6 Chinese Cosmetics Market *37*
2.7 World Division in Market Regions *38*
2.8 Global Cosmetics Market Size *39*
2.9 Trends for Future Development *45*
2.10 Largest Cosmetics Manufacturers Worldwide *45*
2.11 Top Five Manufacturers *48*
2.12 Learnings *56*
References *56*

3 Cost Structure of the Cosmetic Products and Their Manufacturers *59*
3.1 Rough Calculation of the Costs *59*
3.1.1 Overview *59*
3.1.2 Development Products *61*
3.1.3 Determination of the Market Price *61*
3.2 Detailed Calculation of the Manufacturing Costs *62*
3.2.1 Costs in the Production *62*
3.2.2 Production Costs Related to Installation and Building *65*
3.2.3 Labor Costs *65*
3.2.4 Energy Costs *71*
3.2.5 Production Cost Dependencies of Capacity, Operation, and Personnel *73*
3.2.6 Raw Material Costs *76*
3.3 Costs in the Companies *79*
3.3.1 Internal Cost Accounting *80*
3.3.2 Direct Costs *81*
3.3.3 Overheads in the Company's Cost and Performance Accounting *82*
3.4 Figures from the Published Annual Reports *84*
3.4.1 Industry-Dependent Cost Structures of the Companies *84*
3.4.2 Profit and Loss Accounts of Global Cosmetics Companies *86*
3.5 Methods for Pricing *91*
3.5.1 Pricing Depending on the Customer *91*
3.5.2 Cost-Oriented Pricing *92*
3.5.3 Demand-Oriented Pricing *93*
3.5.4 Competition-Oriented Pricing *95*
3.5.5 Influence of the Brand *97*
3.5.6 Summary of Pricing *98*
3.6 Learnings *98*
References *99*

4 Scientific Descriptions of the Skin *103*
4.1 Tasks of the Skin *103*
4.2 Structure of the Skin *103*
4.3 Concepts for Penetration of the Stratum Corneum *108*
4.4 Some Experiments on the Penetration of Lipophilic Substances *114*
4.5 Penetration of Agents into the Skin *120*
4.6 Gender Differences in the Structure of the Skin *126*
4.7 Learnings *127*
References *127*

5 Composition of Creams for Skin Care *131*
5.1 General Structure of a Skin Care Cream *131*
5.2 Modules of a Cream *132*
5.3 Excipients *140*
5.3.1 Emulsifier for Macroemulsions *140*
5.3.2 Emulsifier for Mini (Nano) Emulsions *149*

5.3.3 Stability of Emulsions *149*
5.3.4 Adjusting the Cream Consistency *155*
5.3.5 Preservations *158*
5.3.6 Antioxidants, Complexing Agents, and Buffer Substances *163*
5.4 Additives for Color and Fragrance *167*
5.5 Aids Such as Liposomes for the Introduction of Substances into the Skin *167*
5.6 Learnings *170*
References *171*

6 Proven Active Ingredients for Various Categories of Skin Creams *175*
6.1 Skin Care *175*
6.2 Cream Categories for Skin Care *176*
6.2.1 Cosmetic Creams (Mainstream) *176*
6.2.2 Natural Cosmetics *176*
6.2.3 Cosmeceuticals *182*
6.2.4 Medicines for the Skin *184*
6.3 Moisture in the Skin *188*
6.3.1 Natural Moisturizing Factor *188*
6.3.2 Moisturizing Substances *191*
6.4 Vitalizing Substances, in Particular Vitamins *196*
6.5 Nourishing Vegetable Oils for Smoothing the Skin *201*
6.5.1 Natural Fatty Acids from Vegetable Oils *203*
6.5.2 Vegetable Oils and Fats in Cosmetic Creams *205*
6.6 Active Ingredients for Antiaging Creams *212*
6.7 Essential Oils *216*
6.8 Extracts from Plant Parts *219*
6.9 Active Ingredients from the Sea *223*
6.10 Origin of the Active Ingredients *225*
6.11 Learnings *225*
References *227*

7 Active Ingredients for Special Products *231*
7.1 Definition of Special Creams *231*
7.2 Antiacne Creams for Blemished Skin *231*
7.3 After-Sun Creams and Lotions for Reddened Skin *236*
7.4 Creams for Baby Skin *237*
7.5 Prophylaxis at Risk of Pressure Ulcers (Bedsore) *239*
7.6 Improving the Appearance of the Skin in the Case of Cellulite *239*
7.7 Chemical Removal of Unwanted Hair (Depilatory Cream) *241*
7.8 Treatment of Eczema *242*
7.9 Cream for the Feet and Against Athlete's Foot *243*
7.10 Cream for Hands *244*
7.11 Antiherpes Cream *245*
7.12 Cream for Removing Thick Horny Layers (Callus) *246*
7.13 Lotions for Body Care *247*

7.14 Cream for Itchy Shins *248*
7.15 Self-tanning Cream *248*
7.16 Sunscreens (UV Protection) *249*
7.16.1 Solar Radiation *249*
7.16.2 Character of UV Rays *250*
7.16.3 Radiation-Induced Damage in the Skin *252*
7.16.4 Sunscreen Substances According to the Cosmetics Regulations Worldwide *254*
7.16.5 Application and Warning Notices *261*
7.16.6 Measurements for the Determination of Sun Protection *265*
7.16.7 Recommended Active Ingredients *268*
7.16.8 Care Creams with Sun Protection *269*
7.17 Comment on Cosmeceuticals *271*
7.18 Learnings *271*
References *272*

8 Proposals for the Formulation of Creams *275*
8.1 General Remarks *275*
8.2 Moisturizers *276*
8.3 Vitalizing Creams *277*
8.4 Creams with Smoothing Properties *282*
8.5 Antiaging Creams *284*
8.6 Acne Creams *287*
8.7 After Sun Creams/Lotions *288*
8.8 Baby Cream *289*
8.9 Bedsore Cream *295*
8.10 Cellulite Cream *295*
8.11 Foot Care Cream Against Athlete's Foot *300*
8.12 Hand Cream *300*
8.13 Callus Removal Cream *300*
8.14 Body Lotion *301*
8.15 Eye Area Formulation with Sun Protection *304*
8.16 Sunscreen Lotion *307*

9 Perfumes *309*
9.1 Importance of the Perfume for Cosmetic Creams *309*
9.2 History of Perfume Oils *310*
9.3 Perfume Composition and Markets *312*
9.4 Extraction of Fragrances from Plants *315*
9.5 Chemical Composition of Natural Fragrances *319*
9.6 Possibilities in Product Design of Perfume Oils *327*
9.7 Personal Care and Other Products *329*
9.8 Safety *330*
9.9 Learnings *332*
References *333*

10 Production of Cosmetic Creams *335*
10.1 Method *335*
10.2 Stirring and Homogenizing Tools *337*
10.3 Laboratory Equipment and Pilot Plant *342*
10.4 Batch Production *345*
10.5 Continuous Production *350*
10.6 Scale-up *350*
10.7 Mini-Emulsions *355*
10.8 Bottles and Filling Lines *361*
10.9 Learnings *366*
References *368*

11 Regulations and Guidelines for the Execution of Hygienic Productions *371*
11.1 Good Manufacturing Practice Rules for the Manufacture of Cosmetics *371*
11.2 EHEDG Guidelines for the Construction of the Facility *379*
11.3 Materials for the Equipment of Cosmetic Plants *383*
11.3.1 Problem *383*
11.3.2 Choice of Material *384*
11.3.3 Stainless Steel *385*
11.3.4 Smoothing the Metal Surfaces *387*
11.4 Cleaning-in-Place *392*
11.5 Learnings *396*
References *396*

12 Assessment of the Quality of Cosmetic Creams *399*
12.1 Options for Quality Evaluation *399*
12.2 Microbial Checks *400*
12.3 Specifying the Quality of Cosmetic Creams by Physical Measurements *403*
12.3.1 Probes and Devices *403*
12.3.2 Moisture and Sebum *403*
12.3.3 Firmness and Elasticity *406*
12.3.4 Wrinkles *408*
12.3.5 Gloss and Color *409*
12.3.6 Support of Advertising Claims *409*
12.4 Example of a Cream Test by Customers *410*
12.5 Learnings *414*
References *415*

13 Product Information File (P.I.F.) *417*
13.1 Provisions of the Cosmetics Regulation *417*
13.2 Requirements for the Product Safety Report According to the Cosmetics Regulation *417*

13.3 Safety Data Sheet *420*
13.4 Structure of the P.I.F. *422*
13.4.1 Product Identification and Description *422*
13.4.2 Composition of the Cosmetic Cream *422*
13.4.3 Toxicological Profile *424*
13.4.4 Production Instruction *426*
13.4.5 Analysis Report of the Produced Cream *426*
13.4.6 Cream Exposure to the Skin *429*
13.4.7 Safety Consideration for Babies and Children *434*
13.5 Example for a P.I.F. (Body Lotion) *437*
13.6 Learnings *452*
References *453*

Appendix A Formulations *455*

Appendix B MSDS Niacinamide *465*

Index *475*

Preface

For skin care, cosmetic creams are indispensable to many people worldwide. Above all, they provide protection against environmental influences, nourish the skin with moisturizing and revitalizing ingredients, and slow down the process of aging. In addition, there are many special products, such as creams for protection against UV rays or to combat acne or for the prevention of pressure ulcers. Therefore, the cosmetic creams have gained great economic importance. Worldwide sales after selling prices reached €92 bn comparable to the sale of detergents and cleaners (€100 bn, 2018). The increasing cream market accounts for about 25–30% of the total cosmetics market. The undisputed world leader is L'Oréal (France).

The field of cosmetics is exciting, diverse, and spans numerous fields of knowledge that requires teamwork. During the work at HENKEL, my idea to develop highly effective creams in the form of mini (nano) emulsions should expand the existing product range with a line for pharmacy sales. With this task, I started my work in cosmetics. I took over the product and process development and at the beginning the project management (later handed over to the marketing). The project members included a physician, a toxicologist, a researcher, a member of the analytical laboratory, and one person of the dermatological test group to determine the effect of cosmetics. Later, a pharmacist and a marketing member joined the project group. Over the years, as a Doctor of Chemistry with an education in chemical engineering and many years of industrial experience as well as due to close contact with the project members, I have learned a lot from the other fields of knowledge. Furthermore, I was able to contribute my knowledge about hygienic production, Good Manufacturing Practice (GMP) and European Hygienic Engineering and Design Group (EHEDG), from the field of biotechnological production (enzymes). After retiring, I helped a start-up company for a few years and created the recipes for cosmeceuticals and safety assessments.

With this book, I would like to pass on my knowledge to younger people. The wealth of knowledge allowed me to look at "cosmetic creams" from all angles, including the markets, pricing of products, macro- and mini-emulsions, ingredients, production processes, their materials, scale-up and guidelines, the physical and medical measurement methods, customer surveys, as well as the creation of safety assessments according to the European Cosmetics Regulations. Of course, the book focuses on the structure and formulation of a cream as well as the effect of each ingredient. As there are many thousands of ingredients, the discussed selection is largely based on own experiences, i.e. proven effects.

The reader must not expect that company recipes will be disclosed here. However, the formulations discussed are partly similar. However, all the described recipes represent free inventions and were neither manufactured nor tested for stability, approval, and toxicology. They should give suggestions for own developments based on natural substances. All formulations have in common that they contain sorbic acid as the preservative (FDA: Generally Recognized As Safe, GRAS) with a pH buffer from citric/citrate, the best tolerated preservative. All information is based on the best of knowledge and belief at the time of writing.

Acknowledgments

For positive contributions, I thank the former members of the working group "Nanoemulsions" at HENKEL, in particular Dr. med. Wolfgang Matthies. Gratefully, most of the mentioned industrial companies have provided images of their machines, apparatuses, and devices. A colleague, Ovid Dicoi, carried out some nuclear magnetic resonance (NMR) measurements and evaluations, integrated in the book (Figures 4.7–4.11). Professor Dr. Dr. h.c. mult. Bernhard Blümich, RWTH Aachen, provided the measuring devices.

My wife supported the two-year book project with great tolerance and patience. Thanks to all.

Düsseldorf, December 2018 *Wilfried Rähse*

Prior Publications

The content of this book is based on several articles, which were mostly published in German.

Rähse, W. and Dicoi, O. (2009). *Produktdesign disperser Stoffe: Emulsionen für die kosmetische Industrie. Chem. Ing. Tech.* 81 (9): 1369–1383.

Rähse, W. and Dicoi, O. (2010). *Produktdesign von Flüssigkeiten: Parfümöle in der Konsumgüterindustrie. Chem. Ing. Tech.* 82 (5): 583–599.

Rähse, W. (2011). *Produktdesign von Cosmeceuticals am Beispiel der Hautcreme. Chem. Ing. Tech.* 83 (10): 1651–1662.

Rähse, W. (2012). *Enzyme für Waschmittel. Chem. Ing. Tech.* 84 (12): 2152–2163.

Rähse, W. (2013). Design of skin care products, Chap. 10. In: *Product Design and Engineering*, vol. 3 (eds. U. Bröckel, W. Meier and G. Wagner), 273–313. Weinheim: Wiley-VCH.

Rähse, W. (2014). *Production of tailor-made enzymes for detergents. ChemBioEng Reviews* 1 (1): 27–39.

Rähse, W. (2014). *Industrial Design of Solids and Liquids, A Practical Guide.* Weinheim: Wiley-VCH.

Rähse, W. (2014). *Industrial product design: materials for the machinery. ChemBioEng Reviews* 1 (3): 117–132.

Rähse, W. (2014). Praktische Hinweise zur Wahl des Werkstoffs von Maschinen und Apparaten. *Chem. Ing. Tech.* 86 (8): 1163–1179.

Rähse, W. (2017). *Ermittlung eines kompetitiven Marktpreises für neue Produkte über die Herstellkosten. Chem. Ing. Tech.* 89 (9): 1142–1158.

Rähse, W. (2018). *Ökonomische Grundlagen der technischen Reaktionsführung.* In: *Handbuch Chemische Reaktoren.* Springer Reference Naturwissenschaften (ed. W. Reschetilowski). Berlin, Heidelberg: Springer Spektrum.

1

General and Legal Aspects of Cosmetics

1.1 Short Look at the History of Cosmetics

The term cosmetic, originated from the ancient Greek word meaning "order or decorate," refers to the body and beauty care. This includes the maintenance, restoration, and enhancement of the beauty of human body. The first sign of cosmetics dated back about 10 000 BCE. The Mesolithic people applied grease and castor oil to soften their skin. They painted tattoos with plant dyes. About 7000 years later (3000 BCE), Egyptian parchment described the use of creams to sooth the skin and reduce wrinkles [1]. In the ancient Near East, men applied oils to their hair and beard. Women used eye paints, rouge, powders, and ointments on their body. In 50 BCE, Cleopatra was not only known as a beauty but also known for her intensive use of cosmetics. She possessed many products from nature such as beeswax, honey, and natural oils as well as products made from fruits, vegetables, herbs, and seeds, besides eggs and milk. She bathed in goat milk for skin regeneration. At that time, there were mirrors, makeup, makeup containers, combs, wash dishes, wigs, as well as tweezers and blades for removal of unwanted hair. Vermilion and red ocher were used for coloring the lips and cheeks; henna was used for coloring the hair, skin, toenails, and fingernails; and the malachite green, gray galena, and finely ground antimony were used for eyes as an eyeliner. The Greek physician Galen (about 200 CE) developed the first cold cream from beeswax, olive oil, and water [1]. The Romans introduced communal baths for noble persons. In the Middle Ages, they used hair dye and makeup, in addition to natural skin care and herbal remedies. During those times, a pale complexion was considered beautiful. With white lead, they achieved a flawless pallor. This substance and other cosmetics are highly toxic and often caused abscesses that did not heal. During the Renaissance, the Venetians dyed their hair using plant colors, fixed with clay, and baked in the sun. When it came to Elizabeth I of England (about 1580) and Catherine de Medici in France, dyeing of the cheeks and lips became popular again. The red lip color came from cochineal, a red dye from the cochineal scale insect. In the eighteenth century, bismuth oxide, mercury oxide, tin oxide, and talc were used to whiten the skin. Red makeup for the lips and cheeks emerged from safflower, cochineal, redwood, sandalwood, and vermilion. In addition, the hair was treated with greasy pomades. The hair powder consisted mostly of wheat or rice starch, partly colored. At present, there are

Cosmetic Creams: Development, Manufacture and Marketing of Effective Skin Care Products, First Edition. Wilfried Rähse.

ranges of cosmetic products that have been tested for their safe use. The aim has not changed in thousands of years. Primarily, cosmetics mean increasing attractiveness by beautifying the body and face. For a closer look at the historical processes, the following references are suitable [2–4].

1.2 Definition of Cosmetics

The legal text of the European Cosmetics Regulation [5] defines what a cosmetic product is and forms the legal basis in the European Community (EC) for the delimitation to the medical and therapeutic agents. This is the text of (EC) No. 1223/2009, Article 1a (quote): "Cosmetic Product agent any substance or mixture intended to be placed in contact with the external parts of the human body (epidermis, hair system, nails, lips and external genital organs) or with the teeth and the mucous membranes of the oral cavity with a view exclusively or mainly to cleaning them, perfuming them, changing their appearance, protecting them, keeping them in good condition or correcting body odors."

A very similar definition is given in the German LFGB regulation of 2005 (Lebensmittel-, Bedarfsgegenstände-, und Futtermittelgesetzbuch [6]). Text of (GER) LFGB §2, Article 5 (translation): "Cosmetic agents are substances or mixtures of substances exclusively or predominantly intended to be applied externally to the body of the human being or in his oral cavity for cleansing, protecting, maintaining a good condition, for perfuming, changing the appearance or to influence the body odor. Cosmetics are not substances or mixtures of substances which are intended to influence the body forms."

Already, the revision of the Federal Food, Drug, and Cosmetic Act (FD&C Act) of 1938 described the intended use of these products in a wording [7, 8] that is valid until now: (USA, Food and Drug Administration [FDA]) …"(1) articles intended to be rubbed, poured, sprinkled, or sprayed on, introduced into, or otherwise applied to the human body or any part thereof for cleansing, beautifying, promoting attractiveness, or altering the appearance, …" Among the products included in this definition are skin moisturizers, perfumes, lipsticks, fingernail polishes, eye and facial makeup preparations, cleansing shampoos, permanent waves, hair colors, and deodorants, as well as any substance intended for use as a component of a cosmetic product.

Essentially, the purpose of applying cosmetics is to increase the attractiveness of the user. This goal is achieved with many cosmetic products: unobtrusive, barely perceptible in the daily cleaning of skin and hair as well as the teeth, or eye-catching for decorative cosmetics, hair styling or coloring. To maintain the good condition, the daily use of cosmetic products on the skin and in the oral cavity is necessary; this applies with restrictions also for the hair (nose and ears are not mentioned in the regulation). The main tasks of cosmetics are cleansing, beautifying, perfuming, protecting, and maintaining a good condition of skin, nails, hair, and the teeth.

Healing and elimination of skin damage, infections, and diseases are the tasks of the therapeutic agents and are subject to other legal regulations. However, many active ingredients that are often used in cosmetic creams form a potential for conflict because they have a therapeutic effect in addition to the cosmetic (see Section 1.10). For example, dexpanthenol is not only a good moisturizing agent but also accelerates healing of wounds (pharmacy product: Bepanthen). Hemp oil, a good skin oil from nature with a skin-like C-chain distribution, is known (in Russia) for its healing effect of inner throat problems. Natural jasmine oil from Egypt has a very pleasant scent. Inhaling of this oil resulted in a proven relaxation of the body and used to facilitate birth in earlier days. There are many substances that show two or three effects, of which at least one is not a part of cosmetics. For cosmetic products, these available effects are not allowed to be labeled on the packaging as an advantage.

1.3 Typical Cosmetic Products

The product groups, which belong to the cosmetics, comprise the areas of hands, nails, arms, armpits, feet, legs, body and hair, face, eyes, lips, mouth, and teeth as well as the external genital areas. A concise overview of products in categories of cosmetics, summarized after application, and some product examples can be found in Table 1.1. Skin creams, which are the main focus here, belong to the first category "Skin cleansing and care."

Table 1.1 Typical cosmetic products.

Category/ Application	Products	Examples
Skin cleansing and care	➢ Soap, cleansing milk, facial fluid, -foam, -oil, mask, cleaning water, -lotion, -oil, hair and body wash, perfume and bubble baths, shower gel, bathing accessories ➢ Eye and face cream, -lotion, -emulsion, moisturizer and antiaging, facial concentrate (serum), eye patches, body lotion and cream, hand and foot cream, gels, masks, lip balm, genital cream ➢ Shaving cream, shaving soap, aftershave ➢ Depilatories ➢ Sunscreen milk, sunscreen lotion, water-repellent lotion, repellents (insects)	
Dental and oral care	➢ Toothpaste with different promises, powder, gel, dental floss, toothbrush, mouthwash, tongue cleaning ➢ Dentures[a]: cleaning by toothpaste, tab or powder and adhesion with a special cream	

(Continued)

Table 1.1 (Continued)

Category/ Application	Products	Examples
Hair treatment	➢ Shampoo, dry shampoo, styling, conditioner, hair care spray, permanent wave, hair gel, fluid, cream, foam, hair oil, mask, spray, hair color, hair tint, powder, hair care, serum, balsam, wax, hair tonic, antidandruff, pomade, hair perfume	
Decorative cosmetics	➢ Makeup, -remover, rouge, powder, -cloth, foundation, concealer, highlighter, primer, mascara, eye shadow, eye gel, eyeliner, eye pencil, eye brow pencil, eyebrow gel, eyelash care, lipstick, lip gloss, lip contour pin, nail polish, -remover, nail top coat, artificial nails, foundations, brushes ➢ Self-tanners	
Scent, smell	➢ Eau de perfume, eau de toilette, eau de cologne, deodorant, anti-perspirant	

a) Medical device.
Source: Courtesy of Douglas.

1.4 Legal Regulations of Cosmetics in Europe

Cosmetic products must meet a number of legal requirements from the European Union (EU) and the national parliaments, before they can be marketed. In Europe, the development, production, and marketing of cosmetic products are regulated by the **Cosmetics Regulation (EC) No 1223/2009** [5]. In addition, in Germany, the Cosmetics Ordinance (Kosmetik Verordnung) [9] and the Food and Feed Code (LFGB) of 2005 govern the trade of cosmetic products (§ 2 (5), § 26–29) [6] and others. All must be observed. Supplementary EC directives exist for the production of cosmetics, namely the GMP (Good Manufacturing Practice) Guidelines and the EHEDG (European Hygienic Engineering and Design Group) Guidelines, which are discussed in Chapter 11.

The most important basis in all EC Member States is the aforementioned Cosmetics Regulation, which has entered into force on 11 January 2012 as the successor to Directive 76/768 EEC. The new version of the German Cosmetics Ordinance has been in force since 24 August 2014. It takes over the EC Regulation and additionally regulates the information obligation; the use of the German language, information, and treatment centers for poisoning; exceptions for importation; and sanctions in the case of violation of the regulation.

Table 1.2 Structure of the Regulation (EC) No 1223/2009 on cosmetic products.

Chap.	Titles	Articles
I	SCOPE, DEFINITIONS	1 Scope and objective; 2 Definitions
II	SAFETY, RESPONSIBILITY, FREE MOVEMENT	3 Safety; 4 Responsible person; 5 Obligations of responsible persons; 6 Obligations of distributors; 7 Identification within the supply chain; 8 Good manufacturing practice; 9 Free movement
III	SAFETY ASSESSMENT, PRODUCT INFORMATION FILE, NOTIFICATION	10 Safety assessment; 11 Product information file; 12 Sampling and analysis; 13 Notification
IV	RESTRICTIONS FOR CERTAIN SUBSTANCES	14 Restrictions for substances listed in the Annexes; 15 Substances classified as CMR substances; 16 Nanomaterials; 17 Traces of prohibited substances
V	ANIMAL TESTING	18 Animal testing
VI	CONSUMER INFORMATION	19 Labeling; 20 Product claims; 21 Access to information for the public
VII	MARKET SURVEILLANCE	22 In-market control, 23 Communication of serious undesirable effects; 24 Information on substances
VIII	NON-COMPLIANCE, SAFEGUARD CLAUSE	25 Non-compliance by the responsible person; 26 Non-compliance by distributors; 27 Safeguard clause; 28 Good administrative practices
IX	ADMINISTRATIVE COOPERATION	29 Cooperation between competent authorities; 30 Cooperation regarding verification of product information files
X	IMPLEMENTING MEASURES, FINAL PROVISIONS	31 Amendment of the Annexes; 32 Committee procedure; 33 Glossary of common ingredient names; 34 Competent authorities, poison control centres or assimilated entities; 35 Annual report on animal testing; 36 Formal objection against harmonized standards; 37 Penalties; 38 Repeal; 39 Transitional provisions; 40 Entry into force and date of application

Source: Data from Ref. [5].

The structure of the Cosmetics Directive is reproduced in Table 1.2. On the one hand, it determines the substance approval, describes in detail the prohibited and restricted use of substances, also of the dyes, ultraviolet (UV) filters and preservatives, all listed in the annexes (Table 1.3). On the other hand, the directive demands some quality checks, a qualified safety assessment of the formulation (see Chapter 13), and production according to the GMP standard (Chapter 11). The elaboration of a qualified safety assessment of the ingredients and the entire formulation needs an academically trained expert, who has gained knowledge by studying in a related field and having a maximum experience in this field.

For all cosmetic products, the current version of the Cosmetics Regulation in accordance with Article 11 requires a "Product Information File" (P.I.F.).

Table 1.3 Annexes of the Directive (EG) Nr. 1223/2009 on cosmetic products [5].

Annex	Titles	Subjects
I	COSMETIC PRODUCT SAFETY REPORT	PART A – Cosmetic product safety information; PART B – Cosmetic product safety assessment
II	LIST OF SUBSTANCES PROHIBITED IN COSMETIC PRODUCTS	1338 identified prohibited substances
III	LIST OF SUBSTANCES WHICH COSMETIC PRODUCTS MUST NOT CONTAIN EXCEPT SUBJECT TO THE RESTRICTIONS LAID DOWN	256 substances which may be used up to a maximum value
IV	LIST OF COLORANTS ALLOWED IN COSMETIC PRODUCTS	153 allowed colorants
V	LIST OF PRESERVATIVES ALLOWED IN COSMETIC PRODUCTS	57 preservatives which may be used up to a limit value
VI	LIST OF UV FILTERS ALLOWED IN COSMETIC PRODUCTS	28 substances of limited concentration in ready for use preparation
VII	SYMBOLS USED ON PACKAGING/CONTAINER	Period-after-opening; Date of minimum durability
VIII	LIST OF VALIDATED ALTERNATIVE METHODS TO ANIMAL TESTING	
IX		PART A – Repealed Directive with its successive amendments; PART B – List of time-limits for transposition into national law and application

This report must be elaborated by an expert and provided at the request of an authority. Quote of Article 11: "…When a cosmetic product is placed on the market, the responsible person shall keep a product information file for it. The product information file shall be kept for a period of ten years following the date on which the last batch of the cosmetic product was placed on the market…." The most important part of P.I.F. represents the Cosmetic Product Safety Report (CPSR) or Safety Assessment. If not included in the CPSR, a product description must be prefixed, and the production according to GMP guidelines must also be confirmed (Chapter 13). A promised, specific effect or performance requires proof of the effect. For example, it is necessary for sun creams to determine the sun protection factor (SPF), which indicates the reliable effect by a certified method (ISO Standard – ISO 24444-2010). The last point of the P.I.F. need not be mentioned further because cosmetic companies do not carry out animal testing. If this is exceptionally not the case, the cosmetic regulation provides accurate information.

1.5 Label Lettering and Trademark

Various information for the consumer must be given on the packaging. Table 1.4 contains the hints for a label that needs to be checked point by point. For example, the label must include a product description in keywords and a list of ingredients in the INCI nomenclature (International Nomenclature of Cosmetic Ingredients). In addition, an instruction for the intended use with possible hazard warnings (for example, restrictions on children) as well as information about the content and durability of the product should be listed. The Directive requires the easy-to-read indication of the responsible company (with address) on the packaging. If products are made according to the (EC) No 1223/2009, this guarantees free movement of cosmetic products within the European market and ensures a high level of protection for human health under normal or foreseeable conditions of use.

Figure 1.1 demonstrates how the label might look on a cosmetic jar, pot, or dispenser. Within the limited space on the label, all legal requirements must be met, with the font clearly legible. On the front of the packaging, the brand name and logo (trademark) in brand-typical colors are usually found, including the function and application of the content. Instructions, a brief description of the likely effect, ingredients, and further details of the product as well as the content (volume or mass) and the manufacturer details can be found on the back of the

Table 1.4 Labeling of cosmetic products according to the Cosmetics Regulation (EC) No 1223/2009.

General instructions ➢ Name and address of the manufacturer or of person responsible for marketing the product; ➢ The nominal contents at the time of packaging, by weight or by volume; ➢ Date of minimum durability indicated for products with a minimum durability of less than 30 months; ➢ Period of time after opening the package for which the product can be used; valid for products with a minimum durability of more than 30 months (indicated with the symbol representing an open pot); ➢ Function of the product and particular precautions for use; ➢ Batch number of manufacture; ➢ Perfume and aromatic compositions and their raw materials shall be referred to by the terms "parfum" or "aroma"; ➢ Special rules for nanomaterials; ➢ Serious product claims;
Some countries require additional statements such as ➢ Storage: 4–22 °C; ➢ No animal tests
List of ingredients: INCI = International Nomenclature of Cosmetic Ingredients ➢ Order of the ingredients according to their mass proportions (highest percentage first) ➢ Ingredients less than 1% in any order ➢ INCI-specification: Parfum or Aroma; however, 26 fragrance allergens must be declared from a certain amount (Section 9.8) ➢ The CI-number specifies the dyes

CI, color index.

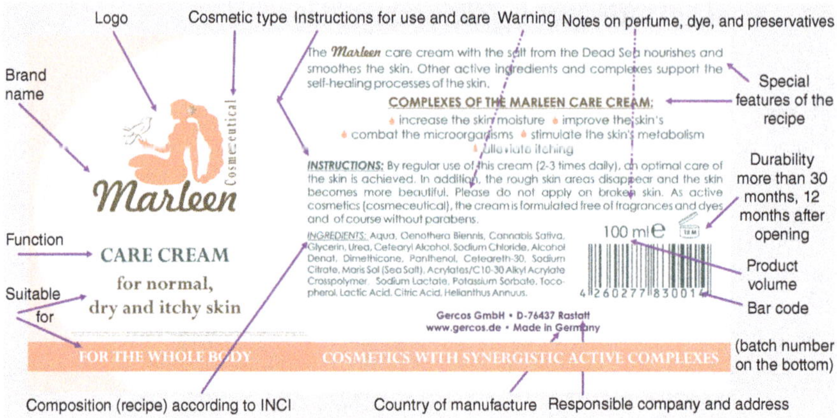

Figure 1.1 Lettering of the label according to the cosmetics regulation; this label is glued on a round dispenser.

packaging. The company name and postal address are sufficient to identify the person responsible for placing the goods on the market. The Internet address (website) provides further information about the product, its application, and the ingredients. Through the imprint of the website, the name of the managing director, the exact address of the company, and the telephone number can be found.

Furthermore, a batch number and the indication of the minimum durability are required on the packaging (e.g. best used before the end of March 2021), except for cosmetic products with a minimum durability of more than 30 months. For such products, an indication of the period after the opening must be given for which the product is safe and can be used without any harm to the consumer. This instruction requires the use of a picture from the Cosmetics Regulation (Annex VII, 2). The image shows an open pot with a figure that indicates the durability in months as shown in Figure 1.1.

In addition to the Cosmetics Ordinance, other laws must also be observed for the information on packaging and for the marketing. Especially in Germany, the Remedies Advertising Act ("Heilmittelwerbegesetz" HWG from 1965, last updated 2015 [10]) restricts advertising claims. This Directive prohibits statements on the label for the detection, elimination, or alleviation of illnesses; ailments; body damage; or morbid complaints. In addition, Directive 2005/29/EC of the European Parliament and of the Council (11 May 2005), concerning unfair business-to-consumer commercial practices in the internal market, provides that the Community works according to the same regulations and contribute to a high level of consumer protection.

New products or product groups sometimes require a new brand. Searching and finding a new brand name and an excellent logo is not easy. The brand should be memorable, easy to pronounce, and distinctive, even in other languages, and should be clearly and positively different from the competition. After finding a good name or a letter combination with logo, a national application is recommended at the Patent and Trademark Office (in Germany: DPMA [11]) or even a European application at the EUIPO (European Union Intellectual Property Office in Alicante, Spain [12]). Protected trademarks may not be used by competitors in the respective countries. The new mark name must be free, i.e. without registration in the own business area. The use of an already registered trade name or a very similar name in the same business area may result in a chargeable warning of the name holder. The owner prohibits the use of his brand name and demands the removal of all market products with this name. The withdrawal of products may result in high costs, which can be avoided. In order to ensure whether a brand name has already been given, it is worth taking a look at the national trademark registries [11], before printing the labels.

In most cases, brands consist of words and/or images or image with letters. However, they can also be three-dimensional, color, and sound brands. The definition of a trademark can be found in Article 2, Directive 2008/95/EC of the European Parliament and of the Council [13]. Quote: "A trade mark may consist of any signs capable of being represented graphically, particularly words, including personal names, designs, letters, numerals, the shape of goods or of their packaging, provided that such signs are capable of distinguishing the goods or services of one undertaking from those of other undertakings."

The national brand register, here the German Patent and Trademark Office (DPMA), precisely shows which mark names and logos are registered and protected in Germany, in the EU, and worldwide. Less significant is the search at the EUIPO because it provides only hits of European registrations. However, it should be considered to complete the search for important brands outside their own national borders. The national trademarks apply to a single country, but EU registrations apply to all member countries. A search for brand names in the DPMA demonstrates the example shown in Figure 1.2, which continues the representation of Figure 1.1. The company that brings the product into the market is mentioned on the label of the packaging.

Before marketing, the company's managing director and his experts must ensure that all measures are implemented in accordance with the EC Directive, especially

- Check all ingredients of the formulation (allowed, maximum application quantity)
- Specify the ingredients in INCI nomenclature and sequence
- The full details printed on the label and package (function of the product, claims, usage, warnings [if necessary], nanoparticles used, volume, date and time of durability, batch number, and manufacturer's address)
- The production according to the GMP standard
- Quality check of the product
- Detailed safety assessment and summary
- Registration of the product at the Cosmetic Products Notification Portal (CPNP) (see Section 1.6)
- Completed P.I.F.

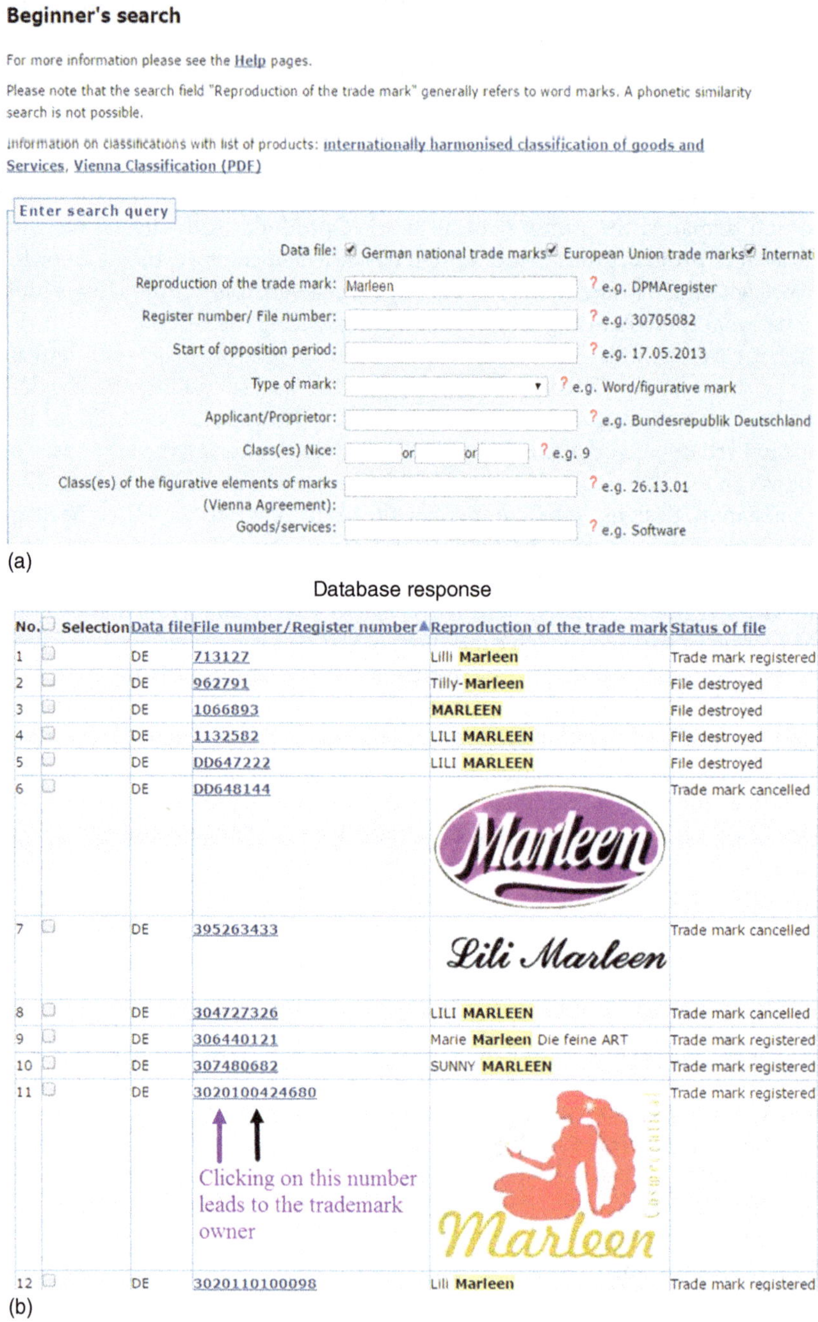

Figure 1.2 Search of the brand name and logo in Germany via the DPMA register, the images show cutouts: (a) start of the search, (b) hits. Source: Data from Ref. [11].

1.6 Mandatory Registration of Cosmetic Products

The Cosmetics Regulation requires the product to be registered with the competent authorities. In Europe, the formulations must be electronically deposited with the CPNP. Quote of the CPNP [14]:

> The **Cosmetic Products Notification Portal** is a free of charge online notification system created for the implementation of Regulation (EC) No 1223/2009 on cosmetic products. When a product has been notified in the CPNP, there is no need for any further notification at national level within the EU. Regulation (EC) No 1223/2009 (Article 13) requires that the responsible persons and, under certain circumstances, the distributors of cosmetic products submit some information about the products they place or make available on the European market through the CPNP.
> The CPNP is making this information available electronically to:
>
> ✓ Competent Authorities (for the purposes of market surveillance, market analysis, evaluation and consumer information)
> ✓ Poison Centres or similar bodies established by EU countries (for the purposes of medical treatment)
> ✓ Cosmetic products responsible persons
> ✓ Distributors of cosmetic products.

1.7 Databases for Ingredients

For a complicated search for individual substances, the Annexes to the Cosmetics Regulation should not be used in the first step because "**CosIng**," the European Commission database [15], allows fast access to individual substances and their possible limitations according to the regulation. The database contains information on all cosmetic substances and ingredients used. CAS and EC numbers that identify the ingredient as well as the INCI names can also be found in CosIng. Permissible maximum amounts of the substance and the wording of warnings, which must be indicated on the packaging, are in the database. A check of the results with the Annexes of the Cosmetics Regulation whereby the search number can be taken from the CosIng answer is indispensable.

How the CosIng files look like is shown by two difficult examples. Sodium fluoride (Figure 1.3) represents the first example. This chemical substance is part of many types of toothpastes. In Germany, experts recommend to use the maximum permitted quantity. Sodium fluoride is toxic, for oral intake suffices 71 mg/kg body weight (LD_{LO}). Accordingly, a person weighing 60 kg may die after taking about 6 g of sodium fluoride. As it had been shown that the substance reduces caries formation, 1500 ppm F is allowed. By brushing the teeth twice a day and total swallowing of the foam, a 60 kg human would take about 6 mg, i.e. 1/1000 of the dangerous amount. However, the amount absorbed is more than a factor of less than 100, as the foam is spat out. The discussed limit values for children less than six years should be 500–700 ppm. The Cosmetics Directive requires that

Substance	Sodium fluoride	INCI-name
CAS #	7681-49-4	CAS = Chemical Abstracts Service
EC #	231-667-8	EC = European Community Number
Name of Common Ingredients Glossary		
INN/ISO/AN		
Regulation	(EC) No 344/2013	(Admentments to the Cos.-Directive)
Other Directives/Regulations		
Annex/Ref #	III/31	Annex III, position 31, same information as here
Product Type, body parts	Oral products	Toothpastes
Maximum concentration in ready for use preparation	0,15 % calculated as F. When mixed with other fluorine compounds permitted under this Annex, total F concentration must not exceed 0,15 %	
Other		
Wording of conditions of use and warnings	Contains sodium fluoride For any toothpaste with compounds containing fluorine in a concentration of 0.1 to 0.15% calculated as F unless it is already labelled as contra-indicated for children (e.g. "for adult use only".) the following labelling is obligatory: "Children of 6 years and younger: use a pea-sized amount for supervised brushing to minimise swallowing. In case of intake of fluoride from other sources consult a dentist or doctor."	Warning on the label
SCCS opinions Scientific Committee on Consumer Safety provides further notes for use of sodium fluoride for children	• 0653/03 - Opinion concerning the Safety of Fluorine Compounds in Oral Hygiene Products for Children under the Age of 6 years • 0882/05 - Opinion on the Safety of Fluorine Compounds in Oral Hygiene Products for Children under the Age of 6 Years • 1214/09 - Clarification on the Opinions SCCNFP/0653/03 and SCCP/0882/05 on the Safety of Fluorine Compounds in Oral Hygiene Products for Children under the Age of 6 Years	

Figure 1.3 Result of the search in the CosIng database for sodium fluoride.

only a small amount may be taken from the adult toothpaste (see warning in Figure 1.3).

There are numbers for the exact identification of ingredients. The well-known CAS numbers, numerical identifiers of chemical substances, are provided by the Chemical Abstracts Service. Another unique seven-digit identifier for substances is the European Community Number (EC Number), which was determined by the European Commission for regulatory purposes within the European Union. The new EC Number comprises three individual substances characterizing numbers, namely the European Inventory of Existing Commercial Chemical Substances (EINECS), the European List of Notified Chemical Substances (ELINCS), and the No-Longer-Polymers (NLP) list. In detail, these are the lists of the EINECS (over 100 000 entries), the ELINCS (more than 4000 entries), and further the NLP-List (NLP-Number with 700 entries). As shown in Figure 1.3, the two numbers, CAS and EC, are now available in CosIng for precise identification.

In the United States, the search engine of the US Association for the Cosmetic and Personal Care Industry is preferred as database. With more substances and information, the database is probably incomparably in the wealth of information [16]. This Cosmetic Ingredient Dictionary provides a comprehensive listing of ingredients used in cosmetic and personal care products for the benefit of consumers. It is authored by the Personal Care Products

Council, the trade Association for the Cosmetic and Personal Care Industry. The combined dictionary/handbook contains more than 13 000 INCI labeling names for the United States, the European Union, and other countries. These are cross-referenced to nearly 60 000 trade and technical names and 3000 suppliers from 91 countries. The U.S. FDA defined the Cosmetic Ingredient Dictionary as the primary source for ingredient names, which are required for cosmetic ingredient labeling. The benefit is the consistency and transparency provided to consumers and scientists as ingredients are identified by a single labeling name regardless of the national origin of the product. Sodium fluoride is also used as an example. In addition to chemical information, the restrictions in application are shown in Figure 1.4. The FD&C allows less fluoride in toothpastes than the European regulation. In Canada, fluoride-containing dentifrices are prohibited.

[Code of Federal Regulations]
[Title 21, Volume 5]
[Revised as of April 1, 2016]
[CITE: 21CFR355.10]

TITLE 21--FOOD AND DRUGS
CHAPTER I--FOOD AND DRUG ADMINISTRATION
DEPARTMENT OF HEALTH AND HUMAN SERVICES
SUBCHAPTER D--DRUGS FOR HUMAN USE

PART 355 -- ANTICARIES DRUG PRODUCTS FOR OVER-THE-COUNTER HUMAN USE

Subpart B--Active Ingredients

Sec. 355.10 Anticaries active ingredients.

The active ingredient of the product consists of any of the following when used in the concentration and dosage form established for each ingredient:

(a) ***sodium fluoride*** --(1) *Dentifrices containing 850 to 1,150 ppm theoretical total fluorine in a gel or paste dosage form.* **sodium fluoride** 0.188 to 0.254 percent with an available fluoride ion concentration =650 parts per million (ppm).

(2) *Dentifrices containing 850 to 1,150 ppm theoretical total fluorine in a powdered dosage form.* **sodium fluoride** 0.188 to 0.254 percent with an available fluoride ion concentration of [gteqt]850 ppm for products containing the abrasive sodium bicarbonate and a poured-bulk density of 1.0 to 1.2 grams per milliliter.

(3) *Treatment rinses.* (i) An aqueous solution of acidulated phosphate fluoride derived from **sodium fluoride** acidulated with a mixture of sodium phosphate, monobasic, and phosphoric acid to a level of 0.1 molar phosphate ion and a pH of 3.0 to 4.5 and which yields an effective fluoride ion concentration of 0.02 percent.

(ii) An aqueous solution of acidulated phosphate fluoride derived from **sodium fluoride** acidulated with a mixture of sodium phosphate, dibasic, and phosphoric acid to a pH of 3.5 and which yields an effective fluoride ion concentration of 0.01 percent.

(iii) **sodium fluoride** 0.02 percent aqueous solution with a pH of approximately 7.

(iv) **sodium fluoride** 0.05 percent aqueous solution with a pH of approximately 7.

(v) **sodium fluoride** concentrate containing adequate directions for mixing with water before using to result in a 0.02-percent or 0.05-percent aqueous solution with a pH of approximately 7.

(b) *Sodium monofluorophosphate* --(1) *Dentifrices containing 850 to 1,150 ppm theoretical total fluorine in a gel or paste dosage form.* Sodium monofluorophosphate 0.654 to 0.884 percent with an available fluoride ion concentration (consisting of PO3 F= and F- combined) =800 ppm.

(2) *Dentifrices containing 1,500 ppm theoretical total fluorine in a gel or paste dosage form.* Sodium monofluorophosphate 1.153 percent with an available fluoride ion concentration (consisting of PO3 F= and F- combined) =1,275 ppm.

(c) *Stannous fluoride* --(1) *Dentifrices containing 850 to 1,150 ppm theoretical total fluorine in a gel or paste dosage form.* (i) Stannous fluoride 0.351 to 0.474 percent with an available fluoride ion concentration [gteqt]700 ppm for products containing abrasives other than calcium pyrophosphate.

(ii) Stannous fluoride 0.351 to 0.474 percent with an available fluoride ion concentration [gteqt]290 ppm for products containing the abrasive calcium pyrophosphate.

(2) *Preventive treatment gel.* Stannous fluoride 0.4 percent in an anhydrous glycerin gel, made from anhydrous glycerin and the addition of suitable thickening agents to adjust viscosity.

(3) *Treatment rinse.* Stannous fluoride concentrate marketed in a stable form and containing adequate directions for mixing with water immediately before using to result in a 0.1-percent aqueous solution.

Figure 1.4 Sodium fluoride in the US database . Source: Data from Ref. [16].

Substance	Thioglycolic acid and its salts
CAS #	68-11-1
EC #	200-677-4
Name of Common Ingredients Glossary	THIOGLYCOLIC ACID
INN/ISO/AN	
Regulation	(EC) No 1223/2009
Regulated By	88/233/EEC
Other Directives/Regulations	
Annex/Ref #	III/2a
Product Type, body parts	(a) Hair products (b) Depilatories (c) Hair rinse-off products (d) Products intended for eyelash waving
Maximum concentration in ready for use preparation	(a) 1) 8% 2) 11% (b) 5% (c) 2% (d) 11% The abovementioned percentages are calculated as thioglycollic acid
Other	(a) General use ready for use pH 7 to 9.5 Professional use ready for use pH 7 to 9.5 (b) ready for use pH 7 to 12.7 (c) ready for use pH 7 to 9.5 (d) ready for use pH 7 to 9,5
Wording of conditions of use and warnings	Condition of use: (a) (b) (c) Avoid contact with eyes In the event of contact with eyes, rinse immediately with plenty of water and seek medical advice (a) (c) (d) Wear suitable gloves Warnings to be printed on the label: (a) (b) (c) Contains thioglycolate Follow the instructions Keep out of reach of children (a) (d) For professional use only: Cotains thiglycolata, folow the instruction.
SCCS opinions	• 1520/13 - Thioglycolic acid and its salts (TGA)
Chemical/IUPAC Name	
Identified INGREDIENTS or substances e.g.	• AMMONIUM THIOGLYCOLATE • CALCIUM THIOGLYCOLATE • CALCIUM THIOGLYCOLATE HYDROXIDE • ETHANOLAMINE THIOGLYCOLATE • MAGNESIUM THIOGLYCOLATE • POTASSIUM THIOGLYCOLATE • SODIUM THIOGLYCOLATE • STRONTIUM THIOGLYCOLATE • THIOGLYCOLIC ACID

Figure 1.5 Results of the search in the CosIng database for "thioglycolic acid and its salts."

As a second example for testing the CosIng database, a chemical substance was selected, which cleaves amide and disulfide bonds. As the cream is applied to hairy areas, the substance acts as a hair-removing agent. The salts of thioglycolic acid, such as potassium thioglycolate, are suitable for cleavage. When searching for "potassium thioglyconate" in the CosIng database (CAS # 34452-51-2, EC # 252-038-4), a reference is made to the Annexes of the Cosmetics Regulation. "Thioglycolic acid and its salts" are found there. Unfortunately, the warnings are missing in the current issue. Therefore, under "thioglycolic acid," a new search in CosIng has to take place, which leads to the goal, if all hints are observed.

Figure 1.5 shows the result of the search. There the maximum permissible amount and the permitted pH range are specified for hair removal (depilation). Furthermore, CosIng disclosed the prescribed wording for the warnings. In this exceptional case, CosIng offers more and more detailed information than the Annexes of the Cosmetics Regulation. It also shows that an intensive search for restricted use substances can be necessary. In addition, there is a reference to the detailed opinion of the Scientific Committee on Consumer Safety (SCCS) on the use of this substance group. Safety instructions for pure potassium thioglyconate in aqueous solution are given in the safety data sheet (SDS) of the manufacturer (example: Bruno Bock). For the removal of the hair, precise instructions for the application and the maximum duration of use are required because the formulations are strongly alkaline (warning: contains alkali). Good formulations contain substances, such as weak acids and a buffer, to correct the strongly alkaline pH as well as skin-protecting and skin-soothing substances, which altogether help to reduce the negative effects of the alkali.

1.8 Regulations in the United States

In the United States, the FDA is the competent regulatory authority, a department of Health and Human Services [17]. The top priority of FDA is the protection of consumers. Within the FDA, the cosmetics are integrated in the Center for Food Safety and Applied Nutrition (CFSAN), which is responsible for regulation and approval of food for human consumption, such as food additives, color additives, and cosmetics. Within the authority, cosmetics are the least regulated products. Sunscreens are subject to the Medicine Act in the United States. The problem is discussed in Section 7.16.4.

The US Federal Food, Drug, and Cosmetic Act (abbreviated as FDCA, or FD&C) is a set of laws passed by Congress in 1938. For cosmetics, the laws are amended in title 21 of the United States Code (21 U.S.C.), Chapter 9. They authorize the FDA to oversee the safety of food, drugs, and cosmetics. For cosmetic products, the Act prohibits the marketing of unsafe or mislabeled cosmetics. Therefore, the FDA does not approve cosmetic products but remove cosmetics from the market that contain unsafe ingredients or are mislabeled. A regulation as in Europe does not exist. The mostly unwritten rules are similar to the European ones, for example, what information must be on the packaging. An examination is only carried out if a violation of the written regulations is found. Then, the FDA imposes high penalties. The FDA can inspect cosmetics manufacturing facilities to ensure quality. The manufacturer of cosmetics is obliged to comply with the regulations. Above all, he must ensure the safety and stability of the products as well as make truthful statements about the product on the packaging.

1.9 Regulations of the Cosmetics Markets in Asia

The ASEAN Cosmetic Directive (ACD, [18]) was created to eliminate restrictions for the trading of cosmetic products among Member States through adoption and implementation of harmonized technical requirements. The Directive represents in large parts a reproduction of the EC Regulation and was signed 2 September 2003 in Cambodia by the Economic Ministers. Since 1 January 2008, the ACD has entered into force in ASEAN Member States after transposing into local regulations. The ACD contains the following subjects, comparable with the European Cosmetics Regulation:

1. General Provisions
2. Definition and Scope
3. Safety Requirements
4. Ingredient Listings
5. ASEAN Handbook of Cosmetic Ingredients
6. Labeling Requirements
7. Product Claims
8. Product Information
9. Methods of Analysis
10. Institutional Arrangements
11. Special Cases
12. Implementation

The Member States (Indonesia, Thailand, Vietnam, China, Japan, Hong Kong, India, Korea, Taiwan, and Philippines) implemented the ACD principles, followed by other states (Brunei Darussalam, Cambodia, Laos, Malaysia, and Singapore). The main principles are as follows:

- Product notification;
- P.I.F. requirement;
- Annexes to control ingredients;
- GMP–GDP (good distribution practice) requirements; and
- Postmarket surveillance.

Annexes II–VI of the prohibited and restricted substances as well as the permitted dyes, preservatives, and UV filters also comply with European requirements [18, 19]. The company or person responsible for marketing the cosmetic products shall ensure that the product will not cause damage to human health when applied under normal or reasonably foreseeable conditions of use. This is actually a matter of course. Before marketing, the product formulation has to be notified to the regulatory authority. Product information and the safety assessment (P.I.F.) must be readily accessible to the regulatory authority. GMP-standard for the manufacture and GDP for the distribution are mandatory. The company should have experience with the legal requirements for cosmetics. Except Thailand, all countries have transposed the ACD into local regulations, with small deviations.

The implementation of the ACD is accompanied by various committees [20], namely the

- *ACC*: The ASEAN Cosmetic Committee coordinated and monitored the implementation of the Directive. ASEAN Secretariat and ASEAN Cosmetic Association (ACA) are composed by representatives of each member state.
- *ACSB*: The task of the ASEAN Cosmetic Scientific Body is to elaborate recommendations for the ACC on safety, technical, and scientific matters.
- *ACTLC*: The ASEAN Cosmetic Testing Laboratory Committee was established as a postmarket surveillance initiative to support the implementation of ACD through establishing and maintaining an efficient quality assurance system in line with international practices and guidelines.

Deviations from the ACD text should be briefly addressed in the case of China and Japan. China's authorities define normal and special cosmetics. Special cosmetics cover the following product categories:

- Sunscreen
- Spot corrector/antipigmentation/whitening
- Slimming
- Breast care
- Hair growth
- Hair colors
- Perms
- Deodorant
- Depilatories

For all imported products and China's special cosmetics, a full dossier and samples have to be submitted to the CFDA (China Food and Drug Administration) for an evaluation by the technical review expert committee. CFDA test them for unauthorized features, toxicology, and sometimes efficacy (sunscreen). The test period for imported specials lasts up to 12 months.

The Japanese Government regulates the cosmetics market through the Ministry of Health, Labor and Welfare according to the Pharmaceutical Affairs Law [21]. Japan published a list of prohibited and restricted ingredients, as well as a positive list of UV filters and preservatives [19]. In Japan, the cosmetics market is divided into cosmetics and quasi-drugs. Products against acne and dandruff or skin chapping as well as for whitening (bleaching) or sterilizing the skin are among the quasi-drugs, furthermore, products for prevention of foul breath or body odor, promotion of hair growth, or removal of hair, hair dyes, and waving of hair. All these quasi drugs need a special, time-consuming registration. After notification, the other cosmetics with ingredients of the positive list can be easily marketed.

1.10 Delimitation of Cosmetic Products

The delimitation of cosmetics from neighboring areas (Figure 1.6) is not uniformly regulated around the world. Therefore, it is necessary to check for each country whether the product group is classified as cosmetics. These delimitations also play a role in the determination of market sizes in the individual countries. In some cases, unfortunately, it is not clear whether certain products are cosmetics or not, this concerns, for example, sunscreens, depilatories or hygienic articles, and many others. Therefore, the published statistics provide partly very different values (Chapter 2).

Products, which do not belong to cosmetics, are subject to other legal regulations. For some cosmetics-related articles, there is another law (LFGB [6]) in Germany. Objects intended to come into contact with the mucous membranes of the mouth as well as objects intended for personal care fall under the LFGB (§ 5, Nos. 3 and 4). This is why in Germany, the toothbrushes belong to the items of daily necessities and not to cosmetics, in contrast to other European Countries. The same applies to nail files, scissors, razor blades and shavers, combs and hairpins, as well as sponges, towels, and washcloths. In most countries, oral hygiene products (toothpaste, toothbrushes, and the like) are considered to be cosmetics, although they correspond to the definition of medical devices. The hygienic articles, which could be assigned to cosmetics, include, for example, the cotton swabs, tissues, and baby diapers but belong to the items of daily necessities and are subject to the relevant law.

The Medical Device Directive (MDD), also called Directive 93/42/EEC [22], regulates the safety and medical–technical performance of medical devices in the European Economic Area together with Directives 90/385/EEC and 98/79/EC. The amending to the Directives became legally binding in the EU on 21 March 2010 in the directive 2007/47/EC [23]. All apparatus, instruments, and aids used for the medical care of humans by the physician are called medical devices, as well

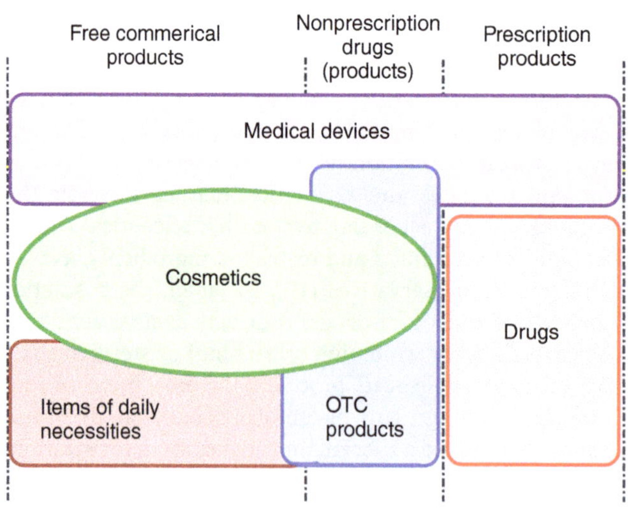

Figure 1.6 Overlaps of the cosmetics with neighboring areas.

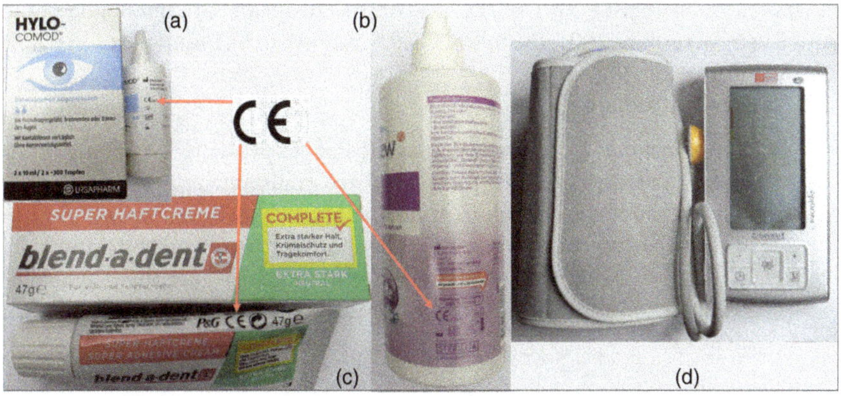

Figure 1.7 Medical products characterized by the prescribed CE symbol: (a) sterile liquid for moistening the eyes in a dispenser for dosing individual drops (OTC), (b) rinsing and storage solution for contact lenses, (c) adhesive cream for dentures, and (d) blood pressure gauge with cuff, the CE mark is on the back.

as all the aids and tools that patients need. On the one hand, the devices enhance or save lives and support healing processes; on the other hand, people need different medical devices (products) to improve their lives. Especially, they are objects or substances, used for hygienic care and medical or diagnostic purposes, and which are generally physically or physically chemically active.

In Europe, medical devices must bear the CE marking before they may be placed on the market or put into service. The CE marking presupposes that the products meet the requirements and that this is confirmed by the prescribed conformity assessment (Figure 1.7). Some concrete examples of the devices, assigned to four classes (I, IIa, IIb, and III), are bandages, medical plasters, support stockings, wheelchairs, disinfectants (for equipment), one-way injection, hearing aids, dental materials, dentures, contact lenses, glasses, respirators and dialysis machines, heart catheters, and breast implant. Not only the contact lenses belong to the medical products but also the cleaning liquid and the storage solution, as well as the denture adhesive cream and special cleansers. In contrast, the toothpaste used for the cleaning of dentures belongs to cosmetics. The category of medical products also comprises the physical contraceptives. According to EU Directive 93/42/EEC, sucking incontinence aids are medical devices/class I, like incontinence pants (in contrast to diapers, assigned to the items of daily necessities). A number of products that can be bought in drug stores or in pharmacies belong to the category of medical devices.

If, because of their ingredients, a product restores, corrects, or modifies physiological functions by exerting a pharmacological, immunological, or metabolic action, the product shall be qualified as a drug (medicinal product or therapeutic agent). However, products that, while having an effect on the human body, do not significantly affect the metabolism and thus do not strictly modify the way in which it functions may be qualified as cosmetic products. The FD&C Act defines drugs, in part, by their intended use, as "articles intended for use in the diagnosis, cure, mitigation, treatment, or prevention of disease" and "articles (other than food) intended to affect the structure or any function of the body of man or other animals." In the Directive 2001/83/EC [24] of the European Parliament and of the Council of 6 November 2001, the following definition can be found (Article 1, § 2):

> Medicinal product:
>
> (a) Any substance or combination of substances presented as having properties for treating or preventing disease in human beings; or
> (b) Any substance or combination of substances which may be used in or administered to human beings either with a view to restoring, correcting or modifying physiological functions by exerting a pharmacological, immunological or metabolic action or to making a medical diagnosis.

Claims stated on the product labeling that descript medical effects are forbidden for cosmetics. Examples for forbidden claims are restore hair growth, reduce cellulite, treat varicose veins, increase or decrease the production of melanin (pigment) in the skin, or regenerate cells.

According to the FA&C Act, a product can be a drug, a cosmetic, or a combination of both [8]. There are products that meet the definitions of both cosmetics and drugs. Examples display Table 1.5. Depending on the country,

Table 1.5 Product examples with allocation to cosmetics or OTC or drugs.

Product/substance	Cosmetic function	Medical function
Anti-dandruff shampoo	Shampooing	Anti-dandruff agent, e.g. Piroctone Olamine
Fluoride-containing toothpaste	Cleaning the teeth, restoration of oral hygiene	Harden the enamel by sodium fluoride or tin fluoride
Deodorant/ antiperspirant	Prevention of sweat decomposition, binding of smell, perfuming	Inhibiting of sweat production, aluminum compounds
Hair growth liquid or spray	Hair and scalp care	Hair growth stimulating substance, biotin, peptides, hormones
Sunscreen cream, foundations	Skin care	UVA and UVB blocker, see Annexes of the Cosmetic Directive
Anti-acne	Low-fat cleansing cream	Anti-inflammatory, disinfectant, keratolytic substances such as allantoin, salicylic acid, and chlorhexidine, benzoyl peroxide
Hormone cream	Skin care	Hormone such as estrogens, phytohormones with estrogen-like effect
Fragrance	Promoting attractiveness	Aromatherapy, support for falling asleep
Dexpanthenol	"Beauty vitamin"	Healing wounds
Jasmine oil	Enchanting fragrance	Relaxation (previously administered to facilitate births)
Evening primrose oil	Skin care	Agent against atopic dermatitis

the product will be assigned and must fulfill both laws in some countries. Based on the superordinate regulations, the regulatory authority of each country determines in disputed questions the allocation of cosmetic and OTC products (over-the-counter). In other cases, the classification also takes place according to the intended use and the claim on the packaging. As the manufacturer wants, special products can be marketed both as cosmetics or medicament. Therefore, the product of Figure 1.1 with the labeling "support the self-healing processes of the skin" is only an effective cosmetic product, although it can smooth neurodermitic skin. Such products are referred to in the literature as "cosmeceuticals," although this term does not appear in laws. If, on the other hand, the mentioned product is advertised with the claim "cream for removing scaly, neurodermitic skin," it would be an OTC product. It is clear that they have to fulfill the corresponding law. These statements also include some products that are called "quasi drugs" in Japan or "special cosmetics" in China.

The description "OTC" is referred to nonprescription medicines, which are preferably sold through the pharmacy [25]. In cases of low, typical symptoms, people perform a self-medication with OTC medicines. The OTC area comprises different product groups depending on the country. Figure 1.8 gives some examples. According to Article 48 of the German Medicines Act, the Federal Ministry of Health classifies medicinal products as nonprescription if they, based on the formulation and experience, cannot endanger the user's health, even without medical supervision. The condition is that they are used as intended. An expert committee develops proposals, whose substances can be released from the prescription obligation or must be subordinated to it.

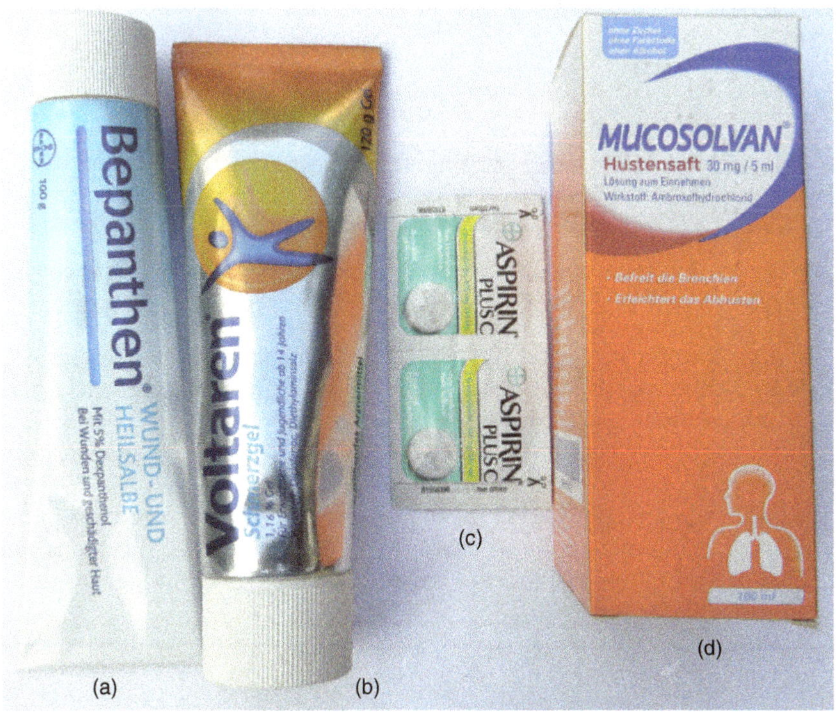

Figure 1.8 Much purchased OTC products: (a) wound and healing ointment, (b) pain gel, (c) headache tablet, and (d) cough syrup.

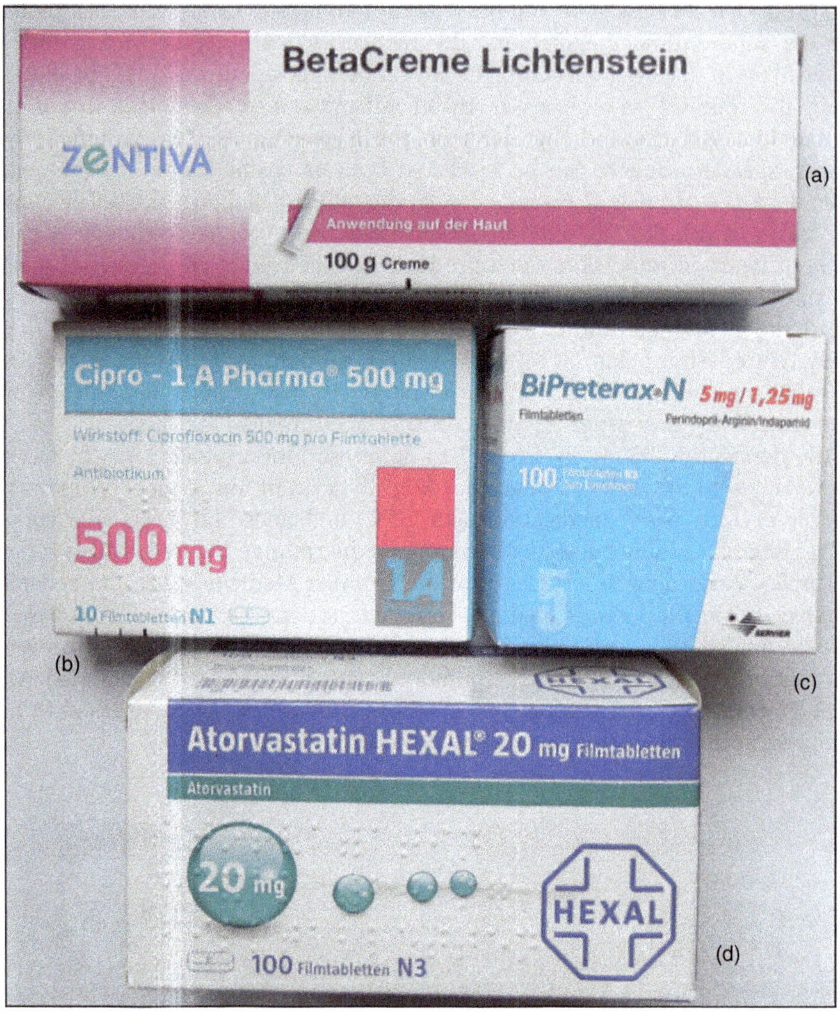

Figure 1.9 Typical drugs: (a) cortisone-containing skin ointment, (b) antibiotic, (c) tablets for blood pressure regulation, and (d) statins for cholesterol lowering.

The manufacturer can optionally change a cosmetic product into an OTC article by complying with the requirements of a drug and allowing only a sale through the pharmacies. Through this way of marketing, L'Oréal goes successfully with the "active cosmetics" series, which also includes the known Vichy skin creams.

The doctor can prescribe medicines for the whole body and for all organs. Some examples of skin, heart, and vein problems as well as bacterial infections can be found in Figure 1.9. In contrast to cosmetics, the drugs act preferably inside the body that means systemic.

1.11 Learnings

- ✓ In Europe, the Cosmetics Regulation (EC) No 1223/2009 clarifies what cosmetic products are and prescribes the ingredients in type and quantity, manufacture and marketing, as well as the responsibilities of the manufacturer.
- ✓ There are very similar regulations in Asia. In the United States, the regulations are not so comprehensive and less stringent.
- ✓ The marketer is responsible for his product worldwide and ensures that all steps, from the formulation to the consumption, follow the guideline.
- ✓ Each manufacturer must provide a P.I.F. for each product/product group that includes the product description and in particular a detailed safety assessment by an expert.
- ✓ The Cosmetics Regulation requires the product to be registered with the competent authority. In Europe, formulations must be submitted electronically to the CPNP.
- ✓ The regulation contains in the Annexes the permissible ingredients, including the colors, preservatives and UV filters, and specifies permissible limits. The restrictions of ingredients can be easily found in the CosIng (**Cos**metic **Ing**redients) database.
- ✓ The manufacturer should usefully protect its brand name in the Patent and Trademark Office.
- ✓ Cosmetic products should be distinguished from items of daily use as well as medical devices and OTC products for which other laws apply.

References

1 Stepanovs, J. (1999). *Skin Saver Remedies*. Australia: Harald W. Tietze Publishing.

2 Umbach, W. (ed.) (2004). *Kosmetik und Hygiene: von Kopf bis Fuß*, 3e. Weinheim: Wiley-VCH.

3 History of cosmetics, from Wikipedia, the free encyclopedia, 2019 https://en.wikipedia.org/wiki/History_of_cosmetics (accessed 6 May 2019).

4 Stewart, S. (2017). *Painted Faces: A Colourful History of Cosmetics*. Amberley Publishing.

5 REGULATION (EC) No 1223/2009 OF THE EUROPEAN PARLIAMENT AND OF THE COUNCIL of 30 November 2009 on cosmetic products. http://eur-lex.europa.eu/LexUriServ/LexUriServ.do?uri=OJ:L:2009:342:0059:0209:EN:PDF (accessed 17 November 2018). http://eur-lex.europa.eu/LexUriServ/LexUriServ.do?uri=OJ:L:2009:342:0059:0209:DE:PDF (accessed 17 November 2018)

6 Lebensmittel-, Bedarfsgegenstände- und Futtermittelgesetzbuch (Lebensmittel- und Futtermittelgesetzbuch - LFGB), 2005. http://www.gesetze-im-internet.de/lfgb (accessed 17 November 2018).

7 Kirk-Othmer (2013). *Kirk-Othmer Chemical Technology of Cosmetics*. Hoboken, NJ: Wiley.

8 U.S. Food & Drug Administration, Is it a cosmetic, a drug, or both?. https://www.fda.gov/Cosmetics/GuidanceRegulation/LawsRegulations/ucm074201.htm (accessed 17 November 2018).

9 Verordnung über kosmetische Mittel (Kosmetik-Verordnung), KosmetikV 2014. https://www.gesetze-im-internet.de/bundesrecht/kosmetikv_2014/gesamt.pdf (accessed 17 November 2018).

10 Gesetz über die Werbung auf dem Gebiet des Heilwesens (Heilmittelwerbegesetz - HWG), 1965. https://www.gesetze-im-internet.de/heilmwerbg/BJNR006049965.html (accessed 17 November 2018).

11 DPMAregister, Beginner's search, 2019. https://register.dpma.de/DPMAregister/marke/einsteiger?lang=en (accessed 6 May 2019).

12 EUIPO, European Union Intellectual Property Office, 2019. https://euipo.europa.eu/ohimportal/en/trade-marks-in-the-european-union (accessed 6 May 2019).

13 Directive 2008/95/EC of the European Parliament and of the Council, 22 October 2008, to approximate the laws of the Member States relating to trade marks. http://eur-lex.europa.eu/legal-content/EN/TXT/?uri=CELEX%3A32008L0095 (accessed 17 November 2018).

14 European Commission, Cosmetic Product Notification Portal (CPNP). https://ec.europa.eu/growth/sectors/cosmetics/cpnp_en (accessed 17 November 2018).

15 European Commission, Cosmetic ingredient database (Cosing). https://ec.europa.eu/growth/sectors/cosmetics/cosing_en (accessed 17 November 2018).

16 Cosmetics Info, Cosmetic Ingredient Dictionary, 2016. http://www.cosmeticsinfo.org/Ingredient-dictionary (accessed 17 November 2018).

17 U.S. Food & Drug Administration, Cosmetics, https://www.fda.gov/Cosmetics/default.htm (accessed 17 November 2018).

18 Health Sciences Authority (HSA), ASEAN Cosmetic Directive, 2017. https://www.hsa.gov.sg/content/hsa/en/Health_Products_Regulation/Cosmetic_Products/Overview/ASEAN_Cosmetic_Directive.html (accessed 17 November 2018).

19 Standards for Cosmetics (Ministry of Health and Welfare Notification No.331 of 2000), Japan. http://www.mhlw.go.jp/file/06-Seisakujouhou-11120000-Iyakushokuhinkyoku/0000032704.pdf (accessed 17 November 2018).

20 Asean Cosmetics Association, ASEAN Cosmetic Directive, 2019. http://aseancosmetics.org/information-center/asean-cosmetic-directive/(accessed 18 November 2018).

21 Information on Japanese Regulatory Affairs Regulatory Information Task Force Japan Pharmaceutical Manufacturers Association, 2015. http://www.jpma.or.jp/english/parj/pdf/2015.pdf (accessed 18 November 2018).

22 COUNCIL DIRECTIVE 93/42/EEC of 14 June 1993 concerning medical devices. http://eur-lex.europa.eu/LexUriServ/LexUriServ.do?uri=CONSLEG:1993L0042:20071011:en:PDF (accessed 18 November 2018).

23 DIRECTIVE 2007/47/EC OF THE EUROPEAN PARLIAMENT AND OF THE COUNCIL of 5 September 2007 amending Council Directive

90/385/EEC on the approximation of the laws of the Member States relating to active implantable medical devices, Council Directive 93/42/EEC concerning medical devices and Directive 98/8/EC concerning the placing of biocidal products on the market. http://eur-lex.europa.eu/LexUriServ/LexUriServ.do?uri=OJ:L:2007:247:0021:0055:en:PDF (accessed 18 November 2018).

24 DIRECTIVE 2001/83/EC OF THE EUROPEAN PARLIAMENT AND OF THE COUNCIL of 6 November 2001 on the Community code relating to medicinal products for human use. http://ec.europa.eu/health//sites/health/files/files/eudralex/vol-1/dir_2001_83_consol_2012/dir_2001_83_cons_2012_en.pdf (accessed 18 November 2018).

25 Over-the-counter drug, From Wikipedia, the free encyclopedia, 2019. https://en.wikipedia.org/wiki/Over-the-counter_drug (accessed 6 May 2019).

2

Economic Importance of Cosmetics

2.1 Cosmetics Market and Distribution Channels in Germany

In Germany, the total market volume of real cosmetic products amounted to about €13.6 billion (bn) in 2016 [1], according to IKW figures (German Cosmetic, Toiletry, Perfumery, and Detergent Association; Figure 2.1, and Table 2.1). The most important product groups are the hair care, facial care, and personal care products. The largest winner compared to 2015 is the decorative cosmetics, followed by men's fragrances, deodorants, and the bath and shower preparations. The overall sales in Germany increased by 1.6%. Another reference, Statista [2], comes in 2016 to $17 bn that correspond to around €15.3 bn. The difference of about 10% is likely to be in the wider product portfolio (see Section 1.7), which includes the hygiene products.

It is necessary to distinguish between the normal and the green cosmetics. Sales of natural cosmetics are rising with 8–9% more strongly than with normal cosmetics. According to figures from the market research institute IRI, certified natural cosmetics could have a market share of 9% and natural cosmetics of 6% in 2015. The shares are included in the figures for the overall market.

Today and in the future, almost every second German customer buys the cosmetics products mainly in the big drugstore markets, which offer a wide selection and reasonable prices. In Germany, the three largest drugstore markets are dm, Rossmann, and Müller. They generate together sales of €16.4 bn (2015) with cosmetics, hygiene products, and daily necessity products, as well as animal feed. The most famous perfumery is Douglas with a turnover of €2.7 bn, which is achieved in 18 countries of Europe. Perfumeries generally sell, as known, mainly decorative cosmetics and perfumes in the high-price and luxury segment. In the consumer markets (Edeka, REWE, Kaufland, and real) and in the discounters (Lidl, Aldi, Plus, and Penny), the consumers buy affordable everyday cosmetics such as soaps, bath and shower preparations, skin creams, and hair and oral care products. Numerous high-priced skin and face creams as well as special products (OTC, over-the-counter) are only marketed by the pharmacies, which have a market share of almost 9%.

The big companies spend an unusually high amount of money on sales promotions and advertising. An estimate gives a value of approximately €1.4 bn for

Cosmetic Creams: Development, Manufacture and Marketing of Effective Skin Care Products, First Edition. Wilfried Rähse.

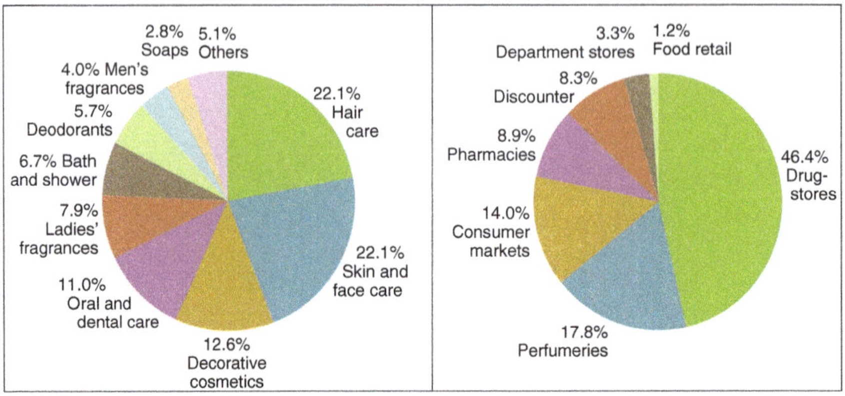

Figure 2.1 Sizes of the market segments by turnover and distribution channels at sales prices in Germany 2016.

Table 2.1 Beauty and body care market in Germany at retail prices [1].

Product category	2015 (Mio. €)	2016 (Mio. €)	2015/2016 (%)
Hair care	3 030	3 003	−0.9
Skin and face care	*2 991*	*3 003*	*0.4*
Decorative cosmetics	1 605	1 710	6.5
Oral and dental care	1 455	1 495	2.7
Ladies' fragrances	1 067	1 074	0.7
Bath and shower preparations	877	914	4.2
Deodorants	745	779	4.6
Men's fragrances	516	538	4.3
Soaps and syndets	370	379	2.4
Foot care	*209*	*190*	*−9.1*
Baby care	*146*	*148*	*1.4*
Aftershave/preshave	141	137	−2.5
Depilatories	*138*	*138*	*−0.1*
Shaving care men and women	*98*	*95*	*−2.6*
Sum	**13 387**	**13 603**	**1.6**

"Italics" are the creams as subject of the book.
Source: Data from Ref. [1].

Germany, which corresponds to 10.4% of the selling price in the markets. If the advertising costs are based on the supplier's price, the percentage increases to about 18–25%. The advertising took place on television, newspapers, magazines, journals, and radio. The amounts earmarked for this can be taken from Figure 2.2. It is not surprising that the four major providers of cosmetics spend the highest amounts on advertising. In the cosmetics segment, they represent less than 30% of the market. The 15 brand manufacturers, who advertise intensively, do not achieve a 50% market share with their medium and premium products. According to the sales channels, consumers buy increasingly "no name" (mass)

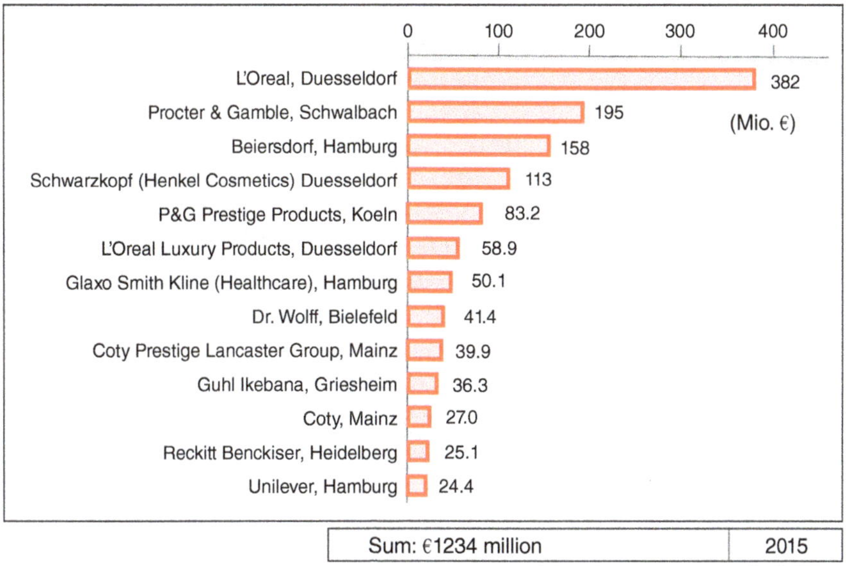

Figure 2.2 Expenditure for sales promotion of cosmetics by advertising in Germany. Source: From AGOF [3], Nielsen June 2016.

Figure 2.3 Estimated growth of the cosmetics market in Germany (February 2017; CAGR: Compound Annual Growth Rate).

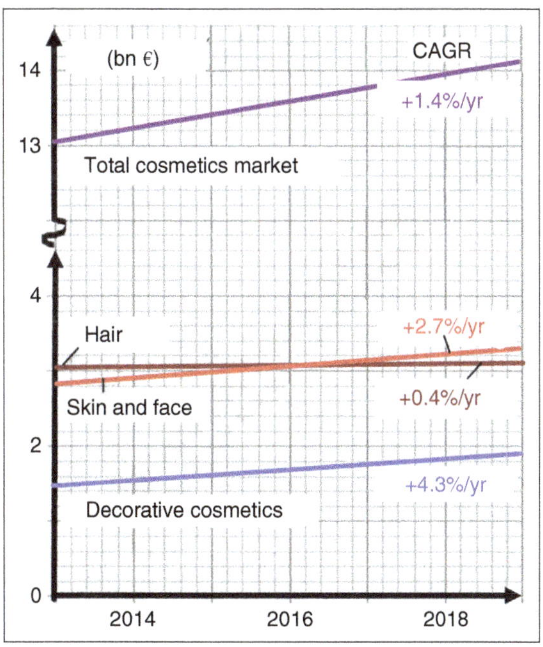

products, which cost much less and in many cases are qualitatively comparable to the brand manufacturers.

For the overall market as shown in Figure 2.3, the individual product categories have different growth rates. In Germany, the decorative cosmetics and also the skin care products grow significantly disproportionately, whereas the products for the hair practically stagnate. The oral care products and soaps also rise above average. In European comparison, German women use less makeup. However, they are closing the gap slowly. The eyebrows of girls and women are more pronounced and nails are more colorful and fashion-conscious. Young women use more lipsticks. Various specialized skin care products represent the second driver on the cosmetics market. The increasing aging of the population enhances the demand for skin care products for men and women, especially with helpful antiaging ingredients. Antiaging as well as creams with special natural oils and cosmetics for men reflects the trend. In terms of value, around one-third of the products are purchased by men, mainly shower gels, deodorants, and shaving water.

2.2 Shopping on the Internet in Germany

Over the last 15 years, a global sales channel has been created, regardless of the location of the cosmetic supplier. By 2015, an online marketing was set up by all providers, so that the customer can order his usual products over the PC, tablet, or smartphone, at his or another supplier. In 2014, products worth €800 million were sold through this channel, representing a 6% (€10 per capita) sales share of the German cosmetic market. Companies are continuously increasing their sales worldwide via the Internet. The online sales potential is enormous [3, 4]. For advertising on the Internet, the cosmetics providers spent €78 million. In retail prices, this value corresponds to about 10% of sales or 0.6% of the cosmetics market in Germany.

Almost half of the online sales refer to skin care products, followed by perfume (20%) and makeup (13%). Products from all distribution channels can be ordered online, regardless of whether they are usually available in the drugstore or consumer market, discounter, or pharmacy. As a rule, the customer gives his product request at Google into the search engine. In the answers, interesting suppliers can be found directly. There, customers often buy from big drugstores. Now pharmacies also use these distribution channels. Alternatively, the search can start at Amazon and Ebay. The product selection is very large online, and the prices are often equal or even below those in the supermarket or the pharmacy. In addition, there is also the possibility to find the cheapest supplier via the "best price comparison search engines," like Idealo. The user of a best price search machine should pay attention to the customer's evaluation of the cheapest supplier. With a good judgment from many customers and a "trusted shop" – valuation, the risk of a faulty purchase is low. If the products did not please without trying them out, they can be returned within two weeks without any problem.

Purchasing via the Internet usually requires an advance payment. Some shops offer a discount of 2–3% by bank transfer in advance. This possibility of payment should only be used in case of contact with well-known companies. For your own security, a payment via PayPal is recommended. Some providers require a 2% surcharge for the PayPal payment. As the delivery of the goods takes place by mail, a fee (postage) for small orders is required below a limited order value (for example, €25, €50, or €100). Therefore, only larger, postage-free purchases are worthwhile. The return of cosmetic products within two weeks is only possible if the content has not been touched. Many articles have an (aluminum-)seal, which must be intact for returned products. A noncommittal test of the goods is not possible, unlike in the perfumery.

If the ordered products are in stock of the supplier, the delivery time of many online shops is between one and three days, maximum one week. In many cases, the package service will inform the customer via e-mail about the day of delivery, sometimes time windows are specified. The great advantages of the convenient online shopping are in the wide selection, possibility of time-saving comparison of different offers, favorable price, and fast delivery directly into the house. Orders can be placed at any time seven days a week. Inquiries via the Internet (e-mail) will be answered by the provider at normal business hours, and also available by telephone. The great advantages convince more and more customers from online shopping. Depending on the region, in 2015, the e-commerce market possesses between 5% and 15% market share and grows worldwide by 20% per year.

2.3 European Cosmetics Markets

The 500 million consumers in Europe mainly use cosmetics to cleanse and care for the body, which contribute to hygiene and well-being and healthy lifestyles. The strength of the European cosmetics industry lies in its mix of both big and small companies. There are more than 4600 small and medium-sized companies, and this number is growing [5]. In the cosmetic industry, real innovations are not short-term feasible. It can take over five years of research, formulation, and testing to bring a new product to the market. That is why the total expenditure on R&D in Europe is estimated to be €1.27 bn, which is 1.65% of the sales. In 2015, the European market amounted to be **€76.4 bn** [2], approximately €77 bn [6]. Of the world's 50 leading cosmetic brands, more than half (26) are based in Europe. The cosmetics industry in Europe employs directly 179 000 people and indirectly a multiple [6]. The European countries, in particular France (€5.8), Germany (€2.9 bn), United Kingdom (€1.4), Italy (€1.4), and Spain (€1.3 bn), export products for €17.2 bn outside the EU 28 [5]. The share of exports from France and Germany is over 50%. Including the export, the European cosmetic industry produces goods for approximately €93 bn (estimated at retail prices). Imports to the European Union (EU) are comparatively low because companies such as Procter & Gamble (P&G) and others are present in Europe for the European market. Anyway, the cosmetic industry thus represents an important economic factor. According to a rough estimate, the cosmetics companies pay several billion in corporate taxes, which benefits individual countries.

The five largest markets [2] are in Germany, secondly in the United Kingdom and at the same level in France, and moreover in Italy and Spain (Figure 2.4). This sequence almost corresponds to the population figures. Therefore, the Polish

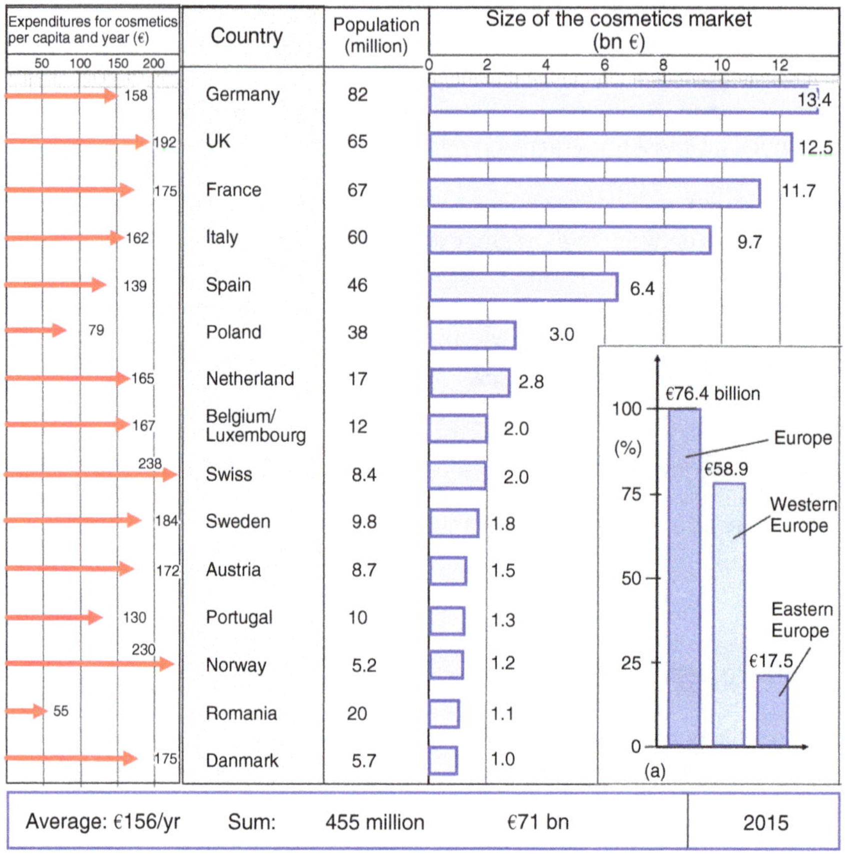

Figure 2.4 European cosmetics market, regions with the largest markets in 2015; (a) Shares of Western and Eastern Europe (Figures from various sources, February 2017).

market is by far the largest "Eastern European" cosmetics market. The figures come from different sources [2, 6, 7]. All are based on the market research, but the scope of "cosmetics" and the investigated distribution channels are not known. Therefore, the numbers contradict each other. For example, according to one source, the British market represents the number 2 in Europe, whereas according to another source, it is the French market. In addition, this could be due to the differences in the exchange rate between the pound and the euro at the time of the calculation.

In the per capita and per year expenditure on cosmetics, the Swiss are clearly ahead, followed by the Norwegians. The British, French, Swedes, and Danes also spend a lot of money on cosmetic products. Among the 15 largest cosmetics markets in Europe, Romania ranks 14th. Of the countries listed, the Romanians expend the least money for cosmetics. Calculated from the 15 largest European cosmetics markets with a total volume of €71 bn, 455 million people have an average annual per capita consumption of €156. This figure is roughly equivalent to the average expenditure in Germany.

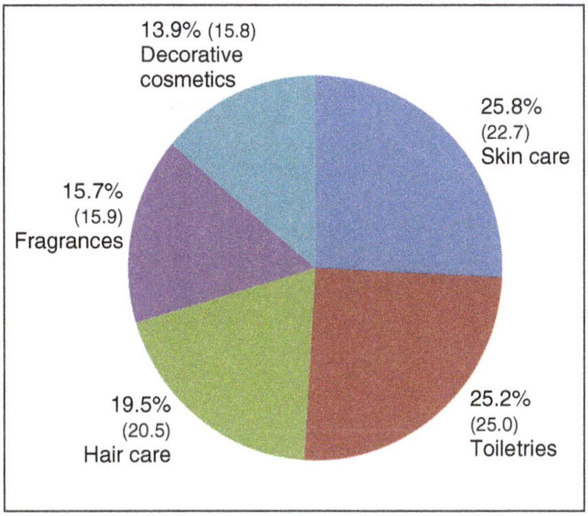

Figure 2.5 European cosmetics market (2015), percentage of product segments, in brackets shares of the United Kingdom for comparison.

Figure 2.5 illustrates the breakdown of the European market into major product segments. From the picture, it can be seen that products for skin care and the usual toilet articles are the most frequently bought, followed by products for the hair, fragrances, and decorative cosmetics. A comparison with the British market results in broad accordance, while the German market differs significantly, apart from skin care. With a corresponding summary of the values for the German market (skin and shaving and foot and baby care products), the skin care segment is 25.3%, hair care 22.1%, fragrances 11.9%, and the decorative cosmetics 12.6%. It is striking that German consumers spend more money for the hair but less on the fragrance, compared to the European average.

According to the figures released by Cosmetics Europe [6], the European cosmetics and toiletries market grew by +3.1% in 2015 (Figure 2.6), compared

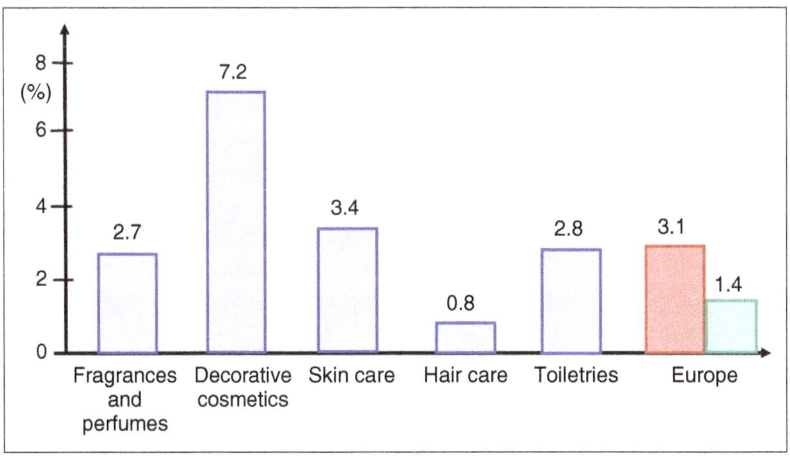

Figure 2.6 Growth of the cosmetic product categories in Europe; changes in 2015 compared to 2014, according to Cosmetics Europe [6]; (according to Statista [2]: total growth = 1.4%).

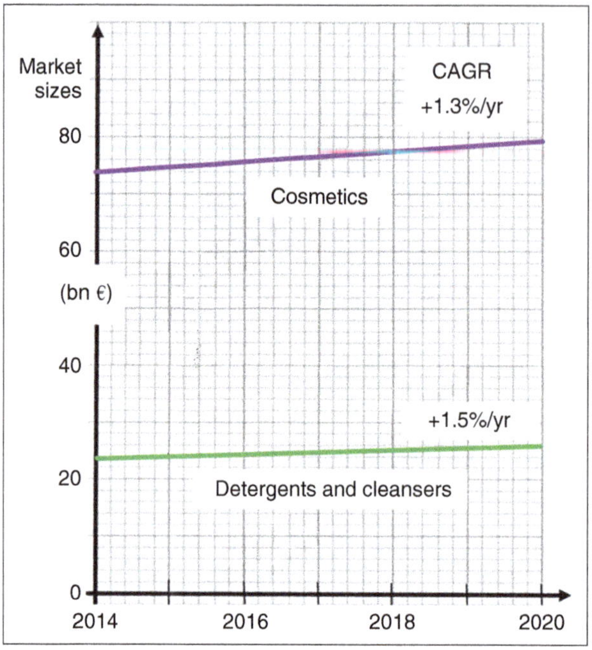

Figure 2.7 Comparison of the European markets for cosmetics as well as for detergents and cleansers (estimated figures over six years). Source: Data from Ref. [2].

to 2.1% in 2014. Growth was driven by a disproportionate increase in sales in various countries, particularly in Switzerland (+12.4%), the United Kingdom (+10.9%), Romania (+8%), and Hungary (7.5%). Regarding the product categories, makeup was the most dynamic (+7.2%), followed by skin care (+3.4%), toiletries (+2.8%), and fragrances and perfumes (+2.7%). Hair care remains the weakest category, but was able to return to the profit zone in 2015 (+0.8% vs. −0.2% last year).

The market growth sees another source, namely Statista, with average about 1.4% per year significantly different. The same value applies to the total growth of 2015, after +0.8% in 2012, −1.4% in 2013, and +0.5% in 2014. It seems as if the lower values are closer to reality (see Germany). The average compound annual growth rate (CAGR) of 1.3% has been determined over six years.

As can be seen in Figure 2.7, the cosmetics market is slightly more than three times (f = 3.18) the size of the market for detergents and cleansers (without air freshener, scented block, etc.). This statement is true not only for Europe but also for the global market (f = 3.01). In 2016, the factor is 2.94 for Germany. The markets, related to sales prices, amounted to €13 603 and €4 624 million, respectively. Over several years, growth of both product groups is between 1% and 2% in Europe, so the distances remain relatively constant. In the longer term, it can be assumed that the overall markets will continue to grow on average, with individual product groups growing faster, stagnating, or even shrinking.

2.4 Cosmetics Market in USA

The United States is the largest cosmetics market in the world with a total revenue of about $62 bn in 2016 (Market Research [8], and Statista [2]) and an annual increase (CAGR) of 3.1%. Euromonitor [9] found that driven by makeup and premium offerings, the sector is expected to rise from $80 bn (2016) to $90 bn by 2020 (CAGR 3.0%). However, the market is still growing. In addition, there are many export opportunities in emerging markets throughout Asia, Latin America, and Eastern Europe. A similar figure ($81 bn projected for 2016) was reported by Shiseidō [10] and is also calculated by multiplying the global market value ($360 bn) by approximately 24% market share for United States and Canada ($86.4 – $6.8 = $79.6). Therefore, a figure of **$80 bn** should reflect reality. According to a graphic, only the premium area grows with 3%, the total market shows an increase of 2% [10].

2.5 Cosmetics Market and Distribution Channels in Japan

Because of strong exchange rate fluctuations between the yen and dollar, the values for the cosmetics market differ widely, depending on the source and year. For 2015, a value of **$32 bn** seems realistic. The value represents an expenditure per capita and year of just over $250, comparable to the United States. As in Japan, the market for premium products is around 30% worldwide.

In terms of the population, the Japanese are the No. 1 in expenditure of money on skin care and decorative cosmetics, as demonstrated in Figure 2.8. In absolute numbers, they are second in skin care right behind the Chinese. United States is in first place in decorative cosmetics, and the Japanese are also right behind them on position 2. Astonishing is the quite high expenditure of the South Koreans for skin care. According to the Japanese, the Americans and British spend the most money on decorative cosmetics per capita. Figure 2.9 shows the market shares of the individual product categories. It is noteworthy that the Japanese use almost no perfume, only about 1/30 of the amount of Europeans. Contrary to this, the proportions of skin care products are clearly disproportionately high, a consistent feature of Asian markets. The cosmetics industry in Japan, dominated by the three companies Shiseidō, Kao, and Kosé, with a market share of 35–40% [11], uses various distribution channels that are rather unusual in Europe. The sale through pharmacies prevails, followed by mail order including Internet ordering as well as

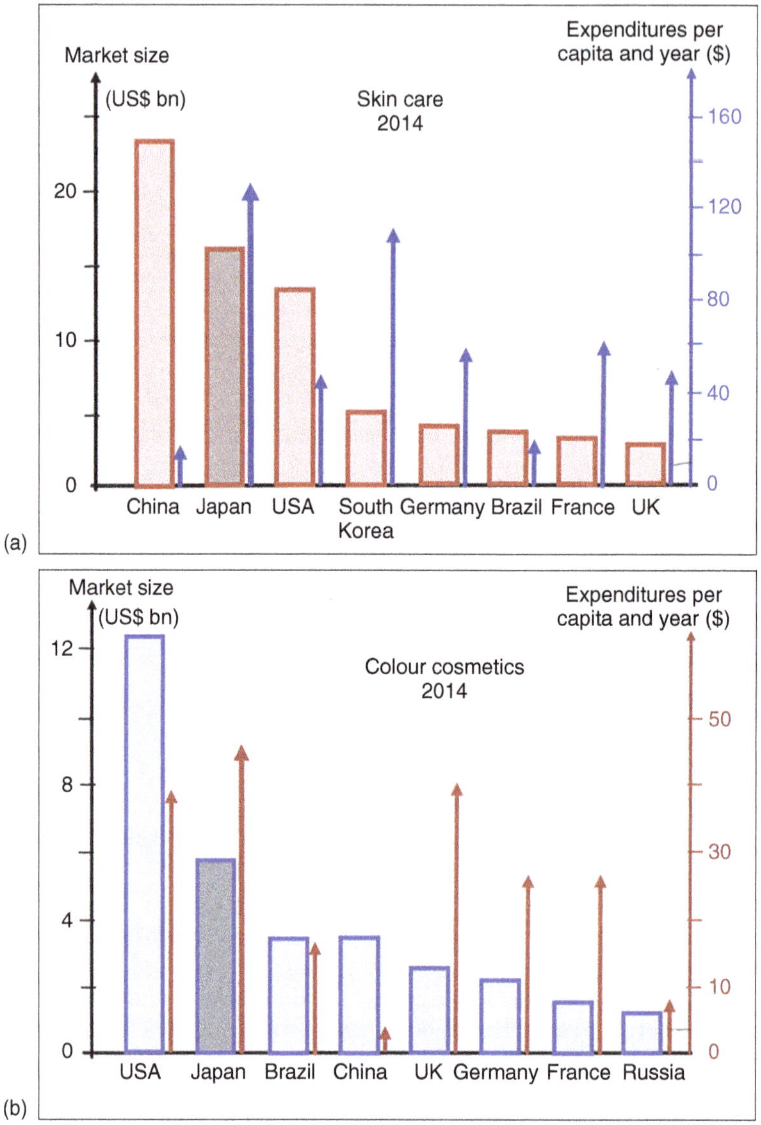

Figure 2.8 (a, b) Global market sizes for skin care and color cosmetic products as well as the positions of the three largest markets USA, China, and Japan. Source: Data from Ref. [10].

department stores and supermarkets [10]. These channels run approximately 75% of Japan's cosmetics sales. Cosmetics purchases by mail order are not common in Germany.

Asians, and especially the Japanese, attach great importance to their skin. The women, as well as the men, pay attention to their skin more and more with increasing age. Young, pure appearance is very important for Asians. Therefore, the Asian markets demand good products for skin care.

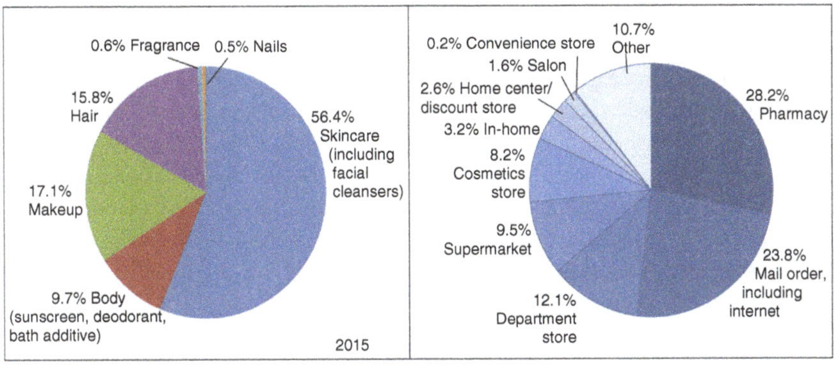

Figure 2.9 Market shares in the Japanese cosmetics market with regard to product categories and sales channels. Source: Data from Ref. [10].

2.6 Chinese Cosmetics Market

In 2015, the cosmetics industry in China probably had realized about **$50 bn** (at retail prices). Again, there are sometimes very different data in the press. By 2022, China is likely to be the world's largest market for cosmetics and toiletries, if the published growth rates are correct and the exchange rates more or less constant. This means, a $100 bn cosmetics market arises, which may be greater than that in the United States. However, the expenditures per person and year remain still long behind the United States, Japan, and the Western European countries.

Especially in China, online trading is growing strongly from 4% in 2010 to already 22% in 2015. The increases are also seen in skin care (2010: 3%, 2015: 19%) and fragrances (2010: 3%, 2015: 11%). Ordering over the Internet focuses on platforms such as Tmall (58% market share) or JD.com (17%) and less on the manufacturers' pages. According to surveys (2016/2017), some global corporations are particularly good performers. Winners represent the brands Estée Lauder, L'Oréal, Lancôme, and P&G with Olay [12].

In emerging countries with high levels of air pollution, such as China, the cosmetic "natural" products with medical effects (cosmeceuticals) are attributed special opportunities. The cosmeceuticals yield more than €2 bn thanks to a growth rate of 10–20%. The market research company Mintel sees in the future billions of growth for dermatological creams, which promote the protection of the body or, in particular, the skin from environmental toxins. The "Cosmétique Active" division of L'Oréal benefits from this trend. With a return on sales of almost 23%, it already has the world's highest profitability of cosmetic product groups in the world [13].

According to the current trends, skin care and makeup products dominated more than half of the market. Interestingly, the fast increase in the mailing of selfies via the smartphones is boosting the makeup business in China. In addition, there are small but interesting niches. Examples are skin care of the children as well as the skin care of the men, both of which are growing fast. The skin care for children is controlled by Johnson's baby care with 80% market share. According to Euromonitor, men's skin care shows an annual growth of 30%. The revenue reached €380 million in 2012.

Customer trust foreign cosmetics brands as they offer quality and recognition. However, those brands are expensive and increasingly bought by young or middle-aged female in big cities. In former times, foreign brands represented with a share of more than 80% China's cosmetics market. Today (2016), their market share in a fast-growing market is slightly below 50%. Big well-known companies from France (L'Oréal), United States (P&G and Estée Lauder), Japan (Shiseidō), and recently also from South Korea are dominating the premium market. For Chinese customers, these brands are already very popular.

2.7 World Division in Market Regions

The commercial world is divided into market regions, which are only partly based on political and geographic realities (Figure 2.10). This division also applies to the cosmetics markets. The breakdown can be found in L'Oréal annual report 2016 [14] and is important for all graphics and statistics (Europe and World). As can be seen on the map, the regions differ considerably in their geographic extent. Western Europe is very small, but of great economic importance. For Eastern Europe, the opposite is true, economically rather small in the field of cosmetics, huge in the geographical extent. North America consists of the United States and Canada. Mexico as well as Central and South America are combined to Latin America. Eastern Europe includes Russia completely and all former Eastern Bloc countries. From India in the west to Japan in the east and via Australia to New Zealand in the south extends the Asia Pacific area. The Middle East covers all countries from Egypt in the west to Pakistan in the east and represents together

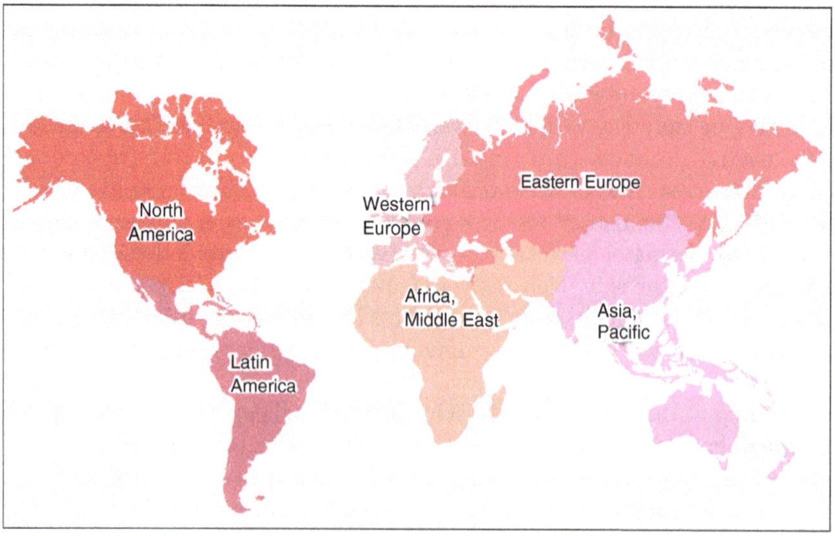

Figure 2.10 Division of the world into market regions according to commercial considerations.

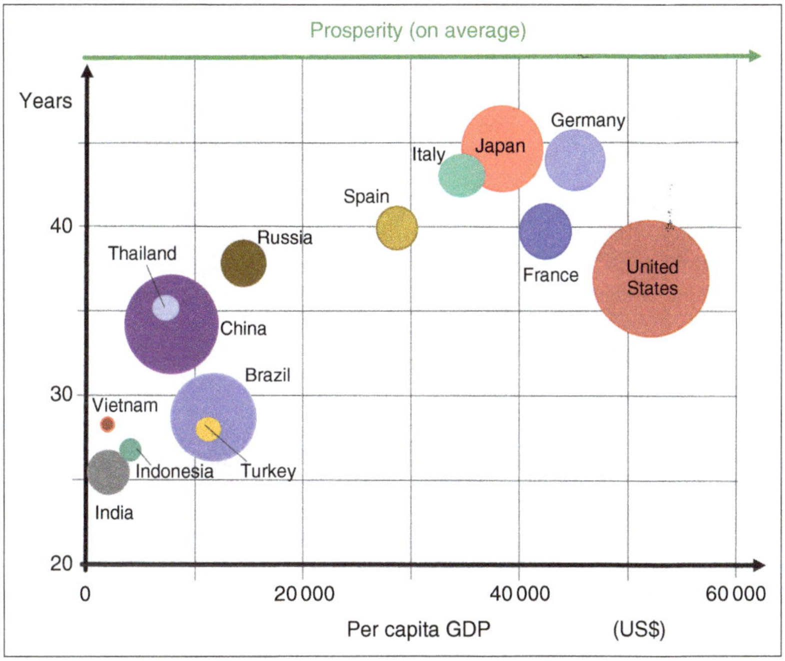

Figure 2.11 Beauty and body care markets in dependence of the GDP per capita and the median age of the population (2014); the market size in each country is expressed by the size of the circle. Source: Data from Ref. [10].

with Africa the Middle East/Africa region. In some cases, India is included there. Then, the region is called IMEA (India, Middle East, and Africa).

Figure 2.11 shows a selection of large cosmetic markets whose size is represented by the area of the circle. In addition, the ordinate shows the mean age of the population, while the abscissa demonstrates the economic strength in the form of the GDP (gross domestic product) per inhabitant. The developed countries are characterized by a relatively high middle age of the population and by a high GDP. On the other hand, there is a country like India with a very young population and a small GDP per capita. Most other countries lie between the extremes. Therefore, in addition to traditional and religious peculiarities, there are country-dependent differences in the requirements for the cosmetic industry, less for cleaning agents, but more for the personal beauty, such as skin care and decorative cosmetics for face and hair.

2.8 Global Cosmetics Market Size

The global market volume is growing steadily (2007–2016 average of +3.8% [14]) and should be around US$**128 bn for skin care**, US$89.3 bn for hair care products, and US$66.9 bn for decorative cosmetics in 2017 [2]. With the percentage of the global product distribution (Figures 2.12 and 2.13), these values lead to a total market between $350 and $390 bn for 2017, including 10.4% hygiene

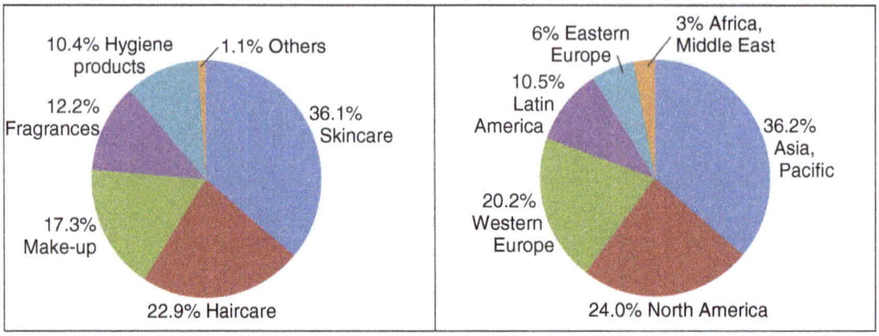

Figure 2.12 Distribution of global markets by products and regions (Basis 2015).

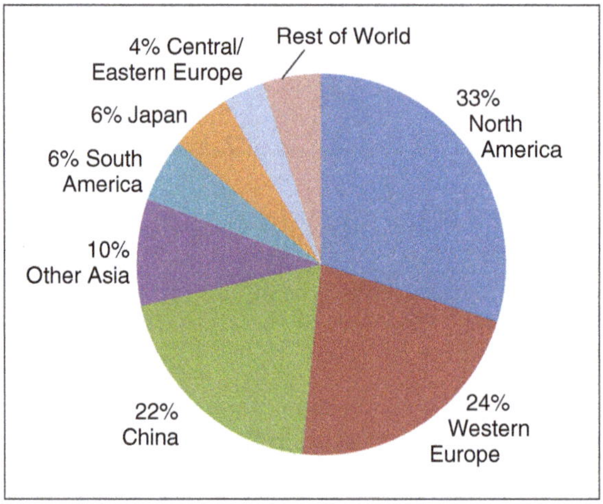

Figure 2.13 World consumptions of cosmetics chemicals by regions, based on value. Source: Data from Ref. [15].

products. Another fixed point is the size of the European market, which was €76.4 bn in 2015. Here, the extrapolation to the global market yields about €300 bn (Figure 2.14). This means, at an exchange rate of €/$ = 1.21 (2014/2015), a value of $363 bn arises.

Calculated on the basis of various sources in dollars, the size of the fragmented market is likely to be around **US$360 bn** in 2015 and should reach a value of $430 bn at a CAGR of 3.7% by 2020. If the strong increase in China will be confirmed, the value could rise to $450–470 bn. Therefore, the 360-bn (March 2017) estimates seem to come close to reality for the global cosmetics market, although other sources mention both downward and upward divergent figures. For the global cosmetics market, a statistic [16] indicates a value of $288 bn (deviation: −$72 bn) for 2015, which will grow to $336 bn by 2020. According to another source, the global cosmetics market is expected to reach $390 bn by 2020, with an expected CAGR of 3.7% during the years 2015–2020 [17]. This leads to just $330 bn in 2015. As this value is already reached by the 20 largest cosmetics markets, the figure is too low. An opposite view represents "business wire" [18] with a value of $460 bn for the year 2014 and $675 bn for 2020. This would correspond

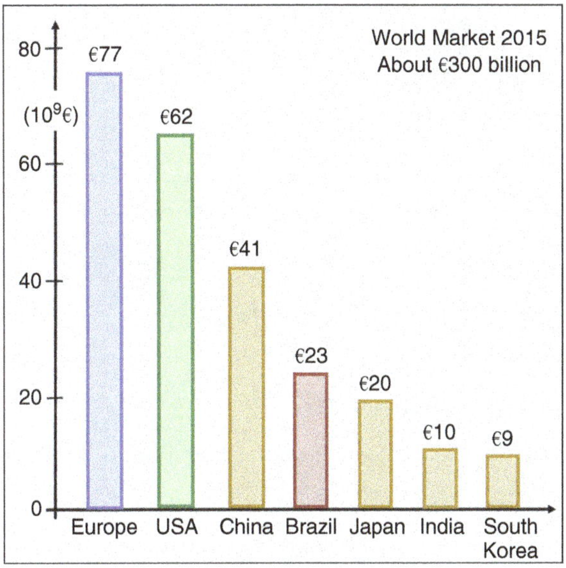

Figure 2.14 Global cosmetics market and the main sales regions (rated by retail prices in €; exchange rate probably about 1.21 = €/$; Cosmetics Europe [6]). Source: Data from Ref. [6].

to an increase rate (CAGR) of 6.6% per year. These values can only be achieved if other large products with strong growth are attributed to the cosmetics market.

Statista [10, 19] reports the size of the global markets and their percentage breakdowns, as L'Oréal does in the annual report 2015 [7]. The figures show that the Asian/Pacific market is by far the largest, followed by Europe and the United States, which are not so far from each other. In the cosmetics categories, the products for the cleaning and care of the skin are in the market share before the hair care. The decorative cosmetics grow and rank already third. These statements apply to a global view of the cosmetics markets. The percentages of each category and the order can differ from region to region (Figure 2.5).

Figure 2.14 displays the regions of the world where most cosmetics are sold. According to the market shares, Europe and North America are ahead, but a merge of the Asian countries gives the largest market (of value and geography). More countries and markets are shown in Figure 2.15. Obviously, the statistical data vary within larger areas according to the source. This demonstrates the comparison between Figures 2.14 and 2.15. The differences arise, on the one hand, from the exchange rate between the euro and dollar and, on the other hand, probably based on the differences in the definition of the cosmetic markets (e.g. with or without hygienic articles). Only the figures for the Japanese market lie outside of acceptable error areas. The numbers, found in the annual report of Shiseidō [10], indicates a significantly larger Japanese market than described in Figure 2.14. However, Shiseidō should know the home market, so the value is trustworthy. Unfortunately, their source shows only selected countries with market shares above 1 bn. For example, Korea with sales about $10 bn is missing; furthermore, Mexico ($10.5 bn), Canada ($6.8 bn), and others must be included in the list under the first 15. The addition of further countries leads to the 20 largest cosmetics markets with a volume of around $330 bn in 2015. The sum of the 20 largest markets represents more than 90% of the world market. To date (December 2017),

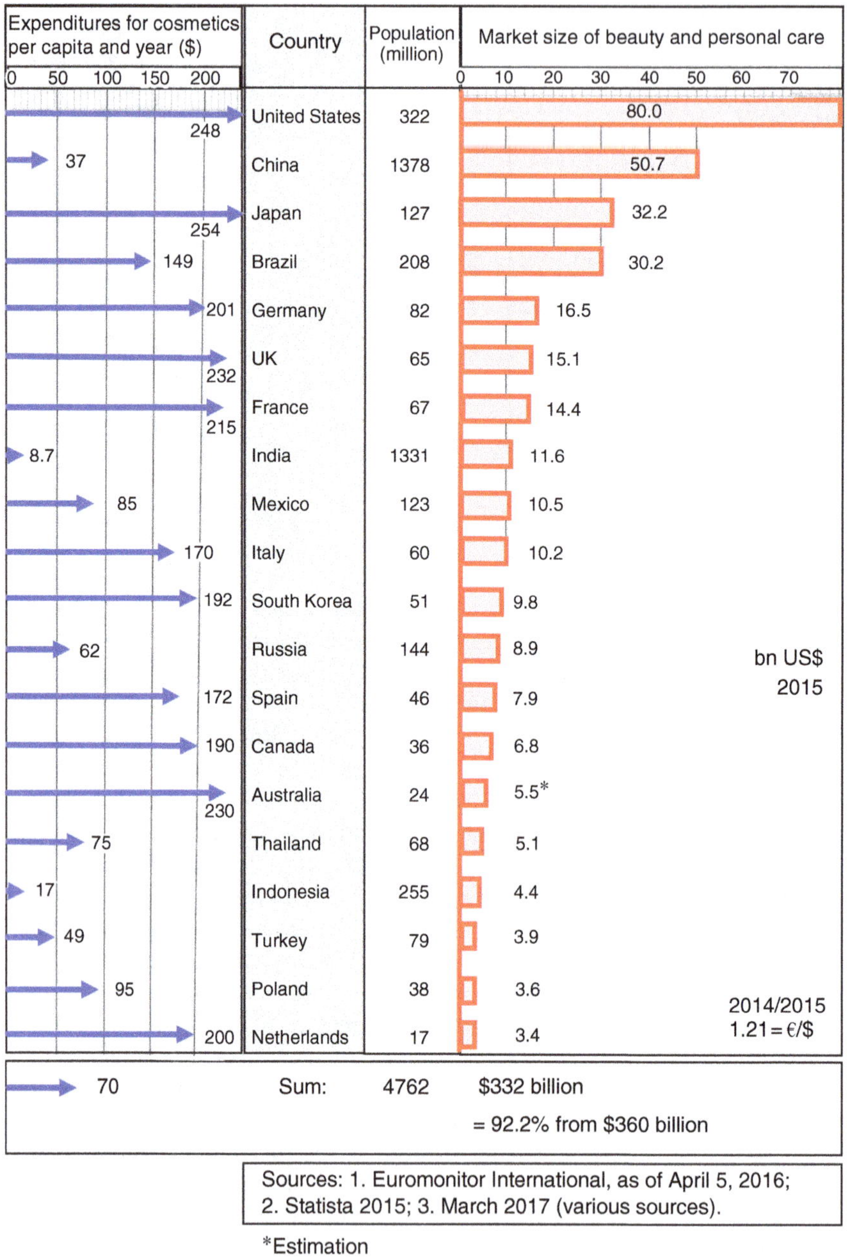

Figure 2.15 Top 20 countries with the largest cosmetics markets in the world (March 2017).

the market would have to be grown on to about \$390 bn, without taking account of exchange rate fluctuations. A strong dollar could lead to an apparent shrinking of the markets, with the market size falling to around \$340–350 bn.

The five largest cosmetics markets are the United States, China, Japan, Brazil, and Germany, which together have a market size of approximately \$210 bn

(58% of world sales). Among the top five markets, China shows by far the smallest per capita consumption. Expenditures per capita and year are highest in Switzerland and Japan, followed by the United States and the United Kingdom. The last place is occupied by India. Japanese spend 29 times more money compared to an Indian and 7 times more than a Chinese. In the wealthy countries, people expend more than \$170 per year for cosmetic products, usually over \$200 to about \$300. Expenditures per capita for cosmetics reflect approximately the prosperity of the country/region. As an underestimated Asian country, South Korea surprises as a large cosmetics producer and market with expenditures per capita roughly equivalent to Western European countries.

The division of the markets into individual product categories [20] is shown in Figure 2.16. Here, a comparison between the global and the German market distribution can be seen. For oral hygiene, Germans spend 10 times as much money as the world average. This large difference at the dental care products is very striking. In addition, remarkable differences at the toilet articles and makeup stand out. The other market shares are quite comparable.

In the "skin cleansing and care" – abbreviated as "skin care" – category, skin, facial, hand, and foot care as well as the baby creams are combined together with the typical cleaning products for bathing and showering as well as for washing with syndet soaps. The size of the pure skin creams market is expected to be between 65% and 80% of the total market for skin cleansing and care, which means that up to 2019, the market for creams will grow to a global market size of over \$100 bn. The estimate is based on the size shares of the product categories in Germany and can be taken from Figure 2.17.

The market for skin care products is very interesting because it is significantly larger and grows twice as fast as for hair care, as shown in Figure 2.18. In six years, the skin care market increases by 34%, the hair care by 16.5%. In addition, the great economic importance of skin care creams is demonstrated by a comparison with the global market for detergents and cleansers. According to Figure 2.19, the global market for skin creams also grows faster than that for detergents and

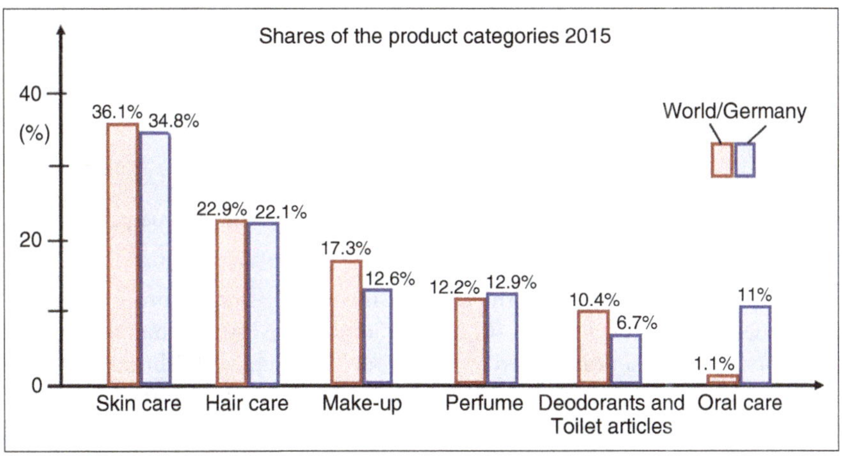

Figure 2.16 Percentage comparisons of the product shares: global cosmetics market vs. German market.

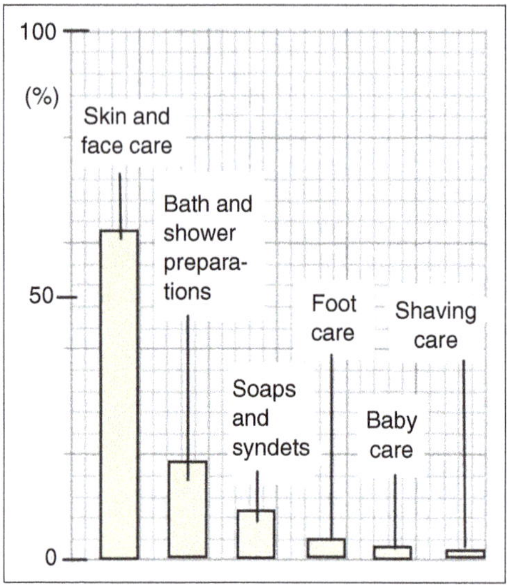

Figure 2.17 Composition of the category "skin care products."

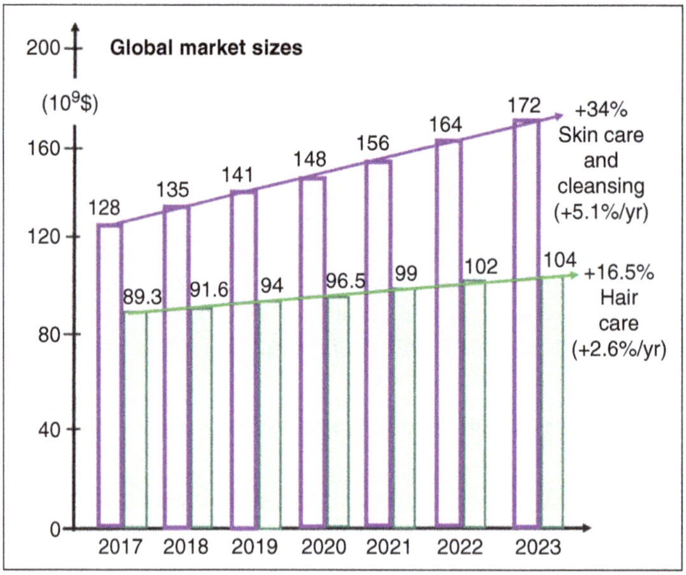

Figure 2.18 Estimated figures for the global market growth up to 2023 (February 2017; CAGR).

cleansers. From the year 2022, the skin creams should have more importance than the detergents and cleansers (€110 compared to 106 bn). The market sizes for detergents and cleansers are based on data from Statista [21] of February 2017 for the European market, adjusted for the results of polishes, room fragrances, and insect sprays. From these data, figures for the global market can be estimated. All mentioned facts prove the great importance of the skin creams that are the focus of discussions here.

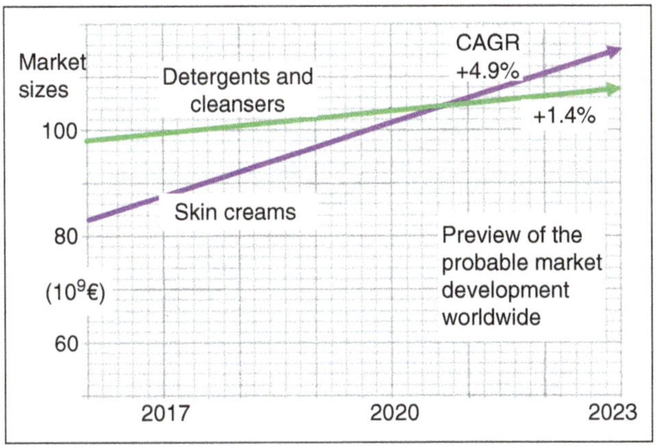

Figure 2.19 Global market growth for skin creams in comparison to detergents and cleansers up to 2023.

2.9 Trends for Future Development

Seven major trends will shape the cosmetics markets of the future and will ensure further growth. Growth rates are moderate (1–2%) in developed countries, but high (3–10%) in the developing countries and especially in the emerging markets. As main development lines are seen:

- Population growth results in more consumers
- Higher disposable income in the emerging and developing countries allows for more money to spend on cosmetics.
- People want to look attractive and also younger in age than they are, therefore developing strong interest in skin creams and antiaging products, particularly in the developed world, as well as in decorative cosmetics.
- Increased interest in premium cosmetics (forecast: $126 bn in 2019).
- Increasing consumer awareness regarding wellness and health creates a further demand for mild and natural cosmetics as well as for cosmeceuticals.
- Enhanced use of personal care products by men.
- Further market segregation is likely for ethnic products, male grooming, silver agers, and adolescents. They may create new market opportunities.

It is to be expected that in some regions, the legal requirements for the safety of cosmetic formulations will be tightened, so that some expensive adjustments have to be made by the companies.

2.10 Largest Cosmetics Manufacturers Worldwide

The largest manufacturers of cosmetic products are listed in Figure 2.20. This listing makes it clear that the three big markets, L'Oréal, Unilever, and Procter & Gamble, occupy top positions in their cosmetics areas worldwide. Looking at the individual countries, the strong position of France with three manufacturers

Country	Company	Turnover (bn US$)
France	L'Oreál	28.0
UK and Netherlands	Unilever*	20.5
USA	Procter & Gamble*	17.6
USA	Estee Lauder	11.1
Japan	Shiseido	7.11
Germany	Beiersdorf*	5.89
USA	Johnson & Johnson*	5.57
France	Chanel*	5.24
USA	Avon Products	5.20
Japan	Kao Corp.	5.04
France	LVMH	5.02
South Korea	Amorepacific Group	4.94
USA	L. Brands*	4.60
USA	Coty	4.28
Germany	Henkel*	4.04
USA	Mary Kay	4.00
USA	Colgate-Palmolive*	2.95
Brazil	Natura Cosmeticos*	2.41
South Korea	LG Household & Healthcare*	2.25
USA	Alticor	2.15

2015

*Cosmetics division.

Number of companies under the top 20	Region	Turnover (bn US$)	Share of the top 20
6	Western Europe	68.69	46.4%
9	North America (USA)	57.45	38.8%
4	Asia	19.34	13.1%
1	Latin America (Brazil)	2.41	1.6%
20	sum: **$148 billion** turnover = 65% of the total turnover of all companies		

Figure 2.20 World's top 20 manufacturers of cosmetics with the country/region of their headquarters. Source: Data from Ref. [22].

under the top 20 stands out with total turnovers of approximately $38 bn. Therefore, France is the home to L'Oréal, the world's largest cosmetics company. With a fivefold larger population and sales of $57 bn, the United States is also heavily involved in cosmetics and represents 2 manufacturers under the top five (third and fourth place) and 9 among the first 20. The United Kingdom and the Netherlands occupy with Unilever the second place in the cosmetic world. In the fifth position, Japan follows with a total of $12 bn under the top 20 and then Germany with almost $10 bn turnover.

Both Europe and the United States export large parts of their production. L'Oréal generates two-thirds of sales outside Europe. A list of regions shows that Europe is ahead of the United States and Asia. With L'Oréal and Unilever, Europe places the two largest players in cosmetics. This is followed by the United States with Procter & Gamble and Estée Lauder. The No. 5 is the first representative of Asia, the Japanese company Shiseidō. In terms of the population, there is a great potential for resident companies in Asia, especially for daily low-cost cosmetics, such as strong growing companies in South Korea and China demonstrate.

Caution when comparing! The list of the top 20 companies is based on the real turnovers (Figure 2.20), which are accessible through annual reports. The accurate figures are published annually, at the end of the company's fiscal year, i.e. generally at a quarter-end or year-end. In contrast to this, the market sizes, based on retail prices, are queried by market research companies in the trade (Figure 2.15) and may differ according to the accuracy of the query and the definition of cosmetics, besides the exchange rate problems. For a comparison on the same basis, the market prices would have to be adjusted to the turnover of the producers, respectively, to the purchasing prices of dealers. In a first approximation, the market size is nearly to halve, so the $360 bn in retail prices would give about **€200 bn turnover** to all cosmetics companies (2015, a figure from L'Oréal [14]; €205 in 2016).

The top 20 represents about 65% of the cosmetics produced. The strongly fragmented residual market with sales of approximately $82 bn is spread over thousands of small companies. A look at the world's top five suppliers of cosmetic products [7] displays that they generate turnovers of more than $84.3 bn, corresponding to 37.0% of the world's turnover at manufacturer prices, approximately 57% of the top 20. Evidently, the top 20 dominate the world market. As they preferably cover the premium and luxury sectors, this applies in particular to higher priced products.

The top five markets benefit from their strong market positions, based on high-quality products and well-known brands, as well as from the worldwide growing demand. The branded item concept, which the major manufacturers all pursue, has at least three basic features [23]:

- Proof of origin and quality guarantee
- Image and valence
- Ubiquity (anywhere to buy).

In the branded article strategy, the achievement of a performance-specific advantage (quality leadership) is used as sales argument. This is replaced in the price–quantity strategy by the low price. The brand manufacturers want to offer their customers products that they value above average and therefore are willing to pay a higher price. This is all the more successful, the better the manufacturer is known and the stronger the brand. Especially effective is a brand name that is identical to the manufacturer's name. Good examples are L'Oréal and Shiseidō. The world market leaders are described in more detail in the following section. Many of their brands are well known; a selection is listed.

2.11 Top Five Manufacturers

No. 1 in the world is **L'Oréal** [7, 14], based in Paris (Clichy), France. According to own data, revenues amounted in 2015 to **\$28.0 bn** (=€25.3 bn, with €4.4 bn profit = 17.4%) and were obtained with 82 900 employees worldwide. The company focuses exclusively on cosmetics with a wide range of consumer and professional as well as luxury products. The company possesses a global share of 12.7% in the cosmetics market. This is divided into the hair care/coloration (>30%), the skin care (28.5%), and with more than 26% in the decorative cosmetics (Figure 2.21). L'Oréal sells its products predominantly in Western Europe and North America, but increasingly also in the Asia/Pacific region. Sales and earnings have risen steadily over the last 20 years (Figure 2.22). The growth of L'Oréal's sales was in the last 10 years with 4.7% CAGR clearly above the average global growth of 3.8%. The company now distributes its products in 140 countries. Founded in 1909, the business originated from a small-scale production of hair dyeing, which the chemist Eugène Schueller took up in Paris 1907. The most important shareholder with 30.8% was Liliane Bettencourt (* 1922, † 2017, * birth and † date of death, signs are significant), the daughter of Eugène Schueller. The Nestlé Group is the second-largest shareholder with 29.6%.

As usual, in many companies, L'Oréal is organized in four divisions, as shown in Figure 2.23. Consumer products account for almost 50% of sales. However, also the luxury division runs very well with a share of over 30%. The more than 30 international, predominantly strong brands of the four divisions can be found in Figure 2.24. This broad positioning supports market growth in all regions.

In 2016, L'Oréal extended its leadership in Western Europe to 20.2% market share with a growth of 2.4%. The group achieved strong market share gains in the United Kingdom through double-digit growth and increased sales also in Germany and Spain. Russia, Poland, and Ukraine drove growth in Eastern Europe. All divisions share the success. Market share in Eastern Europe: +10.4%, growth 12.8%. In the United States, the largest beauty market in the world, L'Oréal is the

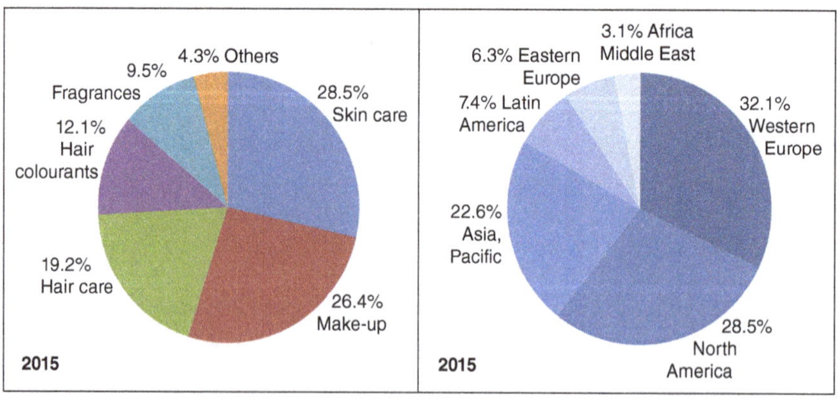

Figure 2.21 L'Oréal's product portfolio and their global sales regions. Source: Data from Ref. [24].

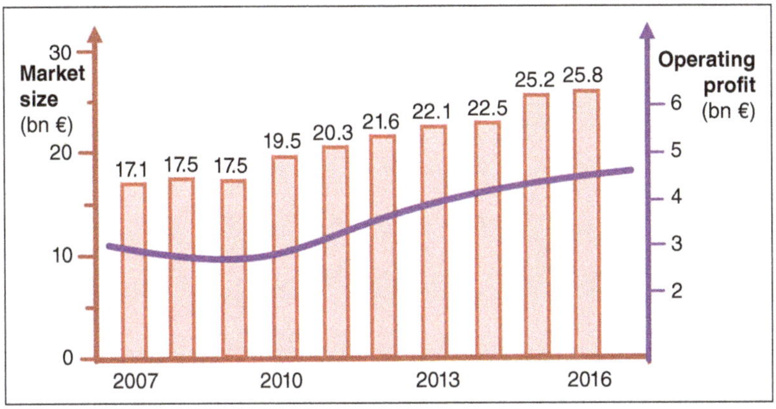

Figure 2.22 L'Oréal's revenues in the past 10 years and their operating profits [14]; the CAGR amounts over 10 years to 4.7% for sales and 3.5% for the operating profit, respectively, over 7 years (2009–2016) to 5.8% and 8.4%.

Figure 2.23 L'Oréal's operational divisions (business units).

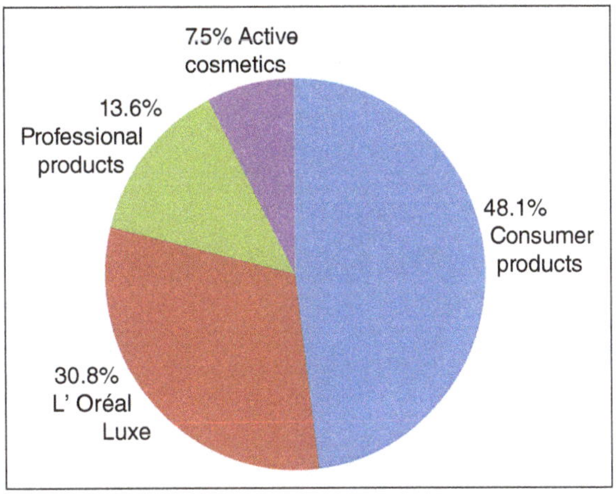

No. 1 beauty company with 14.0% market share and a growth rate of 5.8%. Drivers of growth are the SkinCeutical products as well as products for sun protection and against acne, both recommended by doctors. SkinCeuticals (American medical esthetic skincare brand) owes its success to a combination of uniqueness and function. The products were developed in collaboration with beauty institutes. The special skin care begins after the completion of esthetic skin treatments with intensive pulsed light, to calm the skin after laser treatment.

In Latin America, especially in Mexico, Colombia, and Peru, L'Oréal is generating substantial market share gains, thanks to the dynamic growth of the makeup brands. In Brazil, the Active Cosmetics Division outperformed the market. Overall, the market share is 8.8% with a growth of 11.1%. In the Asia-Pacific region, L'Oréal performed well thanks to buoyant markets in Indonesia, Thailand, and Australia. Taiwan and South Korea also posted good growth. In China, L'Oréal Luxe achieved double-digit growth, but the Consumer Products Division weakened. Market share in the region is 9.5% (without Japan)

Figure 2.24 L'Oréal's brands, sorted by operational divisions [14].

with 3.6% growth rate. The market share in Africa and Middle East is 13.0% with a growth of 7.9%.

No. 2 is **Unilever**, a Dutch–British company, with headquarters in Rotterdam (the Netherlands), as well as in Wirral and London (UK). The personal care division achieved revenues of **$20.5 bn** in 2015. According to own data [25], the turnover was €20.1 bn and the operating margin was 18.1%. Together with the foods (€12.9 bn), home care (€10.2 bn), and refreshment (€10.1 bn) industry, the company employs 172 000 people, who generate sales of approximately €53.3 bn in more than 100 countries. Unilever is one of the world's largest consumer goods manufacturers. There are two roots. On the one hand, at the end of the nineteenth century, two Dutch entrepreneurs founded companies for the production of Margarine (Margarine Unie). On the other hand, the Lever brothers built a soap factory (Sunlight Soap) in 1885, using natural palm oil instead of the customary tallow. In 1929/1930, Margarine Unie and Lever Brothers Ltd. joined together to Unilever.

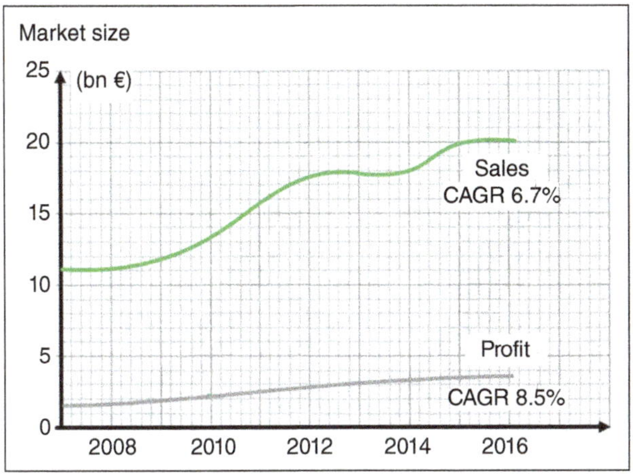

Figure 2.25 Sales and profits of Unilever cosmetics division.

Brands of the cosmetics division: Axe, Baba, Dove, duschdas, Fissan, Impulse, Lux, Pond's, Rexona, Signal (in some countries, e.g. Austria, and Italy: Mentadent), Sunsilk, TIGI, Timotei.

Figure 2.25 shows the strong growing sales and profit of Unilever Cosmetics. The relative operating margin is highest in the United States, followed by Europe. That means, in the United States, the relative operating margin is significantly higher than the sales share with 32%. In Europe, the relationship seems to be balanced. In the rapidly growing Asian market, which already accounts over 40%, profits per unit are the lowest. Especially thanks to strong growth in Asia, also compared to the competition, Unilever Cosmetics has almost doubled its turnover in 10 years (until 2016) and more than doubled the profit in the same time.

The **Procter & Gamble** Company (worldwide No. 3), represented in 70 countries, is a US-based consumer goods group, headquartered in Cincinnati, OH. The sale of cosmetic products achieved 2015 a turnover of **$17.6 bn** [26], corresponding to 27% of the total sales. Probably, the market researchers have summarized the beauty with a part of the health care (oral care) and the grooming to the cosmetics sector. Based on the Annual Report 2016, the share of beauty is $11.5 bn (without the oral care). In 2016, the company generated a total turnover of $65.3 bn (2015; $70.7 bn) with 105 000 employees in the field of consumer goods (Figure 2.26 and Table 2.2). The operating income was 20.6% of the net sales and is above that of comparable companies, probably because of the strong market presence in the United States.

In 1837, Procter & Gamble was established in the United States by two emigrants from the United Kingdom and Ireland. These were the English-born candle maker William Procter (1801–1884) and his brother-in-law, the Irish soap-maker James Gamble (1803–1891). Procter & Gamble is the pioneer of brand management. They always place the brand in the foreground, while the manufacturer is hardly recognizable on the products. The company is focused on North America. This is reflected in the high market share of 44%, whereas Europe only accounting

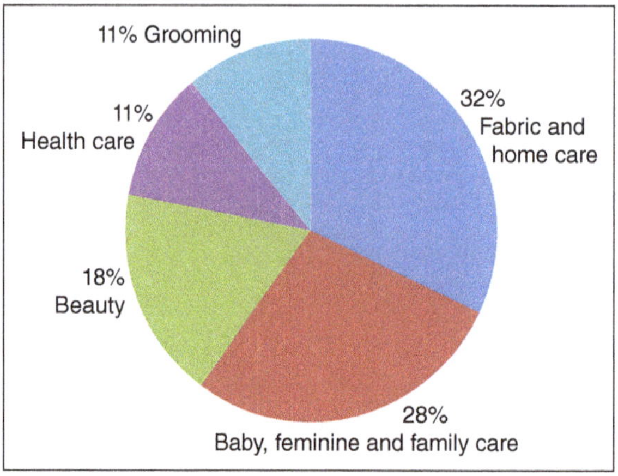

Figure 2.26 Shares of the Procter & Gamble's divisions.

Table 2.2 Global business units of P&G 2016 [26].

	Fabric and home care	Baby, feminine, and family care	Beauty[a]	Health care	Grooming
Net sales	**32%**	**28%**	**18%**	**11%**	**11%**
Net earnings	**27%**	**26%**	**20%**	**12%**	**15%**
	– Fabric care Laundry detergents, fabric enhancers, laundry additives	– Baby care Diapers and pants, baby wipes	– *Skin and personal care* *Skin care, antiperspirant and deodorant, personal cleansing*	– Personal health care Gastrointestinal, respiratory, rapid diagnostics, vitamins/minerals/supplements, other personal health care	– Grooming Male blades and razors, female blades and razors, *pre- and post-shave products, appliances, other shave care*
	– Home care Dish care, air care, surface care, P&G professional	– Feminine care Feminine care, adult incontinence	– *Hair care* *Shampoo, conditioner, styling aids, treatments*	– *Oral care* *Toothbrushes, toothpaste, other oral care*	
		– Family care Paper towels, tissues, toilet paper			

a) Cosmetics in italics.
Source: Data from Ref. [26].

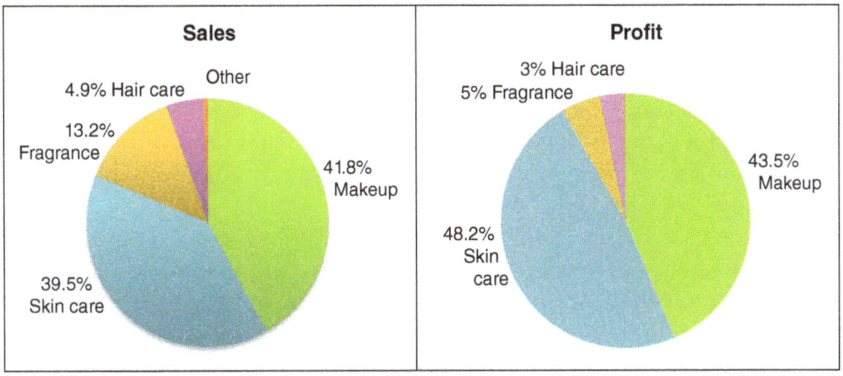

Figure 2.27 Estée Lauder's product portfolio, sales, and profits (2016).

for 23% as well as the Asia/Pacific region for 17% (including 8% China). Latin America and IMEA are for P&G regions with the lowest sales, both with 8%.

Well-known P&G-brands in the field of cosmetics: Avril Lavigne, blend-a-dent, blend-a-med, Blendax, Crest, Head & Shoulders, Herbal Essences, Oil of Olaz, Oral-B, Pantene, Puma.

Estée Lauder Companies (No. 4 in the world) is an American cosmetics business founded in 1946 by Estée Lauder, a US-American cosmetics entrepreneur (born 1906). Headquartered in New York City, the company generated sales of **$11.1 bn** (2015) with 10.1% profit [27]. The net earnings are significantly lower than the usual 17–20% for the top three. Estée Lauder Companies sold 42% of products in the Americas, 39% predominantly in Europe, and 19% in Asia/Pacific. The company has great cosmetic competence, especially in the area of decorative cosmetics (see Figure 2.27), but also in the skin care. The two areas are characterized by an above-average profit, the skin care business is particularly positive. The cosmetics company used the brand name Estée Lauder consistently. Under this brand and the brands of subsidiaries, care and decorative cosmetics as well as perfume for women and men are marketed in over 150 countries worldwide. The Group owns more than 30 other brands, partially internationally well known, either established or purchased by Estée Lauder Companies.

Brands overview (next to Estée Lauder): AERIN Beauty, Aramis, Aveda, BECCA, Bobbi Brown, Bumble and bumble, Clinique Darphin, DKNY Fragrances, Donna Karan Cosmetics, Editions de Parfums Frédéric Malle, Ermenegildo Zegna Parfums, GLAMGLOW, Jo Malone London, KILIAN, Kiton, La Mer, Lab Series, Le Labo, MAC, Michael Kors Beauty, Origins, RODIN olio lusso, Smashbox, Tom Ford Beauty, Tommy Hilfiger Toiletries, Too Faced, Tory Burch Beauty.

K.K. Shiseidō/Japan is worldwide No. 5 with sales of **$7.11 bn** (own data: $6.47 bn [10] in the report period 2015, ended March 2015) in 71 countries and employs 33 000 people. Shiseidō sells cosmetics for the luxury segment as well as other skin, hair, and personal care products. The company produces a wide range of products for body, face, and hair care. In addition, there are famous decorative cosmetics from the company and known perfumes. In Europe, the products are marketed under the brand name Shiseidō. In Japan, also in China, there are a number of products under different, here unknown brands for the local market. Exceptions are the perfumes with the international brands such as Issey Miyake, Jean Paul Gaultier, and Narciso Rodriguez. Shiseidō wants to grow again after a longer stagnating phase until 2013 and years without or with only small operating margin. The operating income in 2015 was about 3.6%. The company has launched a program to increase profits. Target is >10% by 2020. The company aims to become the No. 1 in the home market Japan by overtaking Kao with a growth of 3% CAGR by 2020. Within the next three years with starting point 2017, the market position in China is to be expanded with a CAGR of 9% and in Asia with 7% as well as a share of 30% of the e-commerce trade. The growth targets for North America are 6% and for Europe 2% (CAGR).

It is considered one of the oldest cosmetics companies in the world. The roots date back to the nineteenth century. The pharmacologist Arinobu Fukuhara founded 1872 in Japan (Tokyo) the first pharmacy in Western style. In 1888, Japan's first toothpaste (instead of powder) was made. The 1897 introduced skin care products "Eudermine" exists in a revised formulation until now. This was the start of the cosmetics business.

It can be seen from Figure 2.28 that L'Oréal and Estée Lauder are growing in sales continuously with a CAGR of 4.7% and 5.4%, respectively, over nine years. Shiseidō stagnated for a long time and has only been growing since 2014. The CAGR is calculated to be 1.4%. Procter & Gamble's total turnover has fallen since the year 2011. From 2008 to 2016, the CAGR is – 3%. At Unilever, some dents can be seen in the course of sales. Overall sales have been positive with a CAGR of 3.0%. Unilever sales of cosmetic products increased by 6.6% (2005–2016, CAGR), which is the highest rate of growth among the top five companies. The increase of profit by 8.5% CAGR exceeds the competitors in the top five by far. Unilever

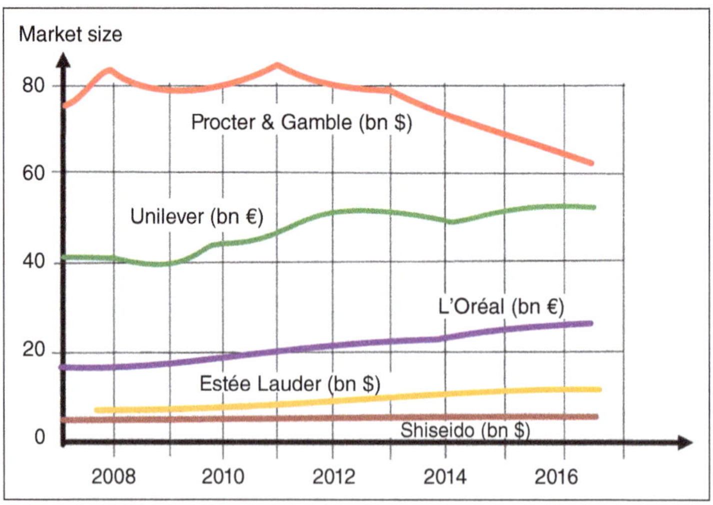

Figure 2.28 Total revenue of the top five companies in the course of 10 years (Unilever and P&G as consumer product companies, the others are cosmetics companies).

Table 2.3 Characteristics of the top cosmetics companies in comparison with food, consumer goods, and pharmaceutical companies (2015).

Company	Country	Founded	Turnover	Number of employees 2015	Revenue per employee
L'Oréal	France	1909	€25.3 bn	82 900	€305 000
Unilever	UK/the Netherlands	1929/1930	€53.3 bn €20.1 bn[a)]	179 000	€298 000
Procter & Gamble	USA	1837	$70.7 bn $17.6 bn[a)]	110 000	$643 000
Estée Lauder	USA	1946	$11.1 bn	46 000	$241 000
Shiseidō	Japan	1872	$7.11 bn	33 000	$215 000
Beiersdorf	Germany	1882	€6.69 bn €5.55 bn[a)]	17 650	€379 000
For comparison					
Nestlé (foods)	Swiss	1867	88.8 bn CHF	335 000	265 000 CHF
Henkel[b)](consumer goods)	Germany	1876	€18.1 bn	49 450	€366 000
Novartis (pharma)	Swiss	1996 fusion of Ciba Geigy (1859) and Sandoz (1886)	$49.4 bn	118 700	$416 000

CHF, Swiss francs.
a) Cosmetics.
b) About 21% cosmetics.

has an operating margin of 14.8% (after 14.1%) for 2016, which is borne by the personal care sector (18.4%) and the food sector (17.4%). Home Care and Refreshment are significantly lower with 9.5% and 9.7%, respectively.

Table 2.3 shows some characteristics of the top five companies (2015). The roots of the most successful cosmetics companies, with the exception of Estée Lauder, date back to the nineteenth or early twentieth century, in contrast to the modern technology companies. The high per capita sales of Procter & Gamble employees, which make up almost three times the Shiseidō value, are striking. At this key figure, L'Oréal and Unilever are lower, but on the same level, whereas Estée Lauder lies clearly below.

In addition to the top five companies, the No. 6 should be briefly mentioned. Beiersdorf has managed to create an umbrella brand with the world-famous name "Nivea" for over 300 different products. The company was founded in 1882 in Hamburg by the pharmacist Paul Carl Beiersdorf, who claimed a patent for a novel manufacturing method for medical patches. Subsequently, the company was taken over by Oscar Troplowitz in 1890. In the year 1911, the first stable fat and moisture cream of the world was developed and marketed under the name Nivea. In 1925, the brand-colors blue and white were introduced for

Nivea. Today, the company mainly sells cosmetics under the brand names Nivea, Eucerin, Labello, and 8x4 as well as Hansaplast as a medical product. About 17% of the turnover is achieved by means of adhesive, in particular adhesive tapes, under the brand name "tesa." According to the 2016 Annual Report, sales of 17 934 employees totaled €6.752 bn, of which €5606 bn (or 83%) are attributable to the "Cosmetics" consumer segment. The earnings before interest and taxes (EBIT) are approximately 15%. In the last 10 years, the CAGR lies just under 2.1%. For research and development, Beiersdorf spent 2.8% of sales.

2.12 Learnings

✓ In 2015, the global cosmetics market has a size of $360 bn at retail prices.
✓ The growth of the global market is 3.7% CAGR in forecasts until 2023.
✓ The largest markets are the United States with $80 bn, followed by China (51), Japan (32), Brazil (30), and Germany ($16.5 bn).
✓ The developed countries show the highest expenditures for cosmetics per capita and year.
✓ China has the highest growth rates.
✓ Calculated in selling prices of producers, the market size is $200 bn (2015).
✓ The top five manufacturers are L'Oréal, Unilever, Procter & Gamble, Estée Lauder, and Shiseidō.
✓ The No. 1 company, L'Oréal, realized $28 bn worldwide (2015) with 17.4% EBIT.
✓ From 2020 to 2021, the global turnover with cosmetic creams will exceed $100 bn (at retail prices), which is thus larger than the market for detergents and cleaners.
✓ The development is driven by the disproportionate use of skin care creams in the Asian countries and of the antiaging products in all parts of the world.

References

1 IKW, Schönheitspflegemittel, Schönheitspflegemittelmarkt in Deutschland 2018 zu Endverbraucherpreisen. https://www.ikw.org/ikw/der-ikw/fakten-zahlen/marktzahlen (accessed 2 May 2017).
2 Statista, Kosmetik & Körperpflege, Marktdaten zum Kosmetik- und Körperpflegemarkt. https://de.statista.com/statistik/kategorien/kategorie/12/themen/93/branche/kosmetik-koerperpflege/ (accessed 20 November 2018).
3 AGOF (2016). Q2 2016 AGOF facts & figures, Parfum & Kosmetik. https://www.agof.de/download/Downloads_FactsundFigures/Downloads_FactsundFigures_2016/Downloads_FactsundFigures_2016_Kosmetik/Q2-2016_AGOF_facts%20figures_Kosmetik.pdf (accessed 9 May 2019).
4 Online Solutions Group, Online Marketing Lösungen: Kosmetik-Branche. https://www.onlinesolutionsgroup.de/online-marketing-kosmetik.html (accessed 20 November 2018).

5 Premium beauty news, The European cosmetics market returns to a more dynamic growth, 2016. http://www.premiumbeautynews.com/en/the-european-cosmetics-market,9946 (accessed 20 November 2018).
6 Cosmetics Europe, The Personal Care Association, Cosmetics and personal care industry overview, 2018. https://www.cosmeticseurope.eu/cosmetics-industry (accessed 20 November 2018).
7 L'Oréal, Annual Report 2015, The world of beauty in 2015, 2016. http://www.loreal-finance.com/en/annual-report-2015/cosmetics-market (accessed 20 November 2018).
8 Global Cosmetics News, 2016. https://globalcosmeticsnews.com/north-america/2276/us-cosmetics-market-to-hit-us-62-bn-in-2016 (accessed 20 November 2018).
9 Euromonitor International, Beauty and Personal Care. https://www.euromonitor.com/beauty-and-personal-care (accessed 20 November 2018).
10 Shiseido, Annual Report 2015 (for the year ended March 31, 2015). https://www.shiseidogroup.com/ir/library/annual/pdf/2015/anu00001.pdf (accessed 20 November 2018).
11 Cosmetic Business, News & Networking for the Cosmetics Industry, Kosmetikmarkt und digitale Kompetenz in Japan, 2015. http://www.cosmetic-business.com/de/neuigkeiten/208981 (accessed 20 November 2018).
12 Cosmetic Business, News & Networking for the Cosmetics Industry, L2 veröffentlicht Digital IQ Index für China, 2017. http://www.cosmetic-business.com/de/neuigkeiten/600292 (accessed 20 November 2018).
13 GTAI Germany Trade and Invest, Chinas Damenwelt treibt Kosmetik-Umsätze in die Höhe, 2018. http://www.gtai.de/GTAI/Navigation/DE/Trade/Maerkte/suche,t=chinas-damenwelt-treibt-kosmetikumsaetze-in-die-hoehe,did=1997890.html (accessed 20 November 2018).
14 L'Oréal Finance, Regulated information, Financial News Releases CLICHY, 9 FEBRUARY 2017, 2016 Annual Results. http://www.loreal-finance.com/eng/news/2016-annual-results-1168.htm (accessed 20 November 2018).
15 HIS Markit, Cosmetic Chemicals, 2016. https://www.ihs.com/products/chemical-cosmetic-scup.html (accessed 21 November 2018).
16 Raconteur, The global cosmetics market, 2016. https://www.raconteur.net/infographics/the-global-cosmetics-market (accessed 20 November 2018).
17 Global Cosmetics Market, Global Cosmetics Manufacturing Industry Revenue, 2016. http://res.cloudinary.com/yumyoshojin/image/upload/v1/pdf/the-beauty-economy-2016.pdf (accessed 21 November 2018).
18 BUSINESS WIRE, A Berkshire Hathaway Company, Research and Markets: Global Cosmetics Market 2015-2020: Market was $460 Billion in 2014 and is Estimated to Reach $675 Billion by 2020, 2015. http://www.businesswire.com/news/home/20150727005524/en/Research-Markets-Global-Cosmetics-Market-2015-2020-Market (accessed 21 November 2018).
19 Statista, Verteilung der Markenwerte von Kosmetikmarken weltweit nach Ländern im Jahr 2018, 2019. https://de.statista.com/statistik/daten/studie/484877/umfrage/weltmarktanteile-von-kosmetika-nach-regionen (accessed 6 May 2019).

20 Statista, Umsatzverteilung im Kosmetikmarkt weltweit nach Produktgruppen im Jahr 2018, 2019. https://de.statista.com/statistik/daten/studie/490169/umfrage/weltmarktanteile-von-kosmetika-nach-produktkategorie (accessed 6 May 2019).

21 Statista, Statistiken zu Wasch-, Putz- und Reinigungsmitteln, 2018, 2019. https://de.statista.com/themen/1473/wasch-und-reinigungsmittel (accessed 6 May 2019).

22 Statista, Markenwert der wertvollsten Kosmetikmarken weltweit im Jahr 2018 (in Millionen US-Dollar), 2019. https://de.statista.com/statistik/daten/studie/484914/umfrage/umsaetze-der-20-fuehrenden-kosmetikhersteller-weltweit (accessed 6 May 2019).

23 Becker, J. (2006). *Marketing-Konzeption, Grundlagen des zielstrategischen und operativen Marketing-Management*, 8e. München: Verlag Vahlen, S. 516 ff.

24 L'Oréal, Annual Report 2016, L'Oréal in figures, A robust and balanced business model, 2017. http://www.loreal-finance.com/en/annual-report-2016/key-figures (accessed 21 November 2018).

25 Unilever Annual Report and Accounts 2016, 2017. https://www.unilever.com/Images/unilever-annual-report-and-accounts-2016_tcm244-498880_en.pdf (accessed 21 November 2018).

26 P&G 2016 Annual Report, 2017. https://www.pg.com/fr_FR/downloads/annual_reports/PG_Annual_Report_2016.pdf (accessed 21 November 2018).

27 http://www.elcompanies.com/~/media/Files/E/Estee-Lauder/investors/earnings-and-financials/annual-reports/elc-ar-2016.pdf (accessed 21 November 2018).

3

Cost Structure of the Cosmetic Products and Their Manufacturers

3.1 Rough Calculation of the Costs

3.1.1 Overview

The profitability of a company in the cosmetic industry can generally be stated. The revenues from the sales of the product must be greater than the accumulated costs, including interest and taxes (expense). The difference between these two figures corresponds to the net income for the financial year. The surplus is published annually in a balance sheet and profit and loss account. The revenues can be put together easily. By contrast, determining the costs of product manufacturing and managing the business is a complex task. The labor, energy, and distribution costs depend heavily on the location of the production facility. The same applies to interest and taxes, whereby the interest rate also depends on the creditworthiness of the company. The comparability of different locations/companies is based on globally valid business key figures.

EBIT stands for corporate earnings before interest and taxes and is referred to as operating income (Table 3.1). The result includes all income and expenses before the financial result, taxes, and extraordinary items. It can be found in the income statement of the annual report. The EBIT enables a high degree of comparability of different companies, also internationally, as differences in financing and tax treatment do not distort the result.

A real example for the construction of a profit and loss account can be found in Table 3.2 for L'Oréal. In the cosmetics industry, the two items "Marketing and sales (M&S)" and "Administration" are often divided into "Advertising and promotion" and "Selling, general and administrative expenses." Assuming 5.5% administrative costs, the very respectable values for marketing and sales are around 45%. This is 1.6 times the cost of manufacturing. No other industry spends so much money on marketing and sales.

In order to capture all costs, the manufacturing costs and expenditures in the enterprise must be registered carefully and summarized. For existing plants, the production manager provides the manufacturing costs. However, these costs do not yet include the costs of the company and the finances. In order to calculate the product costs, the individual and overhead costs of the company determined in accounting are distributed among the products and added to the manufacturing costs.

Cosmetic Creams: Development, Manufacture and Marketing of Effective Skin Care Products, First Edition. Wilfried Rähse.

Table 3.1 Calculation of the EBIT and the net income.

The basis of the profit and loss account is the calculation of EBIT according to the cost of sales method (in Germany analogous to § 275 (3) HGB)

Net sales
– Cost of goods sold
= **Gross profit**
– Marketing and sales
– Administration
– Research and development, R&D
(+ Other company income;
– other operating expenses)
= **EBIT (earnings before interest and taxes)**

EBIT (operating profit)
– Interest expense
+ Interest income
= **EBT (earnings before taxes;** profit of common business operation)
– Tax expense
+ Tax income
= **EAT (earnings after taxes,** net income)

Table 3.2 Profit and loss account for L'Oréal.

€ millions	2016	2015
Net sales	25 837	25 257
– Cost of sales	–28.4%	–28.8%
Gross profit	**71.6%**	**71.2%**
– Research and development	–3.3%	–3.1%
– Advertising and promotion	–29.0%	–29.1%
– Selling, general and administrative expenses	–21.7%	–21.5%
– (Other income and expense)	–2.1%	–0.8%
Operating profit (EBIT)	**15.5%**	**16.6%**
Finance costs, net	0.03%	0.1%
– Other financial expenses	–0.1%	–0.2%
+ Sanofi dividends	1.3%	1.3%
Profit before tax and associates (EBT)	16.7%	17.9%
– Tax	–4.7%	–4.8%
Net profit (EAT)	**12.0%**	**13.1%**

Promising development products from Research and Development (R&D) must be calculated in detail before the deciding to take the product into production. In addition to the investment and manufacturing costs, the prime costs of the product are particularly required because they form the basis for an estimate of the possible market price, which enables the fixing of a profit margin.

3.1.2 Development Products

For the competitiveness and further development of large cosmetics companies, a continuous improvement of their own market products is necessary. On the other hand, innovations promote market position (image) and sales (market shares). For the first in the market, the so-called innovator, novel products with high customer benefit are primarily profitable in the introduction phase and therefore require rapid, intensive development in all departments involved [1].

If there are physically and microbiologically tested laboratory samples in high quality corresponding to the sales product, first an expert creates a safety assessment. In the second step, the tolerability is tested at sensitive probands. At the same time, the process development starts the implementation into production and marketing asks potential customers according to their assessment of the new development. In addition, employees from the plant engineering have to plan a new production plant and building or an adaptation of existing plants in their own company or with a contract manufacturer. For major products, marketing can commission a market analysis in market research. In this way, it is possible to determine the market outlook in the important countries considering the competition. As soon as representative products are available from the pilot plant, the product compatibility must be checked on a large scale, also the product acceptance and benefits as well as the personal relevance of the customers [2]. This is examined in small scale by personal conversations and broadly over sample distribution and questionnaires. Particularly important is the willingness of the target group to buy at the suggested price. This applies to the countries in which the product is to be sold.

Before the start of extensive tests, market opportunities for innovative product ideas can be roughly estimated in the fields of chemistry, biotechnology, pharmacy, cosmetics, as well as in the food and consumer goods industry on criteria known from the literature [3]. Product use, quality (performance, convenience, and esthetic), and the size of the regional/international markets influence the possible sales. In particular, the market price, the brand strength, and the know-how of the company as well as the competition situation determine future market shares or market share gains. Figure 3.1 shows 14 criteria that the top management discusses before releasing funds for an investment in new products. As a rule, market outlooks (rather than the level of investment) are the main reasons for launching the market.

3.1.3 Determination of the Market Price

The market price must be carefully considered and adapted to regional market conditions. It is of crucial importance for the introduction of the product. The

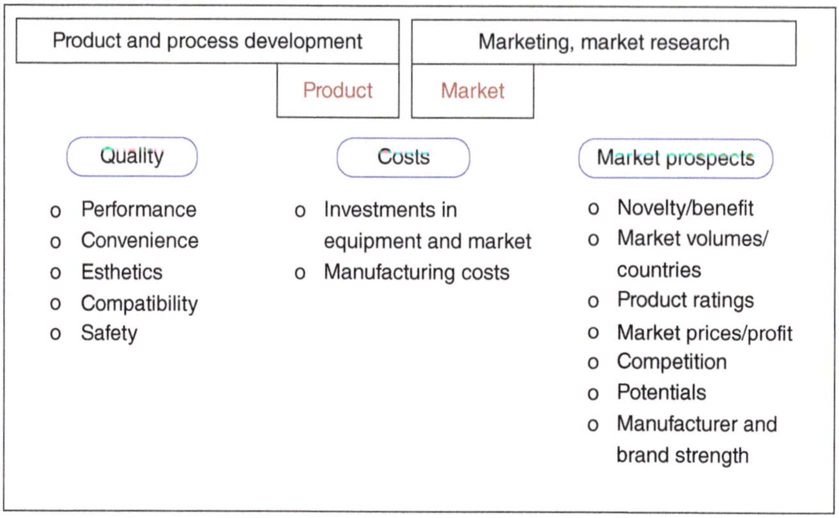

Figure 3.1 Decision criteria for a product launch.

market price arises as shown in Figure 3.2 by adding the manufacturing costs with the direct and overhead costs. As the overhead costs are known in the company and the direct costs can be estimated with sufficient precision, the manufacturing costs represent the only unknown quantity. These comprise the production and raw material costs. This proceeding applies in particular to completely new products with expensive ingredients, made in a new plant to be built entirely or in parts. Therefore, the calculation of the production, raw material, and manufacturing costs are described first, which are performed by the technical department. Subsequently, an explanation of the costs in the company takes place as well as the design of the market price by marketing.

3.2 Detailed Calculation of the Manufacturing Costs

3.2.1 Costs in the Production

For determining the production costs (process costs), a balance sheet over the entire process is used. This includes all costs for carrying out the process that includes the depreciation of the process plants as well as expenditures for all energies and the operating staff. Depreciations mainly arise from the investments in the plant and the building as well as expenses for maintenance, whereby the depreciation period influences the amount of costs per year. In addition, a larger balance sheet includes the raw materials and enables the calculation of the manufacturing costs (see Figure 3.3).

In an early planning phase, the investments for the plant and for the building can be estimated by a simple precalculation over factors on the sums of the machines and apparatus (MaAp) as well as on complete plant units (e.g. packaging units) [4].

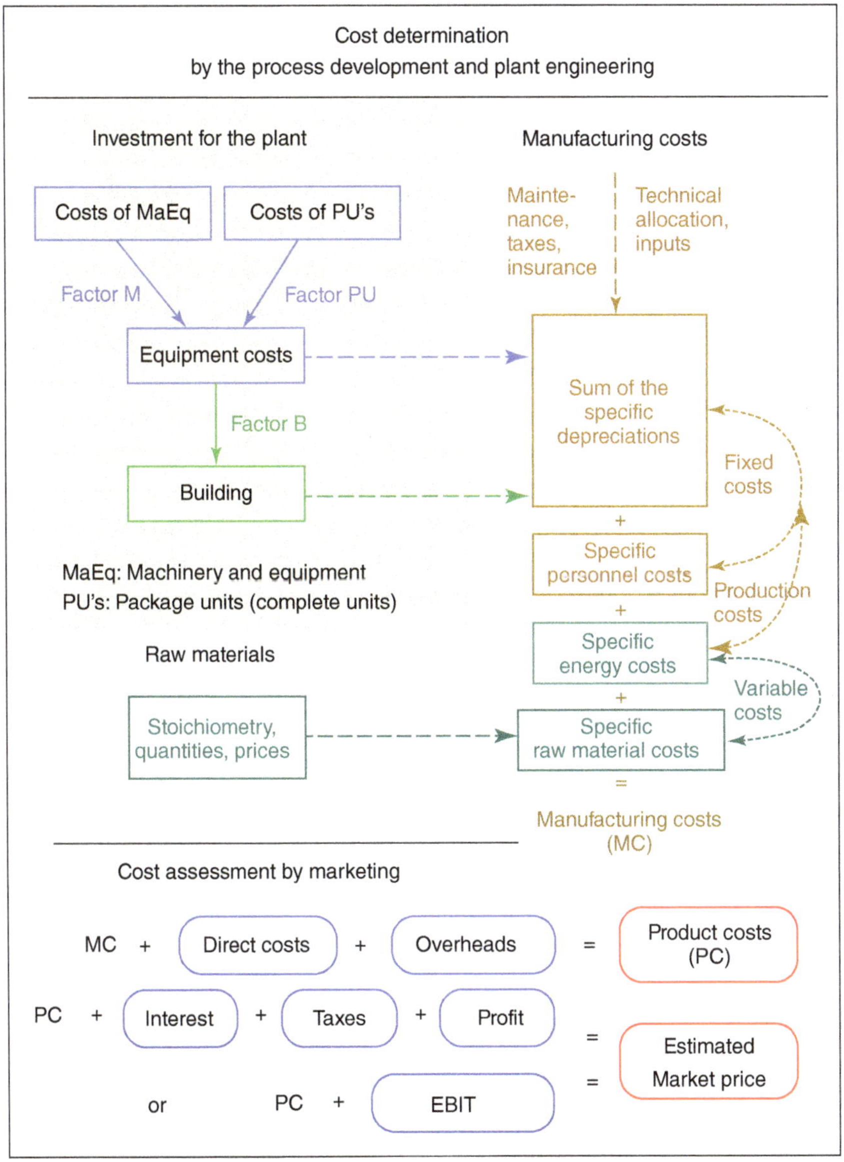

Figure 3.2 Determination of the product costs and possible profit from the sum of manufacturing costs as well as the direct costs of the product and overhead costs of the company.

The balance sheet includes all inputs, which are solid, liquid, and gaseous raw materials, as well as the excipients and the cleaning solutions. The cost allocation of the excipients can be found in Table 3.3. The exiting stream consists of the final products because there are no by-products and residuals for cosmetic creams. In addition, wastewater and exhaust air as well as non-recirculated cooling water leave the balancing field. Inside of this field, the energy and labor

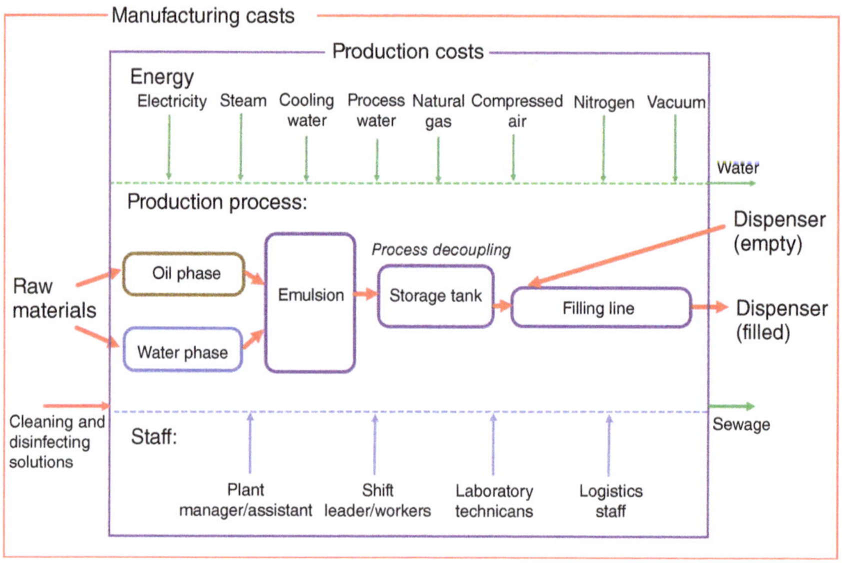

Figure 3.3 Balance sheets for a cream production process, consisting of the production costs (energy, process, labor) and in addition the raw materials to obtain the manufacturing costs.

Table 3.3 Auxiliary materials for carrying out the process and costs for the ecology.

Part of the plant costs	**Part of the raw material costs**		
Input materials for unit operations	**Excipients for**	**Cleaning solutions**	**Disposal costs**
Ion exchanger	Precipitation	Nitric acid	Residues
Adsorption resins	Filtration	Phosphoric acid	By-products
Activated carbon	Floatation	Sodium hydroxide	Wastewater (central cleaning)
Membranes	Emulsification	Surfactant solutions	Exhaust air (cleaning or combustion)
Sieves covers	Granulation	Oxidants	Recycling of off-spec products
Catalysts	Milling	Disinfectant cleaner (alcoholic solutions)	

costs can be found. All process stages require energies, such as electricity, steam, water, natural gas, and compressed air. Their expected consumption and costs are determined during planning. Trained staff supervises the process including filling/packaging/palletizing and carries out the quality control as well as the logistics of raw materials and products. In the following, some examples, calculated and with German labor and energy costs, are presented for clarification. In addition, all calculations are based on full loads, so that the capacities and operating hours correspond to the plan figures.

3.2.2 Production Costs Related to Installation and Building

A calculatory, straight-line depreciation of the investment usually takes 10 years. In very cost-intensive processes, depreciation may be carried out in 15 years or in the case of fast-moving products in 7 years. Buildings are usually amortized over 40 years. The cost of maintenance depends on the process (s, s/l, l, and with/without gas in continuous/batch mode) and on their individual process stages. The filling of finished products into glass jars with a screw cap, a dispenser, and packages requires a relatively high maintenance effort of 7% and more, whereas large continuous plants with high throughputs need 4–5%. The percentage of 5–6% is common for consumer goods. In the cosmetic industry, maintenance of 6.5% is included in the calculation because of the weighting of the filling lines. If the company has type-specific values from the post-calculations, these are always preferable. Table 3.4 shows the cost components depending on the investment costs in typical percentages, which add up the sum of depreciation for wear (SAfa) and thus enter into the calculation of the production costs. The figure of 20.5% mentioned here varies between 16% and 25% depending on the process.

3.2.3 Labor Costs

The composition of the German average labor costs in the chemical and related industries is shown in Table 3.5. The employee usually only perceives the gross annual salary. For the employer, average labor costs are 32.5% higher because of social costs, occupational pensions, and other additional costs.

The shift staff consists of semi-skilled employees as well as trained chemical technicians with a usual classification in E3–E7 pay groups or in E1–E3 pay groups for the filling lines. Therefore, their salaries are below the average. The calculations take place with negotiated wage rates for the chemical industries (Table 3.6) to achieve accurate results. In addition, the German employer pays

Table 3.4 Composition of the total depreciation.

	Investment costs	
Cost component	**Plant (SAK = 100%)**	**Building (SGK = 100%)**
Depreciation of the plant/building	10%	2.5%
Maintenance	6.5%	1.5%
Technical allocation	2.5%	0.5%
Licenses, taxes, insurance	1.5%	
Input materials for unit operations	0–1.5%	
Allowance for depreciation (Afa)	**20.5%**	**4.5%**
Sum of depreciation (based on plant costs SAK; approximate value)	SAfa = **21.54%** γ = **0.2154**	

Table 3.5 Average annual chemical labor costs in Germany (as of 2014, cited from [5], and extrapolated to 2018).

Payments by the employer in the chemical and related industries	Costs (€/a)
Direct remuneration for work actually performed	47 412
Remuneration of nonworking days such as holidays, public holidays, and in case of illness	10 431
Special payments	5 585
Gross annual salary of the employee	**63 428**
Social insurance contributions of the employer	10 381
Retirement provisions	7 389
Other staff costs such as further education and canteen subsidies	2 819
Labor costs in the chemistry	**84 017**
(Factor of gross annual wage without special payments $f = 1.4525$)	
Extrapolated for the year 2018 (+7.9%)	**90 654**

Source: Cited from Ref. [5], and extrapolated to 2018.

Table 3.6 Monthly salaries for technical employees according to the rates of remuneration of the chemical industry of North Rhine-Westphalia for 2018.

Pay group	After start in the job (€) mo/yr salary	After six years in the job (€) mo/yr salary	Job position: manufacturing (M), bottling (B)
E1	2 567/30 804		B
E2	2 717/32 604		B
E3	2 786/33 432		M, B
E7	3 099/37 188	3 656/43 872	Shift leader (E5–E7)
E9	3 274/39 288	4 423/53 076	Laboratory technicians (E9–E10)
E11	4 130/49 560	5 295/63 540	Assistant of the plant manager (E11–E13)

IG BCE = German trade union for mining, chemistry, and energy, annual salary calculated.

surcharges (Table 3.7) for shift, regular night, Sunday, and holiday work [6, 7]. They are dependent on the shift mode and capacity utilization of the facility. The probable labor costs indicated in Table 3.8 are calculated from the basic salary and special payments, social and retirement contributions, and additional personnel costs. All figures refer to Germany; in other European and some global countries, the percentage deviations of labor costs for the manufacturing sector are displayed in Table 3.9.

In a year, employees work about 217 days, calculated from 365 days minus 104 Saturdays/Sundays and minus 14 legal holidays as well as minus 30 holidays.

Table 3.7 Tariff surcharges for shift work in Germany.

Shift operation	Reasons for the extra pay	Percentage surcharge (%)	Sum of the surcharges (%)
4/5 shifts (nonstop, except Easter, Whitsun, Christmas, and New Year)	Shift bonus	10	36.0
	Regular night work	15	
	28 Sundays (+60%)	8.4	
	6 legal holidays (+60%)	1.8	
	1 legal holiday (+150%)	0.8	
3 shifts (Monday 6.00 a.m. to Saturday 6.00 a.m.)	Shift bonus	6	21.9
	Regular night work	15	
	3 legal holidays (+60%)	0.9	
2 shifts (Monday to Friday: 6.00 a.m. to 10.00 p.m.)	Shift bonus	6	6

Calculated from the data of the trade union IG BCE, the percentages may vary slightly depending on the working load.

Table 3.8 Estimated labor costs in 2018 for calculating the personnel costs.

Pay group	Shift modus, working period	Salary (€/yr)	Labor costs (€/yr)
E1	By day	30 804	44 743
	2 shifts	32 652	47 427
	3 shifts	37 550	54 542
E2	By day	32 604	47 357
	2 shifts	34 560	50 199
	3 shifts	39 744	57 729
E3	By day	33 432	48 560
	2 shifts	35 438	51 474
	3 shifts	40 754	59 195
	4/5 shifts	45 468	66 042
E7	By day	37 188	43 872
	2 shifts	39 419	57 257
	3 shifts	45 332	65 845
	4/5 shifts	50 576	73 461
E9		53 076	77 093
E11		63 540	92 292

Table 3.9 Effective labor costs per hour in the world compared to Germany (100% correspond to 40–40.5 €/h in 2018 after an extrapolation; European figures: IV/2016; global figures 2015).

Country (examples)	Costs compared to Germany (%)	Country (examples)	Costs compared to Germany (%)
European Union (EU 28)	68.2		
Swiss	*149.1*	Spain	58.9
Norway	126.4	*Japan*	*58.7*
Denmark	112.1	*Korea*	*56.7*
Belgium	111.9	Portugal	29.2
Sweden	108.3	Slovakia	27.4
Germany	**100**	Poland	19.9
France	97.2	*Brazil*	19.5
Austria	91.7	*Turkey*	*15.6*
Netherlands	90.7	*Mexico*	*13.8*
USA	*87.1*	*Russia*	*12.7*
Italy	70.5	Romania	11.5
UK	65.6	*Philippines*	*5.6*
		Ukraine	*4.6*

Italics for all countries outside the EC and bold type for Germany as reference.

Taking into account the average illness days (approximately 14 days) and the 4/5 shift-bonus of 3 days, 200 effective working days remain. On the other hand, a process operated in a fully continuous alternating shift runs at full load, for example, 338 days (365 days −10 days for Christmas, −4 days for Easter, −3 days for Whitsun, −10 days for maintenance). Thus, 7 additional employees are required for 10 persons (+70%), which must be taken into account in the cost calculation. An exemplary calculation of the labor costs in various shift operations is given in Table 3.10. Depending on the processes, the number of employees can be estimated well and the labor costs are known from publications of the trade unions. Therefore, the cost calculation of the staff is largely error-free.

In the examples, the personnel requirement has been adapted to a batch system consisting of three reactors in three lines with two filling lines downstream. For useful comparison, the cleaning/disinfecting durations must be fixed, whereby the relative times increase significantly with a decreasing shift number. A more accurate adjustment does not provide a fundamentally different assessment. In all variants, the filling is always done in two shifts. As shown in Table 3.11, the lowest specific labor costs arise in both continuous and discontinuous operation for the three-shift mode, compared to the same outputs per batch (same reactor volume). Under these boundary conditions, the number of shifts naturally delivers significantly different plant capacities per year. In a fully continuous alternating shift, the additional surcharges for night/weekend working affect the

Table 3.10 Example of calculating personnel costs in a labor-intensive factory depending on the shift mode (estimated figures for Germany in 2018).

Staff	Production days per year	Number of persons (present per day)	Total number of employees (needed in the year)	Estimated labor costs per person (€/a)	Labor costs for all employees (€/a)
6 shift workers/4 shifts	338	24	41	66 042	2 707 722
6 shift workers/3 shifts	200	18	18	59 195	1 065 510
6 shift workers/2 shifts	200	12	12	51 474	617 688
6 workers/daytime	200	6	6	48 560	291 360
5 shift workers/2 shifts	338	10	17	50 199	853 383
For the filling line	200	10	10	50 199	501 990
5 shift workers/daytime	200	5	5	47 357	236 785
1 shift leader/4 shifts	338	4	7	73 461	514 227
1 shift leader/3 shifts	200	3	3	65 845	197 535
1 shift leader/2 shifts	200	2	2	57 257	114 514
1 shift leader/daytime	200	1	1	43 872	43 872
Employees daytime	338	3	5	47 357	236 785
	200	3	3	47 357	142 071
Technicians	338	4	7	77 093	539 651
	200	4	4	77 093	308 372
Manager assistant/marketing	338	1	2	92 292	184 484
	200	1	1	92 292	92 292

(Continued)

Table 3.10 (Continued)

Staff	Production days per year	Number of persons (present per day)	Total number of employees (needed in the year)	Estimated labor costs per person (€/a)	Labor costs for all employees (€/a)
Plant manager		1	1	145 000	145 000
Sums for production (without filling)					
4 shifts	338	37	63	68 696	4 327 869 = 83.5
3 shifts	200	30	30	65 026	1 950 780 = 37.7
2 shifts	200	23	23	61 736	1 419 937 = 27.4
Daytime	200	16	16	63 935	1 022 967 = 19.7
Sums including filling in 2 shifts					
4 shifts	338	47	80	64 766	5 181 252 = 100%
3 shifts	200	40	40	61 319	2 452 770 = 47.3
2 shifts	200	33	33	58 240	1 921 927 = 37.1
Daytime	200	21	21	59 988	1 259 752 = 24.3

Table 3.11 Capacities of a small factory for cosmetic creams depending on the operating shifts of the manufacturing and with filling lines in 2-shift mode (labor costs from Table 3.10).

Production mode	Working days	Cleaning period (estimated); startup	Capacity (t/yr)	Specific labor costs (€/kg)	Specific labor costs for 5400 t/yr (€/kg)
(a) Continuously operated plant with a capacity of 1.25 t/h					With adapted reactor sizes
4 shifts	338	3 times per week = 12 h; 3 h	9230	0.561	0.960
3 shifts	200	2 times per week = 8 h; 2 h	5500	0.446	0.454
2 shifts	200	5 times per week = 20 h; 5 h	2750	0.699	0.356
(b) Batch-operated plant with a capacity of 18 t that means three reactors of 6 t in 12 h (batch time: 16 h)					With adapted reactor sizes
4 shifts	338	Every 12 h for 3 h; 1 h	9120	0.568	0.959
3 shifts	200		5400	0.454	0.454
2 shifts	200		3600	0.534	0.356

result negatively. Also, in the four-shift operation, the highest number of employees is needed, the least with the day shift. For comparability, the specific personnel costs were additionally calculated for the same production level (5400 t/yr) in the two- to four-shift operation. In the two-shift mode, less employees are working, but with larger reactors. Therefore, the labor costs are on a low level. They can even be lower if the process is running in the daytime. However, this should only be possible for batch times of a maximum of eight hours. As expected, the labor costs increase with the number of shifts, but at the same time, however, the value for depreciation drops considerably.

Processes in a partly alternating shift (two or three shifts) run either 190–210 days, so that no additional personnel costs are required. At the full capacity of 234 days, the surcharge is 15%. Needed additional staff can be internally covered by other, underutilized facilities or by leasing personnel. The absolute labor costs in the three-shift operation are less than 50% of the fully continuous shift modus. However, the three-shift plant has to be 65–70% larger for the same annual output with an adapted number of employees.

3.2.4 Energy Costs

The industry-specific energy costs in Germany, specified in Table 3.12, are based on Refs. [8, 9]. They depend on the location and clearly on the level of consumption, as is shown by the example of the natural gas. Large customers with an annual consumption of more than 100 GWh pay 2.29 cents/kWh (second

Table 3.12 Estimated costs for various energies in Germany (2014/2015, energy prices have not changed significantly until 2016).

Energy	Specific costs
Electrical current	0.09–0.10 €/kWh
Steam	
Low pressure (4 bar)	19.- €/t
Medium pressure (20 bar)	21.- €/t
High pressure (42, 75 bar)	23.- to 28.- €/t
Water	
Cooling water	0.05–0.08 €/m^3
Process water	0.30–0.60 €/m^3
Drinking water	1.40–2.80 €/m^3
Deionized water	2.00–3.00 €/m^3
Natural gas	0.023–0.029 €/kWh
Compressed air	10 € per 1000 N/m^3

half of 2014, [10]) with a term of two years, other customers with consumption between 10 and 100 GWh 2.51 cents/kWh and small customers below 10 GWh 2.86 cents/kWh. The large industrial consumer therefore pays 20% less than the small business.

In every company, the energy consumption of numerous MaAp is available from post-calculations, which can be taken over or extrapolated. If there are no usable values, the energy consumption can be estimated using simple equations [9, 11]. This applies in particular to the consumption of steam, cooling water, process water, and natural gas. The consumption is indicated specifically per year or per effective operating hour, or preferably based on a mass unit (kg, 100 kg, t, and piece). A flow sheeting program, such as Aspen Plus® [12], simplifies plant costing and also delivers energy consumption that can easily be converted into specific costs.

Electricity consumption is a major cost burden in many companies because practically all machines are driven electrically. However, as the power consumption of large machines is dependent on many parameters, the time-consuming calculations are based on various assumptions. For example, consider a mixer where all parameters affect consumption: the design of the machine, type of the electric motor and throughput, the medium as well as all operating conditions (speed, temperature, pressure, degree of filling, and density) influence the power consumption. For quick treatment, it is recommended to use estimates. Power requirements can be found, for example, in the brochures and price offerings of the MaAp manufacturers and should be multiplied by 0.8 to determine the average consumption. This accuracy is sufficient for an estimation of the power consumption. As there are always high as well as too low values, the error of the total energy consumption is usually in the range of ±30%.

3.2.5 Production Cost Dependencies of Capacity, Operation, and Personnel

If the planned installation can start and run in a relatively short time and the production may be interrupted at the weekend, as is customary in the cosmetics industry, the question arises about the shift mode to be selected. The choice determines the specific production costs, more precisely the fixed costs. The calculation must consider the fact that the fully continuous alternating shift (four shifts) requires less investment but higher labor costs compared to the three-shift operation. The differences to the two-shift mode are even more pronounced.

For the production and filling of cosmetic creams, some plant sizes and operation modes are illustrated as typical examples. Based on estimates, Table 3.13 provides such calculations for the specific production costs depending on the operating conditions for the same capacities per year. These can only be achieved if the reactor sizes vary accordingly. In four shift batch operations, the size of the reactors is only 59%, and in two shifts 150%, compared to the three-shift operation. In all cases, the two filling lines are used in the two-shift rhythm, as usual in Germany, and therefore equally expensive.

Specific raw material costs depend heavily on the order quantity and strongly influence the manufacturing costs. Usually, fixed quantities must be purchased depending on the packaging. This is more difficult the smaller the batches are. Often, there is a wide range of products with completely different formulations, which make it difficult to prepare the batches and store the raw materials. Residual quantities cannot be consumed in some cases and must be disposed of if stored too long. In addition, the calculated manufacturing costs assume full utilization, which is not given in most cases. Under-utilization due to lack of orders, too small batches, downtime due to waiting for analysis results, and further cleanings, as well as plant adjustments to the bottling lines, are common and driving up costs (see in Table 3.13 the figure for 50% utilization). The same applies to constant recipe changes. Especially, small orders increase the costs considerably because of the surcharges on raw materials (up to a factor of 2, as shown in Section 3.2.6).

The roughly estimated costs of the production equipment (8 Mio. €) cover the three lines with 6 m^3-emulsifier containers and two premix-stirred tanks and one storage container as well as the metering devices (see Section 10.4). On the other hand, the costs for the filling lines (6 Mio. €) include the bottling and closing of the containers, furthermore the labeling, cartoning, and palletizing (hints in Section 10.8). The facilities for hygiene (3 Mio. €) enclose water treatment, plant cleaning and disinfection (cleaning in place [CIP], Section 11.4), air filters, and fans as well as further equipment for premises according to the Good Manufacturing Practice (GMP) guidelines (Chapter 11).

The costs for the different reactor sizes can be determined using exponents of 0.7 for the up-scaling and 0.6 for the down-scaling [4]. The large reactors for two shifts are nearly twice as expensive as the small ones for four shifts. However, as the specific labor costs for four shifts are high compared to the depreciations, this leads to unattractive costs for the production of cosmetic creams. The four-shift case should therefore not be considered further. Surprisingly, the

Table 3.13 Dependency of the specific production costs from shift operation at constant annual capacities.

Shift modes for manufacturing/ filling lines; capacities and depreciations	4/2 shifts	Quantities and costs (4/2)	3/2 shifts	Quantities and costs (3/2)	2/2 shifts	Quantities and costs (2/2)
Working days	338		200		200	
Capacity per year		5400 t		5400 t		5400 t
Plant output per hour (average)	59.2%	0.666 t/h	100%	1.125 t/h	150%	1.688 t/h
Investment[a]						
➢ Plant[b]		5.84 Mio. €		8.0 Mio. €		10.6 Mio. €
Filling line		9.0 Mio. €		9.0 Mio. €		9.0 Mio. €
Afa	20.5%	3.04 Mio. €	20.5%	3.485 Mio. €	20.5%	4.023 Mio. €
		0.563 €/kg		0.645 €/kg		0.745 €/kg
➢ Building[b]		2.56 Mio. €		3.5 Mio. €		4.65 Mio. €
Afa	4.5%	115 T€	4.5%	158 T€	4.5%	209 T€
		0.0213 €/kg		0.0292 €/kg		0.0387 €/kg
➢ Sum	86.8%	0.585 €/kg	100%	0.674 €/kg	116%	0.784 €/kg
Labor costs (from Table 3.8)	211%	0.959 €/kg	100%	0.454 €/kg	78%	0.356 €/kg
Energy costs[a]	100%	0.10 €/kg	100%	0.10 €/kg	100%	0.10 €/kg

Plant capacity (t/yr)	Specific production costs (€/kg) 4 shifts	3 shifts	2 shifts
5 400	1.644	**1.228**	1.240
Difference	+33.9%	—	+1.0%
Money per year	+2.25 Mio. €	—	−64.8 T€
For 70% utilization = 3780 t	2.275	1.683	1.698
10 800	1.061	**0.878**	0.914
For 70% utilization = 7560 t	1.442	1.182	1.233
For 50% utilization = 5400 t	1.971	1.606	1.678
3 600	2.192	1.578	**1.555**
For 70% utilization = 2520 t	3.058	2.182	2.148
1 800	—	2.52	**2.38**

Boundary conditions of the calculation in these examples: batch-operated facilities with identical capacities per year; batch time: 16 h; three production lines; for the 3-shift mode: $3 \times 6\,m^3$, in other cases, the reactor volumes are adjusted; two filling lines always in 2-shift mode; possible plant investment for 3 shift operations 8 Mio. €, filling lines including facilities for hygiene 9 Mio. €, building 3.5 Mio. €.

a) Estimates.

b) Factors: 3 shift = 1; 4 shift = $0.592^{0.6} = 0.7301$; 2 shift = $1.5^{0.7} = 1.3282$.

specific production costs in two and three shift operations are close to each other, for small capacities in favor of the two shifts, for large ones, the three-shift operation offers advantages. These remarkable results, which concern the cosmetics industry in essential parts, are in contrast to the chemical industry because the chemistry generally requires high investments and relatively low labor costs. The figures discussed are only valid if the plants work as planned. The results also show that a less used large-scale plant can even produce as cheap as a fully operating small plant. In the cosmetic industry, the process costs for manufacturing and filling are generally not related to 1 kg product, but to a packaging unit, i.e. a dispenser with 250, 100, 50, or 30 ml content. At the assumption of a 100 ml dispenser, the production costs of 1.7 €/kg change to 17 cents/unit without raw material and bottle costs.

In the manufacturing costs [13], a distinction is made between fixed and variable costs. The fixed production costs consist of several summands of the depreciations (see Table 3.4) as well as the labor costs. Fixed costs always accrue, even when the system is at a standstill. The variable costs only occur during production. They represent the raw material and energy costs as well as the credit notes and direct debits for by-products/residues and wastewater/exhaust air. All costs are referred to a mass unit.

As the capacity of the plant increases, the absolute figures for the depreciations and for the labor costs increase, while the specific values decrease. The lower the extra effort for the capacity increase, the more the specific costs fall. If the plant capacity is doubled in the planning by erecting two plants of the same size, the specific costs remain constant and are identical to one plant. This is only useful when they are built and operated far away from each other. Erecting a twice as large facility at a single site, the larger MaAp require a lower investment as the multiplication by a factor of 2 results. The costs can often be approximated by an exponent $m = 0.7 \pm 0.2$ [4], which means $2^{0.7} = 1.62$. Significant savings can still be achieved for the larger plants because the number of employees required for the higher capacity does not increase at all or only grows under proportionally (exponent $n = 0.5 \pm 0.2$). Some activities are independent of the size of the plant (control room); others can require little more staff.

In Figure 3.4, the calculated fixed costs from Table 3.13 are graphically represented and into depreciations and labor costs broken down. It turns out that on the one hand, the costs of two-shift operation over a wide range are comparable to three-shift operation. On the other hand, for lower capacities, a shift of the production to cheaper countries can be attractive, as absolute staff costs decrease. However, the additional transport and logistics costs must be considered and rated separately.

Under other conditions, for example, in the case of uncomplicated recipes, the batch takes only 8 instead of 16 hours. Then, the day shift can be compared with the two-shift operation. This is done here by the example of a small factory, which produces in two shifts exactly as much as in Table 3.13 indicated, but in 9 m^3 total reactor volume instead of 18 m^3. Then, the three reactors would have a volume of 1, 3, and 5 m^3 and the filling lines would be identically sized as in

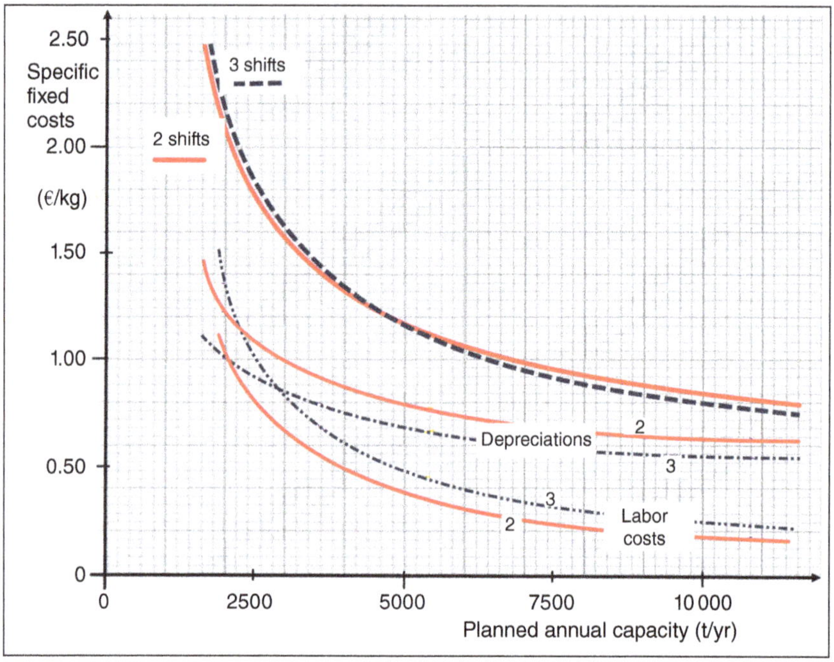

Figure 3.4 Composition of the specific fixed costs and their dependence on the planned capacity (for example, exactly only for the figures from Table 3.13).

the examples before. In day shift, the same facilities provide half the amount, as shown in Table 3.14.

In this exemplary comparison between the two-shift operation and day shift, the cost-effectiveness of shift operation should be shown. The day shift yields comparable results for a yearly capacity of less than 1000 t/a. On the whole, between about 1000 and 4500 t/a, the most favorable production costs are obtained for the two-shift mode. Up to 11 000 t/a, the additional costs in comparison to the three-shift operation lie below 4% and are in most cases acceptable. Thus, the two-shift cycle offers an economical mode of operation for many production sites with comparable boundary conditions. For quantities that are clearly above 11 000 t/a, production is recommended in a three-shift operation with the fillings in usual two shifts. As the real boundary conditions deviate from the examples, each company should calculate its own optimal operation. However, the two shift rhythm seems to be a good solution. Cosmetic factories must have a minimum size to achieve competitive manufacturing costs, which, however, make up only a small part of the product costs. For economic reasons, the production should show a capacity above 6000 t, more preferably above 10 000 t/a.

3.2.6 Raw Material Costs

Although in chemistry the calculation of the raw material costs is based on the stoichiometry, in cosmetics, it is sufficient to add the cost components of the

Table 3.14 Comparison of the 2-shift with daytime operation.

Shift modes for manufacturing/ filling lines; capacities and depreciations	**2/2 shifts**	**Quantities and costs**	**1/1 shift day-time**	**Quantities and costs**	**1/1 shift day-time**	**Quantities and costs**
Working days	200		200		200	
Capacity per year		3600 t		1800 t		3600 t
Plant output per hour (average)	100%	1.125 t/h	100%	1.125 t/h	200%	2.25 t/h
Investment						
➢ Plant		5.28 Mio. €		5.28 Mio. €		8.6 Mio. €
Filling line		7.06 Mio. €		7.06 Mio. €		11.4 Mio. €
Afa	20.5%	2.53 Mio. €	20.5%	2.53 Mio. €	20.5%	4.10 Mio. €
		0.703 €/kg		1.405 €/kg		1.138 €/kg
➢ Building		2.47 Mio. €		2.47 Mio. €		4.00 Mio. €
Afa	4.5%	111 T€	4.5%	111 T€	4.5%	180 T€
		0.0309 €/kg		0.0617 €/kg		0.0500 €/kg
➢ Sum	50.0%	0.734 €/kg	100%	1.467 €/kg	81.2%	1.188 €/kg
Labor costs (from Table 3.8)	76.3%	0.534 €/kg	100%	0.700 €/kg	50%	0.350 €/kg
Energy costs	100%	0.10 €/kg	100%	0.10 €/kg	100%	0.10 €/kg

	Specific production costs (€/kg)			
Plant capacity (t/yr)	**2/2 shifts**	**1/1 daytime**	**1/1 daytime**	**1/2**
3600	**1.368**		1.638	1.455
1800	**2.142**	2.267		
1200	**2.847**	**2.887**		

Boundary conditions for this example: three manufacturing lines of 5, 3, and 1 m^3, batch time 8 h; two filling lines; costs from Tables 3.10 and 3.13.

formula. This is presented in an example from the natural cosmetics (Table 3.15). The composition discussed there is suitable for a high-quality moisturizer and contains, in addition to natural moisturizing factors (NMFs) such as hyaluronic acid and panthenol, valuable natural oils for the skin. The prices are from the Internet (June 2017) and apply for small batches of 50–100 l in the pilot plant.

The quantities and qualities of the oils depend on harvest and certification. Therefore, the exact quality and price are only valid for a short time. From practice, it is known that in particular, the prices of the oils depend strongly on the quantity purchased. Cosmetics companies that buy large amounts of each raw material for a few tons of products, however at least one barrel (200 l), pay much lower prices, which may be 15–70% of the stated values. For example, the almond

Table 3.15 Calculation of the raw material costs for a natural NMF cream over the cost of the individual raw materials (status: June 2017, purchased in small amounts, costs for certificates not always included).

No	Substance	INCI name	Function	Proportion in the recipe (%)	Price (€/kg)	Costs (€)
1	Water (demineralized)	Aqua	Solvent	54.09	—	—
2	Sweet almond oil	Prunus Amygdalus Dulcis Oil	Lipid	16	10.30	1.65
3	Evening primrose	Oenothera Biennis Oil	Lipid	6	39.90	2.39
4	Shea butter	Butyrospermum Parkii Butter	Moisturizer, UV-protection	4	12.15	0.49
5	D-Panthenol 75%	D-Panthenol 751	Moisturizer, skin healing	4 (3% active)	73.00	2.92
6	Glycerin	Glycerin	Moisturizer	3	2.45	0.07
7	Sorbitol	Sorbitol	Moisturizer	2	10.95	0.22
8	Bergamuls® ET-1	Beta-Glucan and Pectin	Emulsifier	2.5	77.00	1.92
9	Imwitor® 375	Glyceryl Citrate/Lactate/Linoleate/Oleate	Emulsifier	2	45.90	0.92
10	Lanette® O; cetyl stearyl alcohol	Cetearyl Alcohol	Nonionic co-emulsifier, consistency generator and emollient (moisturizer)	1	16.20	0.16
11	Trisodium citrate dihydrate	Sodium Citrate	pH-regulation	0.85	4.20	0.04
12	Tocopherol	Tocopherol/Helianthus Annuus (70 : 30)	Vitamin E	0.7	37.00	0.26
13	Lavender oil	Lavandula Angustifolia Oil	Preservative, fragrance correction	0.4	68.30	0.27
14	Xanthan	Xanthan Gum	Consistency generator	0.35	8.00	0.03
15	Potassium sorbate	Potassium Sorbate	Preservative	0.25	5.96	0.02
16	Citric acid monohydrate	Citric Acid	pH-regulation	0.16	2.40	0.01
17	Hyaluronic acid	Hyaluronic acid	Moisturizer	0.1	350.00	0.35
Sum for small product amounts						11.72
Estimated sum for large product amounts						c. 5.1–6.5
Estimated sum for large product amounts and conventional emulsifiers						c. 3.9–5.1

INCI, International Nomenclature of Cosmetic Ingredients.

oil in a 200 l barrel does not cost 10.30 €/kg, but rather less than €3. For the evening primrose oil, the absolute difference is even higher. A purchase of certified products usually requires the payment of surcharges. For the emulsifiers, however, the differences are less extreme. When considering the total costs, in the case of large production quantities, the indicated formulation should not cost €11.70, but roughly estimated at less than €6. If a conventional instead of a natural emulsifier system was selected, the costs can be reduced once more by about 30%.

For costing, the use of the Excel Microsoft Office program is recommended, so that the financial impact for each change of the formulation is immediately visible. Furthermore, the technical release criteria can be included below the costing table, such as the pH value, viscosity, the stability of the emulsion, and the effect of the preservation.

3.3 Costs in the Companies

A globally operating company consists of several product and customer-oriented business units (divisions). For example, L'Oréal is divided into the divisions "Consumer Products, L'Oréal Luxe, Professional Products and Active Cosmetics" ([14], and Section 2.11). As the second example, Henkel has three independent divisions with "Laundry & Home Care, Beauty Care, and Adhesive Technologies" [15]. The concept in other companies is similar. It is therefore necessary to differentiate between the costs of the entire company and the costs of each business unit.

In principle, all costs of the company must be borne by the product sales. In a simplified presentation, the ongoing financing of the company's expenditures (such as personnel, energy, and materials) are carried out through the revenue from product sales. In internal billing, the factories are thus not only burdened by the directly attributable costs (direct costs) of the product for the production, quality control, and marketing. They must also bear overhead costs of the business unit and the entire company. This considerably increases product costs beyond the manufacturing costs.

The calculation of the costs for the product is based on the manufacturing costs (see Figure 3.2), which consist of the relevant direct costs. An equivalent term, the "production costs," means here only the process costs without raw materials. Process-independent costs, emerging in the company, do not belong to the manufacturing costs. However, they can be allocated to the product, such as sales costs. By adding the company-dependent individual and overhead costs, the self-costs of the product result. In the published annual accounts, all costs are grouped together in each business unit and in the company as a whole and assigned to production, administration, research and development, as well as marketing and sales.

3.3.1 Internal Cost Accounting

In addition to legally required financial accounting, which considers the flow of money as well as the profit and loss account, large companies work in the field of company accounting with the cost and revenue calculation, in short, cost accounting [16, 17]. The voluntary cost accounting carried out over the last decades takes into account all company activities necessary for product production and marketing; it generates the operational monetary performance. Although the profit and loss account refers to the last year, the current and future-oriented cost accounting is used to control company activities.

The cost accounting deals with the clarification of the questions, which costs (cost types), where (cost center), and for what (cost objects) incurred [18]. For the purposes of cost accounting, each department or department group of the company captures the monthly expenses based on employee hours. The hour cost rates depend on the department (workplace costs) and the hierarchy.

All costs of the externally sourced materials and services (primary costs) are recorded on the cost center, such as raw materials, machinery and equipment, use of external companies, business travel, maintenance contracts, leasing fees, or training seminars. The same applies to all internal purchased materials and contracted services (secondary costs). This includes, in particular, internal orders, such as workshop or MSR work, toxicological tests, analyses, or training courses. After the addition of the related items, a monthly or quarterly target/actual comparison takes place between the costs incurred in the current year and the approved funds via the cost center accounting. In particular, the individual project costs, the travel cost, and investment costs are considered.

Each responsible department, for example, a development unit for processes or products, has resources through a cost center, which can be consisted of several cost subcenters. Together with the marketing management, the responsible R&D-managers define the new and updated development projects (products and processes) for the next year with the corresponding priorities. The funds, each approximately €200–1500, are then released for the individual projects (cost centers). In the following year, the development unit indicates for each project every month the achievements and costs. Basis is the data of the employees. They write their work per hour on the respective project. The costs for this are billed together with the internal and external expenditures monthly from the approved project funds.

The target/actual comparison are carried out by a controller for the responsible cost center manager. The data reveal timely the deviations in time and money and enables corrections. If the sum lies below the planned value, the work can be intensified with additional staff, while in the case of overdrafts, a reduction in the number of personnel employed allows meeting the project costs by the end of the year. In addition, for an important project, the transfer of funds from another project of lesser priority, in coordination with the clients, is a sensible method of controlling.

3.3.2 Direct Costs

Direct costs [19] represent all costs directly attributable to the products, which are above an individually defined monetary level in the company. Overheads include the purchase of low-value goods as well as other small cost items. Direct costs can be characterized by the fact that they run constant and have a term of several years. These enclose, on the one hand, production costs, i.e. the fixed depreciation and labor costs, as well as the variable costs for energy and raw materials. On the other hand, further fixed direct costs are incurred, which must be added to the manufacturing costs. Examples represent the costs for securing the quality as well as the development of the products (application technology), further costs relate to product sales (marketing/sales) and material procurement (purchasing) as well as the personnel management and operating accounts (administration). These individual costs do not belong to the process, but they are product-specific (see Figure 3.5). The product-accompanying work of the product and process development can also be understood as direct costs.

Fixed and variable marketing and distribution costs include not only labor, equipment, storage, and freight as well as customer visits but also costs for advertising, market research, and sales promotion. Overall, they amount to 5–25% [8], often 10–30%, and for cosmetics even 25–65% of sales (see Figure 3.9), usually assigned to a product group or brand, and make up a large share of the additional costs. Nonassignable small products and their sporadic advertising are subject to the overhead costs of the business division.

Costs beyond the manufacturing costs represent nonprocess-related costs [20] and depend heavily on the company, as well as the size and global importance of the business, its market position, the brands, and the competition situation. The

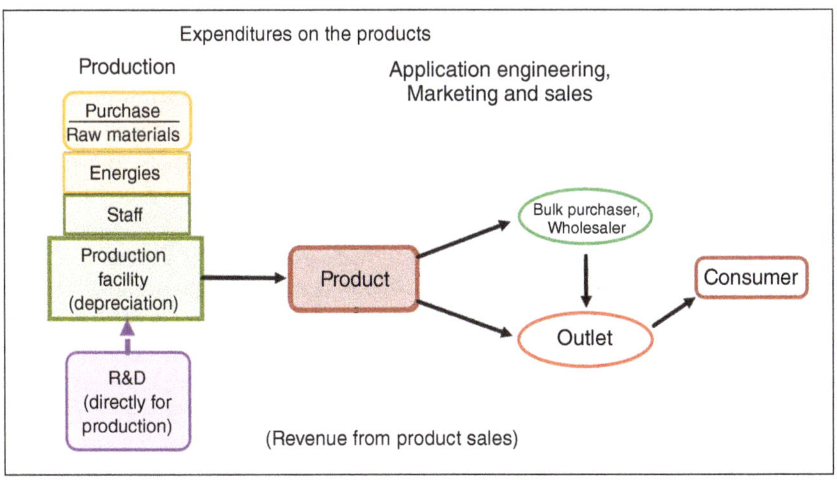

Figure 3.5 Direct costs (attributable to the product) in the production and in the sales (possible outlets in Figure 2.1).

differences in the direct costs are clearly reflected in the market price. Small companies, for example, that exclusively supply a large product to a discounter, need only a small application technology. Furthermore, travel expenses are low. These companies require R&D as well as marketing and sales not or only to a limited extent. As a result, in global companies, the prime costs (product costs) are significantly higher than that of small companies or production sites. The same applies to research-based cosmetics and pharmaceutical companies compared to imitators as well as to brand products compared to trade names and no name products.

3.3.3 Overheads in the Company's Cost and Performance Accounting

The monitoring and steering of a large enterprise in the environment of many competing companies require some management levels and various service departments, which have no or only a weak connection to production and sales. The costs required for this represent the overheads [19, 21] of the company and the company divisions. The indirect overhead costs exist for

- Top management
- Administration
- Services (such as R&D)
- Partially marketing and sales (M&S)
- Infrastructure (establishment and maintenance) and
- Small cost items

M&S are marketing activities that cannot be directly attributed to a product/product group or brand.

It is necessary to split the overheads. On the one hand, they affect the entire enterprise, such as for the infrastructure, fire department, medical care, finance, library, legal department, and public relations, and on the other hand, they concern a business unit of the company such as the costs of R&D, management, and administration. The joint cost of the company can be allocated to the individual business units with a percentage key. In large companies, business units operate as large independent units, which also generate their own overheads and thus have to burden additionally their products. Some costs are made on the basis of actual utilization, so that allocated costs for the different product groups/brands can differ by percentage.

Typical overheads of a company are payments to executives at the top management level, which means for the management board and division management, as well as the proportionate costs for company administration. Additionally to corporate control, the costs are caused by purchasing, accounting/controlling; personnel, legal, and patent departments; and information technology. Also, expenditures on research and development count to the overheads. They are used for new products and processes, as well as all other services (ecology, toxicology, and dermatology). Further all small cost items, such as office materials, are included. The cost structure of a company can be simplified in the form of a house, whereby the foundations and the pillars represent the necessary structures of a company (Figure 3.6). Wide and deep foundations strengthen the company and contribute to the relative reduction of overhead costs.

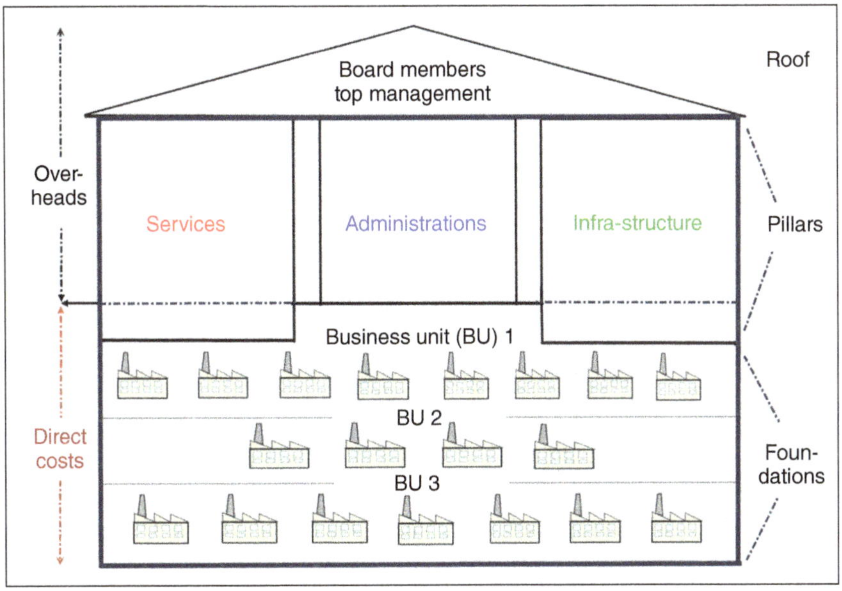

Figure 3.6 Schematic representation of the cost structure of a global company.

The amount of the overhead costs depends on the business as well as on the type, size, and global position of the company. It is necessary to distinguish between large international companies with thousands of products and high overheads and small enterprises with just a few products and low costs. On the one hand, the high overheads of large companies result from the global positioning as well as the desire to bring innovative products onto the market and to take social responsibility. Their costs include the own research and development, testing and ensuring product safety, working in committees, and securing jobs, granting occupational pensions and open public relations. The headquarters of global companies have numerous internal service providers. For all these reasons, global companies show high specific overhead costs. On the other hand, there are regional companies that produce only a few products in a simple production facility with a flat hierarchy. They produce with small direct costs and overheads, not only in absolute terms but also, in particular, relatively after distribution to the products.

In some aspects, the distinction between direct costs and overheads should also likely differ from company to company. Administrative costs are recorded as overheads and distributed over a key, for example, by the number of persons. Production-accompanying R&D and application technology as well as proportionate marketing costs can possibly be considered as overheads of the business unit (not the company) or as direct costs of the products. It is also possible that newly launched products will be freed from overheads to gain market share, until they have established themselves in the marketplace.

In a batch multipurpose plant, as sometimes used for cosmetics, many individual costs are declared to real, process-related overheads because substantial costs, such as the use of changing plant parts or the intensive cleaning, can hardly be allocated fairly. In contrast to the continuous or discontinuous production of a product/product group, multipurpose plants can have important plant parts that are only used for individual products and otherwise stand still. Moreover, simple processes, in contrast to more complex processes, use less energy, less staff, and a simpler process control system. Usually, a breakdown of the costs is made according to a fixed key for all products of the multipurpose plant.

3.4 Figures from the Published Annual Reports

The big companies publish the profit and loss account of the past year in the annual report. From the figures in the annual reports, the cost structures of the companies can be derived. They also provide information on the profits achieved. Operating profit [16] is calculated from the sales proceeds by subtracting the costs, which include all internal and external expenses. All costs are allocated on the one hand to the manufacturing and on the other hand to the administration research and development as well as the marketing and distribution. In addition, there is a position for "others."

3.4.1 Industry-Dependent Cost Structures of the Companies

The cost structure of the cosmetic products is unusual. This is only visible when compared to other branches of industry. Therefore, the cost structure of the chemical, consumer, and pharmaceutical industries will be discussed first. Global chemistry and pharmaceutical companies show extreme different cost structures. To understand the cost structures, both companies are presented in detail. Very high raw material costs characterize the structure of the chemical companies (Figure 3.7), furthermore low administrative and marketing costs. The last two together with R&D constitute the "Overheads and Direct Costs." Because of the level of raw material costs, only a small margin for the profit remains out of the net sales. The profit is still acceptable in height but is clearly below the consumer goods and food industry. As an example, BASF's broad-based chemistry has a unique cost structure because it produces chemicals often in a quantity of more than 100 Tt/yr. Their cost structure changed significantly from 2014 to 2016 because the cost of oil and gas decreased to half (Wintershall).

The second example in Figure 3.8 deals with the typical situation in the pharmaceutical industry. Low raw material costs (<30%) and high research costs, which are often above 18% for international corporations, mark the pharmaceutical cost structures. Marketing and distribution costs are in the typical range around 25% (±5). With 15–30%, the pharmaceutical industry shows unusually high EBIT. In contrast, the food and consumer goods industries use raw material costs around 50% of the net sales. The administrative costs range from 4% to 6%, R&D costs from 1.5% to 3.5%, and marketing and sales costs from 20% to 30%.

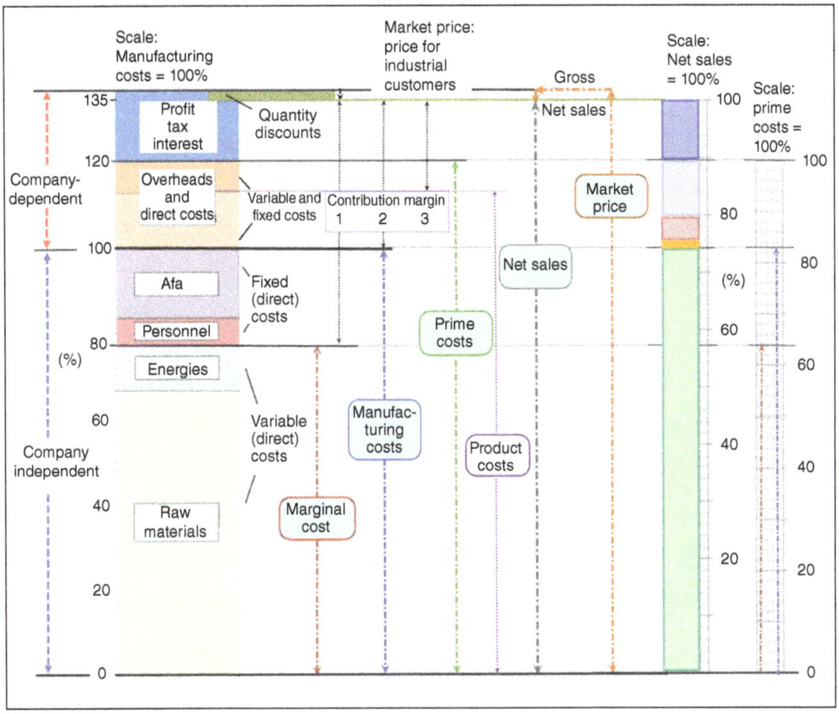

Figure 3.7 Example for the composition of the net sales for a global chemical company (BASF 2014).

In many cases, the cost-oriented market price is based on the total product costs (prime costs). It results from a surcharge on the prime cost, which takes into account the company-specific and/or competition-oriented profit margin. The specific market price is the price that the producer demands from large buyers (trade). Multiplied by the product volume sold, the gross proceeds. The manufacturer's net profit [22, 23] is calculated by subtracting the granted quantity discounts and bonuses.

The contribution margins (CMs) are calculated from the proceeds. The CM 1 [24], i.e. the sum to cover the fixed costs as well as to generate a profit, follows from the sales (=net proceeds) minus the variable shares of the production costs (marginal costs, see Figures 3.7 and 3.8). If the manufacturing costs are subtracted from the revenues, the CM 2 arises. CM 2 consists of the contributions from direct costs, overheads, and EBIT. The CM 3 covers the overheads and the EBIT.

For a plant that is not fully utilized, it may be useful to accept a production order, even if the revenues are below the prime costs. Because of the additional order, it is possible to get a part of the fixed costs, which arise anyway. Not covered direct costs and overheads must carry the main product. This leads there to an increase in the specific costs. Although the main product subsidizes the additional order, the total cost is then lower than without an order.

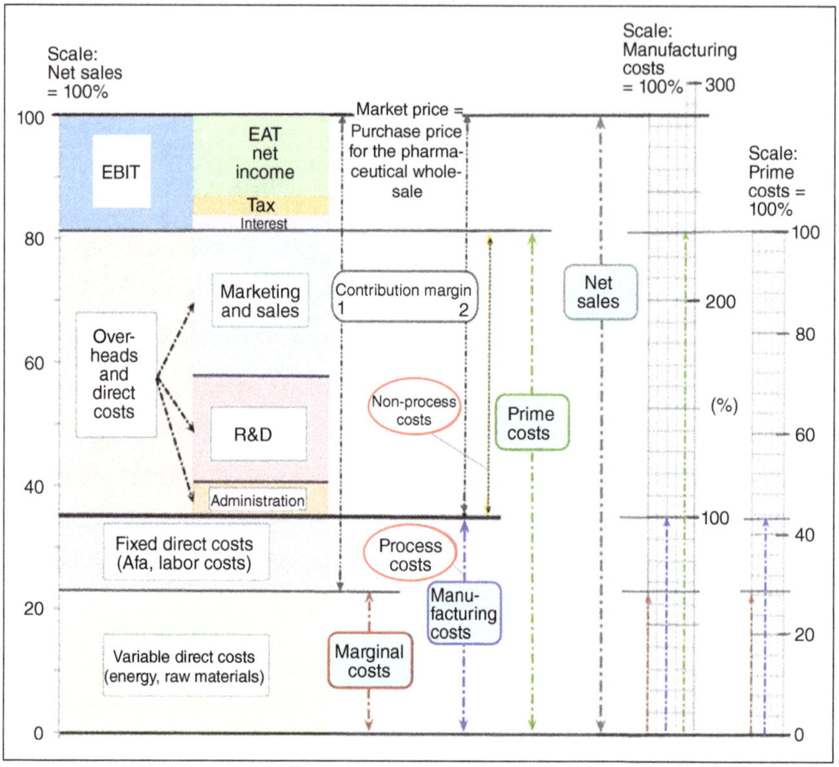

Figure 3.8 Example for the composition of the net sales for a global pharmaceutical company (Novartis 2014).

3.4.2 Profit and Loss Accounts of Global Cosmetics Companies

Figure 3.9 indicates the cost structures of companies from different industries. At first glance, it is noticeable that the cosmetics manufacturer differs significantly. In cosmetics, the manufacturing costs are comparatively low and the marketing costs are very high. In this example, the EBIT is in the normal range. All figures can be found in detail in the annual reports 2016. Table 3.16 shows the profit and loss account of three well-known cosmetics companies in comparison to the chemical and pharmaceutical, consumer goods, and food industries. For this comparison, the absolute figures were converted as a percentage of the net sales. The figures exhibit typical differences in the individual cost blocks.

Four outstanding facts are apparent when looking at the figures: Firstly, cosmetics companies need exceedingly high spending on marketing (see also Figure 3.10). Secondly, their manufacturing costs are low and lie in the same range or below the pharmaceutical industry. Third, the pharmaceutical companies spend a lot of money on research. At last, the chemical industry appears to be characterized by high manufacturing costs.

The more detailed discussion of the values in Table 3.15 shows that the manufacturing costs in the bulk chemicals (BASF) account for 65–75% of the revenues, whereas in pharmaceutical and cosmetic companies, they can be below

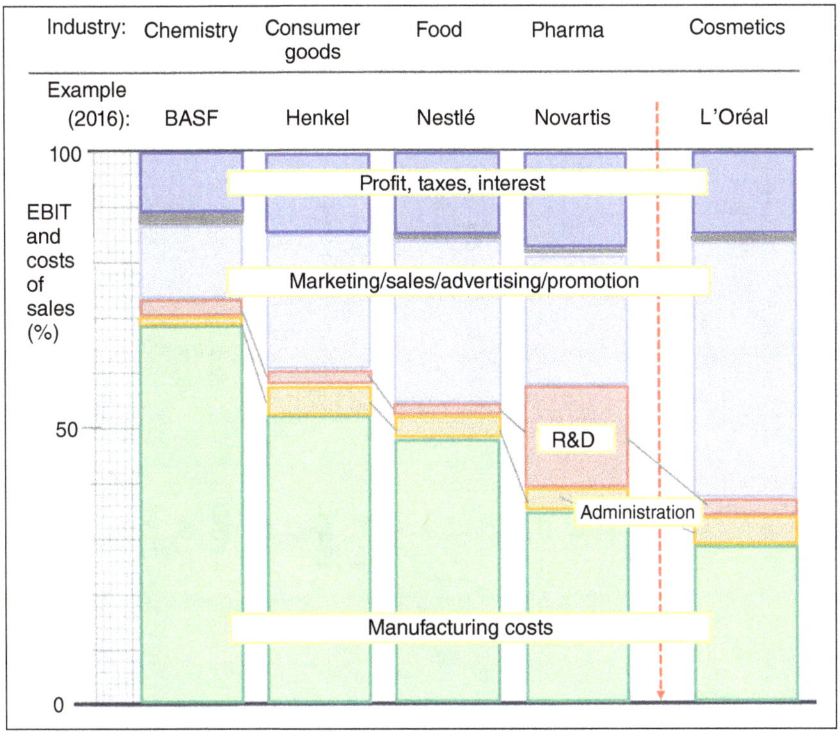

Figure 3.9 Comparison of different industries with the largest cosmetic manufacturer based on the cost breakdown of net sales (2016).

20% (Novo Nordisk, Esteé Lauder). Companies for the production of chemical specialties, consumer goods, and foodstuffs need about 45–55% of the revenues to cover manufacturing costs. The large-scale chemical industry requires about 2%, the others use 4–6% of the revenues for the administration. Nestlé spends 1.9% on research and development, well above the average of 1.1% of the German food industry. Esteé Lauder uses 1.7% for R&D, whereas L'Oréal for the greater product diversity and for the long-lasting peak position needs 3.3%.

The main difference of cosmetics companies to all other industries lies in the high costs for selling, advertising, and promotion. They spend almost twice as much on the total "Marketing and Sales" as the others. This item is stated in the annual reports of the cosmetic companies as a sum or divided into "advertising and promotion expenses" and "selling, general and administrative expenses." The proportion for "Advertising and Promotion" is between 22% and 25%: L'Oréal, however, is significantly higher at 29%. L'Oréal and Beiersdorf contribute to the position "Selling, general and administrative expenses" about 15%, whereas Esteé Lauder and Shiseido spend more than the double with 30–40%. It is striking that Shiseido and especially Chanel (Table 3.17) ultimately earned almost nothing because at least interest and taxes are still subtracted from EBIT. The Shiseido situation can clearly be seen from Figure 3.10 and has improved significantly in recent years.

Table 3.16 Net sales, costs of sales, and expenditures of the various industries; percentage values based on figures of the annual reports 2016.

Company	BASF	Novartis	Henkel	Nestlé[a]	L'Oréal[b]	Esteé Laude-[b]	Shiseido[b]
Industry	Chemistry	Pharma	Consumer goods	Food	Cosmetics	Cosmetics	Cosmetics
Year: 2016							
Net sales (NS) + other revenues (Mio.)	57 550 (€)	49 436 ($)	18 714 (€)	89 786 (CHF)	25 837 (€)	11 262 ($)	6 466 ($)
Costs of sales (CS/NS × 100); (Manufacturing costs)	−68.2%	−35.4%	−52.1%	−49.2%	−28.4%	−19.4%	−25.2%
Expenditures for							
➢ Administration	−2.3%	−4.4%	−5.7%	−9.0%[c]			
➢ R&D	−3.2%	−18.3%	−2.5%	−1.9%	−3.3%		
➢ Marketing and sales	−13.5%	−24.3%	−24.8%	−23.9%[d]			
➢ Advertising and promotion expenses					−29.0%	−25.1%	−23.4%
➢ Selling, general and administrative expenses					−21.7%	−40.0%	−47.1%
➢ Others	−5.4%	−4.7%	−0.8%	−0.8%	−2.1%	−1.2%	−1.2%
Other operationally earnings	+3.6%	+ 3.9%	+0.6%	+0.1%			
Operating profit (EBIT)	**10.9%**	**16.7%**	**14.8%**	**15.2%**	**15.5%**	**14.3%**	**3.6%**
Comparison with some estimated values							
Expenditures for							
➢ Administration	−2.3%	−4.4%	−5.7%	−4.0%	−5.5%	−5.5%	−5.5%
➢ R&D	−3.2%	−18.3%	−2.5%	−1.9%	−3.3%	−1.7%	−1.8%
➢ Marketing and sales (sum)	−13.5%	−24.3%	−24.8%	−28.9%	−45.2%	−57.9%	−63.9%

a) With the assumption of approximately 4% administrative costs, just under 29% would be needed for marketing and sales.
b) Estimated administrative costs approximately 5.5%.
c) Distribution costs.
d) Marketing and administration.

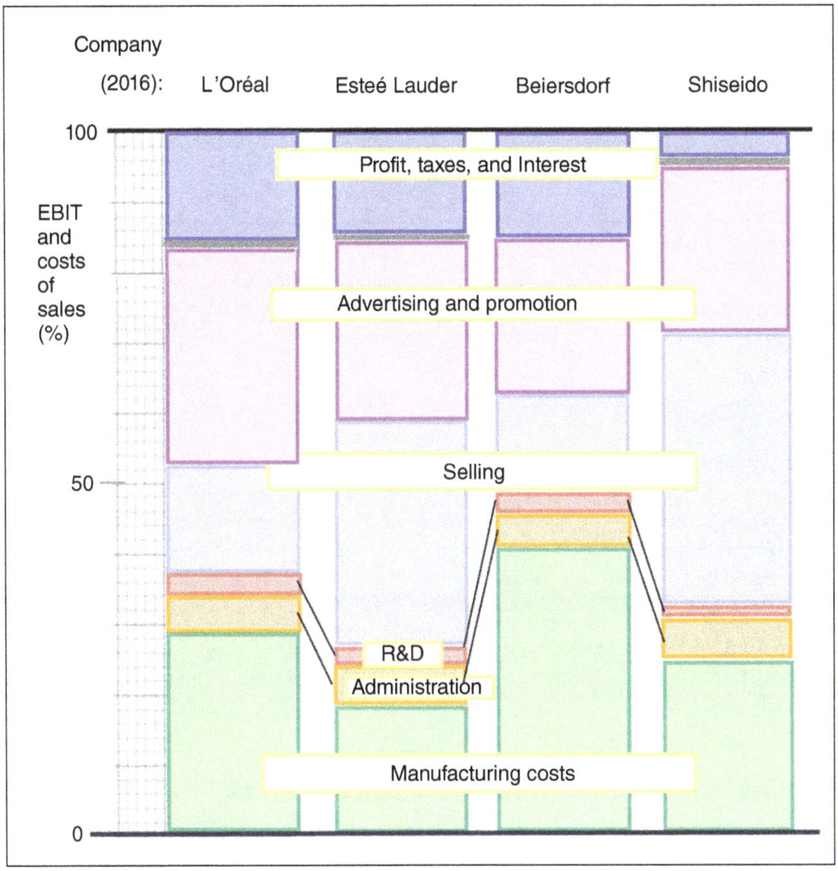

Figure 3.10 Cost structure of four world-leading cosmetics companies.

Depending on the billing regulations of the cosmetics companies, R&D and administrative costs can be included in the "Selling, general and administrative expenses." (Special way only for the cosmetics.) The real R&D costs are published elsewhere in the annual report, while the administrative costs for the comparison had to be estimated to get a real figure for the "Marketing and Sales" costs. For this, the values are between 35% and almost 65% (Shiseido). Well-known cosmetic companies, which however only own a business unit "cosmetics," are illustrated in Table 3.17. Again, the comparatively high marketing costs are recognizable, and in relation to "pure" cosmetics companies, the manufacturing costs are higher.

Estimation shows that the individual summands of the manufacturing costs do not have a major impact on market prices of cosmetics. If the total cost of manufacturing is 30% of net sales, however, the raw material costs thereof amount to approximately 70% (=21%). Labor costs could be around 10% (=3%) and energy costs below 3% (=1%). The remaining 17% (=5.1%) is attributable to depreciation. A relative increase in labor costs by 10% affect the market price by 0.3%. On the other hand, the same rise in Marketing and Sales costs would lead to a price increase of 2.5%.

Table 3.17 Cost structure of companies with a significant business unit "cosmetics" (2016).

Company Cost item	Unilever With 38.3% cosmetics	Beiersdorf With 17% adhesives	Avon With 26% home and fashion	Chanel With fashion, watches, jewelry	Johnson & Johnson With 81.5% pharma and medicine
Net sales (Mio.)	52 713 (€)	6 752 (€)	5 579 ($)	2 319 ($)	71 890 ($)
Costs of sales (manufacturing costs)	−57.4%	−41.1%	−39.5%	−54.7%	−30.2%
Expenditures for					
➢ Administration	−4.4%[a)]	−5.4%	−6.0%[a)]		−4.0%[a)]
➢ R&D	−1.9%	−2.8%	−0.9%[c)]		−12.7%
➢ Selling (general and administrative expenses)[b)]		−13.4%			−20.4%
➢ Advertising and promotion expenses		−22.3%			−3.3%
➢ Marketing and sales (sum of selling and advertising)	−21.5%[b)]	(−35.7%)	−48.0%[b)]	−43.9%	(−23.7%)
➢ Others					
Operating profit (EBIT)	**14.8%**	**15.0%**	**5.6%**	**1.4%**	**29.4%**
For cosmetics	**18.4%**	**13.9%**			

a) Estimates.
b) Position is reduced by administration and research costs.
c) Percentage related to cosmetics = 1.3%.

For the chemical industry, the operating results before interest and taxes (EBIT) are often between 6% and 11%. Other companies (food and consumer goods) range from remarkable 12–20%. These figures surpass the globally active top 10 pharmaceutical companies with an average of 28.1% [25] in 2014, Roche with an EBIT of 29.7%, and Novo Nordisk with 38.8%. To determine the annual net profit from the EBIT, additional (extraordinary) income and expenses, such as interest and country-specific taxes (Germany < 30%), must be taken into account (see Table 3.1).

The EBIT of well-established cosmetics companies/cosmetics divisions is often between approximately 13% (Beiersdorf) and about 20% (Procter & Gamble). Companies with a weak EBIT are trying to catch up. Shiseido, for example, has launched a four-year program to increase the EBIT. As shown in Table 2.3, Shiseido suffers from the low ratio of turnover to the number of employees. It is interesting to note that in the United States, the companies relatively earn the most, followed by Europe and then by Asia. For this reason, companies with a strong mainstay in the United States are particularly successful.

3.5 Methods for Pricing

The market price of a product is based on the site-specific prime costs and arises from the summation with the interest and taxes as well as the variable profit. The margin usually depends on the country/region, the quantity negotiated, the size of the packaged/filled units, and the importance of the buyer as well as the competition situation. Depending on the boundary conditions, there are therefore different prices for the direct customers. These may be employees of industrial enterprises (subsidiaries and foreign partners), wholesalers, and buyers from trade chains. In many department stores, every big cosmetics company has its own shop (shop in shop). Pharmacists buy the cosmetic products either through the wholesale or, if possible, cheaper directly from the manufacturer. The manufacturer can only negotiate market prices with direct customers. Indirect customers are retailers who shop at wholesalers as well as consumers who go shopping in different stores. Pharmaceutical products represent the exception. The way to the consumers usually leads via the wholesale and always through the pharmacy. In Europe, the (fixed) prices for drugs are country specific.

The manufacturer would like to achieve the preset profit targets and therefore put through the highest possible price, taking into account the competition situation. On the other hand, the customers are either interested in a low price with sufficient quality or in an attractive price/performance ratio. Some customers buy deliberately items from high-priced luxury brands with international reputation, to which they trust and value their image. In some cases, the consumers pay higher prices for special qualities/colors/functions. In practice, the brand manufacturer uses three different methods [26–28] for pricing, which are used alone or often together. In theory, the methods are primarily based on cost price. However, for some consumer goods, the expected demand is also important and, of course, the competition. The size and image of the manufacturer as well as the strength of its brands play an important role in determining the market price.

In the real world of cosmetics, the prices to be paid by the customer depend on the point of sale. The companies grant all buyers from trade chains, wholesalers, and businesses different, partially high discounts. Therefore, price comparisons should be made separately for each product. In addition, comparisons of the performance between several products of different manufacturers are difficult and complicate the price comparisons by the customer as well as the correct pricing by the cosmetics company.

3.5.1 Pricing Depending on the Customer

Once a year, the price is negotiated between the cosmetics companies and their big direct customers. The price achieved is a compromise, taking into account long-term business relations (preferred supplier), delivery reliability, and competition. To offer the consumer a wide selection and ensure continuous supply, the trade chains often buy comparable products also from other companies. The negotiated price usually refers to a minimum quantity over the year or to defined quantity ranges with differentiated discounts, whereby larger quantities lead to a higher price reduction. The direct customers often buy the products over many years, sometimes over decades from the same company.

Cosmetics items are offered in small and large selections in various markets. These include the retail trade markets, supermarkets, and consumer markets; department stores; cosmetics stores (Douglas, airports); drugstores; pharmacies; mail order; and the Internet. Some of them buy the goods centrally. They play out their market power and push the prices in the annual negotiations. Discounters are particularly known for this. Depending on the result achieved, products in the assortment are continued or discontinued. After payment of a shelf fee, new products are added. Small manufacturers have to give way in the price and may get into an existential dependency. Large price concessions in negotiations with big companies lead to foreseeable conflicts in price negotiations with other customer.

An indirect sale of the goods to the final consumer takes place through two different strategies. On the one hand, some unknown local manufacturers, usually in cooperation with a discounter, work with the price–quantity strategy, which is characterized by a low-price or discount concept. On the other hand, global brand manufacturers offer their products at higher prices in the preference strategy [27], an upscale concept for branded products with differentiated, country-specific prices. Some branded products can also be successfully sold by the discount drugstores because the opposing goals between manufacturers, the sales market, and customers are not insurmountable opposites. Normally, the consumer accepts a somewhat higher price for branded articles, if the product design and brand image are convincing.

Consumers often buy inexpensive cosmetic products of daily life such as cleaning agents for the body and the hair. For the beauty of their faces, they are ready to spend more money. For this purpose, from time to time, high-quality creams, perfumes, liners, and makeup are bought in a luxurious environment, especially from a luxurious global brand. In the shop, the presentation and the brand are often the deciding factors for the purchase.

3.5.2 Cost-Oriented Pricing

In the market economy, there exist no direct correlation between the costs and the sales price from a scientific point of view. In practice, however, the cost-oriented determination of the market price is often used for consumer goods. The cost prices arise from the cost accounting. As the lower limit, the product must bear its own cost. For this reason, the price calculation is based on the full cost basis including the total marketing and sales after adding the profit margin required by the company. In terms of corporate objectives, pricing is then successful when the planned profits (revenues minus prime costs) can be realized. In order to cover corporate risk, the target is a percentage of at least 15%, preferably 20% for the EBIT.

In the case of consistently adhering to the method of setting the price over the full costing, it can be criticized that the price would have to go up with a decline in sales. The decline would lead to an increase in specific costs because proportional direct and overhead costs must be distributed to fewer products. This would require a price increase or otherwise reduce the profit. In the phase of a drop in sales, however, a higher price would certainly be the wrong decision because it leads to less sales and thus sets a downward spiral in motion.

The partial cost calculation [18] allows reacting within narrow limits. It only contains the variable costs as the absolute lower price limit. When applying this method, there is a risk that medium and long-term costs will not be fully covered through the customer because of an excessive price discount, thus causing losses at this location. Such losses can only be justified for a limited time or for strategic reasons. Both mentioned methods have in common that they do not adequately consider the driving factors of consumer good markets. Thinking from the market and setting the price is the purpose of the target cost management.

3.5.3 Demand-Oriented Pricing

Via supply and demand, the pricing of established products takes place because the price can be adjusted by the demand intensity. Strong demand allows for smaller price increases, and weak sales enforces price concessions (price elasticity) [27] and/or additional investment in the market through marketing. In order to stimulate sales, for example, campaigns with the trade and nationwide advertising via television or magazines are suitable. In some cases, these campaigning measures require special offers, which means that not only advertising costs are incurred but also less profit per unit. In the long term, this strategy would lead to a drop in the market price, which is undesirable. In the campaign period, however, sales can usually be significantly increased, so that the absolute revenues are higher. Despite price concessions, the percentage profit margin may rise as the fixed costs of production and the non-process-related costs will be distributed to more products and thus fall per unit.

For newly developed cosmetic products, target costing [18, 29] allows the determination of a price that is achievable or enforceable in the market. The question is how much customers would be willing to pay a suggested price for the offered product. In order to establish a realistic price, the known alternatives in the market must be considered according to the price/performance ratio. The determination of the price corridor via surveys is carried out by external market research agencies. The results of the surveys depend heavily on the region; therefore, people in different cities/countries are asked.

Initial valuations of a product take place in discussion groups, called focus groups (12 participants of the target group). It is possible that in a preliminary phase, the panelists know neither the brand nor the manufacturer; it is a "blind" test. However, for the indication of a price quotation ("price meter"), the panelists learn the brand and the manufacturer as well as all facts about the product. They see the cream in front of them and can test it briefly. Subsequently, a price range is specified in which the product is located. The participants indicate a price for which they would probably buy the cream. The evaluation of the interviews yields the "price meter" [30], usually in the form of a cumulative distribution curve with the price as an abscissa. The price meter depends on the region in which the survey was carried out. The companies are therefore trying to choose a "typical" region for the whole country.

In Germany, selling prices for therapeutic products are relatively high compared to many other countries. For over-the-counter (OTC) products, there are suggested retail prices, to which the pharmacies sell with the exception of special offers. An example should be briefly discussed. Bepanthen, a wound and healing ointment from Bayer, is a well-known, widely sold OTC product that costs €13.84 in German pharmacies for a 100 g tube. On the Internet, according of the price, search engine is the cheapest price €6.41 (without shipping costs), which means a saving of €7.43 (−53.7%). In an estimate, the distribution of the costs might look like this:

Price in the pharmacy (all prices including VAT)	€13.84
Price in the cheapest online shop	€6.41 (−53.7%)
Purchase price of the pharmacy at the wholesaler	€7.80
Purchase price of the wholesaler	€5.13
Purchase price of the online retailer directly from Bayer	€5.13 (−20%)
Composition of the purchase price	
EBIT Bayer	€1.54 (−30%)
Marketing/sales	€2.05 (−40%)
Administration/R&D	€0.38 (−7.5%)
Manufacturing costs (100 ml tube, including VAT)	€1.16
Manufacturing costs (including packaging, without VAT)	€0.94

The high margins of the trade are due to the low sales quantity per customer. The price margins of low-selling products might be even higher, but for mass products (soaps, toothpastes, and simple creams), however, the margin drops significantly. With some products, almost nothing is earned anymore. This applies above all to mass products in regions with a strong competitive environment. This is not (yet) true for many skincare products.

An example of pricing represents the survey to an antiaging cream. The curve in Figure 3.11 shows a possible result for sales in the pharmacy. Only customers who buy their cosmetics mainly in the pharmacy take part in the survey. The curve arises after an intensive discussion and presentation of the product. The participants answer the question how much money they would spend for a 50 ml dispenser with this cream. The curve demonstrates that 95% of respondents accept a price up to €14.50. No one will pay more than €19.00. Therefore, it would make sense to set the price around €14.50. This price can be easily realized with the calculated prime costs of €5.80 by direct purchase. If the pharmacy needs to buy from a wholesaler, the price is €17.34, which only a minority accepts. This review

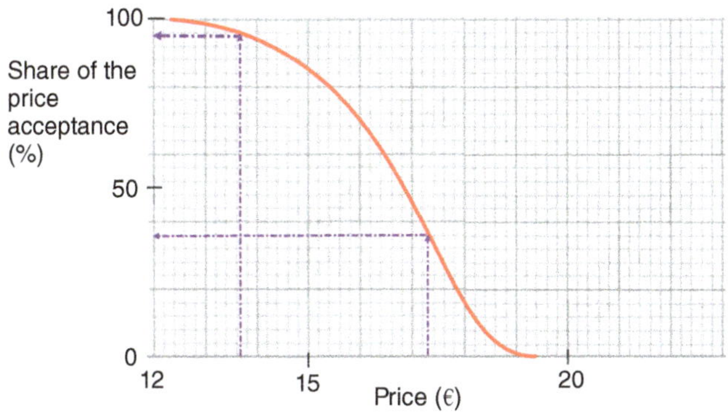

Figure 3.11 Share of consumers who accept a defined price for a 50 ml dispenser with a high-quality antiaging cream ("price meter"; points of the acceptance share results from Table 3.18).

is important to ensure the own profit and is part of the method "target costing." Marketing can also set a price of €16.80 (recommended retail price), which 50% of respondents would pay. In this area, the product is price-sensitive, as the number of buyers halve already at €17.50 (+4%). It is more promising to choose the low price in this case.

Table 3.18 allows an insight into the exemplary cost composition of the featured antiaging cream in a 50 ml dispenser. In case of a direct purchase from the cosmetics manufacturer, the pharmacist can easily sell the cream at a price of €13.70 with a reasonable profit. At a price over €17.00, the product becomes a nonseller. For comparison, the possible pricing of drugstore discounters (trade chains) are displayed. Discounters take centrally large amounts and get discounts of 20–30% and more. Therefore, the prices are lower. Brand manufacturers usually sell products either in the pharmacy or drugstore, never at the same time in both.

3.5.4 Competition-Oriented Pricing

Competitor products affect pricing, whereby quality, market shares, and brand strength as well as their outlets are taken into consideration. In the case of consumption products, also for cosmetics, the market is often divided into two, three, or more (quality/brand/price) segments. The product must be allocated to the "right" segment (=price corridor) in order to survive in the market. Decisive are the own brand strength, the quality of the new product, and the presence and classification of the own market products. In most cases, brand manufacturers possess several brands that they allocated to the luxury segment, or the upper, middle, or lower segment. The new products receive a suitable brand, which determines the price corridor. Figure 3.12 shows an example of possible price segments for creams.

Table 3.18 Determination of selling prices for an antiaging cream (50 ml) in the pharmacy at direct purchase and buying through wholesalers; only for comparison: possible discounter calculation, an arbitrary example.

	Costs of the Pharmacy (branded product)		Costs of the Drugstore (trade chain; "No name"product)
	Brand manufacturer		Small local brand free manufacturer
Item	(a) high sales	(b) low sales	(c) high sales
➢ Raw material (€/kg)	15.00		
➢ Production, filling, packaging	5.00		
➢ Marketing and sales	43.00	47.00	18.00
➢ EBIT	23.00	33.00	11.00
Costs of the cream (€/kg)	86.00	100.00	49.00
➢ Costs for the content of a 50 ml dispenser	4.30	5.00	2.45
➢ Dispenser (€/50 ml)	0.80	0.80	0.80
➢ Label	0.50	0.50	0.50
Costs of the cream and the package (€/50 ml)	5.60	6.30	3.75
Margin wholesale	4.08	—	—
Margin shop	4.89		
Margin shop by direct order		5.18	2.75
Sum	14.57	11.48	6.50
VAT (19%)	2.77	2.18	1.24
Price in shop (€/50 ml dispenser)	**17.34**	**13.66**	**7.74**

For example, L'Oréal (see Figure 2.14) assigns its brands for creams in particular to the luxury segment with the brands Lancôme, Yves Saint Laurent, and Biotherm as well as the upper segment with L'Oréal Paris and Vichy. There can also be rankings within the segment. Lancôme is likely to be the top brand in the field of care creams. On the other hand, Beiersdorf occupied the middle segment with the Nivea brand and the upper segment with Eucerin. In the lower segment are mainly no name manufacturers and local, hardly known brands, which produce for drugstores and discounter.

In the case of competition-oriented price determination, the marketing is geared to the superior market leader, who has the most market shares in the selected segment. In the case of an approximately uniform distribution of the market shares between two or more brands, a renowned company is usually recognized as the price leader, and within the price corridor all others follow. Significant price undercutting would result in a decay of prices in a short period of time, resulting in lower revenues or even losses for all market participants.

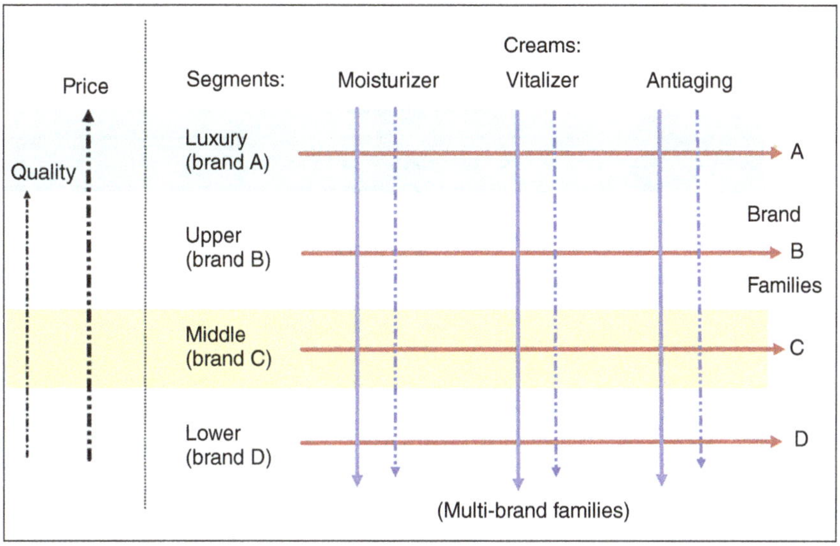

Figure 3.12 Typical segmentation of the cosmetics market using the example of the skin creams (brand families: same brand, different products; multibrand families: comparable products with the same function but of different brands).

Therefore, brand manufacturers (without agreement) stay within the presumed price corridor.

3.5.5 Influence of the Brand

The branded article concept has at least three basic features [27]:

- Proof of origin and quality guarantee
- Image and mark perception of the consumers
- Ubiquity (available everywhere).

The low price in the price-quantity strategy is replaced here by a higher price for the quality leadership. For this, the branded article manufacturers must generate performance-specific advantages, mostly worked out in the R&D department. They want to offer products that their customers value above average and therefore are willing to pay more for it. This is all the more successful, the better the reputation of the manufacturer and the stronger the brand. Consumers want to buy a proven and safety-tested product in high, consistent quality. This is ensured by the brand manufacturer. For these reasons, the brand manufacturers are trying to identify the customer requirement using questionnaires before production. The aim is the market launch of products that achieve the highest possible customer satisfaction at an accepted price.

The calculated prime costs, the market segment, which is defined with the chosen brand, and the expected price-dependent demand usually determine the market price. Furthermore, the determination takes place on the one hand considering the own product performance in comparison with the performances and

Table 3.19 Use of three pricing methods by global companies taking into account the attractiveness and reputation of the own brand.

Methods	Factors influencing pricing
1. Cost orientation	o Contribution margins o Existing sales structures o Customer's acceptance and discounts o Prime costs o Profit margin
2. Demand orientation	o Offer and demand (country dependent) o Product novelty/selling position o Survey and product testing o "Price Meter" (region, customer dependent) o Price positioning for the intended market (region/customers)
3. Competitor orientation	o Competition strength o Price/performance/classification o Product segment o Brand choice o Price corridor
In addition: Brand	o Brand strength of the product family o Reputation/image of the manufacturer o Trustworthiness of the brand for the product

prices of other market products. On the other hand, the brand sympathy (image) of the product/product family as well as the reputation of the manufacturer plays an important role for the positioning of the product and thus the possible profits. Within a segment, the strong brands usually achieve the highest prices. A measure of the strength of a brand [31] is the brand perception, measurable by the number of repeated purchases. In the case of new products, it expresses itself in the expected demand and finally in the shelf space in the store.

3.5.6 Summary of Pricing

The three methods commonly employed for price-fixing can be found in Table 3.19. As a rule, all aspects of the product sale are discussed and considered in the company. When pricing, one's own strength in terms of brand, company's size, sales organization, and market access must be observed. Important factors represent the prime costs and the potential profit in the expected demand at the proposed price.

3.6 Learnings

✓ The manufacturing costs consist of fixed costs (depreciations and personnel) and variable costs (raw materials and energies). In height, they are predominantly determined by the cost of raw materials.

✓ In many cases, production in two-shift operation is the most cost-effective option.
✓ Administration, R&D, and Marketing and Sales (M&S) represent the costs of the company.
✓ The addition of the costs of the company to the manufacturing costs leads to the cost prices (prime costs) of a product.
✓ Cosmetics companies are characterized by very high M&S costs and low manufacturing costs.
✓ With 35–65% of sales, more than twice as much as the usual M&S costs, the cosmetic industry occupies a special position.
✓ Methods of pricing in international companies with strong brands:
  - Cost-oriented pricing through the use of the cost price and a profit margin.
  - Demand-oriented pricing in the target management through the use of a "price meter," obtained by interviewing relevant consumers.
  - Competitor-oriented pricing through an analysis of the competitive environment.
✓ In addition, the price results from considering of brand strength, company size, sales organization, and market access.
✓ The cosmetics market is divided into several price segments. A new product must be assigned to the right brand and segment.

References

1 (a) Rähse, W. (2012). Komprimiertes Entwickeln und Umsetzen von Innovationen. *Chem. Ing. Tech.* 84 (5): 588–596. (b) Rähse, W. (ed.) (2014). Compressed development and implementation of innovations, (Chapter 4). In: *Industrial Design of Solids and Liquids, A Practical Guide*, 39–56. Weinheim: Wiley-VCH.

2 (a) Rähse, W. and Hoffmann, S. (2002). Produkt-Design – Zusammenwirken von Chemie, Technik und Marketing im Dienste des Kunden. *Chem. Ing. Tech.* 74 (9): 1220–1229. (b) Rähse, W. and Hoffmann, S. (2003). Product design – interaction between chemistry, technology and marketing to meet costumers needs. *Chem. Eng. Technol.* 26 (9): 1–10.

3 Rähse, W. (2007). Bewertung der Innovationshöhe und der Marktattraktivität (Chapter 15). In: *Produktdesign in der chemischen Industrie: Schnelle Umsetzung kundenspezifischer Lösungen*, 225–232. Berlin: Springer-Verlag.

4 Rähse, W. (2016). Vorkalkulation chemischer Anlagen. *Chem. Ing. Tech.* 88 (8): 1068–1081.

5 Meyer, D., Bundesarbeitgeberverband Chemie (BAVC) (2014). Chemiearbeitskosten erneut gestiegen. *CHEManager* 21–22/2014 (13 November–10 December), S. 6.

6 § Arbeits-vertrag.org (2019) Schichtzulagen: Wann erhalten Arbeitnehmer mehr Geld? https://www.arbeitsvertrag.org/schichtzulagen/ (accessed 4 May 2019).

7 Chemische Industrie, WSI-Tarifarchiv, 2011. https://www.boeckler.de/pdf/ta_2012_stb_chemische_industrie_nordrhein_2011.pdf (accessed 3 May 2019).

8 Baerns, M., Behr, A., Brehm, A. et al. (2013). Wirtschaftlichkeit von Verfahren und Produktionsanlagen(Chapter 13). In: *Technische Chemie*, 2e, 481–494. Weinheim: Wiley-VCH.

9 Storhas, W. (2013). Wirtschaftlichkeitsbetrachtungen(Chapter 9). In: *Bioverfahrensentwicklung*, 2e, 695–719. Weinheim: Wiley-VCH.

10 E-Control, Ergebnisse Industriegaspreise, 2019. http://www.e-control.at/de/industrie/gas/gaspreis/industriegaspreise/energiepreis (accessed 15 November 2018).

11 Worthoff, R. and Siemes, W. (2012). *Grundbegriffe der Verfahrenstechnik*, 3e. Wiley-VCH.

12 Aspentech, Aspen Plus, maximize profits using a plant-wide simulation solution that combines unparalleled accuracy and engineering collaboration with time-saving workflows, 2019. http://www.aspentech.com/products/aspen-plus.aspx (accessed 6 May 2019).

13 Vogel, H. (2012). *Verfahrensentwicklung: von der ersten Idee zur chemischen Produktionsanlage*. Weinheim: Wiley-VCH.

14 L'Oréal Annual Report 2016, brands overview, a global flotilla of complementary brands, 2017. http://www.loreal-finance.com/en/annual-report-2016/brands-overview (accessed 11 November 2018).

15 Henkel Annual Report 2016, creating sustainable value, 2017. http://annualreport.henkel.com/blob/739562/7adc0b40ea3690105caf799fed39091b/data/2016-annual-report.pdf (accessed 1 August 2017).

16 Joos, T. (2014). Kostenträgerrechnung (Chapter 7.3). In: *Controlling, Kostenrechnung und Kostenmanagement: Grundlagen – Anwendungen*, 5e, 185–218. Springer Gabler.

17 Preißler, P.R. and Preißler, G. (2015). *Entscheidungsorientierte Kosten- und Leistungsrechnung*. Berlin: Walter de Gruyter.

18 Murjahn, R. (2005). *Kostenmanagement in der chemischen Produktentwicklung*. Springer.

19 Zingel, H. (2004). *Lehrbuch der Kosten- und Leistungsrechnung: KLR in Theorie und Praxis*. Birmingham: Goyang Media Ltd.

20 Steinbach, A. (2013). *Ressourceneffizienz und Wirtschaftlichkeit in der Chemie*. Weinheim: Wiley-VCH.

21 Dutta, M. (2005). *Cost Accounting: Principles and Practice*, 2e. Delhi, India: Pearson Education.

22 Möller, H.P., Zimmermann, J., and Hüfner, B. (2005). Erlös und Kosten(Chapter 2). In: *Erlös- und Kostenrechnung*, 67–134. München: Pearson Studium.

23 Plinke, W., Rese, M., and Utzig, B.P. (2015). *Industrielle Kostenrechnung: Eine Einführung*, 8e. Berlin: Springer Vieweg.

24 Preißler, P.R. (2008). *Betriebswirtschaftliche Kennzahlen*. München: Oldenbourg Wissenschaftsverlag, S. 77 ff.

25 Investor Magazin, 2019. https://investor-magazin.de/0313top-20-pharmaunternehmen-u-a-bayer-gilead-novo-nordisk-novartis-roche/ (accessed 6 May 2019).

26 Rähse, W. (2017). Ermittlung eines kompetitiven Marktpreises für neue Produkte über die Herstellkosten. *Chem. Ing. Tech.* 89 (9): 1142–1158.

27 Becker, J. (2006). *Marketing-Konzeption, Grundlagen des zielstrategischen und operativen Marketing-Management*, 8e. München: Verlag Vahlen, S. 516 ff.

28 Meffert, H., Burmann, C., and Kirchgeorg, M. (2014). *Marketing: Grundlagen marktorientierter Unternehmensführung, Konzepte-Instrument- Anwendungsbeispiele*, 12e. Springer Gabler, S. 482 ff.

29 Clifton, M.B., Townsend, W.P., Bird, H.M.B., and Albano, R.E. (2005). *Target Costing: Market Driven Product Design*. New York: Marcel Dekker.

30 Rähse, W. (2007). *Produktdesign in der chemischen Industrie*. Berlin: Springer-Verlag, S. 180 ff.

31 Köster, L. (2006). *Markenstärkenmessung unter besonderer Berücksichtigung von Konsumentenheterogenität*. Wiesbaden: Deutscher Universitätsverlag.

4

Scientific Descriptions of the Skin

4.1 Tasks of the Skin

First, the skin is involved in the metabolism of the whole organism, such as the synthesis of vitamin D and biotransformation of some chemicals. Second, the skin delimits the body from the environment and performs a number of important tasks for the body. These include the protective functions against all exogenous noxae, especially mechanical and chemical influences as well as microorganisms (especially pathogenic microbes), allergens, and also UV rays (sun), and furthermore against cold, heat, and dehydration. Third, the skin regulates the temperature via constriction/dilation of the blood vessels and by sweating as well as its water balance because of evaporating of water on the surface. However, absorption of water from the outside through the skin is not possible; supplemental water comes from inside of the body. The body can excrete pollutants and excess protein via sweat. The sebum is produced in the skin and delivers the important, protective hydro lipid mantle on the skin. Even skin has tissues for the temporary storage of fat, glucose, water, and salts [1]. In addition, the task of the skin is also to protect the organism by perceiving and forwarding external stimuli such as temperature, touch, pain, vibration, pleasure, and pressure [2, 3]. Furthermore, the skin reacts to feelings such as horror, anxiety, and panic, especially visible in the face because of redness and paleness as well as sweating and ruffling of the hair on the arms and legs.

4.2 Structure of the Skin

The skin represents an area of about 1.5–2 m^2, with a weight of 3.5/10–14 kg (strongly varying values, probably dry/moist) and a thickness of 0.3–5 mm, depending on the body site, age, and gender. In younger years to around the age of 65 years, the skin of men is thicker than that of women. Later, the skin thicknesses are similar because men's skin becomes thinner than women's skin. Over 4 million receptors of the skin, the outer sensors of the nerves, provide us with information about cold and heat, pain, and pleasure. One square centimeter of skin includes about 2 heat points, 13 cold spots, and 200 pain points. The skin mass accounts for about 16% of the total body weight. In 100 days, the

Cosmetic Creams: Development, Manufacture and Marketing of Effective Skin Care Products,
First Edition. Wilfried Rähse.

skin grows by 0.2 mm. It contains a quarter of the water stored in the body (total mass = 70% of body weight). On average, in 1 cm^2 of skin, there are about 6 000 000 cells, 5000 sensory cells, 4 m of neural pathways, 200 pain sensors, 100 sweat glands, 1 m of blood vessels, 15 sebaceous glands, 12 cold receptors, 5 hairs, 2 heat receptors, and 150 000 pigment cells [4]. The skin thus represents the sensor and control center for many sensory perceptions [5].

In addition to the many tasks, the largest organ of the people characterizes the appearance. On the surface, the hydro-lipid coat prevents drying out and reduces roughness of the skin. The constantly renewing fat and acid layer is formed from the secretions of the sweat and sebaceous glands. Because of its low pH, the layer acts an antimicrobial and inhibits the growth of many bacteria and fungi. Therefore, this protecting coat is also called as acid mantle, which must be available on the skin in the right composition and thickness, because the skin is daily exposed to many stress factors. These include exposure to too much sun, intense cleanliness, too dry ambient air, environmental influences during leisure time and at work. The strain under prolonged exposure damages the film on the skin. Therefore, a skin care, which strengthens and renews the hydro-lipid film, is of great importance for the well-being of young and, above all, elderly people, today and in the future.

The skin on the body is clearly different from the facial skin. As it is not directly exposed to the environment, the cell metabolism is slower than on the face. The lower desquamation leads to the thicker body skin. The epidermis in the face is about 0.12 mm thick, and the body average is about 0.60 mm. In the upper layer of the facial skin, the cells are smaller compared with other parts of the body. The smaller cells lead to a higher loss of moisture in the face because the barrier is thinner there than on the body.

The skin is divided into three layers (Figure 4.1a), from the inside out into the subcutis, the dermis, and the epidermis. Both the entire **subcutis** and dermis are well supplied with blood. The subcutis represents the lowest of the three skin layers and is considered as a shifting layer, which absorbs tangential forces. The water and the fat in the layer make the skin firm and can cushion shocks. The subcutis layer consists of loose tissues, which connect the upper skin layers with the underlying structures (periosteum and fascia), via partitions (retinacula and septa). These connections can be so strong at some parts of the body, as on the scalp, where the subcutaneous tissue with its underlay is grown together to a uniform, nondisplaceable structure. Between the connective tissue septa, fatty tissue islands are embedded. The fat is either taken up from the blood into the cells or formed from carbohydrates directly in the cell. The fat content of the subcutaneous tissue depends on hormonal influences of sex and constitution.

The **dermis** consists mainly of connective tissue fibers that anchor the epidermis [6]. Here, the finely capillarized blood vessel system supplies the border zone to the epidermis. On the one hand, in this layer, there are sweat glands and sebaceous glands as well as hair follicles and hair muscles. On the other hand, most of the sensory cells for touch stimuli, nerve cells, and neural pathways are located within the dermis. The lower dermis contains smooth muscles and blood vessels

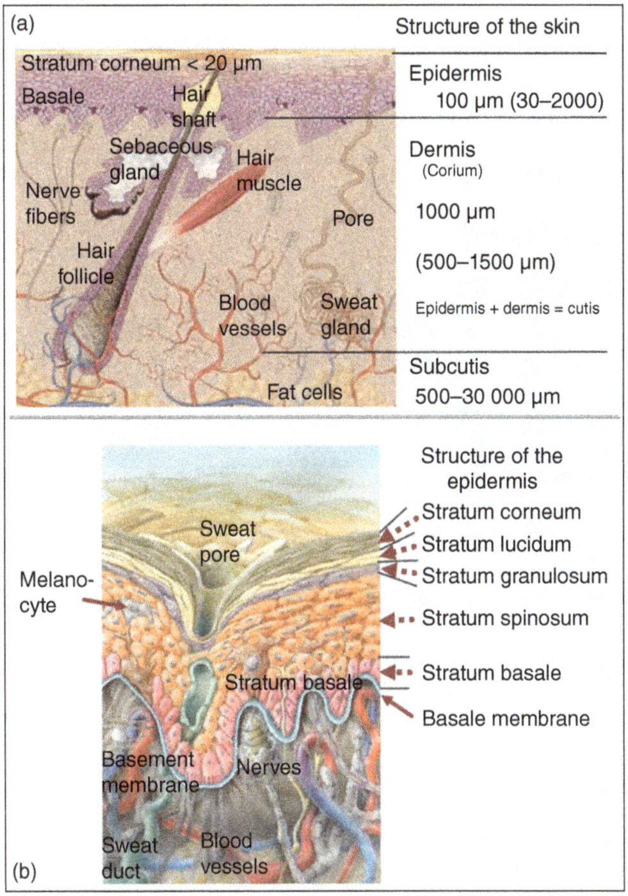

Figure 4.1 View into a cross section of the skin: (a) overview and (b) structure of the epidermis. Source: Courtesy of COGNIS, now BASF, Skin Care Forum 29, 34.

that are important for temperature control. The dermis consists of a thick, elastic, yet firm middle layer of the skin, which can be divided into two layers:

- The upper layer (stratum papillare) is sharply delimited by the basal membrane, the transition to the epidermis.
- The lower layer (stratum reticulare) represents the deeper, thicker layer, which forms the flowing transition to the subcutis.

The elasticity of the skin depends primarily on the three-dimensional mesh of the collagenous and elastic fibers of the dermis.

As the outermost layer of the skin, the **epidermis** represents the actual protective cover and a direct connection of humans to their environment [2, 4]. The epidermis is free of vessels and has no nerves. Their thickness varies greatly and depends on the degree of their mechanical loading. In unloaded areas (such as

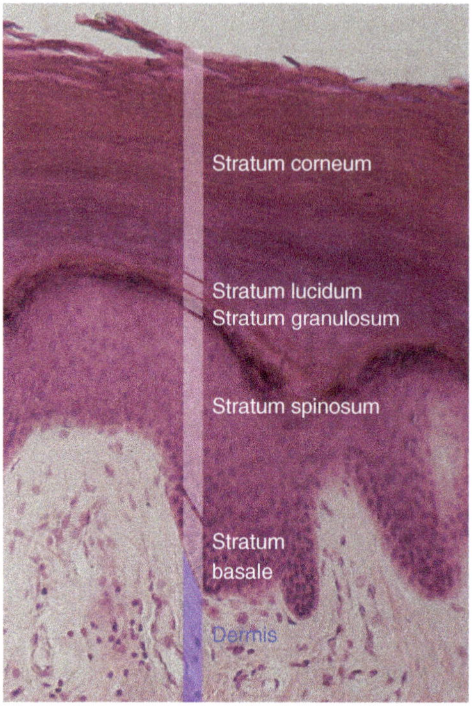

Figure 4.2 Histologic image of human epidermis in thick skin (layers of the epidermis). Source: Wikipedia, GNU Free Documentation License, Mikael Häggström, based on work by Wbensmith, File: WVSOM Meissner's corpuslce.JPG.

eyelids), the epidermis is only 0.03–0.05 mm thick, whereas in areas of high stress (palms and soles of the feet), the thickness amounts up to 2 mm. This in turn is divided into several interlocking layers, which differ in cell size, form, and function. Between the superficial horny layer (stratum corneum) and the lower germ layer (stratum basale) lie the glossy stratum (stratum lucidum), the granule layer (stratum granulosum), and the prickle cell layer (stratum spinosum) (Figure 4.1b). The epidermis is totally separated from the dermis by the basal membrane, which is located directly below the stratum basale and above the dermis cells. Figure 4.2 shows a histological cut through a thick part of the skin. The statements of the drawings shown in Figure 4.1 are confirmed by an image from the natural skin.

The basal layer is the source of epidermal stem cells. In these single-layered cells of the **stratum basale**, continuous new layers are formed by cell division, moving upward. The journey of the more or less round cells from the basal to the horny layer takes about two to four weeks, and another two weeks through the stratum corneum until desquamation. Each basal cell splits every 200–400 hours. In the process, a cell remains in the basal layer, whereas the daughter cell moves in the direction of the horny layer and provides replacement for desquamated horny cells. On the way up, the cells change their size and shape because of the reduced amounts of phospholipids. They become more and more flat. Outer horny layers (stratum disjunctum) gradually disappear by abrasion. In the stratum basale, predominantly basal cells occur, interspersed with 5–10% melanocytes and Merkel cells, which absorb pressure stimuli and transmit them to the brain. In the melanosomes, a part of the melanocytes, the brown dye is synthesized. The formed melamine is passed on to adjacent keratinocytes, giving the skin a uniform, genetically determined color.

In the **stratum spinosum**, also known as "spiny cell layer," the cells are only connected by desmosomes, small spines. Lymphatic fluid is present in the interspaces. Because of this structure, this layer can absorb tensile and compressive forces. This is where keratinization begins; in other words, the process of cornification. In the stratum spinosum, there are also immune cells of the lymphatic system, which are called Langerhans cells. Stratum basale and stratum spinosum together are referred to as stratum germinativum (synonym: germ layer, regeneration layer).

The granule cells of the **stratum granulosum** are larger, flattened, and spindle-shaped. There, they lose their nucleus. Instead, they contain small granules that synthesize an oily substance (keratohyalin). This oil soaks the layers, keeps them supple, and forms keratin in the horny layer. The thickness of the grain layer adapts to the stresses of the skin. **The stratum lucidum**, also known as the "glossy layer," is pronounced only in the inguinal skin of the hands and feet. The layer consisting of flat, coreless cells can be easily moved tangentially and thus reduces the consequences of shocks. Because of the oily keratohyalin, there are only small light refraction differences; the layer appears more or less transparent. The keratohyalin granula forms the semiliquid fat and the protein-rich, acidophilic Eleidin. During cornification, the process where living keratinocytes are transformed into nonliving corneocytes, the cell membrane is replaced by a layer of ceramides. These are covalently linked to structural proteins.

The **stratum corneum** represents the body's important external barrier against chemical, mechanical, microbially, immunological, and radiation effects [7]. The layer is not only a waterproof barrier but also a reservoir for topically applied substances. Because of low metabolic activity of the cells, a depot can be formed for diffusing substances. Forced by the concentration gradients, the stored substances gradually penetrate into the deeper layers of the skin.

The horny layer consists of coreless, compressed flattened dead keratinocytes, containing keratin and filaggrin, and cell debris (scales). Their thickness depends on the load and lie between 0.01 and 0.5 mm (10–500 μm, mostly 10–40 μm), usually at a thickness of 15–20 constantly renewing cell layers. The keratinized cells do not contain cytoplasm, but a dense network of keratin and filaggrin, and are additionally surrounded by a coating. Each keratinocyte has an approximately 10 nm wide envelope consisting of several proteins called the cornified envelope [8]. The proteins keratin and filaggrin inside the cells keeps the skin hydrated by absorbing water and preventing water evaporation. The water content of the horny layer is about 15%, with by far the most water in the keratinocytes, while the corium shows a content of 71.5% water.

In addition, the horny layer is responsible for the elasticity of the skin. After shifting, a weak glutenous protein bond pulls the skin back to the natural shape. In the intercellular space between the corneocytes, there is the lipid matrix (kit) that connects the cells to each other. The kit consists of each 30% ceramides and free fatty acids as well as 40% cholesterol and its derivatives and accounts for about 10–15% of the mass of the stratum corneum. More detailed information can be found in the literature [9]. These intercellular lipids form a bilayer through which various lipophilic active substances can pass through the horny layer. High ceramide content ensures a strong skin barrier. By contrast, this layer blocks the hydrophilic substances dissolved in water. It also prevents severe transepidermal water loss (TEWL) that means a significant evaporation of water through the skin [10]. Transepidermal water comes from deeper layers and reaches the surface of the skin by diffusion. There, it disappears through evaporation. The water content of the healthy horny layer is 15–20%; less than 10% is referred to as dry skin. Within the cornea, there exists a pronounced pH gradient toward less acidic values. In the stratum corneum disjunctum, a pH of 4.5 ± 0.2 occurs in males and 5.3 ± 0.5 in females and increases to 6.9 in the stratum granulosum. This corresponds to a jump of about 2 pH units over a layer thickness of 15–30 μm. For this reason, in creams, a pH of 5 or approximately 5 is recommended.

4.3 Concepts for Penetration of the Stratum Corneum

Many people experience the positive functions of the skin barrier in their daily lives. When showering, the skin does not absorb water, even when using a shower gel. After applying a good lotion or cream, the water runs off from the skin in the form of drops for days. Bathing in chlorinated water in a swimming pool is not dangerous because sodium hypochlorite does not penetrate the skin. The same is observed when swimming in the sea with a salinity of about 3.5%. Sodium chloride, a cytotoxin in higher concentrations, cannot overcome the acid mantle of the skin and the skin barrier. Even with higher salt concentrations, such as when bathing in the Dead Sea with 28% (predominantly magnesium chloride), no salt intake is observed. People with eczema swimming in the Dead Sea also use the high salt content to kill the microorganisms on the diseased skin. Even in ill skin, no salt penetrates.

Conclusion: Neither water nor salts dissolved in water can pass through the skin barrier, also not in case of damaged skin.

Another interesting case that proves the effect of the skin barrier is the bathing of a naked woman in champagne (12.5% alcohol) for 30 minutes. Subsequently, a blood sample is taken and the alcohol content in the blood is determined to be 0.00%. Moreover, surgeons disinfect their arms and hands with pure alcohol before any surgery (>99%). Alcohol is highly antimicrobial, and because of the degreasing effect, it disturbs the acid mantle and the upper part of the lipid matrix, but cannot penetrate.

Conclusion: Unlimited water-miscible organic liquids are retained by the skin barrier, even if they eliminate or overcome the acid mantle.

Nonpolar organic substances show a completely different behavior. At the end of the 1980s, there was a lot of excitement and discussion about a press release. The press reported that in the mother's milk of a young woman, musk could be detected. The discussions focused on the possible consequences for the baby. A few years later, scientists analyzed the urine of a woman who perfumed her neck 30 minutes before. Beyond any doubt, they found components of the perfume in the sample. The applied perfume distributes on the skin. One part evaporates within the first hour, another part evaporates slowly within hours, and a third part penetrates the skin. Although the amounts of perfume oil absorbed are low, they reach the entire body via the bloodstream. Therefore, the selection of the components of a perfume is a responsible task and it is better to abstain from perfuming. Even gasoline, which at the end of the refueling of a car possibly flows over the hand, penetrates immediately into the skin. The same applies to paint remover (brush cleaner).

Conclusion: Nonpolar organic substances with relatively low molecular weights (such as terpenes) can pass the skin barrier immediately and without any problem.

All of these findings will be considered later in the discussion of substance penetration.

Penetration in the other direction, namely out of the skin, takes place only for pure water (except sweat) and is measurable by the TEWL. The TEWL is used to determine the rate of evaporation, which averages between 300 and 400 ml/d. The value represents a diffusion process in which water moves from the lower part of the skin (high concentration) to the skin surface (low concentration). As water loss and substance absorption overcome the same barrier in the form of the stratum corneum, both values should be related. The more damaged the horny layer barrier of the skin, the higher TEWL and the greater the penetration of topically applied substances. The TEWL also increases in the absence of important fatty acids, especially linoleic acid.

The extraordinary mechanical stability of the stratum corneum and the penetration of some dissolved substances cannot be explained by the old "brick-mortar-model" by P. Elias [11]. Nevertheless, this model is briefly presented here because it is memorable and reflects essential processes correctly. In the concept, it is assumed that the horny layer is built up like a wall (Figure 4.3a). The bricks represent the keratinocytes and the mortar the lipophilic, waterproof kit [12]. In this old model, the bricks, with their hydrophilic contents, are falsely not connected to each other; hence, hydrophilic substances reach only the upper cell layer from the outside [13]. In contrast, lipophilic substances can pass through the mortar into the lower part of the epidermis. Dry and diseased skin has more or less deep cracks (Figure 4.3b). Through this, water diffuses from the inside through the epidermis and evaporates at the surface, detectable by high TEWL levels. Conversely, lipophilic ingredients penetrate the skin from the outside through the cracks. In this way, some hydrophilic molecules also pass the barrier. Furthermore, microcracks in the lipid kit (mortar) are discussed, which should allow the penetration of topically applied ingredients of the creams.

In 2007, the model conceptions were significantly improved by precise, electron microscopic examinations of the epidermis [14]. It was shown that the corneocytes have hook-like structures on the outside, so that the cells interlock with each other and the cell structures get high stability. In addition, corneodesmosomes create 400–600 connection points per cell, thus forming a multiplicity of cell-to-cell junctions (Figure 4.4). Presumably, some substances dissolved in water (such as urea) reach the deeper layers of the epidermis via these connection sites through the stratum corneum. The research showed that the integrity and stability of the skin barrier is realized by four structures. These are the hooks on

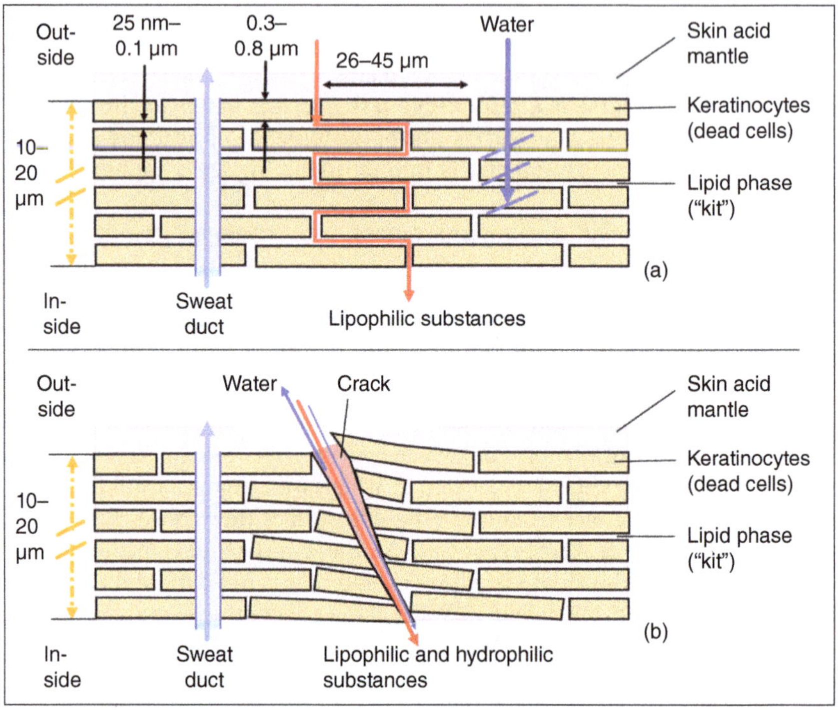

Figure 4.3 Structure of the stratum corneum analogous to the brick mortar model: (a) healthy skin and (b) damaged skin with a crack.

the corneocytes, the junctions via the corneodesmosomes, and the lipid bilayer (kit) of the stratum corneum, as well as tight junctions at the transition from the living to the dead skin layer [15].

Today, four pathways are considered possible, through the active ingredients of a cream, to pass the horny layer and enter the skin. The first possible and most important passage leads through many intertwined intercellular pathways in the lipid matrix (mortar) of the stratum corneum. It is practicable for non-polar, lipophilic substances because the water amounts are normally too small for a hydrophilic diffusion. Microcracks in the lipid bilayer, through which even hydrophilic substances can diffuse, are in the discussion (see Figure 4.3b). However, there are really a sufficient number of microcracks in healthy skin? This should be visible in high TEWL values, but healthy, well-groomed skin indicates low levels. This approach is conceivable for hydrophilic substances only in dry or diseased skin.

The lipid matrix is made up of several bilayers, containing some water in-between. Ceramide 1 shows a very long C-chain, which protrudes not only through the entire hydrophobic region of the bilayer but also into the adjacent bilayer with the linoleic acid chain. This stabilizes the entire multilamellar system ("bolted" layers; Figure 4.5). The chemical structures of the ceramides are shown in Figure 4.6. Between the lipophilic chains, there is a hydrophilic area in the matrix, partially filled with water. Only one monomolecular water layer or less is

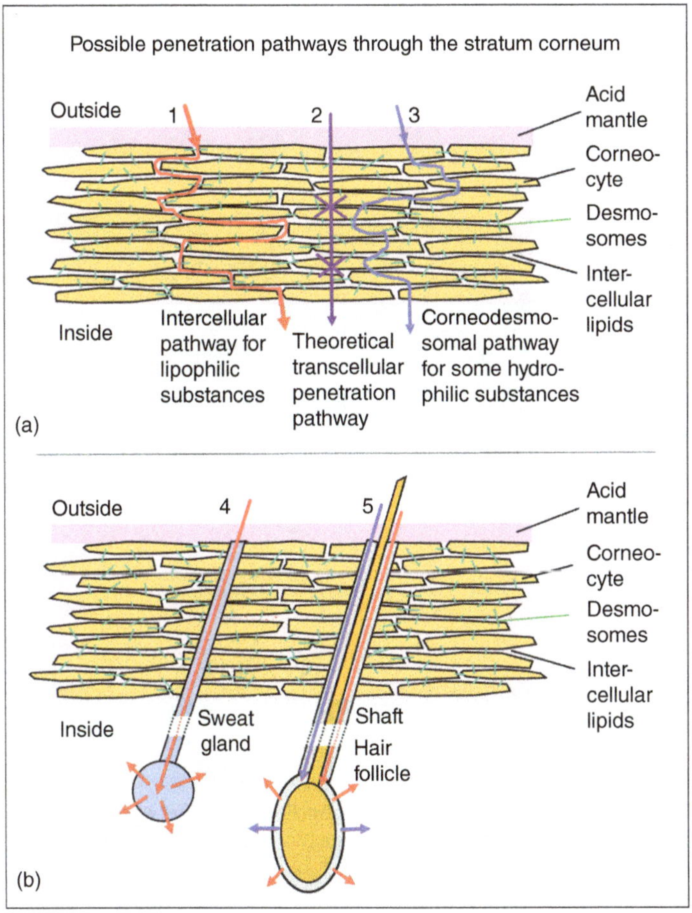

Figure 4.4 Modern concept of stratum corneum with the connections of the corneocytes via desmosomes: (a) intercellular pathway for lipophilic and corneodesmosomal way for hydrophilic substances and (b) pathways through skin adnexa, preferably way 5.

located there, but the low quantity might be insignificant for the water balance. The second imaginable way passes transcellular through many corneocytes and the intercellular lipids that surround every dead cell. Scientists do not believe in this possibility. For hydrophilic substances, the third pathway may be suitable. According to new findings, it leads through the horny layer via the corneocytes and the bridge-forming desmosomes (Figure 4.4a).

The other two ways lead directly through the skin appendages. In addition to the skin, the epidermal complex also includes nails, hairs, and sebaceous glands, free sebaceous glands, as well as sweat glands and scent glands along with their excretory ducts. In their entirety, the appendages are referred to as adnexa. As direct passages through the horny layer, only the sweat ducts and the hair shafts come into consideration. The vertical mass transfer via the excretory ducts of the sweat glands (transglandular) or via the hair follicles and the associated sebaceous

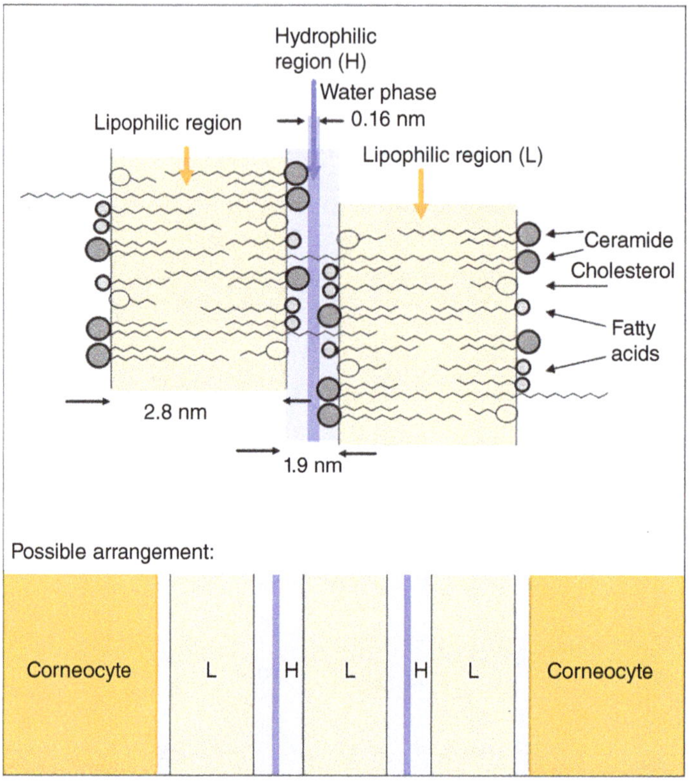

Figure 4.5 Narrowest arrangement of the components in the lipid matrix between the corneocytes, presumably one of the upper elements occurs two to six times next to each other. Source: Data from Glombitza 2001 [16].

glands (transfollicular) should be possible for both lipophilic and hydrophilic substances. The direct molecule penetration of the stratum corneum along the adnexa is referred to as the "shunt pathway" [17]. As a separate epithelial lining is missing for the sweat duct in the stratum corneum, the secretion is carried out through extended intercellular tubular gaps between the keratinocytes. The transfollicular pathway is the more plausible one, discussed especially for long-chain molecules.

For a long time, the skin appendages were not considered to be a route of penetration because of their small proportion of the skin surface, assuming 0.1% of the total skin surface area. However, recent studies showed a higher hair follicle density and because of larger follicle diameter an available penetration area of up to 13.7% (forehead). The epithelium there also shows, compared with regular interfollicular epithelium, smaller corneocytes. At these points can be concluded on a weakened barrier function. Recent studies show a surprisingly large influence of the skin appendages on the penetration kinetics of topically applied substances. For example, the absence of any hair follicle reduces the penetrated amount of hydrocortisone by a factor of 2–4. Especially, in the initial phase of penetration, the transfollicular route is important for drug entry. Later in the penetration

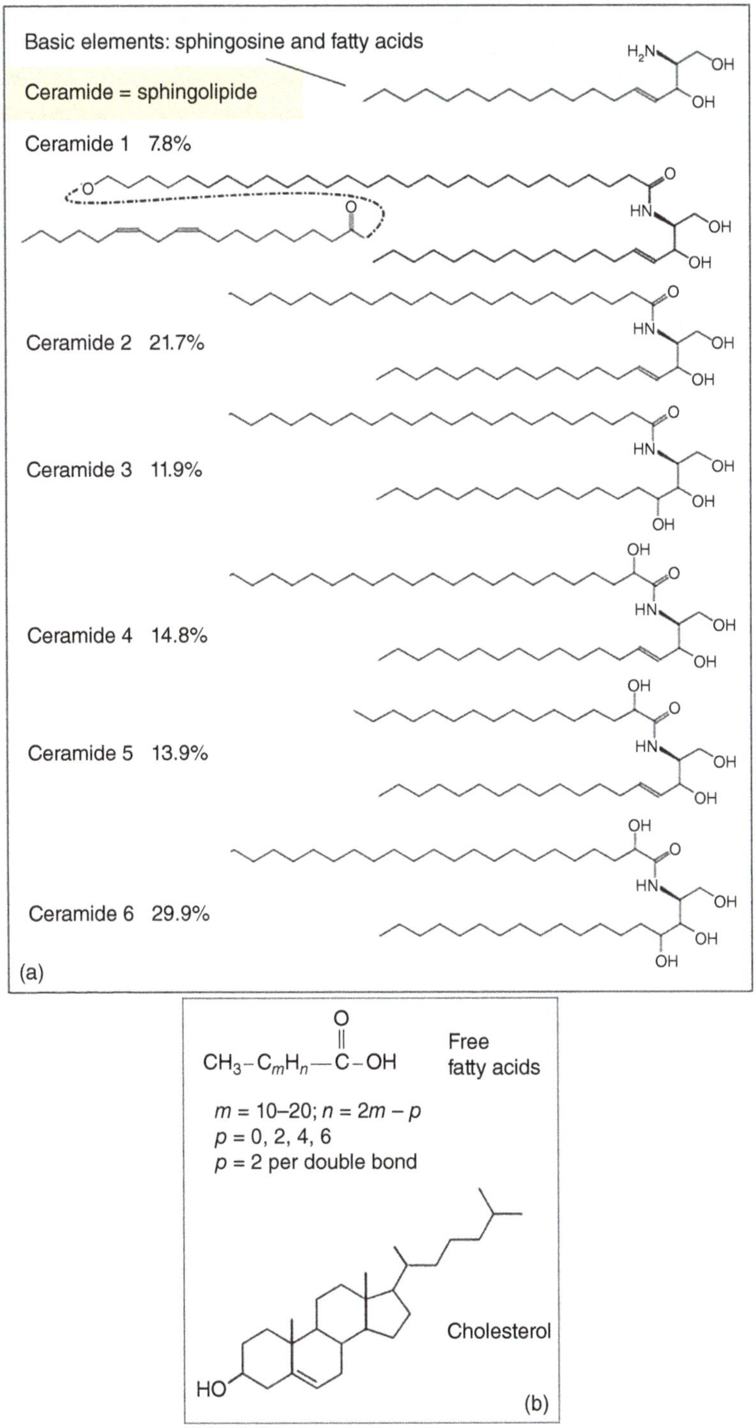

Figure 4.6 Chemical structures of the lipid matrix in the stratum corneum: (a) the ceramides 1–6 and their composition and (b) natural free fatty acid with even C numbers and cholesterol. Source: Panel (a) Data from Glombitza 2001 [16].

process, however, they only play a role for extremely hydrophilic components or for high-molecular-weight substances, as these substances can only overcome the stratum corneum in this way. The sebum may be another obstacle on the trans-follicular way. It should also be noted that the molecular size can be a limiting factor for the penetration rate.

4.4 Some Experiments on the Penetration of Lipophilic Substances

A specialized method of the NMR (nuclear magnetic resonance) spectroscopy [18], performed in vivo on volunteers, can track the diffusion of liquid natural oils and cream components into the skin. Particularly suited is the mobile NMR-Mouse® [19] in conjunction with a newly developed and assembled device to hold up the arm. The equipment changes the position of the measuring arrangement in small steps according to the desired measuring depth. As the skin surface cannot be determined precisely because of the penetration depth of the rays, this position has to be defined. It seems most logical to choose the highest value of the oil proton resonance as the exact point on the skin surface after determining the curves.

The measuring areas used are on the left and right forearm. The measuring system produces data approximately every 15–30 μm over the skin depth. Each measurement takes about 1 minute (65 seconds) to complete. More recently, both the step size and the measurement times have been reduced. Evaluation of the NMR measurements succeeds, on the one hand, by calculating the proton densities, which are based on the maximum value of the untreated skin in the epidermis. On the other hand, the response time t_2 of the signal identifies, after calibration, the substance (water: 12.2 ms; natural oil: 88.1 ms). Each position on the skin provides different readings. Therefore, the relevant test results are from curves of the treated minus untreated skin at the same position and stand for the penetrated substance quantity.

The comparison test between young and old as well as male and female show large differences in the proton resonance on the left forearm, as can be seen for untreated skin in Figure 4.7. The largest differences appear in the dermis because there the proton resonances are much higher in the young female skin than in the older man. One example for supplying the skin with pure natural oils (mixture of almond oil with borage) is displayed in Figure 4.8. After 10 minutes, the supernatant oil on the skin was removed. The measured values [20] indicate that these oils diffuse very quickly in considerable amounts into the epidermis. About 40 minutes after application, there is still a significant effect in the epidermis. In the meantime, most of the oil has entered the deeper layers of the dermis and subcutis, where they leave the skin via the blood vessels and get into the entire body (systemic effect). All further measurements are done on the skin of the older man.

Figure 4.9 shows measurement results of a strong penetrating macroemulsion from the market. The substances diffuse very quickly through the horny layer, but then remain in the epidermis for a longer time (possibly a reservoir effect).

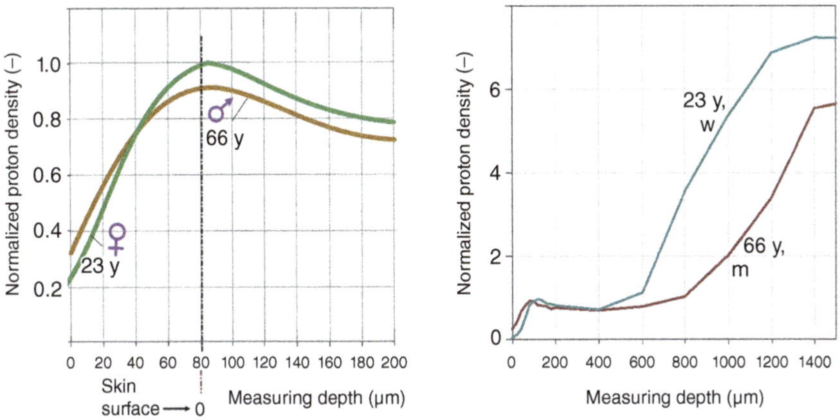

Figure 4.7 NMR results in untreated skin for a young woman in comparison with an older man. Source: All NMR measurements and their evaluations were carried out by Ovid Dicoi.

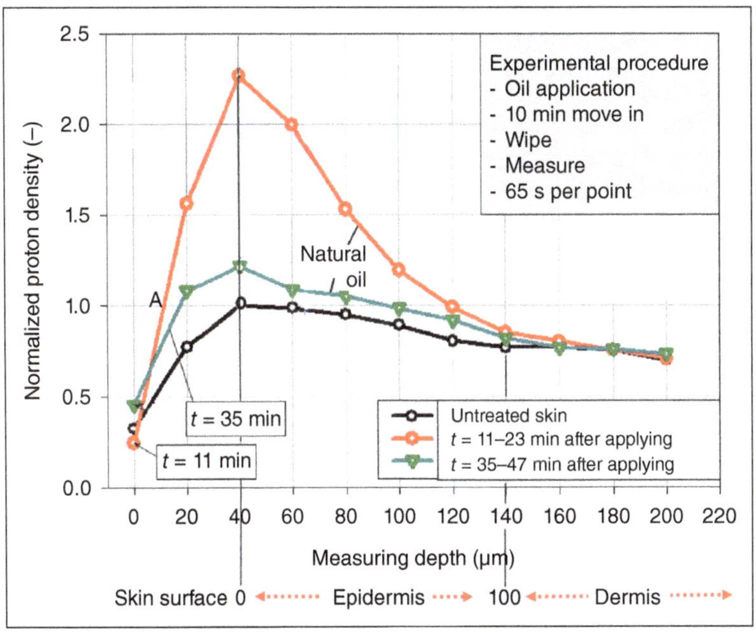

Figure 4.8 Normalized proton densities (differences are relative measures of the oil concentrations) as a function of the measured depth in the skin after treatment with a mixture of natural oils (triglycerides).

Other creams display a different behavior and diffuse deeper into the dermis, as shown by the example of a miniemulsion in Figure 4.10. The typical response signals, firstly for the untreated skin and secondly after applying a miniemulsion, must be subtracted to get the real size of the effect. In sections, calculated according to Fick's law, the curves emerged under the assumption of thermodynamic equilibrium at the basal membrane. Already after a few minutes, high

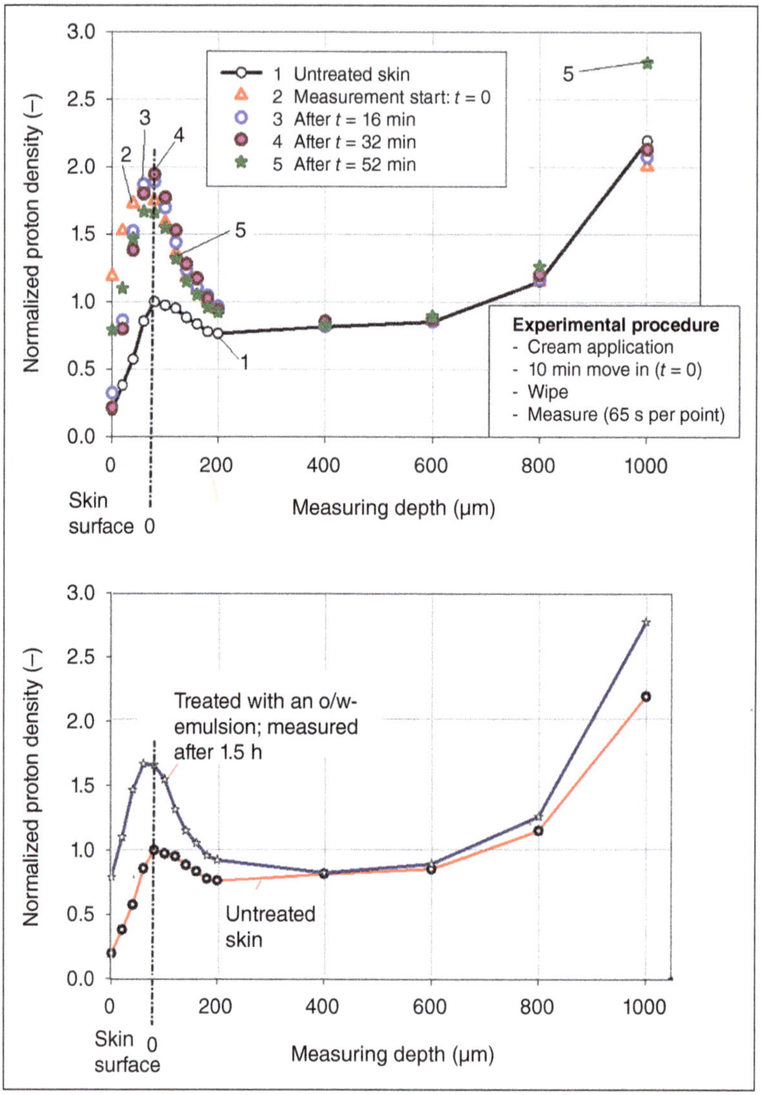

Figure 4.9 Measurement after application of a cream that remains for a longer time in the epidermis (cream from the market, $d_{50} = 9\,\mu m$).

values of the oil appear in the epidermis and in the bottom layers of the dermis. In the lower layers of the dermis, the values increase for some hours before the oil phase values slowly drop to normal levels. The measurements give reproducibly the same patterns, in the case of miniemulsions ($d_{50} = 120\,nm$) with open, as in practical application, as well as with covered application without water evaporation for creams with occlusive substances. Comparable values for the untreated skin originated from magnetic resonance imaging (MRI) measurements [21]. The examined inter- and intracellular water contents in the epidermis depend on the measuring depth in the skin [22].

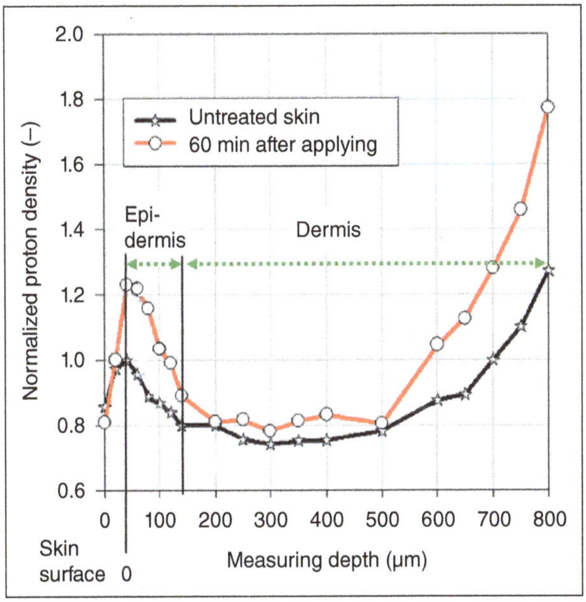

Figure 4.10 Treatment of the skin with an o/w miniemulsion; differences between the two curves caused by diffused lipids after one hour on the left forearm are clearly visible in the epidermis and in the deeper layers of the dermis.

On covering the skin area after applying the cream (occlusion), for several formulations, both macroemulsions (d_{50} about 2 μm) and miniemulsions (d_{50} about 120 nm) show by and large comparable values of the oil content in the skin. Of course, the effects largely depend on the formulations that are similar here. An illustrative comparison of comparable emulsions is shown in Figure 4.11. From the differences between treated and untreated skin, it can be seen that there are clearly differences in the epidermis in favor of the miniemulsion. It can be assumed that the nanosized oil droplets enter the stratum corneum. In practical applications, both emulsions disintegrate on the skin, but at different speeds, because of the stability and water evaporation. Miniemulsions have superior stability. The velocity of the water evaporation on the skin surface, also measurable by using the NMR-Mouse [23], was not determined.

The three investigated cases of penetration differ dramatically in practical application: Pure oil is unpleasant on the skin and penetrates into the skin too quickly to be curative. Better effects are produced with the creams. The application of creams is easy; both fat- and some water-soluble substances diffuse in a controlled manner into the skin without a greasy feeling. The lipophilic phase should contain key ingredients for the skin, such as mono-, di-, and triglycerides with high levels of double or triple unsaturated acids (omega-6 fatty acids, C_{18}), possibly also squalene (triterpenes). Thickeners of a cream reduce more or less the rate of diffusion of the lipophilic phase compared to pure oil. This is desirable if the penetration rate does not decrease too much. A depot in the epidermis, which gradually discharges by diffusion of the ingredients into the dermis, is

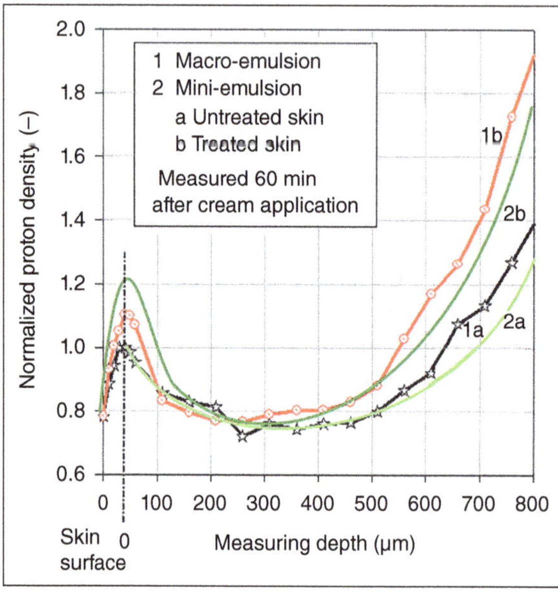

Figure 4.11 Penetration of effective ingredients into the skin: Comparison of a macro- and a miniemulsion in an example. Evaluation of the effects: 1b–1a vs. 2b–2a.

quite useful for the development of maximum effects. The next application of the cream, 12 or 24 hours later, recharge the empty depot again. For effective antiaging agents, this might be the optimal treatment for the skin. When using too much wax, the diffusion can be totally suppressed, which is not favorable for many applications. An optimization leads to the right consistency of the cream, which the dispensers pump out and that can be easily distributed.

The penetration behavior of major water-soluble substances cannot be assessed because of lack of appropriate in vivo investigation methods. The task of developing specific effects for skin care products can be solved chemically through an optimization procedure because there are unlimited possibilities to select and mix components both in type and quantity. According to Figure 4.12, preoptimization of the technology starts with a basic formula in combination with proven emulsifiers and viscosity regulators in the first cycle. During further optimization, it is advisable to examine from time to time the emulsification by measuring the droplet size distribution, testing the stability with a centrifuge, and checking the preservation by microbiological tests. Physical measurements on the skin and simultaneous application tests allow feedback concerning the use of active ingredients in terms of type and quantity (cycle 2, Figure 4.12). Furthermore, there is the possibility of setting the properties of the developed skin care product [12]. Using this optimization method in combination with known active ingredients, several skin properties are ultimately better than the market standard in terms of skin moisture, elasticity, barrier function, surface structure, and other. Targeted product designs presuppose laboratory equipment, measuring apparatus, and internal and external resources to test the skin properties of probands. In addition, such development is not possible without the knowledge and skills of different specialists (Figure 4.13) working synergistically in this process. In the

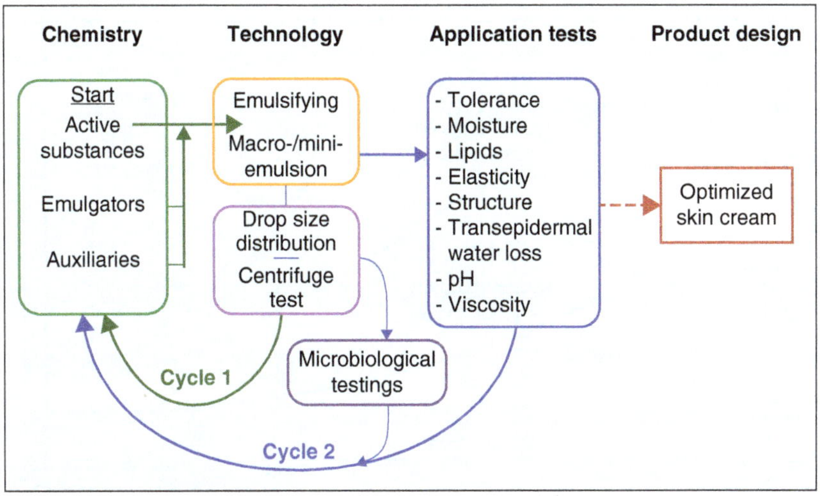

Figure 4.12 Cycles for optimizing product designs.

Figure 4.13 Cooperation with various departments in developing new formulations for skin care products (in addition, toxicology for the safety assessment).

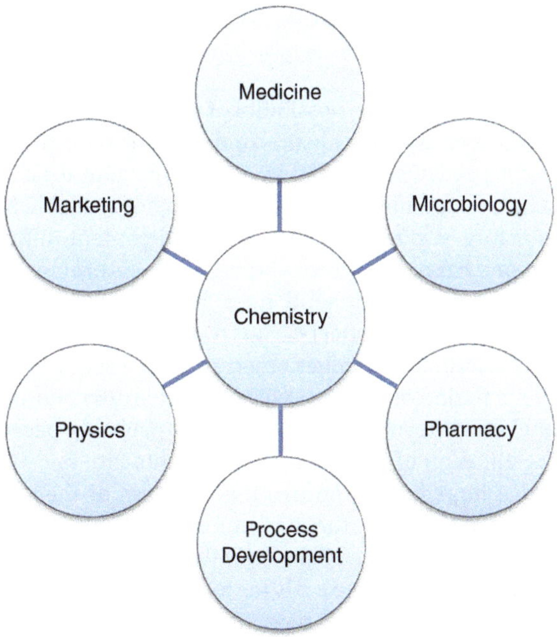

end, the product goes to the market, and for the next cream, the same optimization process is started.

Comparison of different skin parameters before and after application and over a longer period proves the performance (power) of a skin cream/lotion. Any distortion of results is avoided through a statistical design with about 15–25 (in special cases 50) persons. First, a skin tolerance test excludes primary irritant properties. This happens either in patch tests or in the open application tests. The test

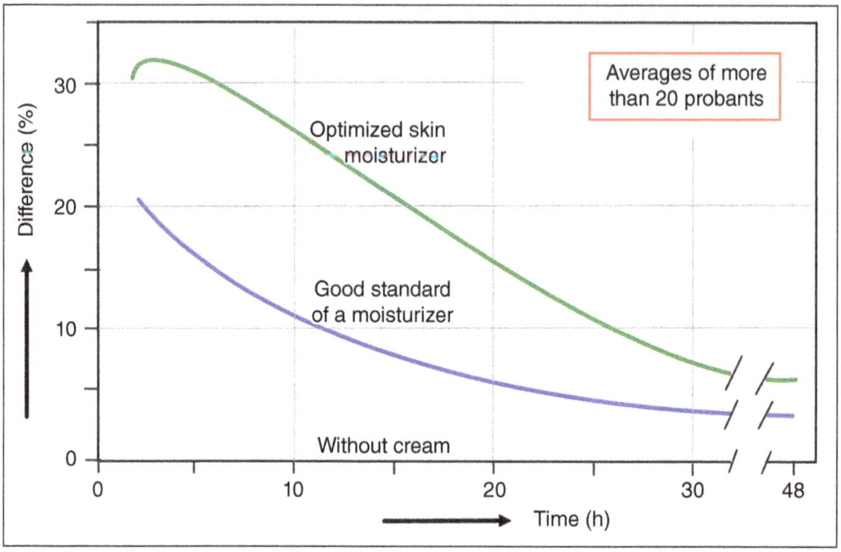

Figure 4.14 Measurement of skin moisture differences (%) with a corneometer; the values clearly demonstrate the effect of creams. The cream was applied at time $t = 0$; capacitive measurements of the treated skin minus measured values of the untreated skin.

aims to exclude the possibility of allergic reactions over a period of one to about five weeks, under realistic conditions or intensified by a provocation test. This test is particularly sensitive and is important for allergic skin types [24]. Various standardized measurement methods are available to assess the barrier function, moisture, surface structure, and lipid content in the skin (see Section 12.3). On the one hand, the values obtained allow a precise statement about the skin condition. On the other hand, the results disclose improvements after using the care product with statistical certainty. Figure 4.14 displays the result of a moisturizer optimization. The values emerged after a single use by capacitive measurements over a period of 48 hours with more than 20 volunteers. As can be seen, the maximal effect result after three hours, probably because of the time required for the diffusion of some active ingredients into the skin. Furthermore, the selected active ingredients improve the moisture of the skin significantly better than a good product from the market for a period of more than 48 hours. The large differences to untreated skin show that some ingredients have penetrated the horny layer and work in the cutis for longer time. The optimized cream should then be applied only every two days; otherwise, the skin is over-supplied.

4.5 Penetration of Agents into the Skin

Two effects of skin creams are undeniable and verifiable. These include the strengthening and renewal of the hydro-lipid coat (acid mantle) and parts of the skin barrier as well as the care, smoothing, and softening of the upper layers of the cornea. Most additional desirable longer lasting cream effects, e.g. elimination of free radicals, strengthening of the skin scaffold, and stimulation of cell renewal, are based on substances that act inside the skin. For this, they must be able to overcome the skin barrier. In the literature for cosmetics, there is little explanation about the penetration of active ingredients. In some cases, the penetration of drugs from the pharmaceutical sector is described. For cosmetic creams, there exist no sound explanation for the penetration and action in the skin. Therefore, some known, probably correct facts are summarized here [25, 26].

The permeability of the stratum corneum depends on the condition of the corneocytes and the lipid matrix, the formulation of the cream, the molecule structure of the ingredient and excipients, as well as physicochemical parameters. First, the diffusion of ingredients depends on the condition of the horny layer:

➢ Acid mantle
➢ Thickness, body site
➢ Density of the cell packing and the intracellular structure
➢ Lipid content and molar composition
➢ Hydration
➢ Age, gender
➢ Microcracks in the lipid matrix
➢ Possibly existing injuries and pathological skin changes (e.g. eczema).

Secondly, during transport through the horny layer, the properties of the active substance used play an essential role. It should be noted that the individual substances can only be assessed to a limited extent because complex multisubstance mixtures are present in the creams. On the one hand, the water phase contains all hydrophilic substances mixed or dissolved. On the other hand, the oil phase predominantly comprises the formulated lipophilic substances dissolved in the oil mixture. When the diffusion into the cornea begins, the process depends mainly on the penetrating substances and on any substances dissolved in it. Therefore, the following parameters have to be considered:

➢ Lipophilicity, amphiphilicity, or hydrophilicity (nonpolar/polar substances with all intermediates)
➢ Spatial structure and molecular size (weight) of the substance
➢ Charges (ions)
➢ Solubilities
➢ Dissolved substances
➢ Viscosities
➢ Presence of functional groups.

To accelerate the transport of substances, chemicals called **enhancers** can be used by acting through different mechanisms [27]. Enhancer and active ingredient may form a solution with completely different mass fractions, which are predetermined by the formulation. Figure 4.15 shows some known enhancers. Further enhancers for cosmetic creams can be fatty alcohols, terpenes, and surfactants [28]. Unfortunately, only a few are approved for cosmetics. Solvents such as alcohols, alkylmethyl sulfoxides, and polyols increase the solubility of the active ingredients and can have a favorable influence on the distribution coefficients. In addition, they more or less extract the upper parts of the epidermal lipids,

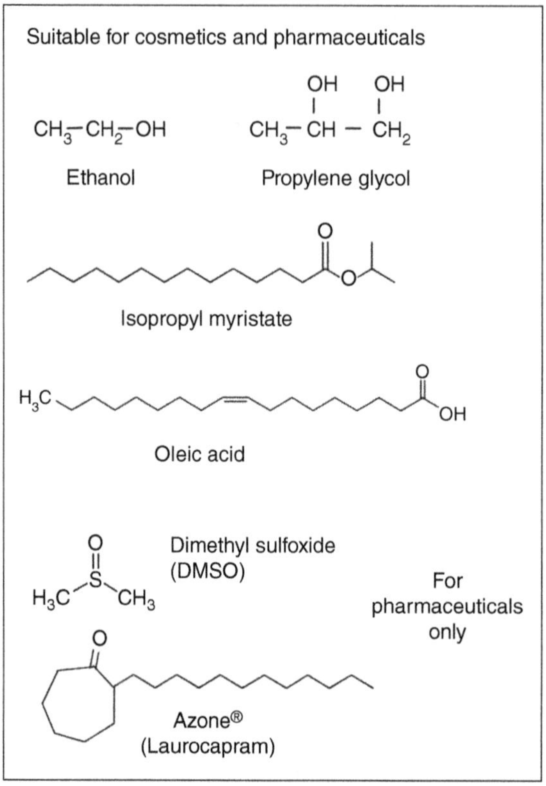

Figure 4.15 Examples for penetration enhancers.

increasing the permeability of the horny layer. The extracted lipids are replaced by the oils and fats of the cream. Effects of the enhancer are mainly based on chemical processes with the lipid matrix and the content of the corneocytes in the stratum corneum:

- Extraction of the hydro-lipid film on the skin surface
- Partial extraction of lipids from the upper part of the stratum corneum
- Disturbance of the structure of the lipid bilayers
- Change in the distribution coefficient by replacing water in some corneocytes with solvents such as propylene glycol or polyethylene glycol (DMSO, especially effective but not approved for cosmetic products)
- Change in the properties of ingredients by solvents (enhancers) and thickeners
- Loosening of the keratin structure in the horny cells
- Delamination of the stratum corneum.

It is believed that substances resembling the constituents of the lipid matrix may be incorporated into the lipid bilayers. These include the fatty acids and the fatty alcohols. As a mechanism of action, a temporary expansion of the channels is assumed. This process accelerates the permeation in the lipid matrix.

The **occlusion** of the respective skin regions provides a common method for increasing the permeability of effective ingredients. Occlusive skin covers block the physiological release of water from the surface of the skin. Presumably by a water uptake within the intercellular lipid structures, the corneocytes swell. In this way, the stratum corneum can increase its water content by maximum 50%. In addition, the occlusion locally causes an increase in skin temperature as well as more blood circulation. All three effects lead to an increase in the percutaneous absorption rates of topically applied substances. The method is used clinically worldwide. In therapeutic use, the occlusion is often performed through a medical bandage or patch or through appropriate substances such as Vaseline. In cosmetics, an occlusion is usually not acceptable because the feeling on the skin is not pleasant. Instead, the evaporation of water is not stopped but reduced by means of suitable ingredients without the user noticing (e.g. long-chain hyaluronic acid), whereby a slight increase in permeability may be achieved.

Sufficient water-soluble, small-molecule substances, in which the hydrophilic and lipophilic properties are balanced (**amphiphilic** and/or zwitterionic **ingredients**), can permeate the skin well in small amounts through the corneodesmosomal pathway 3 (see Figure 4.4). These include in particular the amino acids, urea, small peptides, and various surfactants (Figure 4.16). It is noticeable that many water-soluble substances that use this path contain one or more nitrogen atoms. Betaines, like the amino acids, show a zwitterionic structure but cannot compensate for internal charges. As an example of nonionic emulsifiers, the group of fatty alcohol ethoxylates is worth mentioning, with the long lipophilic C-chain and the hydrophilic EO-units. Molecules with hydrophilic/lipophilic balance (HLB)-values (see Section 5.3.1) between 9 and 11 are well-balanced and therefore probably well-penetrating the horny layer. An example represents

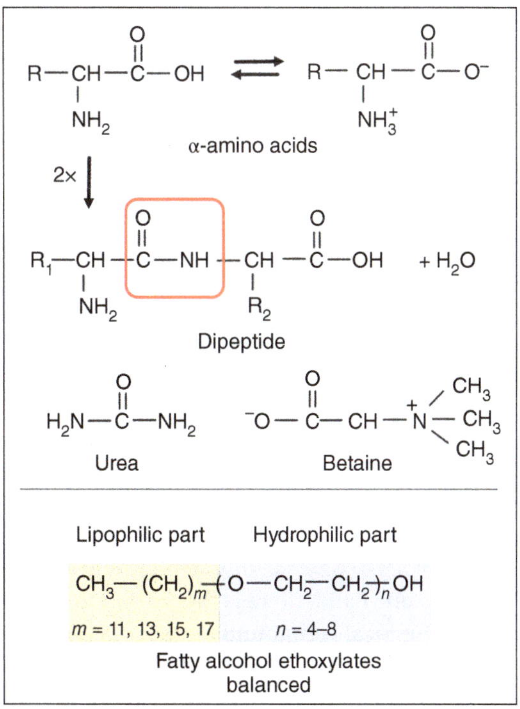

Figure 4.16 Examples for amphiphilic molecules that likely penetrate the stratum corneum via the corneodesmosomal pathway.

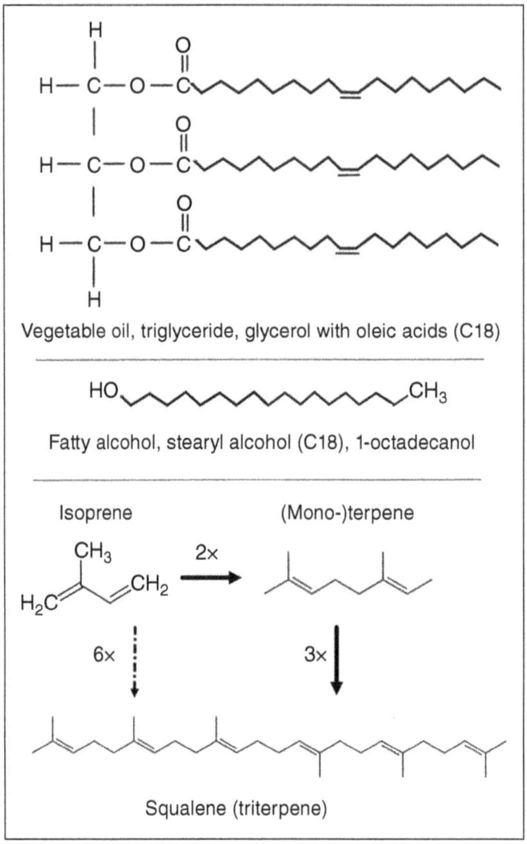

Figure 4.17 Chemical structures of some lipophilic ingredients that penetrate the horny layer through the lipid matrix (pathway 1).

Ceteareth-6 with a HLB-value of 10. The surfactants, which are clearly on the hydrophilic side, are unlikely to penetrate the skin, while the highly hydrophobic emulsifiers possibly diffuse into the matrix via pathway 1. Up to a molecular weight of some hundred Dalton (Da), the mentioned substances should diffuse relatively quickly, but from 1000 Da upward, permeation slows down.

The **lipophilic substances** have a whole series of permeable substances, especially those similar to the components of the lipid matrix. As the experiments show, these include liquid, low viscosity vegetable and essential oils, presumably also their hydrolysis products and fatty alcohols. Examples are shown in Figure 4.17. Essential oils are a mixture of terpenes, which in turn comprise of at least two and a maximum of six isoprene units (C10–C30). The limit for the size of the molecules that can penetrate the horny layer should be above 1000 Da, depending on the spatial volume of the molecule probably even over 10 000 Da. Small molecules use pathway 1 (see Figure 4.4). For them, pathways 4 and 5, with a share of a few percent should play only a minor role.

On the one hand, there is the 500 Da rule in the literature for the skin penetration of chemical compounds and drugs [29]. However, even the well-penetrating vegetable oils have molecular weights of more than 800 Da. On the other hand, the scientists proved the penetration of hyaluronic acid with molecular weights between 20 000 and 300 000 Da [30] and the pharmacists of heparin with 4000–40 000 Da. For polymers with molecular weights below 50 000 Da, the transfollicular pathway is suitable for slow diffusion over hours and therefore important not only for cosmetic effects. Obviously, the limit of permeability depends only to a limited extent on the molecular size and spatial structure. Decisive for the penetrating amount is the permeation rate. Although small molecules can pass the barrier in just a few minutes, macromolecules take many hours until an effect can be detected.

Solid products, such as the fatty alcohols, dissolve in the oil and thus can penetrate the skin barrier. The total penetrating quantities are not very high; the larger parts probably remain on the skin. However, even small amounts (such as vitamins) can have a remarkable effect in the cutis. Higher viscosities and additional solutes decrease the rate of diffusion of the vegetable oils. Active ingredients in almost solid creams will remain on the skin surface. Therefore, slightly more liquid lotions are preferable, if positive antiaging effects in the skin should occur. Pure plant and essential oils penetrate the skin layers so fast that they have already reached the blood after 5–30 minutes. In this short time, the oils generate no effect. Hence, an increase in viscosity makes sense, so that the diffusion process slows down and lasts for hours. Then, the hydrolysis of the triglycerides can take place and the important free fatty acids and glycerol are released. A general relationship between the polarity, diffusion rate, and molecular size, roughly estimated by the described facts, is shown in Figure 4.18.

Caution: Nonpolar solvents such as toluene, benzene, chloroform, carbon tetrachloride, hexane, octane, and other with all dissolved substances immediately penetrate the skin barrier. These and similar solvents are not approved for cosmetic creams.

The diffusion of active substances proceeds along a concentration gradient from high to low concentration. Very polar substances are water soluble. They

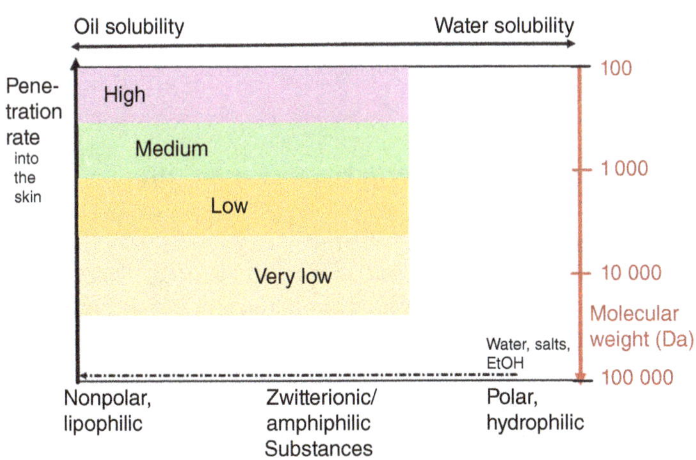

Figure 4.18 Schematic representation of the correlation between penetration rate, molecular weight, and polarity of the molecules.

diffuse so slowly into the horny layer that no penetration can be determined. The barrier works reliably. Therefore, suitable carriers (vesicles) must be used for interesting polar substances, which overcome the barrier via pathway 3 because of their smallness in the range around and under 100 nm as well as by their special surface (see Section 5.5). Thus, the transdermal application of active ingredients can be carried out via so-called "drug carrier" systems. In this case, spherical membrane hollow bodies act as carrier elements for hydrophilic active substances. The vesicles used are usually **liposomes** (Greek: lipos = fat, soma = body), which possess at least one lipid bilayer enclosing an aqueous core with dissolved active ingredients. Liposomes consist of lipids (e.g. phosphatidylcholines) that spontaneously form bilayer aggregates after suspending in water. To increase the stability of the liposome vesicles, cholesterol or other bilayer stiffening substances can be incorporated into the lipid membrane. The diffusion of some liquids or solutes can be influenced by the drop size. Nanodroplets with diameters around 100 nm pass through the stratum corneum relatively quickly (production in Section 10.7).

A newer carrier, the **transfersomes**®, are complex and highly flexible lipid vesicles prepared from liposomes by the addition of surfactants and/or membrane-softening substances. They easily pass the cornea by slipping through skin pores. Advantageously, the transport is independent of the concentration of the drug in the transfersomes. Compared with the usual liposomes, the transfersomes are characterized by a much more flexible membrane. This deformability enables the transfersomes to pass through pores within the stratum corneum that are one-tenth smaller than the diameter of a liposome. Transfersomes allow the transport of macromolecules such as insulin through the skin. Moreover, because the membrane of the transfersomes is much more hydrophilic and swells more easily, the particles are less prone to agglomerate or fuse.

Another variant represents the **ethosome**. The multilamellar vesicles consist of phospholipids, ethanol, and water. Ethanol enhances penetration, making these carriers superior to ordinary liposomes. **Niosomes**, another form of vesicles, resemble the liposomes but have nonionic surfactant structures on the surface. The vesicles can also overcome the stratum corneum via pathway 5, the follicular penetration.

4.6 Gender Differences in the Structure of the Skin

In the skin, there are many similarities between the two sexes, such as the tasks of the skin and the skin diseases, but also gender typical differences. The woman skin with invisible pores looks mostly tender, whiter, and shining through. The hormone testosterone determines the properties of male skin and provides a different structure compared with women. Men's skin is thicker, on average about 20%, and also greasier, and they age in other ways. Men have more and larger sebaceous glands and thus more and larger pores than women. Men are less prone to dry skin than women because men's skin produces about twice as much sebum. This makes them greasier and shiny leading to blemishes and acne. The skin contains more collagen and has a stronger, firmer look.

Differences between the sexes are noticeable in skin aging. With age, the collagen content of men's skin decreases continuously. In woman's skin, the collagen degradation starts later, usually after the menopause. Skin aging in men becomes noticeable later, but then increases rapidly. The reason is muscle atrophy, which manifests itself in sagging skin on the body and face and forms wrinkles on the face. The wrinkles that occur later in lives of men compared with women appear more visible and pronounced. Another difference is that men shave once or twice a day and thereby impair large parts of facial skin. This can lead to irritation and other skin problems (burning and pimples). Otherwise, the skin care of men and women is essentially the same and can be done with identical ingredients. Differences exist in the fragrance notes and partly in the color as well as in marketing. For example, men would reject a reddish-colored cream. They need creams with less oils and with more pore-reducing components. Women should start well before menopause (i.e. around 45) with effective antiaging creams, which they also need to use daily in order to achieve tissue strengthening on time.

4.7 Learnings

- The skin protects against all exogenous noxae and regulates the temperature and water balance.
- It perceives external stimuli such as temperature, touch, pain, pleasure, and pressure.
- The skin consists of three layers: epidermis, dermis, and subcutis.
- In the epidermis, a distinction is made between stratum corneum (horny layer), stratum lucidum, stratum granulosum, stratum spinosum, and stratum basale.
- In the stratum basale arise the new cells, which need about four weeks until the desquamation on the skin surface.
- Various important effects of creams are achieved only by active ingredients that penetrate into the skin.
- The penetration rate depends on the composition of the cream and the condition of the skin.
- Vegetable and essential oils penetrate the stratum corneum within a few minutes.
- Some substances, called enhancers, facilitate the diffusion through the skin.
- Three main paths lead through the horny layer: (i) lipophilic substances diffuse through the lipid matrix, (ii) amphiphilic substances via the corneodesmosomal pathway, and (iii) long C-chain molecules via the transfollicular pathway.
- Especially small, spherical liposomes and transferosomes can transport hydrophilic substances through the stratum corneum.

References

1 Scott, A.S. and Fong, E. (2016). Integumentary system (Chapter 5). In: *Body Structures and Functions*, 13e, 65ff. Cengage Learning.

2 Baranoski, S., Ayello, E.A., and Tomic-Canic, M. (2008). Skin: An Essential Organ, In: *Wound Care Essentials: Practice Principles*, 2e (eds. S. Baranosli and E.A. Ayello), 47–63. Wolters Kluwer Health.

3 Montagna, W. and Parakkal, P.F. (eds.) (2012). *The Structure and Function of Skin*. Academic Press, Inc.

4 Principles of dermatological practice, structure of the epidermis, worksheets (2008). https://www.dermnetnz.org/cme/principles/structure-of-the-epidermis (accessed 16 December 2017).

5 Freinkel, R.K. and Woodley, D.T. (eds.) (2001). *The Biology of the Skin*. Parthenon Publishing Group.

6 Shimizu, H. (2006). Structure and function of the skin. In: *Shimizu's Textbook of Dermatology*. http://www.derm-hokudai.jp/shimizu-dermatology/pdf/01-05.pdf (accessed 18 December 2017).

7 Fluhr, J.W., Elsner, P., Berardesca, E., and Maibach, H.I. (2005). *Bioengineering of the Skin: Water and the Stratum Corneum*, 2e. CRC Press.

8 Brannon, H. (Updated 2017). The anatomy of the stratum corneum. https://www.verywell.com/stratum-corneum-anatomy-1069189 (accessed 18 December 2017).

9 Grundlagen, Das Organ Haut. https://sundoc.bibliothek.uni-halle.de/habil-online/01/01H036/t4.pdf (accessed 18 December 2017).

10 Elias, P.M. and Feingold, K.R. (eds.) (2006). *Skin Barrier*. Taylor & Francis Group.

11 Elias, P.M. and Friends, D.S. (1975). The permeability barriere in mammalian epidermis. *J. Cell Biol.* 20: 1–19.

12 Rähse, W. (2013). Design of skin care products. In: *Product Design and Engineering, Formulation of Gels and Pastes*, Chapter 10, vol. 3 (eds. U. Bröckel, W. Meier and G. Wagner), 273–313. Weinheim: Wiley-VCH.

13 Neubert, R.H.H. and Wepf, R. (2017). Struktur und Morphologie einer Barriere. *Pharm. Ztg.* (Online) (17/2007). https://www.pharmazeutische-zeitung.de/index.php?id=2957 (accessed 18 December 2017).

14 Neubert, R.H.H. and Wepf, R. (2008). Das Stratum corneum, Struktur und Morphologie einer hoch effizienten Barriere. *Dermopharmazie* (4/2008): 21–28.

15 Neubert, R.H.H (2008). New insights into the morphological and molecular structure of the stratum corneum. Charts from the Institute of Pharmacy of the Martin Luther University, Halle-Wittenberg. http://www.pharmtech.uni-halle.de/downloads/newinsights.pdf (accessed 20 December 2017).

16 Glombitza, B. (2001). Lipidsysteme als Stratum corneum Modelle, Charakterisierung und Eignung für Permeationsuntersuchungen. Dissertation. TU-Braunschweig. https://d-nb.info/962768324/34 (assessed 22 November 2018).

17 Bhowmick, M. and Sengodan, T. (2013). Mechanisms, kinetics and mathematical modelling of transdermal permeation – an updated review. *Int. J. Res. Dev. Pharm. Life Sci.* 2 (6): 636–641. https://image.slidesharecdn.com/drugpermeation-150530185013-lva1-app6892/95/drug-permeation-1-638.jpg?cb=1433011846 (assessed 22 November 2018).

18 Blümich, B. (2005). *Essential NMR for Scientists and Engineers*. Berlin: Springer.

19 Blümich, B. and Blümler, P. (1999). Verfahren zur Erzeugung von Messsignalen in Magnetfeldern mit einem NMRmouse-Gerät. DE 199 39 626.4. 20 August 1999.

20 Rähse, W. and Dicoi, O. (2009). Produktdesign disperser Stoffe: Emulsionen für die kosmetische Industrie. *Chem. Ing. Tech.* 81 (9): 1369–1383.

21 Agache, P. and Humbert, P. (eds.) (2004). *Measuring the Skin*. Berlin: Springer.

22 Kemenade, P.M. (1998). Water and ion transport through intact and damaged skin. PhD thesis. CIP-Data Library Technische Universiteit Eindhoven.

23 Dicoi, O., Walzel, P., Blümich, B., and Rähse, W. (2004). Untersuchung des Trocknungsverhaltens von Feststoffen mit kernmagnetischer Resonanz. *Chem. Ing. Tech.* 76 (1–2): 94–99.

24 Oakley, A. (2008). DermaNet NZ: baseline series of patch test allergens. https://www.dermnetnz.org/topics/baseline-series-of-patch-test-allergens (accessed 18 December 2017).

25 Dayan, N. (2005). Pathways for skin penetration. *Cosmetics & Toiletries* 120: 67–76. http://www.lipochemicals.com/system/files/articles/1916992660_040105-C%26T_Dayan_PthwysSknPntrtnSM.pdf (accessed 22 December 2017).

26 Dragicevic, N. and Maibach, H.I. (eds.) (2016). *Percutaneous Penetration Enhancers, Chemical Methods in Penetration*. Berlin: Springer.

27 Trommer, H. and Neubert, R.H.H. (2006). Overcoming the stratum corneum: the modulation of skin penetration. *Skin Pharmacol. Physiol.* 19: 106–121. http://nanobiotec.iqm.unicamp.br/download/Trommer_skin%20penetration-2006rev.pdf (accessed 22 December 2017).

28 Smith, E.W. and Maibach, H.I. (2006). *Percutaneous Penetration Enhancers*, 2e. CRC-Press LLC.

29 Bos, J.D. and Meinardi, M.M. (2000). The 500 Dalton rule for the skin penetration of chemical compounds and drugs. *Exp. Dermatol.* 9 (3): 165–169.

30 Essendoubi, M., Gobinet, C., Reynaud, R. et al. (2016). Human skin penetration of hyaluronic acid of different molecular weights as probed by Raman spectroscopy. *J. Controlled Release* 232: 175–187.

5

Composition of Creams for Skin Care

5.1 General Structure of a Skin Care Cream

A cosmetic cream provides the skin with the needed substances, which ensure a well-groomed, beautiful skin after a long-term usage. Except for a few exceptions, the ingredients of a cream are based on renewable plants. They come from plant parts after extraction, isolation, and purification with chemical technology processes (unit operations). An example represents the oil extraction (triglycerides) from sunflower seeds by means of a press followed by filtration. In order to obtain the required fatty alcohols (FAs) and subsequent products, either the oils are split with water vapor into the fatty acids or the oils transesterified with methanol to form fatty acid methyl ester. In both cases, valuable glycerol is obtained as a by-product. Fatty acids and methyl esters react catalytically under pressure with hydrogen to obtain the FA that is purified by vacuum distillation. FA of different plants forms the starting material for many ingredients, in particular for the thickeners, coemulsifiers, and after further reaction (e.g. ethoxylation) for the emulsifiers. All cosmetic creams are based on the chemistry, including the natural cosmetics. This applies to almost all ingredients and excipients that undergo several stages of isolation, often also of chemical modification or synthesis.

Creams as emulsions offer the best physical requirements for skin care products to achieve the effects on and in the skin. Thus, it is at the same time possible to supply the skin with lipophilic and hydrophilic active substances in larger quantities. The preparation happens preferably as oil in water (o/w) emulsion because some oil droplets of 1–10 μm size can penetrate into the skin, the smaller the faster. These emulsions usually leave no disturbing oily film on the skin. However, applying water in oil (w/o) emulsions, oil remains for several minutes often as a sticky film on the skin, which is undesirable in cosmetics for daytime products. In positive words, their lipid action lasts longer, which is desired for some special products as well as for night and therapeutic creams. Out of the applied cream, water evaporates slowly from the skin, leaving a thin film with the nondiffusible substances and reinforcing the hydro-lipid film (acid mantle). The substances smooth the skin and provide protection against microorganisms with the acid mantle and a suitable, partially water-soluble preservative.

The creams/lotions are thickened emulsions. They comprise excipients, additives, and active ingredients. The nature, number, concentrations, and

Cosmetic Creams: Development, Manufacture and Marketing of Effective Skin Care Products,
First Edition. Wilfried Rähse.

interactions of these substances determine the product design (performance, handling, and esthetics) and can be influenced in large parts. In order to simplify the development of creams for a wide range of applications, it is proposed to divide the formulations into modules, comprising basic and active ingredient modules.

5.2 Modules of a Cream

A possible composition of creams is shown in Figure 5.1. To simplify the developments and reduce the number of tests, a modular design is suggested. The cream consists of the variable, active ingredients that make up 10% and the fixed formulation of the remaining 90%. The main part, cosmetic water, forms the basis of the emulsion. Lipids, vitamins, and moisturizing substances as well as preservatives and emulsifiers/thickeners are part of the fixed recipe. Useful is the addition of alcohol (about 0–5%), which supports the dissolution of sorbic acid, causes a cooling effect on the skin, and acts as an enhancer.

All substances, needed for manufacturing, stabilization of the cream, and application on the skin, are excipients. The enumeration of their functions in Table 5.1 shows that a stable cream needs at least seven agents, in some cases up to the double. These include emulsifiers, preservatives, solvents, enhancers, spreading

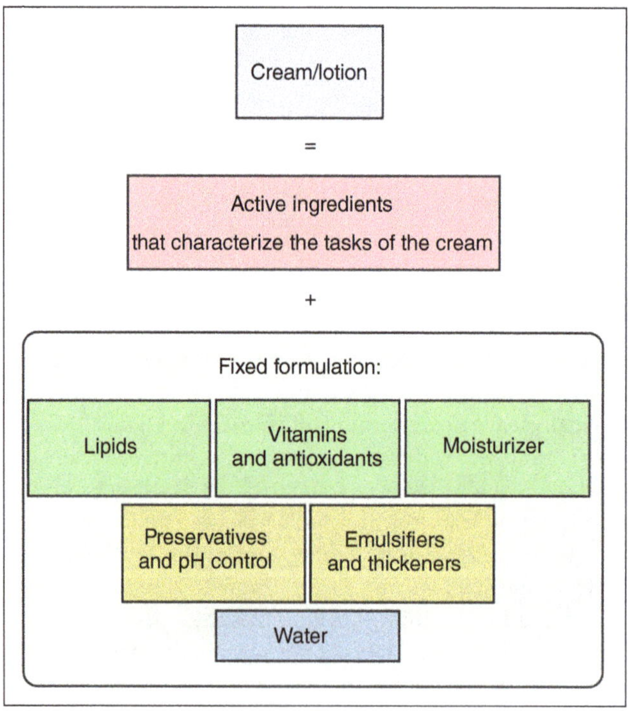

Figure 5.1 Modular compositions of creams (bottom-up).

Table 5.1 Excipients of a cream according to their functions.

	Excipients	Number of common substances[a)]	Number of required/recommended substances	Used concentration ranges[b)]	Typical substances (examples)
(a)	Consistency (viscosity control)				
	• Organic phase	>20	1	2–5%	Fatty alcohols (FA), waxes, glyceryl stearate
	• Aqueous phase	>40	1	0.2–3%	Xanthan, polyacrylates
(b)	Emulsifiers				
	• o/w macro emulsions	>80	1	0.7–4%	Ethoxylates of FA and of glycerides
	• w/o macro emulsions	>50	1	2–6%	Sorbitan ester, glycerides
	• Mini (nano) emulsions	Many combinations possible	2	0.5–3%, together 2.5–6%	Two emulsifiers with different HLB values (9–16, 5–8)
(c)	Preservatives	56	1	0.1–0.4%	Sorbic, salicylic acid, parabens
(d)	Complexing, chelating	12	0 or 1	0.2–0.6%	Sodium citrate, sodium glyconate, cyclodextrins
(e)	Antioxidants	15	1	0.2–2.5%	Tocopherol, ascorbic acid, cysteine
(f)	pH control	31	2	0.1–1.2%	Citric acid with sodium citrate, lactic acid/lactates
(g)	Solvents, solubilizer	>14	0 or 1	0.5–4%	Ethanol, glycols, some emulsifier
(h)	Spreading, antifoam	>17	1	0.5–3%	Silicones, mono- and diglycerides of fatty acids
(i)	Enhancer	About 10	0 or 1	1–5%	Ethanol, propylene glycol
(j)	Odor neutralizer	About 10	0 or 1	<1%	Cyclodextrins
(k)	Liposomes, ethosomes	>5	0 or 1	<2%	Lecithin

a) Estimated.
b) Material and application dependent.

agents, and thickeners for the two phases. In addition, the cream requires one or two pH regulators to set the skin's pH to 5, preferably through a buffer system (acid/base). The buffer and complexing agents as well as antioxidants, the absence of oxygen, moderate temperatures, and ultraviolet (UV) light impermeable dispensers prevent undesirable chemical reactions. The excipients in the table are given in groups with usual amounts.

The composition and amounts of the modules are shown by way of examples. Each developer can change the given scheme by decreasing or increasing the number of ingredients and fill it with own formulations.

Basic module "preservative": Preservatives and pH regulation consisting of 0.25% potassium sorbate, 0.64% trisodium citrate dihydrate, and 0.16% citric acid monohydrate.

Basic module "emulsifier": Emulsifier/coemulsifier and consistency generator consisting of 3.5% fatty alcohol (FA), 1.5% FA-ethoxylates, 1.5% silicone oil, and 0.3% xanthan gum.

In some cases, it is advantageous to add a small amount of a w/o emulsifier. For this purpose, 0.5% glycerol monostearate with emulsifying and thickening properties can be used. The combination FA ethoxylate (INCI [International Nomenclature of Cosmetic Ingredients]: Ceteareth-30) with FA (INCI: Cetearyl Alcohol) has been successfully tested in amounts of 0.7/3.0% to 1.5/4.0%. Levels of less than 1% of ethoxylates can certainly be sufficient for the emulsification (see Section 8.8). As often tested, the formulations here in most cases have quantities of 1.5/3.5%.

Of course, each module can be composed differently. For example, an alternative 1 for natural cosmetic formulations represents the combination of 4.5% "Lamecreme" (INCI: Glyceryl Stearate, Glyceryl Stearate Citrate) with 0.5% lysolecithin (for fatty phases >30%). Alternative 2 may be 2% Imwitor 375 (INCI: Glyceryl Citrate, Linoleate, Oleate) with 2.5% BergaMuls ET 1 (INCI: Beta-Glucan and Pectin) and 0.5% Cetyl palmitate. Alternative 3 is Polyglyceryl-3 Dicitrate/Stearate (4%) with Myristyl myristate 4% (INCI: Myristyl Myristate) as a thickening agent. Other examples can be found in Chapter 7, also for w/o emulsions. In contrast to the emulsifier system with the ethoxylated FA emulsifier, however, the alternatives were not tested.

The emulsifier system is crucial for the stability of an emulsion and for the effect on the skin. This consists of one or two emulsifiers/coemulsifiers and viscosity regulators for both phases. The stabilization of the emulsion occurs not only by emulsifiers but also by the thickeners as viscosity regulators are at least two substances desirable, a lipophilic and a hydrophilic. For adjusting the viscosity of the water phase, there are suitable numerous natural/semisynthetic polymers, such as xanthan and synthetic polymers, like modified polyacrylates. The high molecular hyaluronic acid thickens the water phase extremely, thus even tiny amounts of this well-known skin ingredient are enough to increase the viscosity. However, this increase is not enough for a cream. Lipophilic compounds of medium molecular weights (MWs), as long-chain fatty alcohols or waxes, set the viscosity of the oil phase. The viscosity, reached in the emulsification process after cooling down, depends on the optimum amount of thickening agents as a function of type and amount of the ingredients, estimated from a series of experiments. The excipients determine the structure of the cream. Furthermore, an extensive experience helps for the choice of suitable thickeners, which also considers the spreadability as well as the haptic feeling.

The pH of the skin (4.5–5.5) should be accurately adjusted in the cream via a buffer system with an acid and the corresponding salt (Figure 5.12) to maintain constant conditions for years. In addition, some preservatives such as sorbic acid are fully effective only in a narrow pH interval; thus, in this case, it is consistent with the usual pH of the skin. The stabilization of polyunsaturated natural lipids requires an addition of antioxidants. Even a slight oxidation of these natural oils results in a clearly perceptible, rancid odor. This undesired reaction, catalyzed by heavy metal ions, should be additionally suppressed by the exclusion of oxygen and the use of complexing agents. To reduce odors, even the typical smell of an unperfumed cream or lotion, it may be useful to add a substance that absorbs odors. Cyclodextrins represent suitable agents for this purpose. In addition, they constitute chelators for heavy metal ions. For an unobtrusive odor coverage, birch oil is suitable.

Many people use a cosmetic cream with a perfume of their choice for a personal note (see Chapter 9). For the creation of a perfume, a perfumer selects 30 to maximal 200 scents with one main fragrance [1]. Because of possible allergenic effects, 26 fragrances should be used only in very small quantities, or better not. According to the annexes of the Cosmetics Directive [2], they must be declared on the product package, when exceeding a certain amount (100 ppm for rinse-off and 10 ppm for leave-on materials). Many customers expect a pleasant scent from a cream and choose cosmetic products for the scent impression. Sensitive people search for well-tolerated, fragrance-free products. After patch test studies, 15–20% of the German population show sensitization to at least one of the most important contact allergens (studies of IDU: Information Network of Departments of Dermatology [3]). More than 7% suffer from perfume allergies, which is a major problem [4]. Additives (Table 5.2) increase the risk of an allergic reaction on sensitive skin. Therefore, substances for the scent and for staining are typical ingredients in "normal" cosmetic products, but usually undesirable in cosmeceuticals (cosmetic creams with pharmaceutical effects).

A number of essential oils show positive effects on the skin. Examples are lavender, chamomile, carrot, rose, rosemary, sage, and cedar. So far, only special products contain these oils. Some of their typical, intense smell is not perceived as

Table 5.2 Additives for color and fragrance.

	Additives	Number of common substances	Number of required/recommended substances
(a)	Fragrances including essential oils	>2500	1
(b)	Perfumes	>1000	0, (1)
(c)	Dyes	145	0

Table 5.3 Substances with specific effects on and in the skin [5].

No	Active ingredients	Number of common substances	Number of required/ recommended substances	Used con-centration ranges	Typical substances
1	Lipids				
	• Vegetable oils	>40	1–6	10–15% 15–30%[a]	Hemp, evening primrose, borage, sweet almond
	• Synthetic oils	12	0/1	<3%	Silicone
	• Natural waxes	>20	0/0 or 1	<2%	Beeswax, wool wax (lanolin), candelilla wax
	• Hydrogenated oils	About 15	0/0 or 1	<2%	Coconut, palm, castor oil
	• Synthetic waxes	>40	0/0 or 1	<3%	Artificial jojoba, compounds of fatty acids with alcohols
2	Vitamins/provitamins and their compounds	>20	2–6	0.2–6%	Vitamins A, E, and their ester (palmitate), B_3, B_5, B_6.
3	Natural moisturizing factors (NMFs)	>30	2–6	0.2–6%	Urea, lactate, amino acids, sorbitol, glycerol, protein hydrolysates, PCA
4	AHA (α-hydroxy acids, fruit acids, salts)	9	1–4	<1%	Lactic, citric, salicylic acid
5	Essential oils from healing plants	>100	0–2	0.1–5%	Lavender, geranium, camphor, sage, chamomile (powder), hamamelis
6	Plant extracts (healing agents), natural or synthesized identical to nature	>100	0–2	0.1–5%	Green tea, aloe vera (juice or powder), hops, allantoin
7	Skin care substances, emollients	>20	0–2	0.5–3%	Sphingolipids, cholesterol, lecithins, phospholipids
8	Antiaging substances	>30	0–5	<3%	Kinetin, vitamin A palmitate, squalene, coenzyme Q10
	• Biotechnological synthesized	>3	0–1	<0.1%	Antarcticine, ectoin
	• Synthetic peptides	>20	0–1	<0.1%	Hexa-, tetrapeptides

PCA, pyrrolidone carboxylic acid.

a) w/o emulsion.

Table 5.4 Suggestions for basic formulations and for the indicated quantities (90% of the total recipe).

Basis module	Ingredient	INCI name	Amount (%)
(a) o/w Creams			
Preservative (and pH regulation)	Potassium sorbate	Potassium Sorbate	0.25
	Trisodium citrate dihydrate	Sodium Citrate	0.64
	Citric acid monohydrate	Citric Acid	0.16
Emulsifier (coemulsifier, thickening agents)	Cetyl stearyl ether 30EO	Ceteareth-30	1.5
	Cetyl stearyl alcohol	Cetearyl Alcohol	3.5
	Glycerol monostearates	Glyceryl Stearate	0.5
	Silicone oil	Dimethicone	1.5
	Xanthan	Xanthan Gum	0.3
Lipids	Sweet almond oil	Prunus Amygdalus Dulcis Oil	4
	Evening primrose oil	Oenothera Biennis Oil	3
	Hemp oil	Cannabis Sativa (Hemp) Seed Oil	3
	Shea butter	Butyrospermum Parkii Butter	3
Vitamins	D-panthenol	Panthenol	2.5
	Nicotinamide	Niacinamide	2
	Vitamin E (natural)/ sunflower oil (70 : 30)	Tocopherol/*Helianthus Annuus*	1
Moisturizing substances	Glycerol	Glycerin	3
	Sorbitol	Sorbitol	3
	Hyaluronic acid	Hyaluronic Acid	0.1
	Allantoin (degradation product of nucleic acids)	Allantoin	0.1
Sum			33.05
	Water/alcohol (up to 90%)	Ethanol	2
		Aqua	54.95
Additional recipe components to fulfill the application purpose (%)			10
(b) w/o Creams			
Preservative (and pH regulation)	Potassium sorbate	Potassium Sorbate	0.25
	Trisodium citrate dihydrate	Sodium Citrate	0.64
	Citric acid monohydrate	Citric Acid	0.16

(*continued*)

Table 5.4 (Continued)

Basis module	Ingredient	INCI name	Amount (%)
Emulsifier (coemulsifier, thickening agents)	Sorbitan stearate	Sorbitan Stearate	4.5 (4–5)
	Dermofeel® Viscolid Palm Oil Free	Hydrogenated Rapeseed Oil	1 (0.5–2)
	Cetyl palmitate	Cetyl Palmitate	2 (1–3)
	Xanthan	Xanthan Gum	0.3 (0.1–0.5)
Lipids	Sweet almond oil	Prunus Amygdalus Dulcis oil	7
	Evening primrose oil	Oenothera Biennis oil	5
	Hemp oil	Cannabis Sativa (Hemp) Seed Oil	5
	Sheabutter	Butyrospermum Parkii Butter	5
Vitamins	D-panthenol	Panthenol	2.5
	Nicotinamide	Niacinamide	2
	Vitamin E (natural)	Tocopherol/Helianthus Annuus (70 : 30)	1
Moisturizing substances	Glycerol	Glycerin	3
	Sorbitol	Sorbitol	3
	Hyaluronic acid	Hyaluronic acid	0.1
	Allantoin (degradation product of nucleic acids)	Allantoin	0.1
Sum			42.55
	Water/alcohol (up to 90%)	Ethanol	2
		Aqua	45.45
Additional recipe components to fulfill the application purpose (%)			10

pleasant, despite good efficacy. In contrast to the normal cosmetics, the cosmeceuticals usually contain no perfume, but usefully often a fragrant, essential oil with healing properties.

In combination with the excipients, the active ingredients (Table 5.3, see also Chapter 6) determine the effects of the cream. Of the active ingredients, vegetable oils are of particular importance. On the one hand, nondry oils with iodine numbers up to 100, such as almond and olive oil or shea butter, show visible skin care effects. On the other hand, semidry oils with iodine numbers up to 170 diffuse into the skin in a short time and form virtually no film (sunflowers, evening primrose, and hemp oil). In general, these two types of oils are mixed in order to obtain both effects. For the day cream, a fast intake is desired, i.e. a nondry/semidry setting with 10–15% oil. For a night cream, significantly more oil is used with 15–30% of predominantly nondry oils. These are processed into a w/o emulsion with the aid of a suitable emulsifier. An example from natural cosmetics represents polyglyceryl-3 polyricinoleate (trade name: Dermofeel PR)/beewax or sorbitan fatty acid ester/FA-alcohol.

When carefully selected, the freely combinable modules can be suitable for all formulations. Here, for o/w emulsions, the **Module "Lipids"** contains essential oils, such as 4% sweet almond oil, 3% evening primrose, 3% hemp oil, and 3% shea butter (Table 5.4a). The vitamins are summarized in the **Module "Vitamins,"** which consist of 2.5% D-panthenol, 2% nicacinamide, and 0.7% natural vitamin E (α-tocopherol). The **Module "Moisturizing substances"** includes 3% glycerol and 3% sorbitol as well as 0.1% allantoin and 0.1% hyaluronic acid with a molecular weight of 1–1.5 MDa. The five modules are supplemented by different active ingredient modules to form the desired product type (see Chapter 7). For the w/o emulsions, the emulsifiers and thickeners have to be adjusted. In addition, approximately 1.5–2 times higher oil input is recommended (Table 5.4b).

From the five modules, the 90% formulation is obtained (Table 5.4). The remaining 10% consists of ingredients that are characteristic for the intended task. This may, for example, concern the foot care, a body lotion, the face cream for antiaging, or a cream for the care of neurodermitic skin. Thus, the entire range of creams can be produced from a single large batch (e.g. 10 t). After

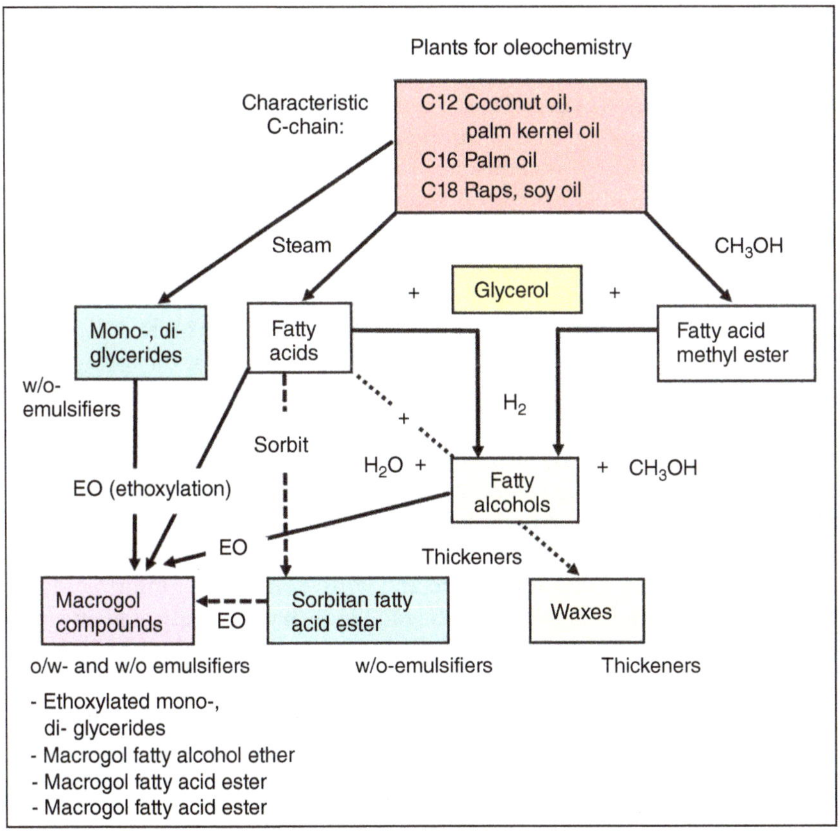

Figure 5.2 Derivatives of vegetable oils for cosmetic creams.

production, the batch is divided into smaller units (e.g. 4, 3, 2, 1, and 0.5 t), to which the remaining 10% are added in each case. This approach is useful if relatively small amounts of each cream are required. Otherwise 100% formulas should be produced as usual.

Many substances come from oleochemistry and are derived from renewable plants. In Figure 5.2, the tree of the follow-up products is shown. Only important substances for creams are indicated.

5.3 Excipients

The emulsifiers/coemulsifiers, consistency regulators (thickeners), preservatives, antioxidants, complexing agents, and pH-adjusting buffer systems comprise the excipients for the preparation of stable creams. In some cases, the formulation demands additional solubilizers, odor absorbers, antifoam agents, or liposomes. Selected excipients preferably meet two or more tasks with beneficial effects on the skin. Some examples are discussed, such as fatty alcohol with coemulsifying and additionally thickening properties. Besides, the coemulsifier provides the skin with fat and makes it soft.

5.3.1 Emulsifier for Macroemulsions

Emulsions are thermodynamically unstable systems of two immiscible liquids that separate after mixing. The oil droplets float up and form an oil phase on the water. In the case of w/o emulsions, the water droplets sink to the bottom; a water phase is formed under the oil. Emulsifiers are surfactants that lower the interfacial tension (surface tension) between nonmiscible liquids. With the aid of an emulsifier, the lipophilic phase is introduced into the hydrophilic, aqueous phase or vice versa, in the form of droplets. By using a suitable emulsifier and a high shear rate mixer, a white liquid (milk) is formed, which is more or less flowable because of the consistency factors and the temperature of the formulation. Depending on the operating conditions of the manufacture, the type and amount of the emulsifier, as well as the components of the formulation, the formed emulsion can be stable for many years. There is a tremendous variety of emulsifiers and coemulsifiers. In a table, several thousands are listed. In addition to the conventional, mostly used o/w and w/o emulsions, other types of emulsions exist, which are rather rare. These include the phase inversion temperature (PIT), Pickering, and the multiple emulsions o/w/o and w/o/w. For further information on these types, which are already used in the food industry, the study of literature is recommended [6].

During the intensive mixing, the emulsifier (surfactant) allows the distribution of the first phase uniformly in the form of small droplets in the other, continuous phase. The liquid is the outer or continuous phase and droplets are the inner or dispersed phase. Oil droplets produce an o/w emulsion and water droplets a w/o emulsion [7]. According to the Bancroft rule [8], the HLB (hydrophilic/lipophilic balance) number of the emulsifier (Figure 5.3) decides the type of emulsion. The

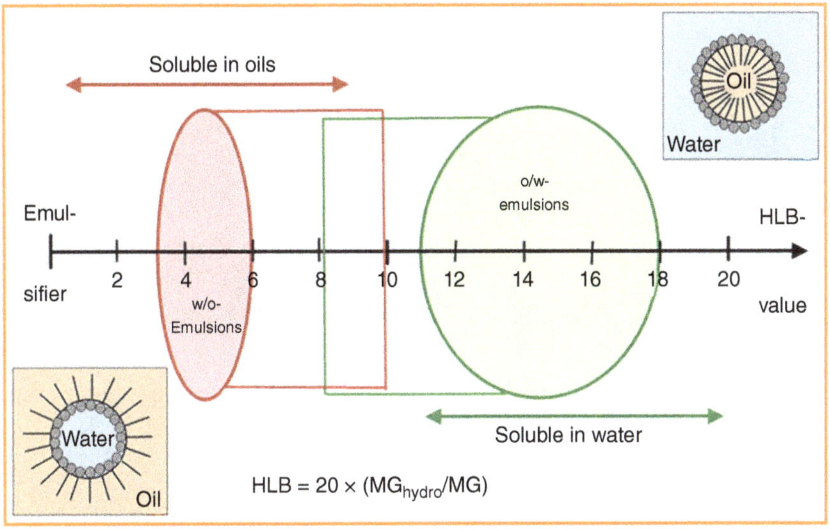

Figure 5.3 Regulation of the emulsion type by the HLB value of the emulsifier (in the little pictures are the lipophilic parts represented by lines, the hydrophilic groups by circles).

HLB value indicates the hydrophilic/lipophilic balance of the system and can be determined analogue to Griffin [9] from the molecular weight of the hydrophilic portion (MG_{hydro}) in relation to the total molecular weight (MW), multiplied with 20. The HLB value is an old aid for the assessment of emulsifiers. There are some suggestions for increasing accuracy. Ultimately, the old value for the practitioner has prevailed, which can be found in product information from the supplier.

Emulsifiers with HLB values between 3 and 8, dissoluble in oils, form (w/o) emulsions. Moreover, surfactants with HLB values of 11–17 produce the (o/w) emulsions. HLB values of 8–11 characterize the transition region between the lipophilic and hydrophilic behavior [10]. In this region arises an o/w or w/o emulsion, depending on the type of production and temperature, but at temperatures below 70 °C, preferably the o/w type arises. The percentage distribution of the phase components (water and oil) does not affect the type. Water droplets in oil are often smaller than oil droplets in water (Figure 5.4).

The stability of the system depends crucially on the emulsifier and used coemulsifier [11]. Some nonionic surfactants stabilize emulsions very effective and are therefore preferred. In the case of the ethoxylated compounds, those with 2 or 3 EO units support the formation of w/o emulsions; from EO 5 to 7 and higher values, usually o/w emulsions arise. The choice of the emulsifier depends on the chemistry of the oil because oil and emulsifier should resemble. High-quality natural oils such as borage, hemp, and evening primrose consist mainly of C_{16} and C_{18} triglycerides (glycerol-esterified fatty acids), which usually contain one, two, or three double bonds. It is therefore self-evident to use for the manufacture of an emulsion nature-based $C_{16, 18}$ emulsifier and coemulsifier, if possible, with double bonds. The emulsifiers should only be based on vegetable oils that regrow and therefore are abundant.

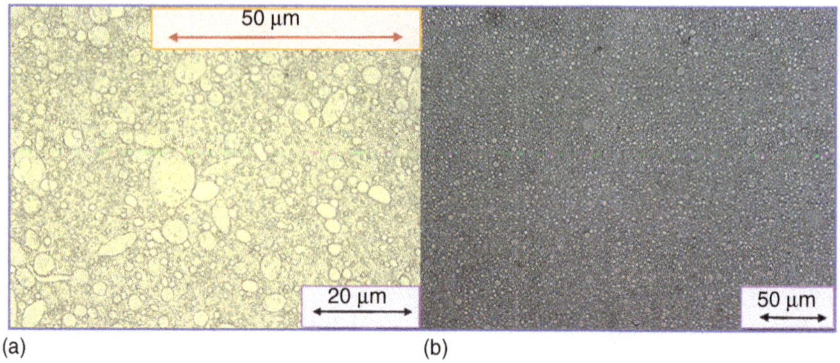

Figure 5.4 Microscopy images of commercial skin care products, bought in the supermarket: (a) o/w emulsion with small and large droplets, (b) w/o emulsion type with some big drops.

Not all surfactants are suitable for the production of skin care products. For instance, cationic surfactants carry a positive charge in water-based solution and cannot be used for emulsifying. On the other hand, anionic surfactants hydrolyze in aqueous solution into a cation (Na, K) and the longer-chain anionic residue, which reduces the surface tension and causes the emulsifying effect. Typical for this are the soaps of the fatty acids as well as the alkyl ether sulfates with a chain length of C_{12}–C_{18}. Because of the possible substance-dependent reduction of skin tolerance and barrier function, the use of these surfactant classes is less advisable. Owing to the good emulsifying effect and the skin compatibility, nonionic surfactants based on fatty alcohols are often used. The chemistry is shown in Figure 5.5, several examples can be found in Table 5.5. Usually, the low ethoxylated compounds support the formation of w/o emulsions; higher degrees of ethoxylation indicate good water solubility.

As the even C-chain compounds based on vegetable oils are better tolerated by the skin than the synthetic ones from petroleum with odd chain lengths, only these should be used for active creams. Owing to the large number of vegetable oils, fatty acids, fatty alcohols and sugars, as well as the possibility of chemical modification, there are far more than hundred well-working emulsifiers from a few thousands, also without synthetic starting materials or animal-based substances. A comprehensive description of the emulsifiers can be found in the literature [12].

Ethoxylated fatty alcohol-based surfactants are preferably used in o/w formulations, especially if they contain a high amount of salts. Their safety is discussed in the literature: "Ethoxylated compounds are used in a great variety of cosmetic applications because of their solubility and viscosity properties and because of their low toxicity. The polyethylene glycols (PEGs), their ethers, and their fatty acid esters produce little or no ocular or dermal irritation and have extremely low acute and chronic toxicities. They do not readily penetrate intact skin" (quoted from [13]). These newer experimental results of this important literature are in contrast to the statements of natural cosmetics that are not adequately scientifically proven.

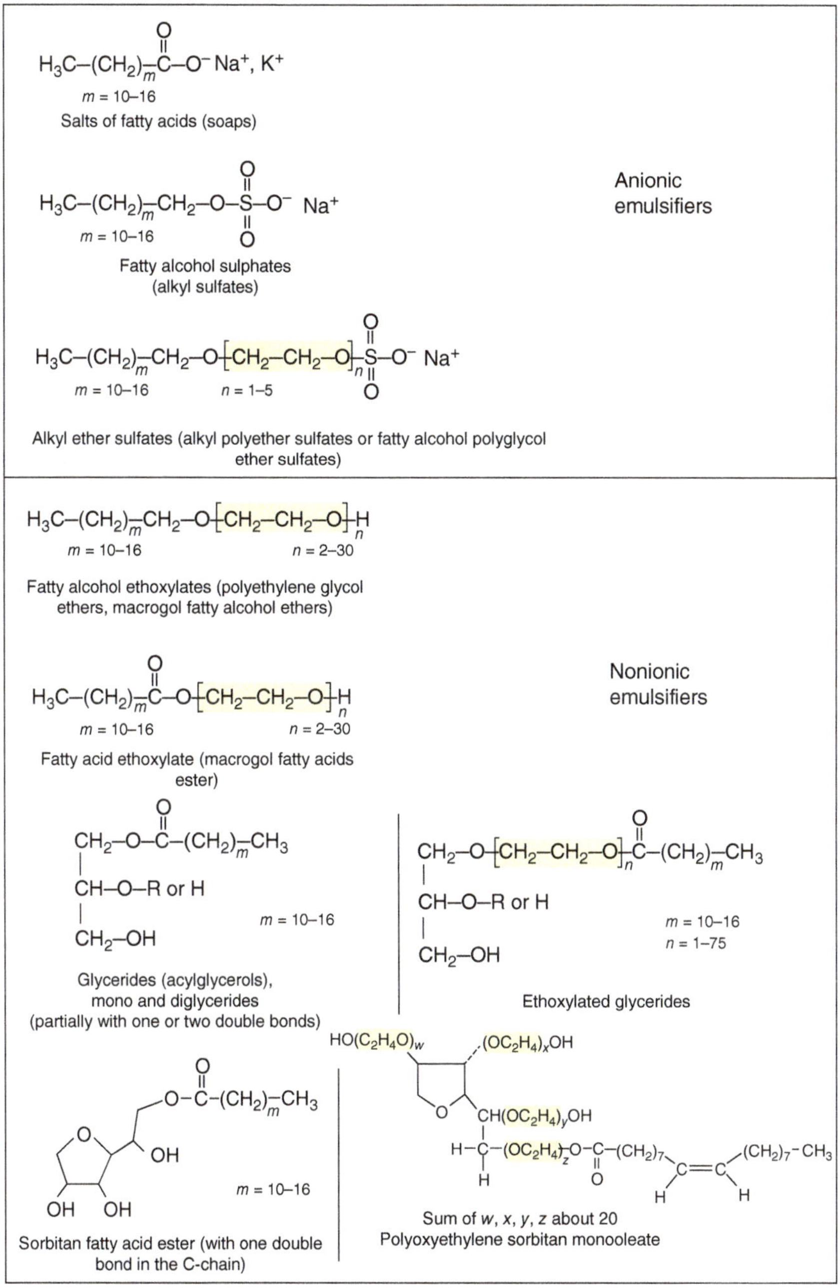

Figure 5.5 Chemical structures of some important emulsifiers.

Table 5.5 Nonionic emulsifiers (the selected examples are based on vegetable oils with C_{12}, C_{14}, C_{16}, and C_{18} chains; polyoxyethylene = PEG in a compound; glyceryl… = glycerol…).

Chemical name	C-chain (oil)	Number of EO units, *X* =	INCI description	Emulsifier type
Macrogol fatty alcohol ethers				
Polyoxyethylene (X) lauryl ether (Macrogol lauryl ether)	12	4, 9, 23	Laureth-4, -9, -23	o/w
Polyoxyethylene (X) cetyl ether (Macrogol cetyl ether)	16	2, 10, 20	Ceteth-2, -10, -20	w/o, o/w, o/w
Polyoxyethylene (X) cetyl stearyl ether (Macrogol cetyl stearyl ether)	16/18	6, 20, 25, 30	Ceteareth-6, -20, -25, -30	o/w
Polyoxyethylene (X) stearyl ether (Macrogol stearyl ether)	18	2, 10, 20	Steareth-2, -10, -20	w/o, o/w, o/w
Macrogol fatty acid esters				
PEG-ester of stearic acid (also as diester)	18	2, 6, 8, 9, 20, 30, 32	PEG-2, -6, -8, -9, -20, -30, -32 Stearate	w/o, o/w
PEG-diester of lauric acid	2 × 10	4, 8	PEG-4, -8 Dilaurate	o/w
Ethoxylated castor oil	3 × 18	7, 36, 40	PEG-7, -36, -40 Castor Oil	
Ethoxylated and hydrogenated castor oil	3 × 18	2, 7, 20, 25, 40	PEG-2, -7, -20, -25, -40 Hydrogenated Castor Oil	w/o, o/w
Glyceryl (mono) esters of natural fatty acids				
Glyceryl monolinoleate	18	—	Glyceryl Linoleate	w/o
Glyceryl monostearate	18	—	Glyceryl Stearate	w/o
Glyceryl monocaprylate	8	—	Glyceryl Caprylate	w/o
Glyceryl monolaurate	12	—	Glyceryl Laurate	w/o

Macrogol (mono- and di-) esters of natural fatty acids				
Polyoxyethylene (X) glyceryl isostearate	18	15, 20, 30, 60	PEG-15, -20, -30, -60 Gyceryl Isostearate	o/w
Polyoxyethylene (X) glyceryl laurate	12	8, 12, 30	PEG-8, -12, -30 Glyceryl Laurate	o/w
Polyoxyethylene (X) sunflower glycerides (also some other oils)	18, 2 × 18	10, 13	PEG-10, -13 Sunflower Glycerides	o/w
Sorbitan fatty acid ester				
Sorbitan laurate	12	—	Sorbitan Laurate	w/o
Sorbitan isostearate	18	—	Sorbitan Isostearate	w/o
Sorbitan trioleate	18	—	Sorbitan Trioleate	w/o
Macrogol sorbitan fatty acid ester (polysorbate)				
Polyoxyethylene (X) sorbitan monoisostearate	18	2, 5, 20	PEG-2, -5, -20 Sorbitan Isostearate	o/w
Polyoxyethylene (X) sorbitan monolaurate	12	10, 20, 40, 44, 75, 80	PEG-10, -40, -44, -75, -80 Sorbitan Laurate	o/w
Polyoxyethylene (X) sorbitan tetraoleate	4 × 18	30, 40, 60	PEG-30, -40, -60 Sorbitan Tetraoleate	o/w
Coemulsifier				
Cetearyl alcohol	16/18		Cetearyl Alcohol	o/w
Cetyl alcohol	16		Cetyl Alcohol	o/w

The sorbitan and glycerides esters proved useful for w/o emulsions. Many emulsifiers are liquid, some solid. In order to keep the temperature load of the oil phase low, the melting point of the emulsifier/coemulsifier should preferably not exceed 60 °C. Altogether, there are a large number of different emulsifiers, mostly based on the fat chemistry. Anyone looking for specialty products will find a wide selection on the manufacturers' pages, such as BASF, Evonik, Croda, Lubrizol, Huntsman, and others. BASF [14] alone sold approximately 30 o/w and 6 w/o emulsifiers. Evonik also offers over 20 emulsifiers as well as the Mani GmbH and others.

About 25–30 years ago, acidic catalysts were used for ethoxylation, which led to the adverse formation of 1,4-dioxane as a by-product. These problems could be solved by the application of suitable alkaline catalysts. Ethoxylated emulsifiers have good skin compatibility. In numerous comments in the Internet is the assertion that the classes of (nonionic fatty alcohol) ethoxylates makes the skin more permeable and supports the introduction of (harmful) substances further mobilize the skin fats. However, no developer formulates a cosmetic cream without a refatting component. With an emulsifier system, consisting of fatty alcohol ethoxylates and fatty alcohol as coemulsifier, the described skin irritations could never be detected, rather the opposite. A quote of Paracelsus (1493–1541) may be the right comment to these findings: "*All things are poison and nothing is without poison, only the dose makes, that a thing is no poison.*" In particular, the low concentrations, but also interactions and influences of the environment (pH, salts), the thickening agent, and especially the lipids modify the possible effect of surfactants on the skin. The ethoxylates have the advantage to emulsify at low concentrations very well, also in the presence of high salt loads (up to 5%), which can be an important prerequisite for the choice of the system. Even with intensive research, a damage of the skin barrier is not detectable (see later). This means that possible disadvantages in the presence of suitable components of the formulation detectable do not occur. Therefore, effects should always be checked with the own formulations.

The boundary conditions must be determined to choose a suitable emulsifier system. They depend, among other things, on the formulations, preparation, long-term stability, and skin conditions. An example can be found in Table 5.6. If these sharp limits are required, the stability and salt tolerance as well as the maintenance of the pH value of the skin are unattainable requirements for many systems.

Personal Note 1: From the almost innumerable emulsifier systems, it is very difficult to find a combination usable for the whole product range. That is why, I let me advice of a commodity specialist, who worked formerly at Henkel, then Cognis, and later at BASF. Therefore, my choice of the system comes from the "Henkel" developments. For o/w emulsions, I decided in 2002 after numerous attempts with different emulsifiers to use the following system: Ceteareth-30, Cetearyl alcohol, and Dimethicone, in the ratio 1 : 2 to 2.5 : 1. The coemulsifier has additional features, acts as emollient, thickened the emulsion, and provided the skin with lipids. The silicone oil is primarily used to suppress foam formation during production. Further properties will be discussed later. This system showed

Table 5.6 Exemplary conditions for the selection of an emulsifier system.

Parameter	Target value
Emulsifier	➢ High skin tolerance (in combination with the coemulsifier) ➢ Low concentration (about <2%) ➢ Relatively low processing temperature ➢ High tolerance for different salts ➢ Effective in the desired pH range
Coemulsifier	➢ Emulsifying and thickening effect ➢ High salt tolerance ➢ (Emoillent, refatting)
Thickener	➢ High salt tolerance ➢ Low necessary quantities ➢ Effective in the desired pH range
Whitening	No or only slight whitening for short time
Stability	
➢ Temperature	4–50 °C
➢ Time	>3 yr (25 °C)
pH value	Range 4.8–5.5; exact value adjustable by a buffer
Salt tolerance	1% to >5%, depending on the formulation
Haptic	Pleasant
Odor	Unobtrusive
Color	White

the desired high salt tolerance, extreme stability with duration of over five years, good skin compatibility, and the suitability for all recipe variants. In particular, the extensively tested parameters of the skin after application of different creams with this emulsifying system show consistently positive results for the transepidermal water loss (barrier function), skin moisture, lipid content, elasticity, and surface structure of the skin. The use of an extensively tested emulsification system for all o/w formulations considerably reduces the development and testing costs.

One of the most effective nonionic surfactants for high salt o/w emulsions include ethoxylated $C_{16,18}$-fatty alcohols, creating optimal results with about 30 molecules of ethylene oxide (MW c. 1600, HLB = 16.6) in the presence of $C_{16,18}$ fatty alcohol as a coemulsifier. Other emulsifiers are in most cases less effective (higher quantity is required, larger droplets or less salt tolerance). A macroemulsion containing high-quality vegetable oils can be stable for several years. Such a stabilized cream, about four weeks old at the time of measurement, is displayed in Figure 5.6. As can be seen, the emulsifier system produces a narrow droplet size distribution with droplet diameters between 0.7 and 7 μm, much better than the o/w emulsion shown in Figure 5.4. This shows the superiority of the chosen

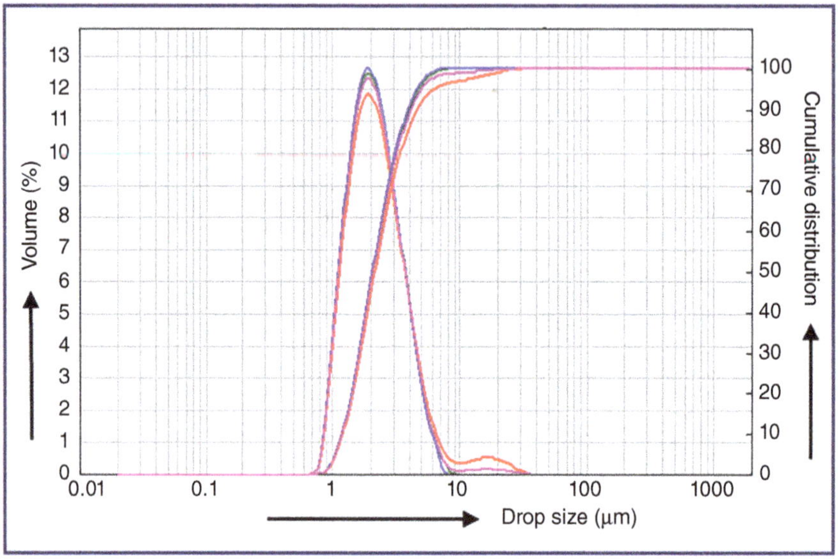

Figure 5.6 Droplet size distribution of a nonionic surfactant-stabilized o/w skin cream with a mean diameter d_{50} of 1.9 μm and triple measurement after dilution by laser diffraction/equipment: Malvern.

system and a reason for extreme stability. Another proof is the successful centrifugal test at 30 000g for 5 minutes at 40 °C without separation of the phases. Surely, there are other systems with similar effectiveness. Every developer knows emulsifiers/systems that have proven themselves and that he prefers.

The coemulsifier $C_{16,18}$-FA promotes the formation of microfoam during the high-energy input in the production. The silicone oil, part of the emulsifier system, prevents the formation of these microfoams. In addition, the dimethicones [12] show numerous positive effects on the skin and are therefore both excipients and active ingredients:

- good skin compatibility
- high spreading capacity because of the low surface tension
- nongreasy
- gas-permeable films on the skin, no heat, and moisture jam
- high thermal and chemical stability
- readily emulsifiable, but not at all soluble in water, soluble in aliphatic hydrocarbons
- water repellent.

Because of the water-repellent and protective properties, they are used in baby creams, decubitus, and sun protection products as well as skin care products up to 5%. Only positive evaluated compounds should be utilized. Some aromatic silicone compounds are not among them. Possible alternatives to the silicone oils may represent in the low-viscosity range neopentyl glycol diethylhexanoate (Soldoc VF 8) and for medium-viscose dimethicones the pentaerythrityl tetraisostearate (Soldoc VF 4/18).

Only a few suitable emulsifiers/emulsifier systems exist for the production of natural cosmetic creams. A list of trade and INCI names taken from the literature can be found in Table 5.7. The incorporation of some emulsifiers requires relatively high working temperatures of 70–80 °C and should therefore not be considered for oxidation-sensitive ingredients or must be handled very carefully in vacuum (inert after nitrogen flushing).

5.3.2 Emulsifier for Mini (Nano) Emulsions

Although the drop size distributions of commercial products are usually in the micron range from 0.8 to 50 μm, nanosized droplets of about 50–600 nm characterize a mini emulsion. Two completely different surfactants are required for the production of such a mini emulsion. The desired emulsion with very small and stable oil droplets is formed by an emulsifier system, consisting of hydrophilic and lipophilic emulsifiers in a ratio of 4 : 1 to 2 : 1 in the presence of a coemulsifier (such as FA) in comparable amounts. There are many suitable w-emulsifiers to supplement the discussed hydrophilic emulsifier systems.

For example, with the lipophilic emulsifier Laureth-2 (2%, Dehydol LS 2, BASF) or preferably Cetyl PEG/PPG-10/1-dimethicone (0.5–1.5%, Abil® EM 90 from Evonik), the system Ceterareth-30/Cetearyl alcohol (BASF)/Dimethicone offers narrow droplet distributions. The search for a very heat-stable w/o emulsifier led to high molecular weights of approximately 10 000 because the temperature stability of the emulsion depends on the molecular weight of the emulsifiers, the higher – the better (see Figure 5.7). With 2% Eumulgin® B 3 and 3.5–4.0% Lanette® O as a coemulsifier and thickener, in combination with a 0.5–1.0% polymeric silicone-based emulsifier (Abil EM 90), which is even suitable for multiple emulsions, temperature-stable o/w mini emulsions arise. Because of its good solubility in water, Eumulgin B 3 does not penetrate the skin, just like Abil EM 90 because their high molecular weights and physical properties prevent diffusion. Besides the emulsifier system, the droplet size distribution also depends on the homogenizer and the operating conditions. With a dual-piston high-pressure homogenizer (Gea Niro Soavi [15]), in three runs, a close and reproducible droplet size distribution arises (d_{10} = 80 nm, d_{50} = 122 nm, and d_{90} = 200 nm; Figure 5.8).

Micro emulsions, used for therapeutica and not for cosmetics, can be clear and transparent. Because of the high amounts of emulsifiers (about 15–35%) needed for the production, a use for skin care seems less appropriate, similar restrictions apply to multiple emulsions. The way to produce a multiple emulsion is the emulsification of an o/w emulsion in oil or a w/o emulsion in water. The additional costs for multiple emulsions (higher amounts of surfactants, manufacturing, and fewer degrees of freedom in the formulation) result in no benefit for the skin care today.

5.3.3 Stability of Emulsions

An essential quality feature of skin care products is time to use after manufacture and after opening the package. The stability of the emulsion [7], the constancy

Table 5.7 Approved emulsifiers for the natural cosmetics [15]; the processing must be carried out according to the manufacturer's instructions, in most cases, a coemulsifier is required.

Trade name/melting point	INCI name	Type/HLB	Amount used	Remarks
BergaMuls® ET 1; Cold to use, fibrous	β-Glucan and Pectin	o/w	>3% 1–2%	Solo emulsifier with thickening properties As coemulsifier
Dermofeel® PR; About 75 °C	Polyglyceryl-3 Polyricinoleate	w/o 4–5	3–5%	For thickened emulsions add 3–5% beewax or comparable substances.
Emulprot®; Dispersed at 70 °C	Sodium Citrate, Hydrolyzed Milk Protein, Xanthan Gum, Cyamopsis Tetragonoloba (Guar) Gum, Magnesium Stearate	o/w 9	2.5–5% 1–1.5%	Solo emulsifier, formulations appear less greasy As coemulsifier (milk proteins can induce sensitization).
Fluidlecithin Super; Liquid	Lecithin, Carthamus Tinctorius Oil, Glycerin, Caprylic/Capric Triglyceride, Alcohol, Glyceryl Stearate, Ascorbyl Palmitate	o/w 7	4% 1–2%	Emulsifier requires 0.2–0.4% thickener such as cetylalkohol, cetylpalmitat, glycerolmonostearat, or cetearylakohol As coemulsifier
Imwitor® 375; Highly viscous liquid	Glyceryl Citrate/Lactate/Linoleate/Oleate	o/w 11	2–3% 0.5–1%	Emulsifier forms low viscous emulsions, requires thickeners. Coemulsifier.
Lamecreme; About 65 °C	Glyceryl Stearate, Glyceryl Stearate Citrate	o/w 7–8	4–5%	Emulsifier for fatty phase higher fatty phases, salt tolerant.
Lanolin 38–44 °C	Lanolin Anhydrid	w/o 4	3–8%	Emulsifier for skin care products: ointments, baby care and skin care creams for stressed skin.
Lysolecithin E 60; Cold to use	Lysolecithin	o/w 9	3–4% 0.5–1%	Emulsifier requires thickener. As coemulsifier.

Montanov™ 68; About 75 °C	Cetearyl Alcohol and Cetearyl Glucoside	o/w 9	5%	Emulsifier for >25% oil phase, further thickeners are not required.
Montanov L; About 75 °C	C_{14-22} Alcohol (and) C_{12-20} Alkyl Glucoside	o/w 10	3–4% 2–3%	Solo emulsifier. In the presence of 0.2% xanthan, further thickeners are not required.
Polyglyceryl-3-dicitrate/stearate; About 55 °C (Tego® Care PSC 3)	Polyglyceryl-3-dicitrate/stearate	o/w 11	2–4%	High salt tolerance. Suitable in the presence of organic acids.
Olivem® 900; About 70 °C	Sorbitan Olivate Wax	w/o 4.7	7%	Emulsifier for >50% fatty phase.
Tego Care PS; About 60–65 °C	Methyl Glucose Sesquistearate	o/w 12	2–4%	Emulsifier for 20–40% oil phase, requires 2–7% thickener, such as cetylalkohol, cetylpalmitat, glycerolmonostearat, or cetearylakohol.
Tego Care CG 90; About 80 °C	Cetearyl Glucoside	o/w 11	1–1.5%	Emulsifier for 20–35% fatty phase requires 3–5% thickener as described and 0.1–0.2% xanthan.
Tegin® Pellets; About 65 °C	Glyceryl Stearate SE	o/w 12	6–8%	Emulsifier with an optimal pH range of 7–8; addition of thickener meaningful.
TegoSMS; About 55 °C	Sorbitan Stearate	w/o 5	4–7%	30–45% fat phase; coemulsifier for o/w emulsions.
Xyliance®; About 70 °C	Cetearyl Wheat Straw Glycosides (and) Cetearyl Alcohol	o/w 8	3–5%	Limits: salts <0.4%, urea <5%, 0.2% xanthan recommended, further thickeners are not required.

Source: Data from Käser 2016 [15].

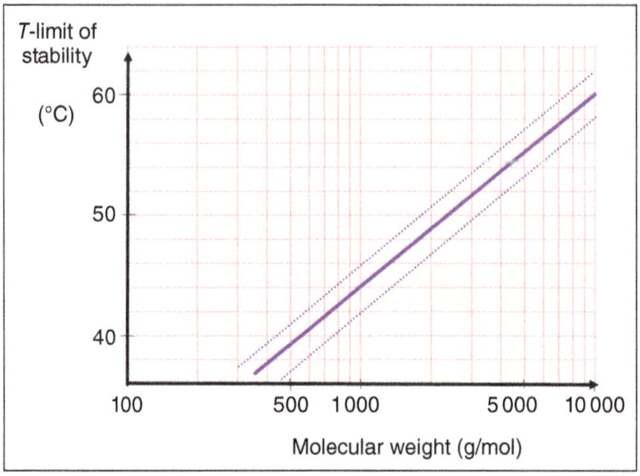

Figure 5.7 Stability of w/o emulsions as a function of the molecular weight of the emulsifier (according to values from a prospectus of Degussa, now Evonik; without the influence of thickeners).

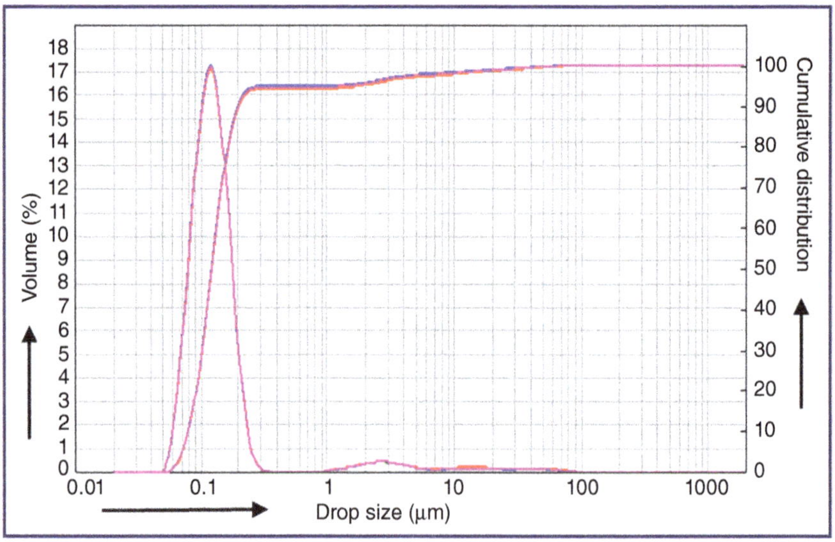

Figure 5.8 Stable cream containing nanodroplets (50–300 nm) after three runs through a high-pressure homogenizer and triple measurement after dilution by laser diffraction/equipment: Malvern.

of color, scent, and consistency (such as immediately after preparation) are the guidelines of the product design. A number of steps, regarding the recipe, the manufacturing and filling conditions, and the packaging material (Figure 5.9), optimize the lifetime of the product.

For oxidation-sensitive ingredients such as vitamins and natural oils, a protection is possible by introducing chemical protective groups and by addition of

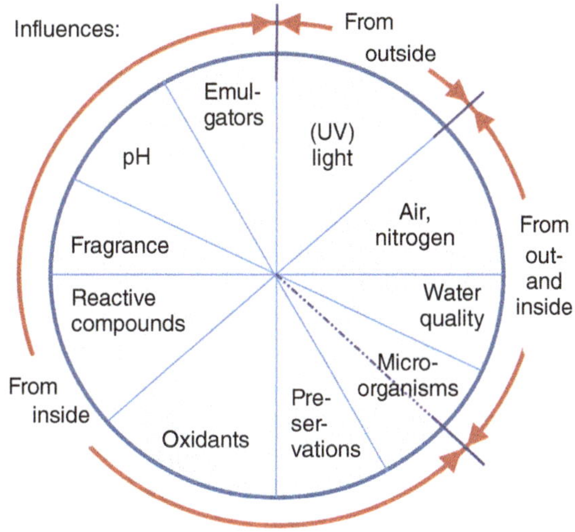

Figure 5.9 Influences on the stability of skin care products.

antioxidants. Vitamin A palmitate and vitamin C phosphate are examples for the chemical modification of the sensitive vitamin A. Tocopherols (especially in the presence of citric acid) protect the precious natural oils with two or three double bonds against rancidity. Weak odors of the basic formulation may, if desired, be absorbed with cyclodextrins or mask with an essential oil. The white color of the emulsions is usually inconspicuous. Large color differences cannot be seen in direct comparison (even if a certain percentage of green hemp oil is processed).

The microbiological stability of emulsions depends on the substantial absence of microorganisms. For the proliferation, the microorganisms necessarily need a preferably continuous water phase and usually the presence of oxygen, a pleasant temperature and nutrients. Therefore, the quality of water is of crucial importance because almost all infections in the product come from the water. Ion exchange columns remove heavy metals and lime-forming ions prior to use. Membrane plants, supported by UV lamps, sterilize the water. Preservatives in the product not only inhibit microorganism growth but also destroy them, depending on quantity. An agent in low concentrations is sufficient, when working in the optimum pH range. Sensitive ingredients also need a stable, constant pH. Various reactions such as hydrolysis, oxidation, esterification, and ester cleavage on the one hand can alter the pH and thus reduce the effectiveness of preservation and on the other hand create a more pleasant environment for microorganisms. Therefore, the use of a buffer system in the range of pH 4.8–5.5 has certain advantages.

When evaluating the stability, the w/o emulsions offer a special feature. Since microorganisms grow only in the aqueous phase, during emulsifying, they can be enclosed in small droplets (about 1 μm) and thus limit the multiplication. Here, a thermal treatment is sufficient or the only very small addition of preservatives. This is an advantage that can be used, for example, for night creams. Other influencing factors include the filling of the creams and the design of dispensers or pots.

During the manufacturing, deviations in the temperature profiles, in the order of recipe ingredients, or in the dissolution of substances as well as an insufficient energy input, can lead to a destabilization of the emulsion. Not only optimal type and amount of emulsifiers, but also reactions that are supported by heavy metals, UV radiation, and/or oxygen; temperature fluctuations; and changes in pH and microbial contaminations over time contribute to the disintegration. In addition, the stability of an emulsion depends on creaming/sedimentation processes as well as the Ostwald ripening [16]. This describes the disappearance of small droplets because the big ones are more stable. However, after some time, aggregations and coalescences stimulate the destabilization.

Careful selection of formulation components, emulsifiers, and preservatives, as well as setting and maintaining a constant pH value provided by a buffer system, guarantees the stability of the cream/lotion. Production and filling run under optimized conditions in clean, disinfected equipment, taking into account the recipe-dependent temperature limits. Recommended is the flush of all containers with nitrogen during the production and bottling. In addition, a packaging (dispenser) is recommended, impermeable to UV light, water vapor, and oxygen, equipped with a removal device without any air intake to hinder product contact with oxygen. Independent of the chemistry and technology, the physically quality (stability) of the emulsion can be seen from the distribution, which should be monomodal. The lower the d_{50} and especially the narrower the droplet distribution, the more stable is the emulsion. The checking of the stability takes place with a centrifuge test (40 °C min, 30 000g, and 5 minutes). All narrow distributed emulsions that pass the test remain stable over a long period (years).

In a few cases, the formulation contains fine solids (powder). The particles settle mainly on the surface of the droplets and assist in the stabilization (Pickering emulsions). These and other types of emulsions, such as the PIT, the micro, and multiple emulsions, are fully described in the literature [1, 17], and are not discussed here. They do not play a major role in skin care.

In order to gain an impression of the emulsion stability, well-known competing creams from the market have been measured with laser diffraction after a few weeks/months after their production. Remarkably, the four investigated samples of different producers show bimodal distributions. The corresponding o/w droplet size distribution shows Figure 5.10 and the associated micrograph (Figure 5.4a) depicts the visible drops. In addition, a measured w/o market product also displays a bimodal distribution (Figure 5.4b). The larger droplets in the emulsions reveal sizes in the range of 20 to about 60 μm (maximum 100 μm); they point to an already incipient destabilization. In other shots, some big bubbles were considerably better visible. Amazingly, several market products start already after a few months with the disintegration. The emulsifiers probably do not stabilize sufficiently and the change to wider distributions starts gradually over many months. The consumer notices the segregation only after the complete disintegration and then at first just pumps water out of the bottle.

When stored before use, creams remain stable at a constant, low temperature (approximately 4–15 °C) for a long time. In the application phase, a temperature of approximately 18–21 °C is optimal. To ensure that no oxygen comes into contact with the product, an airless dispenser should be selected.

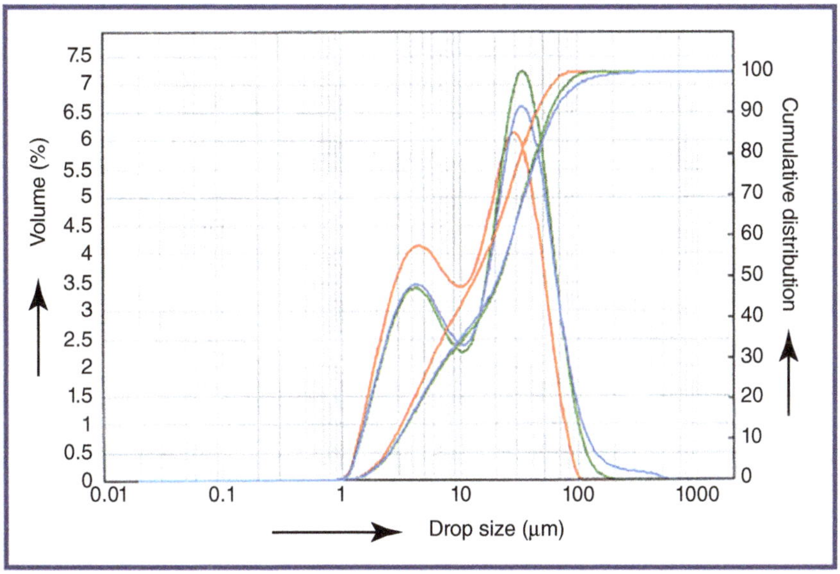

Figure 5.10 Droplet size distributions of a commercially available o/w skin cream, few weeks/months after production, and triple measurement after dilution by laser diffraction/equipment: Malvern.

5.3.4 Adjusting the Cream Consistency

For an easy application, especially for a light distribution on the skin, creams require a viscous, cream- or gel-like, or almost firm consistency, which can be adjusted by different substances ("thickener, viscosity, or consistency regulator"). This task is carried out primarily by oil-soluble substances, which are solid at room temperature. These include in particular fatty alcohols (FA) and their partial esters with glycerol but also a variety of esters with fatty acids (synthetic waxes). As fatty alcohols ($C_{14,16,18}$) and their glycerol esters still have other positive properties in addition to the thickening effect, such as softening of the skin (emollient) and refatting, they are preferably used. Furthermore, they exhibit the properties of a coemulsifier so that the amount of emulsifier can be significantly reduced. In emulsions, the fatty alcohols and esters form liquid-crystalline gel frameworks that bind and store water within the gel framework, i.e. in particular thicken the water phase. This ability significantly increases the viscosity, especially of the o/w emulsions. Furthermore, these groups of substances have moisture-binding properties. A disadvantage of the FA is that they promote the formation of microfoam during production. Without additives, this property leads to a cream that exhibits a white effect on the skin during application, which disappears after a few seconds, when the cream is further distributed. For this reason, the addition of an antifoam agent such as silicone oil or polypropylene glycol is recommended. As a rule, without the use of one or more thickening agents, no sensible cream can be produced. Some useful lipophilic thickeners are shown in Table 5.8. Because of the occlusion effect, facial day creams and body lotions should be formulated completely or largely without waxes.

Table 5.8 Lipophilic thickeners (selected examples, [12, 15]).

Substance	INCI name	Melting point	Characteristics and effects
Fatty alcohols (FA)			
➢ 1-Octadecanol (Lanette 18), C_{18}	Stearyl Alcohol	57 °C	Emulsion-stabilizing (coemulsifier), opacifying, foam-enhancing, refatting, viscosity-consistency-regulating (thickener), softening effect; for all mentioned fatty alcohols
➢ Mix: Octadecanol/ hexadecanol (Lanette® O) $C_{16,18}$	Cetearyl Alcohol	49–56 °C	
➢ 1-Hexadecanol C_{16}	Cetyl Alcohol	49 °C	
➢ 1-Tetradecanol C_{14}	Myristyl Alcohol	38 °C	
➢ 1-Docosanol C_{22}	Behenyl Alcohol	64–67 °C	
Glycerol ester (mono- or diester with fatty acids)			
➢ Glycerolmonostearate, Glycerol ester of stearic acid (Cutina® GMS V)	Glyceryl Stearate	58–59 °C	Viscosity-consistency-regulating (thickener), softens the skin and smoothens, refatting, moisture-binding; for all mentioned glycerol ester
➢ 9-Octadecenoic acid-glycerol ester (1 : 1), oleic acid glycerol esters (Dermofeel® PO)	Glyceryl Oleate	35–38 °C	
Synthetic waxes (ester of alcohols and fatty acid)			
➢ Cetylpalmitat $C_{16/16}$, Ester of Cetylalkohol and Palmitic acid (CutinaCP)	Cetylpalmitat	45–52 °C	Viscosity-consistency-regulating (thickener), moisture-binding, emollient, support for the formation of emulsions, pleasant skin feels, for all waxes
➢ Myristyl myristate ester of Myristyl alcohol and Myristic acid $C_{14/14}$ (Tegosoft® MM)	Myristyl Myristate	37–44 °C	
➢ Isopropylpalmitate and other	Isopropyl Palmitate	10–15 °C cloud point	
Hydrogenated (hardened) oils/fats			
➢ Dermofeel viscolid palm oil free	Hydrogenated Rapeseed Oil	55–60 °C	Emollient, emulsifying, viscosity-adjusting
➢ Castor wax	Hydrogenated Castor Oil	86 °C	

(*continued*)

Table 5.8 (Continued)

Substance	INCI name	Melting point	Characteristics and effects
Natural waxes			
➢ Cocoa butter, deodorized	Theobroma Cacao (Cocoa) Seed Butter	30–35 °C	Concentration 0.2–5%, refatting properties; all consistency generators and weak emulsifiers, normally used as a further thickener in small amounts for o/w emulsions (<0.5%), in w/o emulsions the double; suitable for hand and foot creams; the higher the melting point, the lower the amount added, for all natural waxes
➢ Berry wax	Rhus Verniciflua (Peel) Cera	About 52 °C	
➢ Beeswax	Beeswax, Cera Alba	61–65 °C	
➢ Candelilla	Euphorbia Cerifera (Candelilla) Cera	68–72 °C	
➢ Carnauba wax	Copernicia Cerifera	82–86 °C	
➢ Japan wax	Rhus Succedanea Cera	44–56 °C	
➢ Rice wax	Oryza Sativa Bran Cera	79–85 °C	
➢ Soy wax	Hydrogenated Soy Glycerides	50–55 °C	

With synthetic and natural waxes, higher consistencies can be adjusted up to a solid state. In lower concentrations, they can act as coemulsifiers and stabilize the emulsion. Wax-formulated creams form a film on the skin surface, which is occlusive, protective, and water-repellent. This can be useful in special applications (work safety for the hands; UV protection with water-repellent effect). Low melting points facilitate the incorporation of temperature-sensitive active substances into the emulsion and are preferred as the waxes are introduced into the lipid phase at their high melting temperature. Waxes are needed in the decorative cosmetics for the design and firmness of the products, for example, for lipsticks and mascara as well as for kajal, brow, and lip contour pins, and in hair conditioners for more gloss and grip of hairs.

Water-soluble polymers can assist the consistency regulation of the lipophilic thickener. They must be effective in the desired pH range (usually 4.5–5.5) as well as with the existing concentration of salts and solids. However, the addition also produces in the water-phase gel-like structures, which are responsible for a further viscosity increase and often gives the emulsion the desired soft feel. Some examples of water-soluble synthetic, semisynthetic, or natural thickener can be found in Table 5.9 (more examples in Ref. [12]). Tailor-made,

Table 5.9 Water-soluble thickeners (selection).

Substance	INCI name	Application concentration	pH	Characteristics and effects
Synthetic polymers based on polyacrylic acid or polyurethane				
➢ Carbopol® Ultrez 21	Acrylates/C10–30 Alkyl Acrylate Cross-polymer	0.1–1.0%	5–11	Low compatibility with salts
➢ Carbopol Ultrez 30	Carbomer	0.1–1.0%	4–12	Medium compatibility with salts
➢ Avalure TM Flex-6	Polyurethane-62, Trideceth-6	0.5–1.5%	3–10	Multifunctional polymer, o/w or w/o emulsifier and consistency regulator for high concentrations of electrolytes
Terrestrial plants				
➢ Guar	Guar Gum	0.2%	2–10	Co-thickening, slimy feeling
Microbial polysaccharides				
➢ Xanthan	Xanthan Gum	0.2–0.5%	2–12	High salt tolerance, temperature stable up to over 60 °C
Semisynthetic polymer				
➢ Ethoxylated ester of 1,2-propanediol and oleic acid	PEG-55 Propylene Glycol Oleate	<1.5%	5–10?	High salt tolerance

?: Value is not safe.

water-soluble polymeric thickener for special application, based on polyacrylic acid (>30), provides Lubrizol [18]. A universally applicable thickening agent, which can withstand higher salt concentrations, represents the biotechnologically produced xanthan [19]. Xanthan does not interact with other ingredients of the formulation, is white, and odorless. This water-soluble polymer is easy to process, improves the haptics, and is highly recommended for all cream formulations. Alternatively, Carbopol can be used. Many positive experiences exist through the application of these polymers, but the dissolution takes up to an hour.

5.3.5 Preservations

Cosmetic creams are not sterile, which means that germs can be present in minimal amounts. The manufacturing occurs at a low level of bacterial counts. For this purpose, before operation, a disinfection of the plant is usually carried out with isopropanol. If disinfected water is used, and the dispensers are flushed with nitrogen during filling, as well as the removal from the dispenser takes place without air inflow, and the cream is intended for prompt consumption, then the production can ensure without preservatives. These few germs are usually not a problem and can be controlled by present substances that suppress germ multiplication, i.e. by active substances with germ-reducing effects. (Caution when using solids: organic solids can introduce large amounts of microorganisms into the water phase.)

After opening the cream container, the contents are contaminated by finger contact and air access, when using an ordinary pump dispenser or a glass pot. Without any preservative, the microorganisms grow in the presence of water and dissolved oxygen as well as nutrients contained in each cream. Preferably, they proliferate rapidly at pH values of 6–7 and temperatures between 35 and 40 °C. At lower pH values and temperatures, growth is clearly slowed. Gradually, the microorganisms make the product useless. It spoils, visible by reduced effectiveness, especially when stored in warm, humid environments, such as bathrooms. This process can cause problems such as skin or eye infections. During spoiling, a drop in viscosity occurs. At the same time, supported by oxidation processes, the color begins to change from white to yellow/brown until the emulsion disintegrates. Moreover, the smell changes very slowly but perceptibly. Therefore, the preservation must be carried out by using an effective substance under optimal conditions. First, in small amounts, the preservative prevents a further multiplication of introduced microorganisms. Formulated in slightly higher concentrations, it is able to reduce the number of germs. The required concentration of the preservative lies mainly in the range of 0.1–0.5% and depends particularly on the efficacy of the recipe and the packaging. The European Cosmetics Regulation (see Section 1.4) defines the permissible limit values.

A preservative for cosmetic creams must fulfill a number of requirements. These are

- antimicrobial efficacy already in low concentrations
- broad spectrum of effects
- adequate solubility in water
- high efficacy in the crucial pH range of 4.5–5.5
- Smell and color not striking
- no interactions with other ingredients
- nontoxic or allergenic
- inconspicuous in the formulation
- desirably the substance should develop positive properties on/in the skin.

In creams, some preservatives are preferred such as the para-hydroxybenzoic esters (parabens), sorbic acid, and phenoxyethanol. Phenoxyethanol represents an ether of phenol with ethylene glycol, has only a moderate activity spectrum, especially against Gram-negative bacteria, and is therefore either formulated in relatively high concentrations (about 0.8%; allowed: <1%) or preferred in combination with other agents. The esterified hydroxybenzoic acid with a methyl, ethyl, propyl, or butyl group works effectively in mixtures (0.1–0.8%), preferably in a combination of methyl- with propyl-4-hydroxybenzoate. The optimal effect lies in the pH range 6–8. The propyl ester is more effective against molds, also hydrolytically stable. From the perspective of chemistry, on the one hand, the hydrolysis under basic conditions with a pH shift, and on the other hand the strong oil solubility (=inactivation) must be considered in the formulation.

Parabens rarely cause allergies, are effective over wide pH ranges, and therefore (still) most frequently used in various cosmetics, usually not as a single substance but in combination with further parabens and/or with phenoxyethanol. The addition of phenoxyethanol is recommended for enhancing the activity against Gram-negative bacteria because here phenoxyethanol works well. In the past, various suspected diseases were examined, which have been traced back to parabens. Former analyses of breast tumor tissues of women revealed traces of benzyl esters. They are in suspicion to cause cancer and may not be used longer [20]. Further, butyl ester possibly triggers hormonal changes in children. Methylparaben may promote the skin aging in sunlight (UV rays). According

Table 5.10 Preservatives for cosmetic creams (maximum allowed quantity according to the European Cosmetic Regulation, Annex V [24]; status 2017).

Preservative	INCI name	Effective pH range	Maximum allowed quantity
Benzoic acid and its salts	Benzoic Acid	<4.5	0.5% (as acid)
4-Hydroxybenzoic acid and its salts and esters	4-Hydroxybenzoic Acid		0.4%
Parabene		4–7	0.8% (as acid) for mixtures of esters
Methylparaben	Methyl Paraben		0.4% (as acid) for single ester
Ethylparaben	Ethyl Paraben		0.4% (as acid) for single ester
n-Propylester	Propyl Paraben		0.14% (as acid) for single ester
n-Butylester	Butyl Paraben		0.14% (as acid) for single ester
Isoester and phenyl, benzyl, and pentyl esters are prohibited			
2,4-Hexadienoic (sorbic) acid and its salts, such as potassium sorbate	Sorbic Acid Potassium Sorbate	3.5–5.5	0.6% (as acid)
Propionic acid and its salts	Propionic Acid	3.5–5.5 maximum 3.9	2% (as acid)
Salicylic acid and its salts	Salicylic Acid	4–6	0.5% (as acid); not suitable for children under 3
2-Phenoxyethanol	Phenoxyethanol	3–10	1.0%

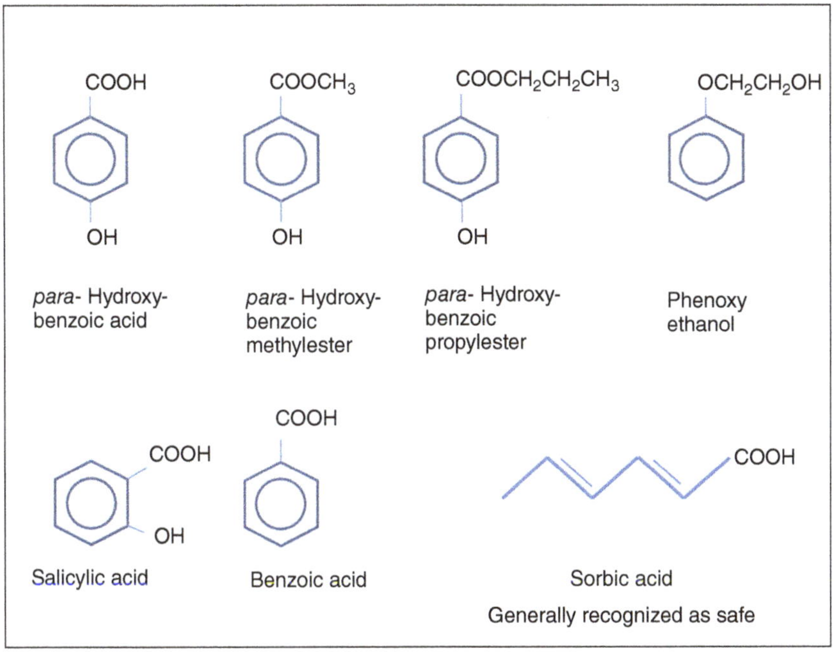

Figure 5.11 Chemical structures of some preservatives.

to the present knowledge, documented in numerous studies, the permissible parabens preserve safe. Although these preservatives are most commonly used, an application should be carefully considered. There are less problematic agents. Save alternatives, however, require emulsifiers and thickeners with medium salt tolerance, i.e. possibly a reformulation of the product.

In 2013, studies by the European Scientific Committee on Consumer Safety (SCCS) have shown that the common parabens are harmless to health below the legally prescribed concentration limits [21–23]. This applies to the methyl and ethyl parabens. For the *n*-propyl and *n*-butyl paraben the permissible maximum amounts were significantly reduced from 0.4% to 0.14% in 2015 (Table 4.10); these two agents can no longer be used in baby products. The use of all other para-hydroxybenzoic acid esters is prohibited.

In the Cosmetics Regulation (Annex 5), the restricted use of 59 preservatives is allowed. Some examples are given in Table 5.10 and Figure 5.11. The effect of the preservation depends on the pH value. As the skin's natural pH is between 4.5 and 5.5, the optimal effect should lie in this pH range. In many experiments, sorbic acid has proven itself. The poor solubility of the acid in water can be improved by adding ethanol and glycerol. This acid is approved for food and by the Food and Drug Administration (FDA) as "generally recognized as safe (GRAS)." An alternative is salicylic acid. Both agents are added in the form of a readily soluble salt and liberated by citric acid. Another alternative represents Rokonsal™ BSB-N, a mixture of benzyl alcohol, glycerin, benzoic, and sorbic acid. This mixture works at pH values <5.2 and may be limited to 0.5% (as acid).

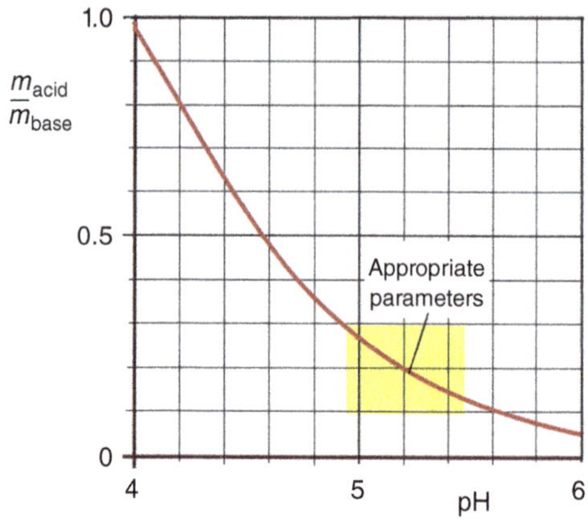

Figure 5.12 Adjustment of the pH in the cream over the mass ratio of citric acid monohydrate to trisodium citrate dihydrate.

In the pH range of skin from 4.5 to 5.5, the physiologically acceptable sorbic acid (hexadienoic) applies as the best preservative. The acid disassembles in the body similar to the fatty acids via the β-oxidation [25]. Above pH 5.5, in equilibrium between the undissociated acid and the salt exists too little free acid so that the effectiveness decreases. The hexadienoic, poorly soluble in oil and in water, dissolves satisfyingly in organic solvents. These are present in cosmetic formulations as solubilizers, and therefore sorbic acid is particularly suitable for cosmetic creams. Sufficient sorbic acid-preserved cosmetic creams pass the microbiological stress test, described in the European Pharmacopoeia [26]. The germ count in the inoculated samples decreases by at least 3 orders of magnitude with the preservative within 7–14 days.

As a suitable salt of the sorbic acid, potassium sorbate is readily soluble in water and is therefore easy to process. The neutral to weakly basic alkali metal salt shows no antimicrobial effect. However, the acid is present in sufficient quantity below a pH of 5.5. The required pH adjustment for the release of the acid from the salt is preferably carried out with citric acid. For preservation, a concentration of 0.05–0.2% is required, which is equivalent to 0.067–0.268% potassium sorbate (selected: 0.25% potassium sorbate and 0.15% anhydrous citric acid or 0.16% of the monohydrate). A combination of potassium sorbate and citric acid in a mass ratio of 2 : 1 is used in numerous new recipe formula (NRF) formulations. The NRF is a collection published by the Federal Association of the German Pharmacists' Associations (ABDA), which includes formulas, standardization, and pharmacy-appropriate manufacturing techniques for medicinal products [27]. There, the corresponding preparation has a pH of about 4.5. The setting of the pH is more meaningful with a buffer. To adjust the desired pH value (as 5.1) exactly, in the discussed example, 0.64% trisodium citrate dihydrate is added in accordance to Figure 5.12. With more or less citrate, another desired pH value can be precisely set.

Sorbic acid is sensitive to oxidation and builds up gradually in the presence of air over time. Three measures will help to suppress the reactions: the exclusion of oxygen, the addition of antioxidants, and stabilizers. Therefore, the cream should run into the container (dispenser) under nitrogen. When using appropriate "airless" dispensers, there is no gas atmosphere above the cream. No growth of microorganisms happens in these buffered, pH-stable creams also in a long-time application. The water-soluble potassium sorbate in combination with the buffer system citric acid/citrate preserves well at the desired pH in the range of 4.5–5.5 [28]. The released sorbic acid is stabilized by the buffer system, especially by the citrate. The buffer provides not only for a constant pH of the cream but also for the correct pH on the skin for a long time, which is important for the effectiveness and durability of the ingredients.

There are a number of natural preservatives such as farnesol, phenylethyl alcohol, propolis, tea tree oil, thymol, and cinnamaldehyde. All substances are out of the question because of their restrictive properties for creams.

5.3.6 Antioxidants, Complexing Agents, and Buffer Substances

Some cosmetic ingredients are easily oxidized by atmospheric oxygen, which are often catalyzed by heavy metal ions. Oxidation develops a lot of heat, hence various reactions can occur. As by-product formed peroxides are characterized by a high reactivity, this can lead to further reactions (autoxidation). The oxidized substances and resulting cleavage products often show an unpleasant smell and partly skin incompatibilities. In particular, the valuable polyunsaturated oils tend to get a rancid odor. To avoid oxidation, the following points should be observed:

- Use of fresh, peroxide-free raw materials (oils, emulsifiers, thickeners, etc.)
- Exclusion of oxygen during storage of the raw materials, production, and filling
- Storage of the products in cool, dry, and dark surroundings without oxygen contact
- Complexation or, the better solution, exclusion of heavy metal ions that catalyze oxidation reactions
- Exclusion of UV rays (sun, light), which can cause peroxide formation
- Ensure adequate preservation.

Cosmetic creams require one or more lipophilic antioxidants because of the mostly lipophilic, oxidation-sensitive ingredients. To prevent oxidation reactions in the cream, antioxidants in quantities of 0.1–0.2% [29] must be added, preferably in combination with complexing agents. Table 5.11 lists the selected substances commonly used. Also, some vitamins belong to the antioxidants (vitamins), described in Section 6.4. Chemical structures can be found in Figures 5.13 and 6.7. Tocopherol, the compound alone or in admixture with vitamin C, is considered as particularly suitable (see Section 6.4 for details). After esterification, vitamins such as vitamins A and C show better shelf life in the product. α-Lipoic acid is a radical scavenger and powerful antioxidant that can regenerate body-consumed antioxidants such as vitamin C, vitamin E, coenzyme Q10, or glutathione. In the case of strongly buffered emulsions and in the presence of complexing agents and in absence of oxygen, as described herein, however, durability should not be a problem. The absence of dyes and perfume also contributes to stability because its ingredients can trigger numerous reactions.

Table 5.11 Free-radical-inhibiting antioxidants for cosmetic creams [12, 29].

No.	Chemical (INCI) name	Chemical formula
1	Vitamin C {L-(+)-Ascorbic Acid}	
	➢ Ascorbic Acid	$C_6H_8O_6$
	➢ Ascorbyl Palmitate	$C_{22}H_{38}O_7$
	➢ Magnesium Ascorbate	$C_6H_6MgO_6$
	➢ Sodium Ascorbate	$C_6H_7NaO_6$
2	Butylated Hydroxyanisole (BHA) is a mixture of 2-*tert*-butyl-4-hydroxyanisole and 3-*tert*-butyl-4-hydroxyanisole	$C_{11}H_{16}O_2$
3	Butylated Hydroxytoluene (BHT)	$C_{15}H_{24}O$
4	*t*-Butylhydroquinone	$C_{10}H_{14}O_2$
5	Cysteine	$C_3H_7NO_2S$
6	Esters of gallic acid	
	➢ Propyl Gallate	$C_{10}H_{12}O_5$
	➢ Octyl Gallate	$C_{15}H_{22}O_5$
	➢ Dodecyl Gallate	$C_{19}H_{30}O_5$
7	Vitamin E	
	➢ α-Tocopherol	$C_{28}H_{48}O_2$
	➢ Tocopheryl Acetate	$C_{31}H_{52}O_3$
	➢ Tocopheryl Linoleate	$C_{46}H_{78}O_3$
	➢ Tocopheryl Palmitate	$C_{44}H_{78}O_3$
8	Coenzymes	
	➢ α-Lipoic Acid	$C_8H_{14}O_2S_2$
	➢ Ubiquinol (part of Q10)	$C_{59}H_{92}O_4$
9	Hydrocarbons	
	➢ Squalene	$C_{30}H_{50}$

The antioxidants in creams must meet a number of conditions:

- ➢ good fat solubility
- ➢ thermo- and UV-stable
- ➢ no interaction with other components
- ➢ high efficacy at low concentrations
- ➢ good skin tolerance
- ➢ toxicologically harmless

Complexing agents must be readily soluble in water because the dissolved heavy metal ions are located there. The first thing to mention is trisodium citrate, which is added for pH adjustment. It also acts as an effective complexing agent for heavy metals. Further agents are then superfluous. Otherwise, the salts of the ethylenediaminetetraacetic acid (EDTA) are widespread, readily soluble in water, and proved as successful complexing agents. Edetic acid itself is insoluble in water and therefore less suitable. However, EDTA can be used in the form of

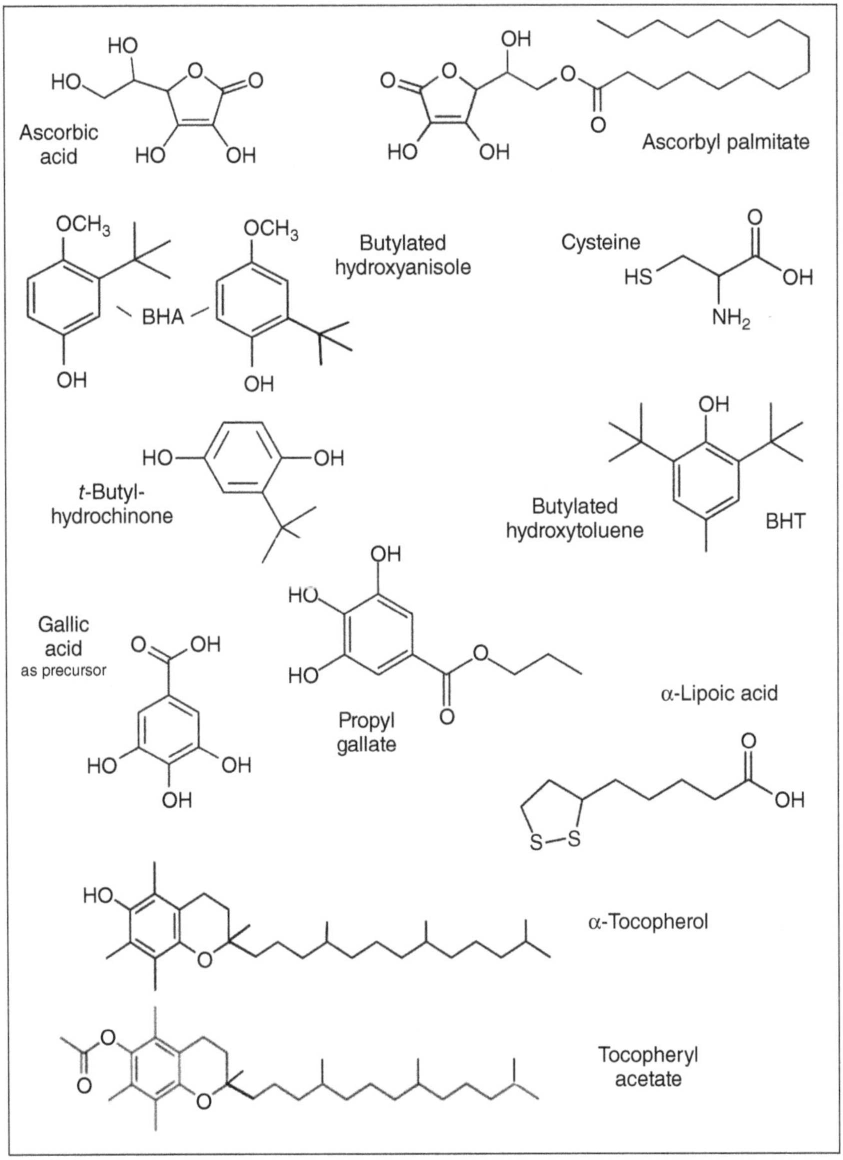

Figure 5.13 Chemical structures of antioxidants.

several well water-soluble salts, notably as di-, tri-, and tetrasodium EDTA and calcium disodium EDTA (Figure 5.14). The substances show their usefulness by the fixing of polyvalent metal ions such as Ca^{2+}, Cu^{2+}, Ni^{2+}, Fe^{3+}, and Co^{2+}, executed with six bindings. The EDTA/metal ion complex remains in solution but shows practically no further reactivity. The concentration of free metal ions is so low, according to the complexation constant, such that no catalysis of the reactions takes place. The hydrophilic complexes cannot penetrate the skin, just as little as sodium chloride it can while bathing in the sea.

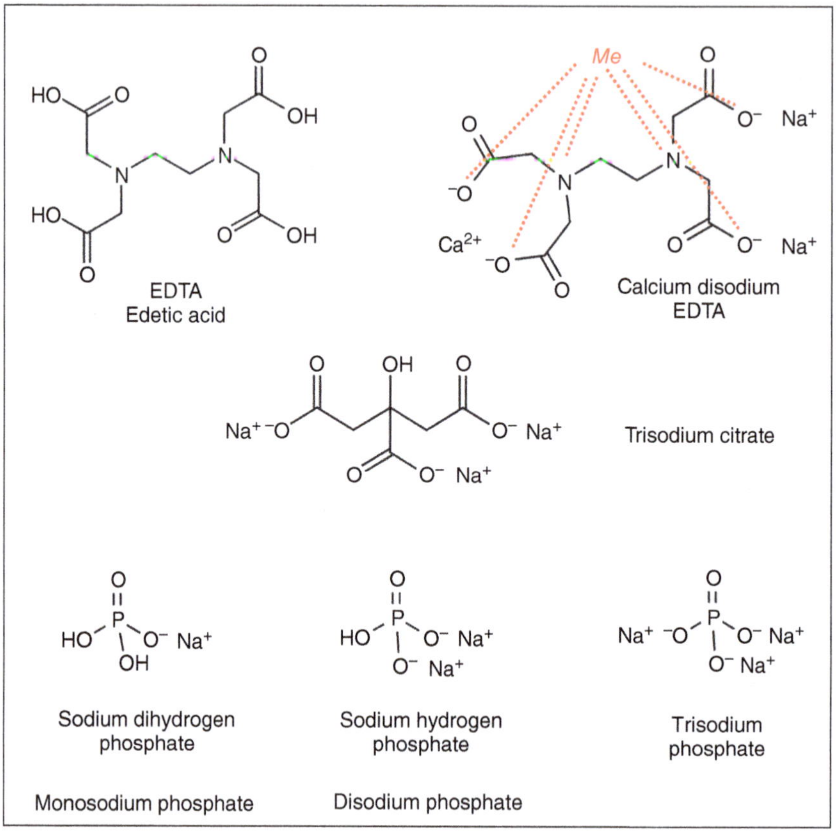

Figure 5.14 Common complexing agents for creams.

Sodium phosphates including monosodium phosphate, disodium phosphate, and trisodium phosphate are approved as antioxidant agents (food additives) in the European Union. The United States FDA lists sodium phosphates as "GRAS." Sodium phosphates can react and form families of condensed anions including di-, tri-, tetra-, and polyphosphates. Most of these salts are known in both anhydrous (water-free) and hydrated forms. The hydrates are more common than the anhydrous forms. In particular, mixtures of sodium phosphates are not only antioxidants but also effective complexing and buffering agents (Figure 5.14).

The effectiveness of the ingredients and the skin compatibility depend on the pH because the pH affects the effect of the substances contained. Changes in the pH value, usually to higher values, can have negative effects on the preservation, the solubility of some substances (possibly precipitations), and their stability. As the pH increases, for example, ester cleavages or similar reactions may occur. Advantageously, the desired pH value can be adjusted in wide areas by the addition of small amounts of acid or base. A constant pH over several years ensures reliable buffer systems, and not only during preparation and a short storage. The system lactic acid/sodium lactate is generally used for cosmetic creams in the pH range of 4–5.2. The substances are also known as moisturizers. The system citric acid/sodium citrate has a stronger buffering action, as described in Section 5.3.5. Both systems require active substances that are stable and effective in this pH range of the skin, which is usually fulfilled for cream ingredients.

Two substances mentioned here act as buffers because of their molecular structure, but not comparable to the acid/base systems. These are the amino acid "glycine" and the salts of edetic acid (see antioxidants), both in actions around the pH of 5. The amino acid "glycine" is also known as a good moisturizing agent. In the literature, a number of other, single substances are counted as buffers [12].

5.4 Additives for Color and Fragrance

In the view of some people, the addition of color to a cosmetic cream makes it more attractive and enhances consumer acceptance. Tinting helps to hide discoloration of colored ingredients. However, the color of active ingredients, such as green for hemp oil and red for carrot and sea buckthorn oil, is barely visible in emulsions, they remain white. Therefore, there is no scientific reason to dye creams. Furthermore, the addition of dyes and perfumes to skin care creams significantly increases the risk of an allergic reaction on the skin, especially for sensitive skin types and for babies as well as for weakened, sick skin.

A total of 153 dyes are listed in the Cosmetics Regulation in Annex IV and can be used under the specified conditions. For identification, they are assigned an order number, known as the color index (CI). On the label of the cream package, the CI number must be indicated in the ingredient list according to INCI.

Perfume plays an important role in cosmetics. Therefore, the fragrances are described in a separate chapter (Chapter 9). The importance arises from the fact that many consumers do not buy the cream after the effect, but after the pleasant odor. As the perfumes are made of many complex fragrances, reactions with other ingredients and changes in the cream are difficult to assess during the formulation. Long-term storage and storage under elevated temperatures increase the possibility of undesired reactions and thereby the allergy risk for sensitive persons. The question arises whether it would not be better to cover the intrinsic odor of a cream by a well-tolerated, natural fragrance (e.g. lavender or jasmine oil). Suitable oils also have a positive effect on the skin. It may be a better alternative for the cream perfume and acceptable for persons who regularly use liquid perfume solutions with the own fragrance.

5.5 Aids Such as Liposomes for the Introduction of Substances into the Skin

With the help of vesicular systems, such as liposomes and ethosomes, the skin penetration of many cosmetic chemicals increases. Liposomes spontaneously arise from lecithin powder by dispersing in water with a high shear agitator. They form ordered, liquid, crystalline-like hollow spheres structures. The walls consist of a lipid bilayer (Figure 5.15). In the cream, the amphiphilic liposomes

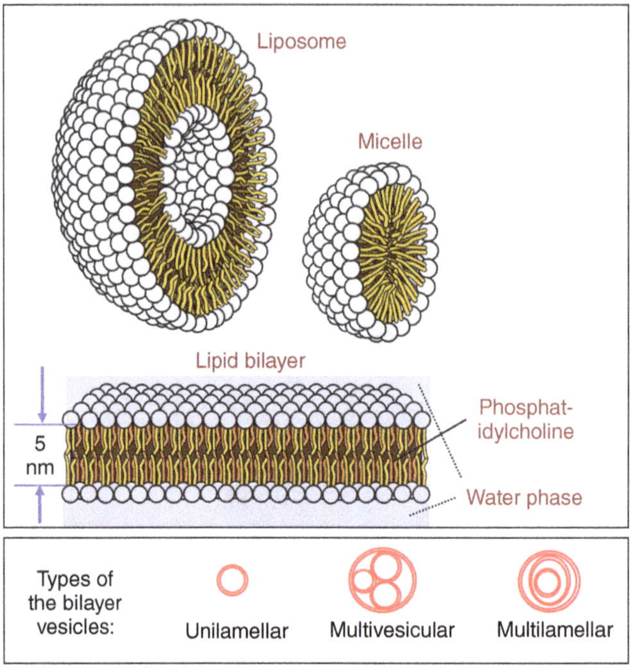

Figure 5.15 Liposome and the double-layer formation. Source: Upper part from the German Wikipedia, in the public domain.

can be used to introduce preferably small amounts of water-soluble active ingredients into the skin. This makes sense in medicine and cosmetics for highly effective, expensive substances, as for some vitamins and synthetic peptides in antiaging applications. Liposomes and other, similarly structured nanoparticles carry different names, which can be recognized by the endings -somes, -spheres, -particles, or -pearls.

Loaded liposomes achieve the following effects on and in the skin:

- transport of hydrophilic active substances into the epidermis
- improved hydration of the horny layer, smoothing of the skin
- drug depot with gradual release
- direct effect at the desired location
- barrier function is improved.

Lecithin usually consists of phosphatidylcholine. This molecule has long lipophilic chains and a hydrophilic part and is therefore a lipid-soluble emulsifier. In addition, this emulsifier is amphiphilic, i.e. the molecule contains a positive and a negative charge. The phospholipids consist of fatty acids, glycerol, phosphoric acid, and choline (Figure 5.16). In addition, there can be shares of phosphatidylethanolamine, phosphatidylinositol, and phosphatidic acid. Lecithin is mainly derived from natural and genetically modified soybeans and from eggs through several chemical purification and isolation steps. For pharmaceutics and cosmetics, lecithin is preferably obtained from the sunflowers

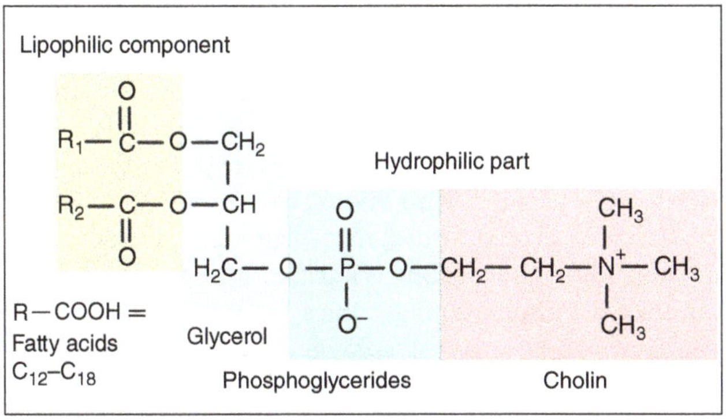

Figure 5.16 Phosphatidylcholine with fatty acid, such as palmitic and oleic acid.

(Non-GMO [genetically modified organism]). Lecithin used in cosmetics is offered as powdered or granulated "pure lecithin," which is prepared by removing the oil and free fatty acids from the native lecithin. The easy-to-dosed solids have a high phospholipid concentration and improved o/w emulsifying properties.

Lecithins not only can transport hydrophilic ingredients in the interior and lipophilic agents in the bilayer but also have other positive effects such as the refatting of the skin. They prevent the skin from drying and help to regulate the pH of the skin as well as support the natural protective coat against aggressive environmental influences. Its high content of linolenic and linolenic acid helps with skin diseases.

In the preparation of the liposomes, lipophilic ingredients dissolve in the shell, and the water-soluble substances are found in the interior and proportionally in the outer water phase. To the mechanism of the penetration into the skin was suggested that the lipophilic liposomes pass intact through the lipid-rich outer layer of the skin to the dermis, where they become localized [29]. Probably, the follicular pathway contributes to the liposomal delivery of drugs or of ingredients into the deeper dermal layers. For the penetration into the epidermis, their size must be less than 200 nm; this means that only the small unilamellar vesicles (SUV) with sizes from 25 to 100 nm are suitable. Large unilamellar vesicles (LUV) exhibit diameters from 100 to 400 nm, whereas multilamellar vesicles (MLV) and multivesicular vesicles (MVV) lie in the range of 100–1000 nm and cannot penetrate.

For the production of the small vesicles, the use of a high-pressure emulsifier with three to five passages is recommended. In this process, at a working pressure of 700 bars and more, the liposomes obtain diameters less than 100 nm. A further process relates the preparation by extrusion. The lipid gel, formed by adding water to lecithin powder and then pre-swollen, is hydrated with vigorous stirring and gradually converted into liposomes. Alternatively, the unilamellar liposomes arise by injection of a lecithin solution in ethanol into the water phase. This is followed by extruding. In the extruder, the gel is pressed through 0.4, 0.2 μm pores and then 20 times through 0.1 μm pores, until the desired diameters are reached. Another method for the laboratory and pilot plant is the 5–15 minute ultrasound treatment of the gel.

Liposomes can be prepared relatively easily from pure lecithin powder. Variants require the use of other surfactants (transfersome), which increase the permeability of the enclosed substances. On the other hand, it is possible to buy already filled liposomes of unknown sizes. The size distribution must be requested by the manufacturer or measured in its own facility. They are available in the market, for example, loaded with vitamin C, curcuma, glutathione, coenzyme Q10, vitamin B12, vitamin D3 and K2, astragalus, *Ginkgo biloba*, nicotinamide, hyaluronic acid, and astragalosides (ActiNovo, [30]). Glutathione is a tripeptide consisting of the three amino acids: glutamic acid, cysteine, and glycine.

When the lecithin is replaced by natural sphingolipids (ceramides), the resulting sphingosomes correspond in composition and action to the liposomes. Liposomes, whose bilayer walls consist of synthetic nonionic surfactants, are called niosomes. Suitable surfactants are alkyl polyglyceryl ethers, macrogol alkyl ethers, and macrogol fatty acid esters [12]. Solid lipid nanoparticles are relatively new [31]. They consist of a solid core, which is surrounded by a surfactant or polymer shell. Embedded in the solid core, the active ingredients are advantageously protected against oxidation. The nanoparticles penetrate well so that the skin does not look greasy. As with the liposomes, a controlled release of active ingredients can also take place, which play an important role particularly for therapeutic agents.

Ethosomes are phospholipid-based elastic nanovesicles containing a high content of ethanol (20–45%). Ethanol is known as an efficient permeation enhancer. The addition of ethanol to the vesicular systems allows to prepare elastic nanovesicles for penetrating small openings. In the delivery of therapeutics through the skin, ethosomes are of great importance. These therapeutics include, on the one hand, highly lipophilic molecules such as testosterone, cannabinoids, and ibuprofen and, on the other hand, hydrophilic drugs such as clindamycin phosphate and buspirone hydrochloride [32]. This technique can also be used for cosmetic ingredients. Studies describe the transdermal and intradermal delivery of peptides, steroids, antibiotics, prostaglandins, antivirals, and anti-pyretics. Ethosomal systems efficiently deliver more substances to the skin, also in the terms of quantity and depth than either conventional liposomes or hydroalcoholic solutions [33].

5.6 Learnings

- ✓ A cosmetic cream consists of excipients and active ingredients, possibly also of additives for color and fragrance.
- ✓ Excipients are the emulsifier/coemulsifier, the preservation system, thickener for both phases, antioxidants, and complexing agents.
- ✓ Emulsifiers and thickeners are mostly based on vegetable oils.
- ✓ The abbreviation HLB (Griffin) stands for the hydrophilic/lipophilic balance of surfactants with values between 1 and 20. The HLB value of the emulsifier determines which type of emulsion is produced (Bancroft rule).

- ✓ Emulsifiers with HLB values >10 dissolve in water and form o/w emulsions, with HLB values <8 they dissolve in oil and form w/o emulsions.
- ✓ A preferred preservation system consists of the water-soluble potassium sorbate as a precursor for sorbic acid and a buffer of citric acid and citrate. The citric liberates the sorbic acid from the salt.
- ✓ The buffer sets the pH to skin's pH and stabilizes the sorbic acid as well as the total cream.
- ✓ Preferred thickeners are the gel forming $C_{16,18}$-fatty alcohols, dissolved in the oil phase, and the water-soluble natural polymer xanthan. Waxes form impermeable films on the skin and thus are restricted in their application.
- ✓ Natural vitamin E is suitable as antioxidants, especially in the presence of citrate and vitamin C.
- ✓ Citrate stabilizes sorbic acid and complexes heavy metal ions, further substances are not needed.
- ✓ With the help of liposomes, water-soluble ingredients can penetrate the skin.
- ✓ For a long lifetime of the cream, the absence of oxygen and microorganisms are necessary in all phases.

References

1 Umbach, W. (ed.) (2004). *Kosmetik und Hygiene*, 3e. Weinheim: Wiley-VCH.

2 Regulation (EC) No 1223/2009 of the European Parliament and of the Council of 30 November 2009 on cosmetic products. http://eur-lex.europa.eu/legal-content/EN/TXT/PDF/?uri=CELEX:32009R1223&from=EN (assessed 26 November 2018).

3 Schnuch, A., Uter, W., Lessmann, H. et al. (2008). *Allergo J.* 17: 611–624.

4 Schnuch, A., Uter, W., Geier, J. et al. (2005). *Allergo J.* 14: 618–629.

5 Rähse, W. (2014). Chapter 13: Industrial product design of solids and liquids. In: *Design of Skin Care Products* (W. Rähse), 319–369. Weinheim: Wiley-VCH.

6 Muschioli, G. and Bunjes, H. (2007). *Multiple Emulsionen: Herstellung und Eigenschaften*. Behr's Verlag.

7 Schuchmann, H.P. (2007). Emulsification techniques for the formulation of emulsions and suspensions. In: *Product Design and Engineering*, vol. 1 (eds. U. Bröckel, W. Meier and G. Wagner), 63–93. Weinheim: Wiley-VCH.

8 Mollet, H. and Grubenmann, A. (1999). *Formulierungstechnik*. Weinheim: Wiley-VCH.

9 Griffin, W.C. (1949). Classification of surface active agents by HLB. *J. Soc. Cosmet. Chem.* 1: 311–326.

10 Mollet, H. and Grubenmann, A. (2008). *Formulation Technology: Emulsions, Suspensions, Solid Forms*. Weinheim: Wiley-VCH.

11 Köhler, K. and Schuchmann, H.P. (eds.) (2012). *Emulgiertechnik: Grundlagen, Verfahren und Anwendungen*, 3e. Behr's Verlag.

12 Ellsässer, S. (2008). *Körperpflegekunde und Kosmetik*, 2e. Berlin: Springer-Verlag.

13 Fruijtier-Pölloth, C. (2005). Safety assessment on polyethylene glycols (PEGs) and their derivatives as used in cosmetic products. *Toxicology* 214: 1–38, Elsevier Ireland Ltd. https://www.ncbi.nlm.nih.gov/pubmed/16011869 (accessed 23 September 2017).
14 BASF, Care Creations, Products, Configure the filter and instantly find the Care Creations® products you are interested in, https://www.carecreations.basf.com/product-formulations/products (accessed 7 May 2019).
15 Käser, H. (2016). *Naturkosmetische Rohstoffe, Wirkung, Verarbeitung, kosmetischer Einsatz*, 5e. Linz: Verlag Freya.
16 Ostwald, W. (1900). *Z. Phys. Chem.* 34: 495.
17 von Rybinski, W. (2005). Chapter 15: Herstellen von Emulsionen nach der Phaseninversions-Methode. In: *Emulgiertechnik* (ed. H. Schubert), 469–485. Hamburg: Behr's Verlag.
18 Lubrizol (2019) Personal care, skin care. https://www.lubrizol.com/Personal-Care/Markets/Skin-Care (accessed 7 May 2019).
19 CP Kelco (2019). Products, Xanthan gum. https://www.cpkelco.com/products/xanthan-gum/ (accessed 7 May 2019).
20 Andersen, K.E., White, I.R., and Goossens, A. (2011). Chapter 31: Allergens from the European baseline series. In: *Contact Dermatitis* (eds. J.D. Johansen, P.J. Frosch and J.-P. Lepoittevin), 560. Berlin: Springer-Verlag.
21 BFR, Bundesinstitut für Risikobewertung (2011). Verwendung von Parabenen in kosmetischen Mitteln, Stellungnahme Nr. 009/2011 des BfR vom 28. Januar 2011. http://www.bfr.bund.de/cm/343/verwendung_von_parabenen_in_kosmetischen_mitteln.pdf (accessed 21 September 2017).
22 European Commision, Scientific Committees, Green Facts, Parabens used in cosmetics. http://ec.europa.eu/health/scientific_committees/docs/citizens_parabens_en.pdf (accessed 21 September 2017).
23 Scientific Committee on Consumer Safety (SCCS) (2013) Opinion on Parabens, Updated request for a scientific opinion on propyl- and butylparaben, COLIPA n° P82, http://ec.europa.eu/health/scientific_committees/consumer_safety/docs/sccs_o_132.pdf (accessed 21 September 2017).
24 Regulation (EC) No 1223/2009 of the European Parliament and of the Council of 30 November 2009 on cosmetic products. http://eur-lex.europa.eu/legal-content/EN/TXT/PDF/?uri=CELEX:02009R1223-20170727&qid=1403001930973&from=DE (EU) Nr. 1004/2014 (accessed 21 September 2017).
25 Lück, E. (1995). Chapter 19: Sorbinsäure. In: *Chemische Konservierungsstoffe*, 3e (eds. E. Lück and M. Jager), 158–174. Berlin: Springer-Verlag.
26 Sigg, J. (2005). Chapter 21: Pharmazeutische Emulsionen. In: *Emulgiertechnik* (ed. H. Schubert), 595–637. Hamburg: Behr's Verlag.
27 Deutscher Arzneimittel-Codex, Neues Rezeptur-Formularium, http://dacnrf.pharmazeutische-zeitung.de (accessed 21 September 2017).
28 Rähse, W. (2011). Produktdesign von Cosmeceuticals am Beispiel der Hautcreme. *Chem. Ing. Tech.* 83 (10): 1651–1662.
29 Kirk-Othmer (2013). *Chemical Technology of Cosmetics*. Hoboken, NJ: Wiley.
30 ActiNova, Nano Vitamins. https://www.actinovo.com/shop/en (assessed 26 November 2018).

31 Our website is under construction, follow us for update now, http://www.mani-gmbh.com/ (accessed 7 May 2019).
32 Ethosome from Wikipedia, the free encyclopedia (2018). https://en.wikipedia.org/wiki/Ethosome (assessed 23 November 2018).
33 Verma, P. and Pathak (2010). Therapeutic and cosmeceutical potential of ethosomes: an overview. *J. Adv. Pharm. Technol. Res.* 1 (3): 274–282. https://www.ncbi.nlm.nih.gov/pmc/articles/PMC3255417 (assessed 23 November 2018).

6

Proven Active Ingredients for Various Categories of Skin Creams

6.1 Skin Care

The main tasks of cosmetic creams are cleansing, beautifying, perfuming, protecting, and maintaining a good condition of the skin. The care cosmetics can fulfill their tasks only with regular, meaningful applications. Chemists and pharmacists formulate tailor-made products for the skin with general or special caring effects. An example represents the slowing down of aging processes in the skin, which occur by removal of the protective mantle on the skin, by drying out, and especially by UV irradiation. On the other side, cosmetic creams support the skin protection and accelerate healing processes with well-chosen substances. The aim is to restore a perfectly healthy and well-conditioned skin. This usually means the recovery of a beautiful skin that everyone strives for. When using cosmetics, the desire for beautification comes first.

The natural and the actual condition of the skin depends not only, among other personal things, on genetic factors, gender, type, age, and weight but also on the nutrition, diseases (e.g. neurodermatitis), and smoking as well as other stress factors in everyday life. The behavior of humans, avoidance of strong sun exposure, regular water supply of the body, cautious washing of the body, and a targeted skin care play an important role, particularly in the elderly. The cosmetic formulations are adapted above all to the part of the body to be treated and to the task. The formulations also differ depending on the thickness of the skin (feet, hands, and face) and the condition of the skin (dry, normal, and bold) as on the intended use (skin texture, moisture, sun protection, and antiaging). Caring creams exist for the whole body; for the face; especially the areas under the eyes, as well as for the face including the neck and décolleté, hands, and feet; and finally for the genital area. Moreover, there are a number of specialty products for babies, for dry noses and aching back (bedsores), for soothing the skin after tattooing or after a sun bath, for the chemical hair removal or to stop unwanted hair growth, as well as for supporting the healing process during treatment with therapeutic agents and afterward. In addition, some special products increase personal safety in the workplace (not part of this book).

Cosmetic Creams: Development, Manufacture and Marketing of Effective Skin Care Products,
First Edition. Wilfried Rähse.

6.2 Cream Categories for Skin Care

6.2.1 Cosmetic Creams (Mainstream)

Dermatological creams are developed for the care, health, and healing of skin, and according to the task, they are classified as cosmetics, cosmeceuticals, and over-the-counter (OTC) pharmaceuticals as well as therapeutic agents. In general, there are five basic human needs for the care and protection of healthy skin, which can be covered with a wide selection of appropriate creams (Table 6.1). These needs consist of providing the skin with moisture and antiaging agents to revitalize the (face) skin, smooth rough and dry skin, and protect the skin against UV radiation (Figure 6.1). In addition, many consumers love the fragrance of perfumed creams and the opportunity to scent themselves discreetly in this way. A specifically formulated cream can meet two or three requirements at the same time. Normally, a main task exists, which is done with several ingredients, and which is presented as an advantage of this cream on the label and in the advertisement. Each cream should contain natural oils for lipid supply and moisturizing ingredients. The promised effect of the cream is achieved by the chemical/biological action, number, and amount of these substances, as well as by additional, specifically added active ingredients (e.g. antiaging agents or vitamins).

6.2.2 Natural Cosmetics

In the seventies and eighties of the past century, a consciousness for the environment, environmental problems, and sustainability developed. An

Table 6.1 Six cream main groups for cosmetics.

No	Cream main task	Offered effect	Main application	Further application
1	Moisturizing	Hydrating moisturizer (rejuvenating, soothing, protective, repairing cream)	Face, eyes	Body, legs
2	Vitalizing	Rejuvenating, refreshing appearance (renewal, repair, shine)	Face, eyes, neck	Body
3	Smoothing	Improves the appearance of rough, dry skin	Body	Face, feet, hands, arms, legs
4	Antiaging	Rejuvenating appearance	Face, eyes, neck	Décolleté, hands
5	UV-protecting	Reduces the harmful effects of sun rays (longer stay in the sun possible)	Body	Face, neck, hands
6	Babies protection	Care under the diaper (protection against environmental influences)	Body	

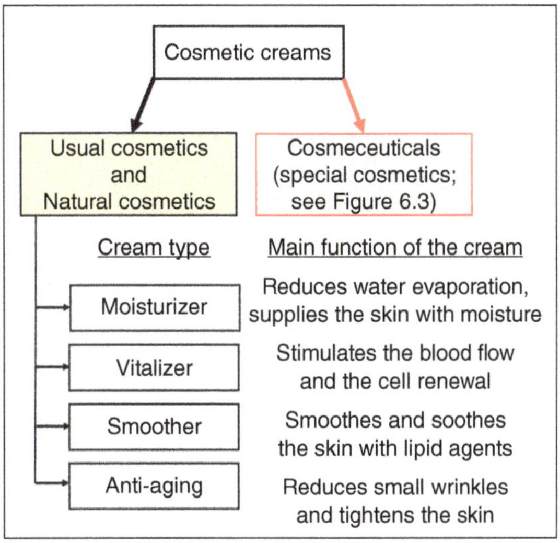

Figure 6.1 Four development directions for skin care creams with high sales.

environmentally conscious customer group wanted to buy "natural," "purely biological" products. Some manufacturers of cosmetic products adapted the trend and offered "bio products," which nowadays are called "natural cosmetics." This term is advertised. For the consumers, "nature" is synonymous with plants. They also think of herbal (homoeopathic) substances, in particular plant extracts. Synthetic substances, preservatives, emulsifiers, dyes, and perfumes are rejected. As some substances, which do not originate from nature, are necessary for the production, the pure doctrine could not fulfill the claims. The conditions that natural cosmetic products have to meet were adapted accordingly.

The organic cosmetics rely on natural ingredients, and in the startup phase, organic cosmetics differed from the normal cosmetics by prohibiting some substances or groups of substances. This includes

- substances based on petroleum (mineral oils),
- pure synthetic ingredients,
- ethoxylated emulsifiers,
- silicones,
- parabens and various other preservatives,
- perfumes with synthetic ingredients,
- raw materials from dead animals, and
- animal experiments.

The introduction of natural cosmetics caused a stronger focus of the entire cosmetic industry on natural ingredients and is therefore to be assessed positively. Therefore, in the past 40 years, the "usual" cosmetics approached the "natural cosmetics," in particular through changes in the usual cosmetics. Nowadays, the differences between reputable manufacturers are often only in some aspects such as the emulsifier. To keep the gap visible, the protagonists of natural cosmetics try to establish new rules and ban more and more substances. This leads to ever-new, stricter regulations and to ever-new "organic" labels (Figure 6.2).

Figure 6.2 Selected ecolabels for natural cosmetics.

The term "natural cosmetics" is not binding defined in Germany or at European level. A proposal for a recommendation on the definition of natural cosmetics was developed under the leadership of the Federal Ministry of Health (Germany). The proposal [1] from 1993 is (shortened):

- Natural cosmetics are products made exclusively from natural products. Natural substances are substances of plant or animal or mineral origin, as well as mixtures and reaction products thereof. Only physical processes, including extraction with water, ethanol, glycerol, or carbonic acid, are authorized for the extraction and further processing. Moreover, enzymatic and microbiological processes are permitted if only naturally occurring enzymes or microorganisms that are not produced by genetic engineering are used.
- In natural cosmetics, it is possible to use those natural fragrances that are named and definitions of the French standard T 75-006 ("Matières premières aromatiques d'origine naturelle Vocabulaire"), as well as those substances listed therein, which have been isolated by physical methods.

- For preservation, the following nature-identical substances may be used (cosmetics regulation Annex VI)

  No. 1 Benzoic acid, its salts, and its ethyl ester
  No. 2 Propionic acid and its salts
  No. 3 Salicylic acid and its salts
  No. 4 Sorbic acid and its salts
  No. 12 4-Hydroxybenzoic acid and its salts and esters (methyl, ethyl, propyl, and butyl esters)
  No. 14 Formic acid
  No. 29 2-Phenoxyethanol
  No. 34 Benzyl alcohol
  In the case of natural cosmetics containing one of these preservatives, the indication "preserved with ...." must be clearly indicated in the immediate vicinity of the indication "natural cosmetics," with the name of the preservative.
- For the production of natural cosmetics, emulsifiers can be used, which are obtained by hydrolysis, esterification, or transesterification from the following natural products:
  fats and oils; waxes; lecithin; lanolin; mono-, oligo-, and polysaccharides; proteins; and lipoproteins.

On this basis, the working group "Nature Cosmetics" of the Federal Association of German Industry and Trade (BDIH) developed a more stringent guideline in 2001. After the certification, it awards the quality label "certified natural cosmetics." The certified products must fulfill the following requirements [2]:

1. Vegetable raw materials should come from controlled organic cultivation (kbA), taking into account availability and quality, and from certified wild harvesting.
2. No animal experiments for raw materials and products. *(This means, no business in China!)*
3. Animal raw materials must come from living animals (beeswax and sheep wool wax).
4. The use of inorganic salts (such as magnesium sulfate) and mineral raw materials (sodium chloride and titanium dioxide) is generally permitted.
5. Ingredients may be used when obtained from the following substances by hydrolysis, hydrogenation, esterification, transesterification, cleavage, or condensation:
   Fats; oils and waxes; lecithin; lanolin (wool wax); mono-, oligo-, and polysaccharides; proteins; and lipoproteins.
6. Prohibition of synthetic organic dyes, synthetic perfumes, ethoxylated substances, silicones, paraffin, and petroleum products.
7. Natural fragrances according to ISO standard 9235 are permitted.
8. Only natural preservation systems and some nature-identical (synthetic) preservatives may be authorized for the granting of microbiological safety: These are benzoic, salicylic, and sorbic acids and their salts, as well as benzyl alcohol.
9. Radioactive irradiation for sterilization is prohibited.

The review of the criteria is carried out by independent institutes. In the case of compliance, the ecolabel is awarded. In addition, the following demands are made:

- Use of an environmentally friendly process.
- Full consumer enlightenment.
- Support for organic farming to preserve the natural basis of life.
- Commitment against genetic engineering.
- Optimally degradable and recyclable packaging.
- Support of the fair-trade.

The certification and marketing of the English version of the ecolabel takes place on a worldwide level through the International Organic and Natural Cosmetics Corporation (IONC GmbH). Many aspects of natural cosmetics are accepted. Some have already influenced the whole cosmetics such as the use of vegetable oils with healing properties. Against a complete takeover of the rules speak three major drawbacks for the customer (not for the manufacturer). From an economic point of view, it must be noted that natural cosmetic products must be considerably more expensive because the raw materials from the controlled cultivation can cost much more. Of course, the support of controlled cultivation and fair-trade makes sense, but the customer has to decide if he wants to pay extra for it. Whether the ingredients have a better or higher quality is an open question (e.g. synthetic alcohol vs. bio-alcohol). In most cases, the qualities are probably comparable, whereby the synthetic substances have fewer by-products and therefore a higher purity. In particular, the organic emulsifiers cost considerably more. Because of the substance restrictions, natural cosmetics in some cases do not reach the quality level of the mainstream cosmetics. An example is the ban on the use of urea because urea is only synthetically accessible. However, this substance is extremely important for the skin. Many products, especially antiaging creams (peptides) and cosmeceuticals such as a cream for decubitus or UV protection, cannot be formulated well in the organic cosmetics. It is further noted that all dermatic medicines break the rules of natural cosmetics. With increasing effectiveness, the quality distance of active cosmetics to natural cosmetics seems to enlarge.

Natural cosmetics have positive aspects, especially the concentration on natural active ingredients. Nevertheless, the campaign of natural cosmetics manufacturers is not always fair because no responsible manufacturer would use toxic or questionable substances; of course, he does not induce any animal experiments or uses substances from dead animals. The quality of a cream depends on the fact that the raw materials are carefully selected and monitored analytically. Furthermore, the cream for the skin must be unproblematic, regardless of the origin of the raw materials. The evidence ensues objectively according to various test methods on at least 12 subjects (see Chapter 12). The rejection of parabens, mineral oils, dyes, and perfumes share many manufacturers. In addition, high melting waxes and the extracts of plants with allergenic potential should be avoided. However, if the customer wants to have perfume in the cream, nothing speaks against an allergy-proven addition. Again, it is a matter of price because almost all perfumes contain higher proportions of synthetic compounds to reduce costs. However, it should be kept in mind that petroleum is also a natural product.

In 2002, five important organizations involved in certifying cosmetics standards came together and agreed to cooperate in the development of a single harmonized standard [3]. They recognized that it is in the best interests of the sales and the consumers to have a single, international standard for certified organic cosmetics. Over the next eight years, they worked out a harmonized standard, coming into force in January 2010. These regulations are called "COSMOS standard," a today's valid standard in Europe, relatively unknown and yet to be accepted in the coming years. The COSMOS standard AISBL (Association Internationale Sans But Lucratif; engl.: international association without lucrative purpose) is a not-for-profit, international association registered in Belgium. It was set up to define and develop the COSMOS standard as an international standard for organic and natural cosmetics. His guiding principles are (quote from [3]):

- Origin and processing of ingredients – describes five categories, their origin requirements, and how they may or may not be treated:
  - Water – must comply with hygienic standards;
  - Minerals and ingredients of mineral origin – must be of natural origin and may be modified with simple chemical reactions;
  - Physically processed agro-ingredients – may be plant, animal, or microbial origin but no GMOs (genetically modified organism), no critically endangered species, only products of (not a part of) animals;
  - Chemically processed agro-ingredients – the same as above, and the chemical treatments must respect the principles of Green Chemistry with the resulting ingredients complying with strict limitations of toxicity and biodegradability;
  - Other ingredients – a very limited list of preservatives and some other ingredients and petrochemical moieties are temporarily allowed and are reviewed on a regular basis, taking into account availability of acceptable alternatives.

Annexes at the end of the standard give further details of the precise ingredients, treatments, chemical reactions, etc., that may be used.

- Composition of total product – including how to calculate the organic content of complex ingredients (e.g. water based and other composite ingredients) and how much organic content is required in products under organic certification (specific percentage limits for physically processed agro-ingredients, chemically processed agro-ingredients, and the total product).
- Storage, manufacturing, and packaging – to ensure adequate cleanliness, hygiene, and traceability throughout all processes and to ensure that packaging respects the environment.
- Environmental management – details the requirements for care of the environment throughout the manufacturing process, and managing, minimizing, and recycling waste.
- Labeling and communication – defines comprehensive requirements for clear product labeling and company advertising, to ensure all necessary information for consumers and no misleading claims.

➢ Inspection, certification, and control – the requirements for all products, their ingredients, and their manufacturing to be certified by a competent body, authorized by independent accreditation. The process is repeated annually to ensure ongoing compliance.

Besides the standard regulations, there are the control manual, the labeling rules, and the technical guide [3]. All companies that comply with the rules can join the standard.

6.2.3 Cosmeceuticals

In addition to the general cream categories, there are specialists who go beyond pure care and help the skin to get healthy. The formulation of a cosmetic cream can be carried out in such a way that it additionally exerts a specific medical benefit. In some cases, the use is in combination with therapeutic creams, for example, with a formulation containing cortisone. Depending on the recipe, cosmetics cure slight dermatological diseases [4]. For such products, between cosmetic care and healing with medicines, dermatologists in the United States created the term "cosmeceuticals." The word, formed from cosmetics and pharmaceuticals, means "active cosmetics" and is no part of the Food & Drug Administration (FDA) classification but well-known among the scientists. The designed skin care products (Figure 6.3) ensure not only beauty but also a healthy condition of the skin with essential ingredients [5].

Cosmeceutical creams or lotions are preferably oil in water (o/w) emulsions. They combine effective cosmetic ingredients with "healing" natural oils and vitamins/provitamins in amounts that provide a rich supply to the skin and show visible success. By using selected ingredients, physiological processes occur more than usual and support the skin's own regeneration. To avoid undesired effects, it is necessary to reduce the excipients to a minimum. This affects not only the amounts of emulsifiers and preservatives but also waxes, synthetic oils, and other synthetic adjuvants. To minimize the risk of allergic reactions, cosmeceuticals should not contain perfumes or dyes. To increase acceptance, some creams contain intensively tested fragrances that do not cause allergic reactions. The possible addition of auxiliary materials and their maximum levels, as well as the identification of substances on the packaging (INCI, International Nomenclature of Cosmetic Ingredients), are subject to the EU Cosmetics regulations (Chapter 1). To meet the demands, cosmeceuticals must contain the present biologically active substances in effective concentrations [6]. The usual cosmetics lack a quality label. If the concentration of the ingredients exceeds 35–40%, such cosmetic creams could bear a seal of quality, for example, a big CC for cosmeceutical cream.

The first group of suitable substances comes from plants/parts, obtained through different processes such as grinding/pressing and extracting with solvents or with the aid of superheated steam. Depending on the origin of applied substances, the mode of action on the skin surface can be described as healing, caring, moisturizing, lipid-supplying, smoothing, refreshing, rejuvenating, revitalizing, promoting circulation, antiseptic, anti-inflammatory, antioxidant, astringent, and UV-protective. The second group includes the vitamins and

Problem	Cream's application area	Main function of the cream
Acne	Face	Support of the benzoyl peroxide-containing therapeutic cream with a nourishing, moisturizing cosmetic cream
After sun	Body	Cools, soothes and regenerates burnt skin
Baby	Body	Nourishes, protects and moisturizes the baby's skin
Bedsore (decubitus)	Back, legs, feet, arms	Prevents and reduces pressure sores
Cellulite	Woman's thighs	Strengthens the tissue, moisturizes and firms the skin; improves the appearance
Depilatory	Armpits, chest, genitals, arms, legs	Removes chemically unwanted hairs
Dry skin sites, eczema	Body	Smoothes and soothes rough skin
Foot care and athlete's foot	Feet	Cares for the feet and reduces fungii
Hands	Hands	Nourishes and protects the hands
Herpes	Face	Reduces the infection
Horny layer	Feet, hands, arms	Removes callus with acids and urea
Body	Body	Nourishes and moisturizes the skin of the body
Self-tanning cream	Face, neck, décolleté, (arms, hands, body)	Tans the skin
Eye area with suncreen	Eye area	Nourishes, moisturizes and protects the skin
Suncreen	Body (face)	Protects against UVA and UVB rays
Water loss (shin)	Legs	Reduces transepidermal water loss of the shins

Figure 6.3 Examples of cosmeceuticals creams (details see Chapter 7).

provitamins. These substances act on and in the skin in many ways. Some work with synergistic effects, especially the antioxidants (vitamins C and E), the stimulators of circulation (vitamin B_3), regulators of sebum production (vitamin B_6), antiaging (vitamin A), and healing ingredients (vitamin B_5). The third group of substances, proteins, and peptides derived from microorganisms or synthesized in analogy with nature provides improvements in skin hydration, firmness, elasticity, and skin protection.

Examples represent products that can reduce small facial wrinkles, improve peripheral circulation, reduce hair growth in unwanted sites, or fight a fungal infection [7]. Cosmeceuticals [8] not only serve for the maintenance of health but also often for the support and supplement of a therapy. These problems require compositions other than those normally used in the cosmetic industry. The active substances work simultaneously in effective concentrations, partially with specific synergistic actions. Among the most important materials are some unsaturated natural oils, which, depending on the amount, slightly intensify the sheen on the skin. If the effect is visible, this might be undesirable for usual, daily cosmetic products but acceptable for active cosmetics. The oils, in combination with moisturizers and vitamins, enable medical actions besides the usual care. The primary target is to maintain or restore a healthy skin. The inclusion of novel peptides [6, 9] and other specialties provides firming effects. A medical application of cosmetics has demonstrated the calming impact on a dry, itchy skin, bringing the skin back to a healthy state. Studies of volunteers and/or the descriptions of many applications document the improvements achieved by cosmeceuticals. Cosmeceuticals have no legally binding rules and are subject to the same laws that apply to the usual cosmetics.

Some of the cosmeceuticals (functional cosmetics) listed here are referred to as "quasi-drugs" in Japan and as OTC drugs (antiperspirants and sunscreens) in the United States. In Europe, the name "cosmeceuticals" applies to cosmetics that have a particular effectiveness, not so far away from therapeutics. The desirable use of the term could be a quality feature, if all cosmetic creams with active ingredient contents above 40% should be referred to as cosmeceuticals. Table 6.2 gives some aspects of the delimitation between general, active, and OTC cosmetics as well as for therapeutic agents.

6.2.4 Medicines for the Skin

Even if they help, cosmeceuticals are cosmetic products and not medications and have to be registered with the CPNP (see Section 1.6). The following marketing requires a qualified security assessment of the formulation, executed in accordance with the EU cosmetic regulations (see Chapter 13). Some brand manufacturers market certain creams only through the pharmacy as OTC products, although these cosmetic products are also subject to the cosmetics regulation. Other OTC products represent therapeutics, which have been freed from prescription because of their safe and problem-free use but must be sold via the pharmacy. They may be authorized by a simplified procedure.

Therapeutics forms the next stage for the treatment of damaged skin. Dermatologists prescribe recipes to treat and heal diseased skin with pharmaceuticals. For these products, however, there are much stricter legal provisions. The Medicinal Product Act (AMG, [11]) contains the definition, requirements, production, licensing, registration, clinical testing, delivery assurance, and quality control aspects. The marketing is subject to approval in accordance with §21 AMG. In Europe, the permit is granted by the competent federal authority or the Commission of the European Communities or the Council of the European Union.

After AMG §1 (abridged), pharmaceuticals are defined as "substances or preparations of substances that are intended for use in the human body and show properties for treating or alleviating or preventing diseases or pathological symptoms ...." The results of clinical studies represent the basis for the

Table 6.2 Main tasks and delimitation of dermatological products in Europe.

	Cosmetics		Medicament
	General and natural cosmetics	Cosmeceuticals and OTC-cosmetics	Dermatics (prescript and free pharmaceuticals)
Legal basis	Cosmetics regulations		Medicinal product act (in Germany: AMG)
Manufacturer	Cosmetic industry		Pharmaceutical industry
Application	Externally on the skin		Exterior and interior
Product design	Creams, liquids, foams		Creams, liquids, and tablets as well as medical syringes
Sale	Everywhere; OTC: available over the counter in pharmacies		Only in the pharmacy, free or on prescription
Tasks	*Basic tasks*: • skin care • cleaning • appearance • odor	*In addition to the basic tasks*: • improve, prevent, and eliminate forms of minimal dermatological diseases; (physiological effects are limited to the skin; no systemic[a] effect detectable)	*Healing*: • heal, alleviate symptoms of diseases and pathological symptoms • fight against harmful pathogens, parasites, and foreign substances; (systemic[a] effects possible)
Competence	*Beauty*: • care • moisture • recovery • scent	*Beauty and health*: • care/prevention/concomitant • therapy • skin rejuvenation/moisture • itching • acne • sun damage • treatment of bad skin	*Healing of skin diseases*: • eczema • dermatitis/psoriasis • seborrheic keratosis • microbial infestation, inflammation • rashes • wounds

(Continued)

Table 6.2 (Continued)

	Cosmetics		Medicament
	General and natural cosmetics	**Cosmeceuticals and OTC-cosmetics**	**Dermatics (prescript and free pharmaceuticals)**
Active ingredients:			
Number	1 (advertised)	>2	mostly 1
Concentrations	Low <30%	Medium to high >40%	Very low, effective-, and age dependent (children, adults)
Additives:	Yes	Usually not in some cases	No
Perfume	Yes		No
Dye	Product dependent yes	No	No
Emulsion type (preferred)	o/w (w/o for night creams)	o/w (w/o for night creams)	w/o (or ointment)
Cream container	• Glass pot • Bottles • Dispenser	• Dispenser • (Glass pot)	Tube

a) Possible effects throughout the body via the blood circulation.
Source: Data from Ref. [10].

approval process of drugs, especially the proof of claimed effects (indications proof). Furthermore, the product must be safe in application. The emulsions or ointments have a healing effect in the case of skin lichen, eczema, inflammations, skin rashes, and nutritional deficiencies.

Dermatics contain the real active ingredient (one, mostly synthesized substance) and a basis formulation for ointments or emulsions (Table 6.3a–d) as well as a preservative. The preservative can be a combination of potassium sorbate and citric acid (0.2; 0.1%). Single-phase systems are predominantly known as an "ointment," which consists in the simplest case of a basis substance (such as Vaseline®) and an active ingredient. Other pharmaceutical creams are formulated as water in oil (w/o) emulsions. A fatty layer on the skin characterizes this type of emulsion after application. To check the effectiveness of therapeutic agents, they usually contain only one medicinal substance, at optimized concentration. A special galenics with suitable excipients guarantees

Table 6.3 Basis formulations for ointments and emulsions according to the German Pharmacopoeia (DAB and DAC) for the incorporation of lipophilic and hydrophilic active ingredients (preservatives must be added to water-containing recipes).

Function	Substance	Proportion (%)
(a) Basics for the wool wax alcohol ointment		
Lipid	Cetylstearyl alcohol	0.5
Lipid	Wool wax alcohol	6.0
Lipid	White vaseline	93.5
(b) Basics for the lanolin ointment/emulsion		
Lipid	Viscous paraffin	15
Lipid/emulsifier	Wool wax alcohol	65
Lipid	Aqua	20
(c) Basic cream DAC (emulsion)		
Emulsifier	Glycerolmonostearate 60	4
Lipid	Cetyl alcohol	6
Lipid	Neutral oil	7.5
Lipid	White vaseline	25.5
Emulsifier	Macrogol-20-glycerolmonostearate	7
Aqua phase	Propylene glycol	10
Aqua	Purified water	40
(d) Nonionic cream DAB (emulsion)		
Emulsifier	Polysorbate 60	5
Lipid	Cetylstearyl alcohol	10
Lipid	White vaseline	25
Aqua phase	Glycerol 85%	10
Aqua	Purified water	50

the pharmaceutical availability [12] of the active ingredient. The pharmacological potency and duration of action depend on the concentration, additives, and the formulation (pharmaceutical chemistry and technology). The active substance exists as a pure or chemically modified natural material, or is synthesized completely, or produced biotechnologically. Except for α-hydroxy- and salicylic acid and urea, the drugs contain completely different agents as formulated in the cosmeceuticals. Examples are the potent cortisones, antibiotics, antifungals, and retinoids.

Many ointments and emulsions are based on the so-called basic formulations. This concept, the use of broadly tested basic formulations, has been successfully applied to cosmetics (see Section 5.2). The one's discussed for therapeutics represent a selection. The composition of the formulations was carried out according to the criteria of compatibility and safety. They must behave absolutely neutral both to the active ingredient and to the skin. Thus, any change in the skin is attributable to the lipophilic or hydrophilic medicinal agent. Its effectiveness can be assessed in this way. The healing ingredients remain not only on the skin surface but also penetrate into the epidermis. Dermatics treat primarily pathological disorders of the skin such as psoriasis, eczema, inflammation, cancer, rashes, and deficiencies, and many others [13]. They are dermatological therapeutics. Adapted galenic compositions support the active material and transport into the skin and release (duration) of the therapeutic substance.

6.3 Moisture in the Skin

Suitable regularly applied cosmetic creams ensure permanent supply of moisture in the skin that leads to a full, firm skin. Such a skin looks healthy and beautiful. This care is the most important task of a cosmetic cream, especially for people over 45. Dry skin is immediately visible and perceived as unpleasant. It is characterized by an excessive water loss. Large quantities of urea and fatty acids are missing in these areas. The symptoms can deepen without treatment. In addition, harmful microorganisms may settle on these sites. Therefore, each cream must be provided with substances that ensure a lasting moisturizing of the skin, by applying once daily. For a deep, lasting moistening, special creams are formulated, the moisturizer. They should contain larger quantities of moisture-binding, natural, or nature-identical substances, such as urea and fatty acids. In this connection, the o/w emulsions have a much better effect than the w/o grades.

6.3.1 Natural Moisturizing Factor

The stratum corneum as the top skin layer serves to separate the human body from the environment. This layer also protects the body from inner moisture loss by regulating the transepidermal water loss (TEWL) and forms a barrier against external chemical, microbiological, and physical influences. The right skin moisture is the basic prerequisite for a well-groomed, supple skin with high elasticity. This applies, in particular, to the stratum corneum (horny layer). This layer is composed of flat dead horn cells (keratin), surrounded by a keratin mantle, as well as a "lipid kit" between the cells. The kit, which consists of 30% ceramides, 30% free fatty acids, and 40% cholesterol derivatives, can be found in the intercellular space and forms a bilayer (see Chapter 4). The hydrophobic lipid kit prevents the passage of water and the water-soluble substances into the skin. It is referred to in its entirety as "skin barrier." Depending on the mechanical stress, the thickness of the horny layer is 10–500 μm, these are 12–200 cell layers (corneocytes). At many body parts, the skin thickness lies between 10 and 20 μm.

In the horny layer, the right quantity of water is 15–20%. The content usually regulates itself. In deeper layers, the skin has a water content of about 40%. Water balances are extremely important and ensured by the skin barrier and the water-storing substances. Mostly located in the corneocytes of the stratum corneum, the group of these substances is referred to as a natural moisturizing factor (NMF). Through diffusion processes, the body supplies the skin from the inside with sufficient moisture through the epidermis to the cornea.

Via the sweat glands, excess water can be transported to the outside through the pores to regulate the body temperature by evaporating water on the skin. The eccrine sweat glands distributed throughout the body can secrete considerable amounts of clear, odorless water (in quantity up to 2–4 l/h) with a pH of 4.5. As a rule, microorganisms do not feel well at this pH and are reduced. The perspiration consists of 99% water, otherwise mainly of electrolytes, such as Na^+, Cl^-, K^+, and furthermore lactate, amino acids, and urea. Sugar and ascorbic acid can also be detected in small amounts. The apocrine sweat glands, which exist only in the hairy body regions of the axillae and genital region as well as in the (hairless) nipples, produce small amounts of a milky secretion, which contains proteins and lipids and is approximately neutral with a pH of 7.2. As microorganisms love this pH value and the temperature of 37 °C, they multiply rapidly and decompose available substances for energy production. The products of decomposition such as butyric acid smell unpleasant.

Substances of the NMF (Table 6.4) are continuously formed by the cleavages of molecules (proteins) during the regeneration of the horny layer by dying cells. The deposited fragments in the corneocytes can bind water directly via hydrogen bonds or in the form of a hydrate shell and to dispense water again, depending on the conditions.

At temperatures below 30 °C, the water of sweat does not matter; the TEWL is determined solely by the water diffusion from the inside (epidermis) to the outside through the horny layer. This passive diffusion is forced by the water vapor pressure gradient between skin surface and environmental air. From the outside, humidity, room temperature, and skin temperature influence the TEWL. Internal factors, such as the thickness of the epidermis, surface structure of the horny layer, and water binding ability also affect the water loss. Owing to extreme external conditions such as low relative humidity, high ambient temperatures, or rapid air movement, furthermore exposure to ultraviolet rays, excessive washing, and many others, the TEWL can clearly increase even with healthy skin. As a result, the water content in the horny layer can drop below 10% and is then visible from the outside by changing the skin surface, which becomes brittle, cracked, rough, i.e. clinically dry. Because of the unfavorable environmental conditions, there

Table 6.4 Composition of natural moisturizing factor.

No	Substance	Amount (%)	Description
1	Amino acids	40	Derived from the hydrolysis of keratin proteins. Because of their small size, amino acids have the ability to penetrate to the deepest horny layers, where they exert their moisturizing action
2	Ions (Na^+, K^+, Ca^{2+}, Mg^{2+}, Cl^-, PO_4^{3-})	18.5	The water inside the corneocytes contains dissolved salts
3	Pyrrolidone carboxylic acid (PCA)	12	The sodium salt of PCA is more hygroscopic than glycerol. Therefore, a PCA cream formulation gives protection against dehydration, especially for bad skin conditions with desquamations
4	Lactate	12	α-Hydroxy acids (AHA) and their salts are highly hygroscopic compounds that boost skin moisturizing, strengthen the skin barrier function, and enhance the horny layer flexibility
5	Sugar, organic acids, peptides	8.5	Moisturizing agents
6	Urea	7	Urea is a major skin hygroscopic agent and comes from the degradation of arginine or from the sebaceous glands. It has mild keratolytic effects, as urea cleaves keratin molecules, thus reducing the thickness of the horny layer in the renewal process
7	Ammonia, uric acid, and other organic acids	1.5	Results of degradation processes
8	Citrate	0.5	The salts of α-hydroxy acids (AHA) are hygroscopic and boost skin moisturizing (see lactate)

Source: Data from Ref. [14].

are probably cracks formed in the lipid kit inside the cornea, in which water is additionally transported to the outside.

The application of a moisturizing cream reduces the TEWL by closing the surface and cracks with polymers such as hyaluronic acid and improves the skin's moisture by penetrating dissolved molecules (NMF). This can be done in different ways. On the one hand, suitable substances can cover the skin surface after application largely air- and water-impermeable (occlusion). This prevents any water from evaporating on the surface of the skin and leads to a moisture accumulation in the horny layer of the skin and consequently to swelling. The increased water supply facilitates the penetration of many active ingredients, well known from pharmaceuticals (drug penetration). On the other hand, small amounts of the NMF and other substances, in particular dissolved urea, can diffuse from the skin surface through the horny layer. The diffusing substances bind water in the stratum corneum, and even in lower layers of the epidermis. In balance, the greater part of these substances is likely to be on the skin surface.

The occlusion can be done with high-molecular substances that remain on the surface. For example, the hyaluronic acid with a molecular weight of 1–1.5 million Dalton (Da) is used in small quantities and forms a very thin film on the skin. The same applies to xanthan that also develops films. Both substances, optimally applied in combination, have further tasks. They thicken the aqueous phase and bind a lot of water, which they slowly release. Hyaluronic acid and its Na salt bind water up to 1000–4000 times of its weight.

6.3.2 Moisturizing Substances

There exist many "moisturizing" substances. Substances that visibly lead to positive effects are mentioned here. Moisturizing ingredients include, on the one hand, substances that diffuse into the horny layer and increase the moisture content for a long time such as substances of the NMF occurring in the cornea

- polyvalent, short-chain alcohols (hydrophilic, nonvolatile), called humectants
- hygroscopic substances
- α-hydroxy acids (AHA).

On the other hand, moisturizing ingredients include film-forming polymers that, because of their highly molecular structure, cannot penetrate the skin but swell and thus create a water reservoir that is sufficient for a few hours, but is not sustainable. Examples are

- proteins
- protein hydrolysates
- hyaluronic acid
- emollients
- natural and synthetic polymers.

Substances that are part of the NMF and often used for moisturizing the skin include individual amino acids, salts, pyrrolidone carboxylic acid (PCA), lactates, sugars, organic acids, urea, and citrates. Many amino acids are relatively small at molecular weight below 300 Da and therefore able to easily diffuse into the horny layer. The penetrated part ensures a long-lasting moisturizing of the skin. **Glycine**, the smallest amino acid, is particularly suitable for this purpose. Other applicable amino acids are alanine, arginine, citrulline, histidine, leucine, lysine, serine, and threonine. Amino acids are chiral compounds, i.e. they exist in two mirror image molecular structures. In addition, the amino acids are mostly present as "inner salt" or zwitterion. The proton of the acidic carboxy group can migrate to the solitary electron pair of the nitrogen atom of the basic amino group (Figure 6.4). These zwitterions are surrounded in aqueous solution by a hydrate envelope as well as salts. Salts dissolved in water do not appear to be able to penetrate far into the stratum corneum at negative diffusion gradients; estimated, a part possibly comes about —three to five cell layers deep.

PCA can be a nature-identical substance synthesized from an amino acid, the glutamic acid. PCA and its Na salt occur in the human skin and protect it by

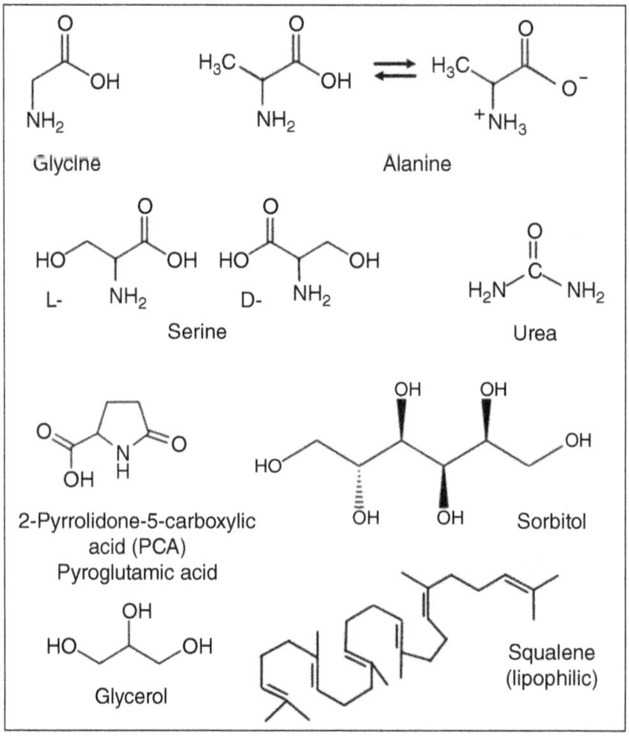

Figure 6.4 Examples for short-chain amino acids and other substances of the natural moisturizing factor (NMF).

excellent water-binding properties before drying out and ensure a high elasticity. The unpleasant skin tension is also significantly reduced. **Sodium lactate** is the sodium salt of **lactic acid** and belongs to the effective hydrating substances, which the skin produces in equilibrium with the lactic acid itself. Hygroscopy is higher than that of glycerol and urea. The weakly acidic pH about 5, which sets itself in the system acid/lactate, inhibits the growth of harmful microorganisms in the skin. As a sugar, **sorbitol** is a good humectant with excellent tolerability. Other less suitable sugars are fructose, glucose, and mannitol. Sorbitol is often used in combination with glycerol (2–5%). **Glycerol** is formed in the skin by the ester hydrolysis of the natural oils. It acts strongly hydrating and barrier-strengthening, in particular in combination with urea. Less important moisturizing substances represent short-chain polyvalent alcohols, such as propylene and butylene glycol as well as macrogols (PEG).

In dermatology, one of the most important NMFs is **urea**. It binds water in the upper layers of the skin, reduces the TEWL, and contributes to the elasticity of the horny layer. Urea acts as an anti-inflammatory, antibacterial substance, and in higher concentrations also as an antipruritic and keratolytic substance. At application on open wounds, this substance – as acids – initiated a short burn. Urea is nontoxic and well tolerated. Essentially, compared to healthy skin, urea is lacking in dry skin by up to 50%, in psoriatic skin up to 60%, and in neurodermatitic up to 70%. These missing amounts must be supplied from outside to the skin, an important task for cosmetic creams. Not only in supporting a therapy but also as a precaution, the active moisturizer should contain 2–5% urea and suitable for little children in the lower concentration.

Squalene is a triterpene and widely common in the animal and plant world. It is synthesized in the human body during the cholesterol biosynthesis. As part of the natural skin barrier, the lipophilic substance penetrates easily, increases the elasticity, and smooths the skin. Furthermore, the TEWL is reduced. Thanks to the antioxidative effect, squalene protects the cell membranes and reduces the damage caused by UV rays in the skin. It slows down degenerative and wrinkling processes. When used, immunostimulatory and healing effects are observed. Squalene helps with atopic dry, irritable, and sensitive skin and is also suitable for baby skin. The technically hydrogenated squalene, called squalane, is resistant to oxidation and used as an ointment base.

The AHA (Figure 6.5) are short-chain carboxylic acids, which carry a hydroxyl group (position: C2) in the immediate vicinity of the acid group (C1). Many of these acids are contained in plants and fruits and therefore called "fruit acids." Of these, the glycolic, **lactic**, tartaric, malic, and **citric acids** as well as their salts and esters are found in various cosmetic creams. AHAs are used for different tasks. On the one hand, dermatologists use the 70% acid for dermal peeling to remove fine wrinkles and pigmentation problems and to stimulate the skin for cell regeneration. After that, the inflamed skin remains for a long time and must be protected with caring medicines. On the other hand, beauticians use highly concentrated fruit acids (40–50%) at pH values <4 for the soft peeling of the face. This causes a peeling of the upper horny layers within five minutes; subsequently they are washed off or better neutralized with sodium bicarbonate solution. After soft peeling, the skin layer is refined for several months. Some consumer creams are applied for a slight peeling. There the concentrations amount to maximum 5% acid or 15% of the well-tolerated salts. So far, the benefit of slight peeling has not been demonstrated. As the AHAs are good moisturizing agents, they are used with 0.1–3% in nourishing

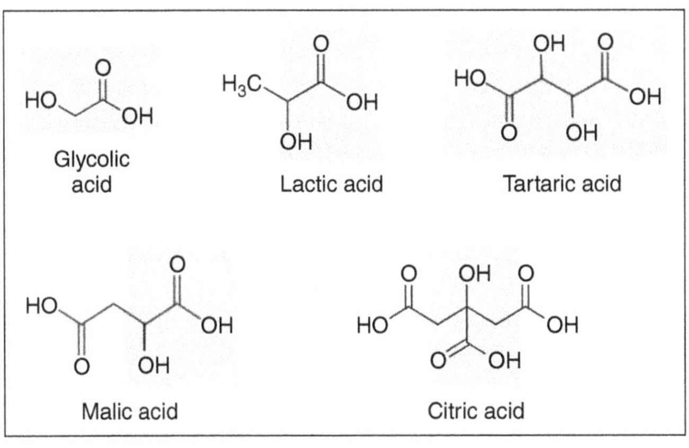

Figure 6.5 α-Hydroxy acids, used for peeling and moisturizing as well as for pH-buffering.

creams, especially **lactic/citric acid** and **sodium lactate/citrate** as substances of the NMF.

Polymers form a film on the skin that is water-rich and protects the skin from drying out. The suitable polymers include proteins consisting of long chains, of many amino acids. They are usually hydrophilic. The proteins known in cosmetics include the collagen and elastin derived from animal skins. These substances ensure the firmness and elasticity of the skin. Because of their molecular weight, however, larger proteins cannot be supplied from the outside, but lead to a smooth, velvety condition on the skin. The chemical or enzymatic cleavage of the proteins yields the protein hydrolysates, for example, the **soybean** or **wheat hydrolysate**. They provide moisture retentive, protective, and sealing films. Other effects of hydrolysates are comparable to the proteins. Animal-based proteins should not be used. They offer no visible and measurable advantages, compared to plant-based hydrolysates.

Hyaluronic acid is a mucopolysaccharide occurring in the dermis. Mucopolysaccharides represent linear acidic polysaccharides constructed from repeating disaccharides. This water-soluble acid can be produced biotechnologically. The film-forming effect is less pronounced than with other polymers but is more effective in the moisture-binding. Through the water evaporation for several hours, the gel film tightens, resulting in a smoothing skin tightening. The effect lasts for several hours and then has to be renewed by cream application. Therefore, the hyaluronic acid and its sodium salt are used in many face creams (Table 6.5).

The fermentatively produced **xanthan** shows moisture-binding properties in addition to water thickening and emulsion stabilization. The film formed on the skin reduces the TEWL. In cosmetic creams, it is not advisable to use for occlusion vaseline or mineral oils; high-melting waxes should also be omitted.

Also, the application of lipids and substances with corresponding properties leads to an occlusion of the skin and causes a slowing of the evaporation. This class of substances is called emollients and comprises all natural oils, for example, the evening primrose oil as well as borage and argan oil, and all esters, especially mono- and di-glyceryl esters. Furthermore, cholesterol, lecithin, and squalene show emollient properties as well as the fatty alcohols.

The re-fatting phospholipids (**Lecithin**) prevent the drying out after washing of normal and especially dry skin. They regulate the pH of the skin and support the natural protective coat against aggressive environmental influences. Their high content of linoleic and linolenic acid has a positive effect on dry and diseased skin. The use of the phospholipids for the production of spherical vesicles, the liposomes, is described in Figures 5.15 and 5.16. Further interesting lipid components, which act as emollient, represent the vegetable oils, discussed in Section 6.5.2. An example represents the shea butter.

Another interesting vegetable substance is **Aloe vera** (desert lily), a stemless plant with 40–50 cm long leaves. The plant is cultivated in many subtropical and tropical regions of the world. Especially in the Mediterranean, India, the West Indies, the Canary Islands, and Mexico, the plants can be found. The well-known

Table 6.5 Moisturizing substances with additional effects in the skin [2, 15].

No.	Chemical name	INCI name	Typical concentrations	Additional properties
1	*NMF* (see Table 6.4)			
	o Lactic acid/ Na-lactate	Lactic Acid/Sodium Lactate	0.4–1%/ 1–2%	pH-regulation, keratolytic effect at pH 5
	o Citric acid/ Na-citrate	Citric Acid/Sodium Citrate	0.1–0.2%/ 0.3–1%	pH-regulation, chelation, stabilization of molecules
	o Glycerol	Glycerol	2–5%	Barrier strengthening and stabilizing
	o Urea (in care creams)	Urea	2–5%	Enhancer (penetration promoting), barrier protecting
	o Squalene (triterpene)	Squalene	0.5–2%	Antioxidant, immune-stimulating, skin protecting, and nourishing
2	*Polymers*			
	o Protein hydrolysate soybean, silk, wheat	Hydrolyzed Soybean, Silk, Wheat Protein	2–5%	Smooths the skin
	o Hyaluronic acid/ Na-hyaluronate	Hyaluronic Acid/Sodium Hyaluronate	0.05–0.5%	Tightens the skin (for hours)
	– High molecular		0.1–0.5%	
	– Medium molecular		0.05–0.2%	
	o Xanthan	Xanthan Gum	0.2–0.5%	Stabilize the emulsion.
3	*Lipophilic vegetable emollients*		10–35%	See Section 6.5 See Table 5.8
	o Natural oils			
	o Hydrogenated natural oils			
	o Lecithin	Lecithin	2–5%	Emulsifier, pH-regulation, skin protection, care of dry skin
	o Shea butter	Butyrospermum Parkii	2–5%	Helps against dry and scaly skin, wrinkles, and skin irritations (sun burns)
4	*Lipophilic compounds* (groups), emollients		*For all*	*For all*
	(a) Fatty alcohols (b) Natural and synthetic ester of fatty alcohols (c) Waxes (d) Paraffines (e) Silicones		0.5–5%	Lipid component, make the skin soft and supple (a–c) thickeners

(Continued)

Table 6.5 (Continued)

No.	Chemical name	INCI name	Typical concentrations	Additional properties
5	*Water-soluble emollients*			
	o Aloe vera (juice)	Aloe Barbadensis Leaf Juice	20–60%	Anti-inflammatory, disinfecting (antiseptic) effect in skin irritation, skin injuries such as wounds and burns
	Powder (1 : 200)	Powder	0.1–0.3%	
	o Hydrolyzed soy proteins	Hydrolyzed Soy Proteins	2–5%	Minimize roughness and chapping of skin caused by dry climates
6	Oil-soluble vitamins		0.3–3%	See Section 6.4

aloe vera gel, obtained from the water storage tissue of the real aloe leaves, may be transferred into the powder by careful spray drying. The gel composed mainly of polysaccharides has a mucous consistency. It consists of sugars such as glucose, mannose, galactose, and xylose, as well as water-soluble vitamins, amino acids, amylase, alkaline phosphatase, lipase, and salicylic acid, as well as glycoproteins and aloenins. Aloe gel is more commonly used in cosmetic creams. It acts moisture-retaining, furthermore anti-inflammatory, and healing, e.g. after sunburn, as well as immunostimulatory. As a drink, it serves as a laxative.

6.4 Vitalizing Substances, in Particular Vitamins

In addition to the moisture-retaining agents, there are a number of other vitalizing substances. Among them, the essential vitamins play an important role. The human body cannot synthesize vitamin; therefore, a daily intake with food is required. On the other hand, however, the body is able to convert provitamins into the corresponding vitamins. The human body requires vitamins only in very small amounts, but continuously. They can be divided into fat-soluble (A, D, E, K) and water-soluble (C, all B, biotin) substances (Table 6.6). As coenzymes, vitamins are involved in many enzymatic reactions, also in the skin. An insufficient amount or even the temporary absence of a vitamin can lead to serious physical damages. Vitamins delivered through the skin have many positive effects. The number and strength of these effects are generated only by the cosmetics and cannot be reached from the inside. In addition to the stimulation of collagen and keratin, a mixture of various vitamins provides the suppression of the radicals. This action is necessary for a vital skin and slows down skin aging.

In the skin, retinol (**vitamin A**) is oxidized in two stages, first to the retinal (retinaldehyde) and finally to the retinoic acid (tretinoin, “vitamin A acid” or “retinol

Table 6.6 Vitamins in cosmetic creams for vitalizing the skin.

No	Common name	INCI name	Typical concentration	Main effects on and in the skin
1	*Vitamin A*			
	Retinol	Retinol	max. 0.3%	All compounds improve dry, scaly, impure skin, against skin aging, wrinkling, and light-induced skin damage; the aldehyde is most effective
	Retinol palmitate	Retinyl Palmitate		
	Retin aldehyde	Retinal		
2	*Vitamin B group*			
	Nicotinamide (old: Vit. B_3)	Niacinamide	1–3 (5)%; acne: 4%; daily care: 1–3%; mature skin: 5%	Stabilizes the skin barrier, reduces the TEWL, improves the skin structure, smooths wrinkles and is anti-inflammatory
	Dexpanthenol (Provitamin B_5)	Panthenol	1–5%	Moisturizing, anti-inflammatory, regenerative, healing; "Beauty vitamin"
	Panthenyl triacetate	Panthenyl Triacetate	1–3%	Stimulate skin metabolism
	Pyridoxine-HCl (provitamin B_6)	Pyridoxine Hydrochlo-ride	About 2%	Regulates excessive sebum production
	Biotin (vitamin B_7)	Biotin	0.5–2%	Improves the skin structure
	Folic acid (vitamin B_9)	Folic Acid	0.05–0.2% (0.5%)	Reduces TEWL, antiaging effect
3	*Vitamin C*		*For all:*	*For all:*
	Ascorbic acid	Ascorbic Acid	0.1–5%	Vitamin C supports collagen synthesis, combat-free radicals and works against age spots, preferably in combination with vitamin E, as well as against acne; the Mg salt is the most stable water-soluble application form
	Ascorbyl palmitate	Ascorbyl Palmitate	Acne 5%	
	Magnesium ascorbyl phosphate	Magnesium Ascorbyl Phosphate	Daily care: 0.2–2%;	
	Sodium ascorbyl phosphate	Sodium Ascorbyl Phosphate	Antioxidative 0.2%	
4	*Vitamin E*			
	Tocopherol	Tocopherol	0.5–5%	Antioxidant, moisture-binding, cell-regenerating and anti-inflammatory effect, reduces skin damage by UV radiation
	Tocopherolacetat	Tocopheryl Acetate	(1–3%)	

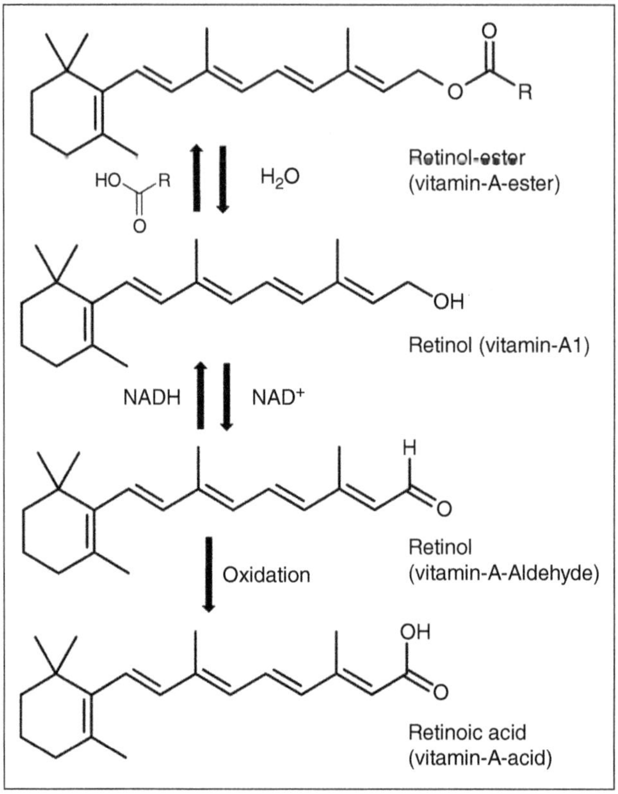

Figure 6.6 Hydrolysis of a retinol ester and oxidation of the retinol via the aldehyde to the active retinoic acid. Source: Wikipedia, public domain.

acid"; see Figure 6.6). The active form of the retinol is thus the tretinoin, which promotes the formation of horny cells (corneocytes) and ensures a visibly smooth skin. It also stimulates collagen production to build a fresh, stable skin structure, and control the cell functions that additionally causes a reduction of the acne. Therefore, retinol is a long-established antiaging drug. Viewed from the outside, retinol reduces wrinkles and improves skin structure. It acts as an antioxidant against the harmful free radicals, alone or in combination with vitamin C and vitamin E, which have proved effective. The skin-compatible retinol esters, such as retinyl palmitates and retinyl linoleates, are converted into retinol in the skin by ester cleavage. The effect magnitude is determined according to the molecular weight fraction. The pure retinol should be used carefully because very high concentrations can cause inflammation, redness, or burning skin. Therefore, the esters are the better alternative in amounts below 1%. According to the recommendation of the German Institute for Risk Assessment, a cream should contain maximum 0.3% pure retinol, a lotion 0.0% [16]. As the vitamins are very effective, small amounts are sufficient for the intended improvement of the skin surface.

The composition of the following substances is particularly effective. **Vitamins A** and **C** play a crucial role in collagen production. The promising combination with **vitamin E or Q10** acts against skin aging by oxidative stress and free radicals. **Hyaluronic acid** stores moisture in the skin and **phosphatidylcholine** starts a natural skin renewing process.

Nicotinamide is the amide of nicotinic acid (niacin and vitamin B_3). It has great biochemical importance as a component of the coenzymes, NAD+ and NADP+, which can reversibly bind hydrogen. Nicotinamide stabilizes the barrier function of the horny layer, measurable in the reduced TEWL and increased skin moisture. It stimulates the production of proteins (such as keratin) and ceramides. In aging skin, topical application of niacinamide improves the surface structure, smooths out wrinkles, and inhibits photo carcinogenesis [17]. It shows anti-inflammatory effects, visible in acne and rosacea. Incompatibility is rarely observed. Because of the measured effects, niacinamide is a suitable component in cosmetic products for use in disorders of epidermal barrier function, for aging skin, as well as for improving pigmentary disturbances.

Dexpanthenol (also known as D-panthenol or provitamin B_5) is a water-soluble compound [18] that penetrates into the skin and reacts there to give pantothenic acid (vitamin B_5). Pantothenic is the main component of coenzyme A, which controls in the skin some metabolic reactions of fats, carbohydrates, and proteins. Therefore, topically applied panthenol enhances the formation of new skin cells and thus measurably promotes regeneration of the skin. Furthermore, this "panacea" improves moisture retention and elasticity of the skin, soothes irritated skin, and shows nourishing and wound-healing properties. D-Panthenol is known for anti-itching, anti-inflammatory, and wound-healing effects; it is also used for the healing of newly engraved tattoos. The active ingredient is present in many skin creams in different degrees of concentrations (typical range: 1–3%). Because of its positive, visible effects on the facial skin, it is also called "beauty vitamin." Panthenol, used in medicines with 5%, speeds up wound healing. It is stable at pH 5 up to 45 °C.

Pyridoxine constitutes the group vitamin B_6 with the derivatives pyridoxal (with aldehyde function) and pyridoxamine (with amino group). The active form of the vitamin B_6 group is called pyridoxal phosphate. It is intended to regulate excessive sebum production and is used in the case of greasy, impure skin and acne.

In addition to its moisturizing effect, **folic acid** can stimulate the formation of fibroblasts [19]. These connective tissue cells are responsible for the formation of collagen, the scaffold of the skin. A stable collagen framework is responsible for a smooth, wrinkle-free skin. In addition to folic acid, vitamin C and retinol are responsible for a healthy collagen framework. The cells in the body are constantly shared by replication of the DNA. External influences, such as the UV radiation, provide free radicals that can attack the DNA. In addition, age increases the error rate during DNA copying so that the skin ages faster. Folic acid supports the repair of the DNA and therefore shows several interesting antiaging effects.

Biotin improves the keratin structure of skin, hair, and nails. It is rather rarely used in skin creams. However, **vitamin C** is often formulated (Figure 6.7). Involved in some enzymatic reactions, it promotes collagen synthesis and neutralizes radicals [20]. Collagen forms the framework of the skin and supplies the necessary strength and tension. Deep wrinkles in the face can be caused by a weakened collagen skeleton. Vitamin C protects the collagen framework against free radicals and helps the skin in the production of new collagen. It provides a stable collagen framework that withstands greater stresses so that the skin remains wrinkle-free for a longer time. In cosmetics, the antioxidant effect is used for the stability of the cream as well as for the protection of the skin against oxidative stress. In combination with vitamin E, it is particularly effective because the reactants regenerate each other. The antioxidant effect of vitamin C helps against micro-inflammation created by acne. Acne is also caused by oxidation processes in which vitamin C counteracts. Therefore, vitamin C should not be lacking in the treatment of acne. As studies show, vitamin C also helps with the increased formation of pigment spots (hyperpigmentation) by inhibition of excessive melanin formation. Thus, the formation of age spots is counteracted. In addition, vitamin C protects our skin from UV radiation. These photoprotective effects also contribute to the reduction of age spots. Vitamin C reduces the formation of radicals in the cornea thus aging. Via liposomes, vitamin C should penetrate deeper into the skin to effect in the epidermis and dermis. It is not sufficiently stable over a prolonged period of time at a pH of 5.

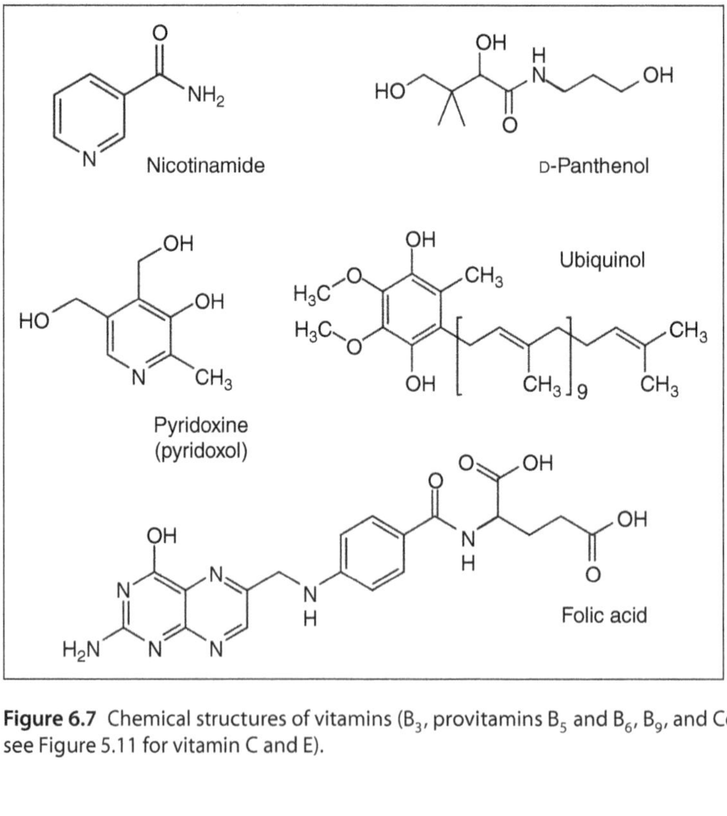

Figure 6.7 Chemical structures of vitamins (B_3, provitamins B_5 and B_6, B_9, and Coenzyme Q10; see Figure 5.11 for vitamin C and E).

The use of the lipophilic ester or, better, the water-soluble magnesium salt is therefore advisable.

Vitamin E is the collective term for different tocopherols, which differ in the substituents on the chroman ring and in the chirality [2]. It is usually processed into creams in the form of the natural α-tocopherol and applied to the skin. As a lipophilic component, it easily penetrates into the cornea in large quantities. These relatively high concentrations, which are not accessible from the inside, have favorable effects on the skin:

- acts as an antioxidant,
- improves the hydration of the stratum corneum,
- strengthens the skin against exogenous noxes,
- eliminates small wrinkles and smooths the skin,
- provides a light protection up to SPF 10,
- protects against photoaging,
- improves wound healing and reduces scarring,
- shows anti-inflammatory properties, and
- reduces the age spots.

Vitamin E, which is also tolerated in high doses, is a valuable substance for all age groups and skin types. Furthermore, α-tocopherol protects the cosmetic preparation from undesired oxidations that lead to color and odor changes. According to recent studies, γ-tocopherol is able to eliminate cancer cells without damaging healthy cells.

6.5 Nourishing Vegetable Oils for Smoothing the Skin

In current cosmetic science and practice, the role of natural vegetable oils is becoming increasingly important because of people's growing awareness for pure nature and sustainability. The same is true, with minor limitations, also for the derivatives, such as fatty alcohols and other. The natural oils from plants have a unique position thanks to their positive effects in the skin and may not be missing in the skin care. Moreover, there are strong interests in beneficial active substances that protect the skin surface and reactivate the metabolism totally without toxic effects or allergy problems.

In addition to moisturizing substances and vitamins, a good cosmetic cream should contain some vegetable oils in an amount of at least 10%. In o/w emulsions, the lipid phase usually has a proportion of 10–20%, in w/o systems of 20–30%. Because of the C-chain distribution in the triglycerides and their accompanying substances, pure natural vegetable oils exhibit positive properties on and in the skin. They prevent water loss through the skin by strengthening the lipid barrier layer, soften and smooth the stratum corneum, and reduce inflammation of the skin. The positive effects depend on the fatty acids present in the range C_{12}–C_{20} (Table 6.7) and on the small amount at existing unsaponifiable substances. These are particularly useful and determine the character of the oil. The unsaponifiable substances include the tocopherols for oxidation protection, furthermore

Table 6.7 Fatty acids for cosmetics from natural oils.

C-chain : double bonds	Chemical name	INCI name	Molecular formula	Description	Melting point
12 : 0	Dodecanoic acid	Lauric Acid	$C_{11}H_{23}COOH$	Antimicrobial, fast-absorbing	43.8 °C
14 : 0	Tetradecanoic acid	Myristic Acid	$C_{13}H_{27}COOH$	Antimicrobial, fast-absorbing	54.2 °C
16 : 0	Hexadecanoic acid	Palmitic Acid	$C_{15}H_{31}COOH$	Low film forming, refatting, restructuring	62.5 °C
16 : 1 (ω-9)	(9*Z*)-Hexadecenoic acid	Palmitoleic Acid	$C_{15}H_{29}COOH$	Skin-soothing, restructuring	1 °C
18 : 0	Octadecanoic acid	Stearic Acid	$C_{17}H_{35}COOH$	Refatting, film forming	69.3 °C
18 : 1 (ω-9)	(9*Z*)-Octadecenoic acid	Oleic Acid	$C_{17}H_{33}COOH$	Caring, slowly absorbing	17 °C
18 : 2 (ω-6)	(9*Z*,12*Z*)-Octadecadienoic acid	Linoleic	$C_{17}H_{31}COOH$	Regenerating of barrier layer, anti-inflammatory	−12 °C
18 : 3 (ω-3)	(9*Z*,12*Z*,15*Z*)-Octadecatrienoic acid	Linolenic Acid (α)	$C_{17}H_{29}COOH$	Stimulates cell growth, anti-inflammatory	−11 °C
18 : 3 (ω-6)	(6*Z*,9*Z*,12*Z*)-Octadecatrienoic acid	Linolenic Acid (γ)	$C_{17}H_{29}COOH$	Anti-inflammatory, antiallergic, against itching	−10 °C
20 : 0	Eicosanoic, icosanoic acid	Arachidic Acid	$C_{19}H_{39}COOH$	Refatting, film forming	76.5 °C

phospholipids, phytosterols, carotenoids (provitamin A), squalene, flavonoids, and others.

6.5.1 Natural Fatty Acids from Vegetable Oils

The hydrolysis of the triglycerides (esters) generates in three steps the fatty acids, which may be the same or different (C-chain, double bonds), as well as the liberated glycerol. For the skin, the C_{18} and partly the C_{16} fatty acids are important, especially the unsaturated compounds. For constant C-chain, the melting point decreases with the number of double bonds. The more double bonds in the oils, the stronger the tendency to oxidation. In the air, these oils become yellow in a short time. However, the ω-6 fatty acids are particularly beneficial for the skin. Therefore, the use of oils with high content of the essential linoleic and γ-linolenic acids requires special measures in the production of such a cream, which will be discussed later.

Because of the lack of compatibility on the skin, free fatty acids should not be present in cosmetic creams. They can be fed in esterified form as oils (triglycerides). The hydrolytic cleavage of the triglycerides takes place gradually in the skin, so that the fatty acids are successively released (Figure 6.8). A lack of fatty acids, which normally does not occur in younger people (<45 years), can have many causes (including incorrect diet and disease) and lead to a disturbance of the barrier function. The TEWL increases and the skin becomes dry, wilted "unhealthy" with poor wound healing ability, and increased number of comedones. By applying a cosmetic cream, the contained natural oils can eliminate these unhealthy states.

Fatty acids from vegetable oils are of great importance in cosmetic industry [21]. The properties and effects of some fatty acids in the skin will be briefly explained. **Lauric acid** (C12 : 0) is rapidly absorbed into the skin. It acts an antimicrobial and fungicidal substance. Lauric acid-containing oils can be distributed well and create a smooth skin feeling. Because of the relatively short C-chain, lauric acid lowers the viscosity slightly. The same applies to **myristic acid** (C14 : 0), which is also antimicrobial and skin softened. **Palmitic acid** (C16 : 0) occurs in human skin sebum. The proportion decreases with age. Therefore, for the mature skin, palmitic acid-containing oils should be used in the creams. In addition, palmitic acid strengthens the acid mantle of the skin. The saturated $C_{14,16}$ fatty acids have a slight refatting effect. Stearic acid (C18 : 0) is contained in most vegetable oils only in the thousandth or single-digit percentage range. It forms a film on the skin, which prevents the evaporation of the water and hinders the sebum flow away. Therefore, the saturated C_{16} and C_{18} oils can be weakly comedogenic and should not dominate. Humans with bad skin should not buy creams high in saturated oils. Monounsaturated **oleic acid** (C18 : 1) are readily absorbed, can easily be distributed, and produce a soft skin feeling. They act as enhancers, which mean that they can make the lipid barrier of the skin more permeable and receptive to lipophilic active ingredients. The application of a cream with high content of oleic acids produces a pleasant, moist skin feeling.

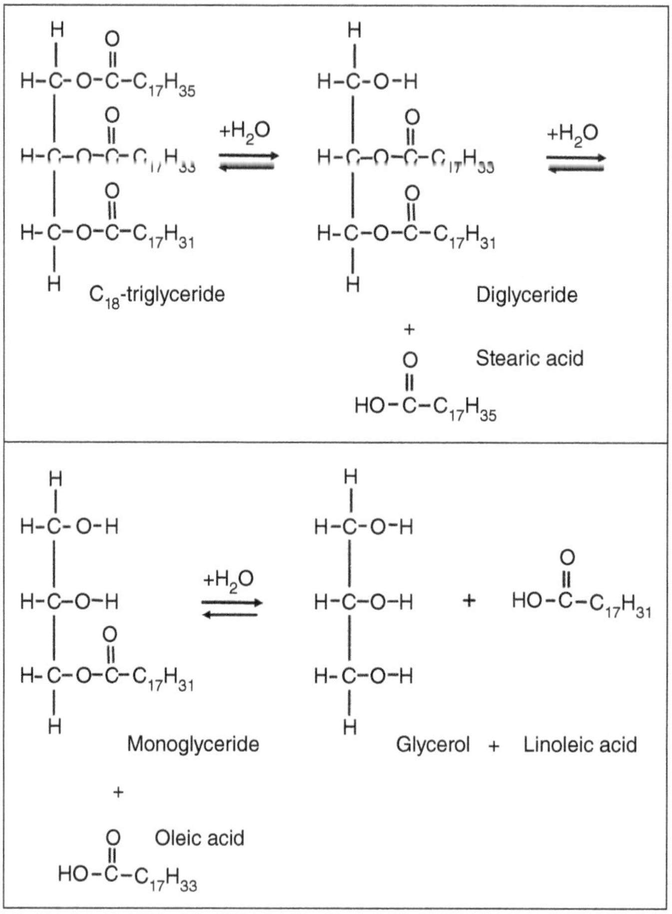

Figure 6.8 Gradual hydrolysis of triglycerides with different C_{18}-fatty acids.

Important natural oils contain bound linoleic and γ-linolenic fatty acids. The human body cannot produce either of these fatty acids. Therefore, they must be supplied from the outside. **Linoleic acid** (LA; C18 : 2, ω-6) is a di-unsaturated fatty acid that represents an important substance in the epidermis. The top layer of skin, the stratum corneum, contains ceramides, free fatty acids, and phospholipids. Ceramides exist as lipid bilayers and regulate the water balance in the skin (Chapter 4). The largest share of the essential ceramide 1 consists of the important LA. Furthermore, LA supports the elimination of skin irritation in contact dermatitis after topical application and reduces light damage to the skin and age spots. LA oils quickly absorb and hardly fat the skin. The natural component of the sebum normalizes the skin metabolism. In acne skin, the LA content is reduced in sebum, resulting in blocked pores and formation of comedons and eczema. The acid improves the work of sebaceous glands, unblocks the pores, and decreases the number of comedons. Moreover, in the skin, LA is needed for the structure of cell membranes and also for the production of intercellular cement of the skin. These two processes are possible, thanks to the enzymatic catalysis in the stratum corneum. On the one hand, LA enhances the skin texture of greasy skin with a tendency to comedones and impurities. On the other hand, the barrier is improved by stabilizing the lipid bilayers in the skin without a greasy film on the skin. The effect can be detected by a reduced TEWL. LA oils are suitable for both dry and greasy skin conditions. In particular, dry, scaly, also diseased skin (neurodermatitis) requires this oil, preferably in admixture with the tri-unsaturated γ-linolenic acid (GLA).

A deficiency of linoleic and linolenic acid leads to a disruption of the barrier function of the skin with a significant increase in the TEWL. Besides the linoleic, at least as important is the triple-unsaturated **GLA** (C18 : 3, ω-6), which is missing in neurodermitic skin and represents a raw material for the group of tissue hormones such as prostaglandins [2]. A lack of these hormones, which are involved in cell metabolism, leads to rough, dry, cracked, and itchy skin. Natural oils such as borage, evening primrose, hemp, and black currant, as well as oil from the seed of pomegranate, supply the skin with linoleic and γ-linolenic. One of these natural oils (triglycerides) in amounts of 3–10% should be included in good skin cosmetics. The composition of fatty acids bound in hemp oil corresponds most closely to the composition of the skin fatty acids. Doctors treat atopic dermatitis with evening primrose oil. GLA has an anti-inflammatory, calming, itching, and regenerating effect. **α-Linolenic acid** (ALA; C18 : 3, ω-3) acts an anti-inflammatory substance. It is particularly useful for mature skin because it promotes cell regeneration and activates the new formation of cells. The chemical structures of the most important C_{18}-fatty acids are shown in Figure 6.9.

6.5.2 Vegetable Oils and Fats in Cosmetic Creams

Unlike the vegetable oils, which consist of the valuable fatty acids and glycerol (triglycerides), the colorless and odorless mineral oils (paraffins) are pure hydrocarbons made by distillation of petroleum. Similar in physical properties, they are composed of paraffinic saturated chain hydrocarbons, naphthenic (saturated ring-shaped) hydrocarbons, and aromatic (ring-shaped with aromatic double bonds) hydrocarbons, including some olefins. As a positive effect of mineral oils in the skin is not known, this class of substance should be avoided. In cosmetics, occlusion on the skin may be achieved with other effective ingredients, if desired.

The differentiation between vegetable oils and fats is based on the aggregate state at room temperature (20–24 °C). Fluids are referred to as oils and nonflowable substances as fats. The harvest, i.e. the oilseed, is first carefully cleaned and dried. The oils are generally obtained by pressing processes (screw presses, seldom extruders), preferably from the seed, but also from the kernels as well as from the fruits, nuts, and bran (Table 6.8). During cold pressing, a temperature of 40 °C must not be exceeded. The resulting press cake can be used directly as a protein-rich animal feed or indirectly after an extraction of further valuable substances. For the hot pressing, the material is collected via riffel presses, conditioned, and heated to 80–100 °C to tune the moisture content precisely to the pressing conditions. This determines the yield during de-oiling. For extraction,

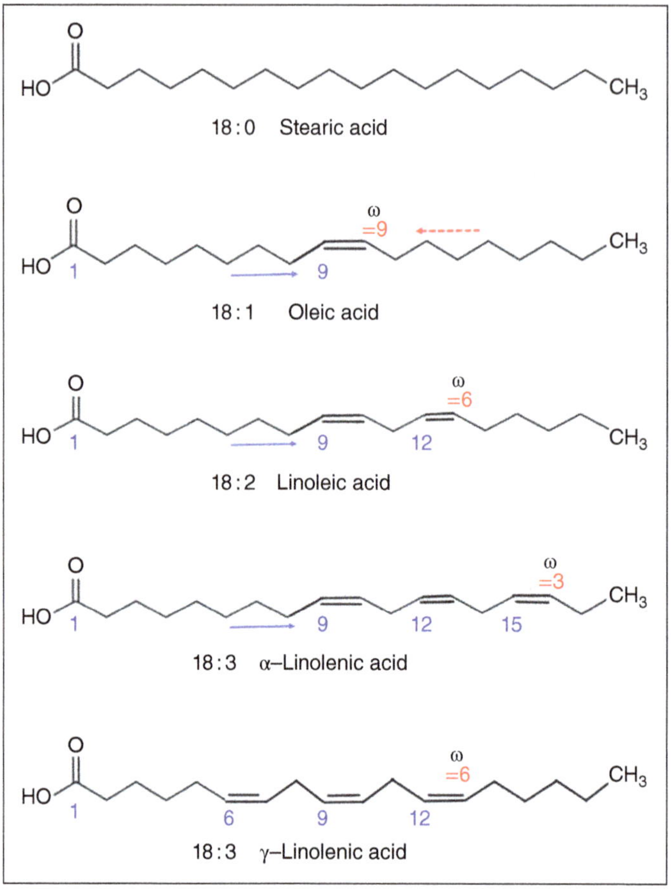

Figure 6.9 Chemical structures of C_{18}-fatty acids.

the oil remaining in the press cake is then mixed with the solvent hexane for extraction.

The oil flowing from the press must then be freed from the accompanying substances. This requires a degumming, which is carried out on the one hand with acidic water (citric or phosphoric acid). The process water is subsequently centrifuged or filtered off. In this way, most of the accompanying substances such as phospholipids, glycolipids, vitamins, soaps, and trace elements can be separated. Subsequently, the water containing some fatty acids is neutralized with sodium hydroxide. On the other hand, a degumming of the vegetable oils, having a higher content of phosphorus, is carried out by the addition of demineralized water. The accompanying substances dissolved with water are processed into lecithin. These substances find their use not only in the food industry for the production of margarines and instant products but also in animal nutrition and in the technical industry. In some cases, the oils can be carefully obtained using hexane or carbon dioxide extraction methods.

Table 6.8 Vegetable oils from different plant parts.

Origin of the oils	Seed oils	Kernel oils/ cores (almonds, nuts)	Fruit oils	Nut oils	Bran oils
Examples	– Black cumin – Borago – Evening – Primrose – Grape seed – Hemp – Jojoba – Safflower – Sunflower	– Argan – Peach – Plum – Soybean – Shea butter – Sweet almond	– Avocado – Coconut – Olive – Rose hip oil – Sea buckthron	– Hazelnut – Inca peanut – Macadamia – Nut – Peanut – Walnut	– Rice germ – Wheat germ

Pressed from plant parts, in particular for cosmetic use cold pressed, well over 100 oils are suitable for cosmetic creams (examples in Table 6.9). In most cases, two or more oils are used. The recipe manager (formulator) ensures that after his experiences, the distribution of the fatty acids corresponds to the task of the cream. Two examples are given in Table 6.10, suitable for wide applications. The selection consists of base oils, for instance sweet almond oil, shea butter, jojoba oil, and several semidry oils. In these, the polyunsaturated acids clearly predominate. The combination of linoleic with GLA is recommended.

Sheabutter [23], part of the module "Lipids" and also known as Karité butter, is obtained from the fruits (nuts) of the Karité tree. These trees grow south of the Sahel zone in the savanna belt of West Africa, from Senegal to Uganda, and reach a height of 10–20 m. The shea nuts belong botanically to the family of berries. From the kernels of fruit, the shea butter is obtained. For centuries, shea butter has been used as a source of edible fat and for body care by the African people, especially as a medical balm. The butter is applied to treat inflammation or skin damage, with astonishing results. Further positive properties are attributed to this substance mixture. Shea butter contains mainly long-chain, C_{18} fatty acids whose main components are oleic (40–55%), stearic (35–45%), linoleic (3–8%), and palmitic acid (3–7%). Compared with other oils, the proportion of unsaponifiable ingredients is between 8% and 11% (otherwise 1–6%). This consists of about 75% triterpenes and triterpene alcohols, phytosterols, and tocopherols. The Karité butter has nourishing substances, which refat the skin and give dry skin their suppleness. The butter regulates the moisture content of the skin. In addition, it shows protective effects against excessive sun exposure and allows the necessary care after sunbathing. The butter also before and after the visit in the sun studio is especially recommended.

Skin care formulations, containing ingredients similar to those in the chemical composition and/or physical properties as those of the skin's surface layer, should be most effective. An example represents **jojoba oil** [24] that is completely miscible with sebum. When applied, there is a very thin, nongreasy lipoid

Table 6.9 List of vegetable oils and butters for cosmetics of a big supplier [22], many of these are also certified for organic cosmetics.

Examples for natural oils and butters		
Acai Oil	Cranberry Seed Oil	Pecanut Oil
Almond Oil	Evening Primrose Oil	Perilla Oil
Amaranth Oil	Grape Seed Oil	Pistachio Nut Oil
Apricot Kernel Oil	Ground Nut oil	Plum Kernel Oil
Argan Oil	Hazel Nut Oil	Pomegranate Oil
Avocado Oil	Hemp Oil	Poppy Seed Oil
Babassu Oil	Jojoba Oil	Pumpkin Seed Oil
Baobab Oil	Kukui Nut Oil	Rape Seed Oil
Black Cumin Oil	Laurel Oil	Rice Bran Oil
Borage Seed Oil	Linseed Oil	Rose Hip Oil
Brazil Nut Oil	Macadamia Nut Oil	Sacha Inchi Oil
Broccoli Seed Oil	Marula Oil	Safflower Oil
Calendula Oil	Meadowfoam Seed Oil	Sea Buckthorn Pulp Oil
Camelina Oil	Milk Thistle Oil	Sesame Oil
Castor Oil	Moringa Oil	Shea Butter
Chia Oil	Mustard Seed Oil	Soybean Oil
Chilean Hazelnut Oil	Olive Oil	St. John's Wort Oil
Cocoa Butter	Palm Oil	Sunflower Oil
Coconut Oil	Palm Kernel Oil	Tamanu Oil
Corn Oil	Palmolein	Tiger Nut Oil
Cotton Seed Oil	Palm stearin	Walnut Oil
Cramble Oil	Passionfruit Seed Oil	Wheat Germ Oil
	Peach Kernel Oil	

layer of jojoba and sebum on the skin. This partially porous layer provides an exceptional transepidermal respiration and moisture control. Therefore, jojoba oil is an excellent moisturizing agent, significantly reducing the TEWL without totally blocking transpiration of gases and water vapor. The oil spreads well and leaves a rich velvety nonoily feel on the skin. Well tolerated by people with skin problems, it helps to drain sebum in blocked pores and balances the production of sebum. Furthermore, jojoba oil has exceptional skin-softening properties and increases the skin's suppleness significantly for several hours. The oil is able to minimize fine lines and wrinkles for some hours, promoting skin suppleness while assisting with the rejuvenation of the skin. The antibacterial effect of jojoba oil was proved. Scientists found that the oil destroys different types of bacteria within two hours. Skin bacteria and certain skin fungi cannot survive in jojoba oil. This oil shows an extraordinarily stability in extreme temperatures up to 370 °C without decomposition of the carbon chain. In addition to thermal stability, the natural oil has unsurpassed oxidative stability. Furthermore, jojoba helps to stabilize oxidative sensitive, natural active ingredients of the formulation. The stability is based on the molecular structure and the presence of antioxidants.

For more than 4000 years, sweet **almond oil** [25] has been used as a skin care product and as a foodstuff. Cold pressed to preserve the valuable ingredients of the fruits of the almond tree, it represents a good, very popular oil for cosmetic

Table 6.10 Distribution of fatty acids of the recommended vegetable oils ((a) modul "lipids," (b) an alternative); the oils come from nature and the information about the content, taken from different sources, therefore differ up to ±10% and more.

(a)

C-chain: double bonds	INCI name of the fatty acids	Prunus amygdalus sativa kernel oil, "sweet almond oil"	Oenothera biennis oil, "evening primrose oil"	Cannabis sativa Seed Oil, "hemp oil"	Butyrospermum parkii butter "sheabutter"
	Iodine number[a]	About 100	About 155	About 155	About 65
	Unsaponifiables	<1.5%	1.5–2.5%	0.5–1.5%	2–12%
	Module "lipids": amount in cream				
	o/w	4%	3%	3%	3%
	w/o	7%	5%	5%	5%
16 : 0	Palmitic Acid	5.7%	6.4%	6.5%	4.5%
16 : 1	Palmitoleic Acid	0.7%			
18 : 0	Stearic Acid	2.5%	2.0%	3.0%	35%
18 : 1	Oleic Acid	71%	6.8%	14.9%	53%
18 : 2	Linoleic Acid	19%	74.8%	56.3%	7%
18 : 3	Linolenic (α)			15.8%	
18 : 3	Linolenic (γ)		9.6%	1.8%	
18 : 4	Stearidonic			0.6%	

(b)

C-chain: double bonds	INCI name of the fatty acids	Triticum Vulgare Oil, "Wheat germ oil"	Carthamus Tinctorius Oil, "Safflower oil"	Borage Officinalis Seed Oil, "Borage seed oil"	Simmondsia Chinensis Seed Oil, "Jojoba Oil"
	Iodine number[a)]	About 130	About 145	About 140	About 85
	Unsaponifiables	<5%	0.5–1.5%	0.5–2%	About 44%
	Module "lipids 2" Amount in cream				
	o/w	4%	3%	3%	3%
	w/o	7%	5%	5%	4.5%
16 : 0	Palmitic Acid	17.5%	7.0%	9.7%	1.2%
18 : 0	Stearic Acid	1.0%	2.5%	4.4%	
18 : 1	Oleic Acid	19.5%	11.0%	17.3%	10.7%
18 : 2	Linoleic Acid	53.2%	77.2%	37.5%	
18 : 3	Linolenic (α)	6.1%			
18 : 3	Linolenic (γ)			21.9%	
20 : 1	Gadoleic (ω-9)				71.3%
22 : 1	Erucic Acid (ω-9)			2.5%	14.6%
24 : 1	Nervonic (ω-9)				1.5%

a) The iodine number (IZ) is a measure of the content of unsaturated fatty acids in the glycerides.

creams. The plant is a 2 or 8 m tall tree or shrub. The almond tree blooms in March or April. The stone fruit ripens in July/August. Today, the almonds come mostly from California (about 300 000 t/a). In addition, there are large cultivation areas in Europe, especially in Spain and Italy (100 000 t/a). The well-tolerated almond oil is rich in valuable ingredients. In addition to the important fatty acids, it contains many important vitamins and minerals in small amounts. These are the vitamins A, some B-types, E, and D as well as the minerals potassium, magnesium, and calcium, which altogether contribute to a smooth and healthy skin. Almond oil protects the skin cells and supports their renewal. The high oleic acid content in the oil nourishes and smooths the skin and gives moisture. Almond oil smells very mild, which supports popularity.

Evening primrose oil [26–28] is an ancient remedy. Native Americans used an aqueous extract of the plant to treat wounds, bruises, and hemorrhoids. The hardy plant, about 1 m in height, can often be found in dry, sunny meadows. Originally native to North America, the plants are today found in almost all of Europe and parts of Asia. The entire edible plant attracts attention because of the pale yellow, phosphorescent blossoms that exudes a concise smell. The seeds of evening primrose that grow in a capsule show numerous medicinal properties. The seeds contain 14–20% of linoleic rich oil, pressed out by a cold process to preserve the oxidation-sensitive ingredients. Evening primrose oil is high in linoleic and GLA, making it to an exceptional skin nourishing oil. These important acids, which the body cannot synthesize, must be supplied through nutrition and topical application. GLA regulates the cell renewal and metabolic process as well as strengthens the barrier function. The essential fatty acids contained in evening primrose oil have a positive effect on the appearance of the skin. The rejuvenation of the skin, visible after prolonged use, may be due to the elimination/repair of damage in the skin. In addition, the oil inhibits bacterial growth and allows our skin to defend against infection and inflammation. Evening primrose oil combats diseases on the skin such as atopic eczema and psoriasis. It helps maintain stratum corneum cohesion and reduce TEWL.

In 2800 BCE, **Hemp oil** was described as a remedy. It owes its green color to the chlorophyll. The hemp plant, also called cannabis, belongs to the cannabaceae, a plant family to which hops also belongs. The roots of this unique plant can reach up to 2 m into the ground and heights between 1.5 and 7 m. Hemp [29] uses sunlight more efficiently than any other plant and requires only a short growth period of one year. The cold-pressed seed oil is high in ω-3 and ω-6 fatty acids and contains the important GLA. In small amounts, it also provides essential vitamins (E, B_1, and B_2) as well as minerals such as potassium, magnesium, calcium, phosphorus, iron, sodium, manganese, copper, and zinc. It comprises 70–80% of the valuable polyunsaturated C_{18} fatty acids. Thanks to a balanced composition, hemp oil relieves many deficiency symptoms and helps to weaken diseases. The stearidonic acid contained in hemp oil (ω-3 fatty acid) is even more effective than linolenic acid against inflammatory processes. This not only applies to inflammations of the skin such as atopic eczema and psoriasis but also to inflammation of the ears, throat, and joints. In Russia, hemp oil is a therapeutic agent for the treatment of throat pain. In cosmetic creams, the high content of unsaturated fatty acids in this oil gives the skin a smooth, healthy look.

For 1500 years, the positive effects of **Borage oil** on the human body have been known. In the sixteenth century, monks cultivated the herb in their monastic gardens because the tea made from it possessed medical effects. The 0.7 m tall borage plant is a wildflower (commonly called the starflower) and characterized by the star-shaped bright blue flowers. The plants are cultivated particularly in Spain and France, but also in western parts of Asia and in the United States. The cold-expressed borage seed oil represents the richest known source of the essential GLA, which contains 20–24% twice as much as the evening primrose oil and therefore very interesting today [30]. The positive properties of borage oil are particularly due to the high GLA content. This acid is an essential component of the tissue hormones (prostaglandins) and takes part in the metabolism. Therefore, the borage oil activates the skin metabolism, improves the skin structure, and helps with skin scaling, itching, and skin dryness, even on diseased skin (atopic eczema and psoriasis).

Wheat is one of the oldest cultivated plants. Its history can be traced back 10 000 years. With growing of the Roman Empire, the wheat crops became very important. During the Dark Ages, only the rich could afford wheat. In the sixteenth century, European colonists brought wheat and other grains to America. Today, wheat is grown in almost every country. From the wheat germ grows the new wheat plant. The germ makes up 2–3% of the whole grain and is nutritionally the richest part of the plant. The cold-pressed **wheat germ oil** contains about 80% of valuable unsaturated fatty acids and in small amounts of tocopherols (0.25%), furthermore phospholipids, phytosterols, and carotinoide [31]. The high-quality ingredients of the oil not only strengthen the barrier layer but also provide a vitalization of the skin. These properties are particularly suitable for pregnant women who use this oil in skin and dam care. The application of the wheat germ oil on the skin shows anti-inflammatory, antioxidant, and pain-relieving properties as well as decreases skin roughness and dryness.

Safflower is one of the oldest cultivated plant originally from Egypt that belongs to the family of basketwort (Asteraceae). Chemical analysis of ancient Egyptian textiles dated to the Twelfth Dynasty (2000 to 1800 BCE) identified dyes made from safflower. Traditionally, the plants were used for coloring and flavoring foods, in medicines, and in making red and yellow dyes until cheaper aniline dyes became available. Today, the safflower is cultivated in around 60 countries, especially in Kazakhstan with 24% of the world production, furthermore India, Mexico, Argentina, and the United States. The crops are 30–150 cm tall with globular flower heads having yellow, orange, or red flowers. Each branch shows one to five flower heads containing 15–20 seeds per head. For the past 50 years, the cultivation of the plant takes place for the valuable vegetable oil extracted from its seeds. This cold-pressed oil is used for food such as olive oil and for cosmetic products. The resulting **Safflower oil** differs from other vegetable oils because of its high proportion of linoleic acid, which is between 75 and 80% [32]. In contrast, olive oil has only 5%, but more than 75% of oleic acid. The α-tocopherol contained in the safflower oil provides for the antioxidant effect. In addition, vitamins K, A, and phytosterols are found in traces. The colorless oil prevents dryness and roughness of the skin and can be used for makeup removal. Its contents make safflower oil also suitable for issues including eczema, psoriasis, and acne. The hydrating properties lend skin a healthy shine, promote elasticity, and reduce the appearance of wrinkles.

Further details on the vegetable oils and the compositions can be found in the literature [33]. In recent years, some "oil-free" creams have been developed. Presumably, some users are bothered by the shine that the oil can leave depending on the formulation. From a medical point of view, it makes no sense to omit the oils and replace them with fatty alcohols. The important di- and tri-unsaturated oils, which the body cannot produce itself, must be supplied from outside when the skin barrier is disturbed. This task is not fulfilled by the oil-free creams, but also many others. Formulations then advertise that they are oil-free. Others contain milk or yoghurt. If a healthy skin needs only to be moisturized, then oil-free creams are suitable and can demonstrate their advantage, namely to cover the skin with a thin rather matte layer.

6.6 Active Ingredients for Antiaging Creams

A skin that is smooth and tight and shows firm contours and looks healthy is considered beautiful across all cultures. According to estimates, the aging process gradually begins around the age of 25, which then becomes visible 10–20 years later, especially on the face. The antiaging ingredients in the creams aim to slow this development as much as possible. The skin aging process cannot be stopped. In the genes, the aging of the body cells is preprogrammed. However, the premature aging of the skin can be slowed down, which reduces the visible signs: the skin looks younger.

Everyone can influence the aging of the skin through right behavior, a proper diet, as well as by use of an effective cream with antioxidative ingredients. Smoking, occupational and private stress, airborne pollutants, bathing in the sun or on the sunbed, extreme temperature changes, and other stressful situations can trigger the formation of free radicals. These cause oxidative processes that damage the skin cells. The uncontrolled oxidations accelerate the breakdown of collagen and elastin fibers, which decreases the firmness of the skin and gradually relaxes the facial contours. Much later, the age spots and redness appear. These processes, which promote the aging of the skin, run slower because of the regular use of a suitable cream. The antiaging research is therefore focused on substances that deactivate the harmful oxidation of cells with radical scavengers. This is done by absorption and reaction with the radicals. Substances that reduce premature skin aging are called antioxidants. These include in particular the externally supplied vitamins C and E. The vitamin-containing vegetable oils, which additionally promote the regeneration of the skin via the active fatty acids, are also effective. Finding substances that are able to stimulate cell production is the second important line of development.

Common antioxidants and their application forms are listed in Table 5.11. With antiaging creams, small wrinkles go back visibly, but they cannot do anything to prevent deep wrinkles. The antiaging active ingredients can be divided into three groups. The substances of the Group 1, led by vitamin A, have proven their effectiveness in double-blind studies (see Table 6.11). Double-blind refers to studies in

Table 6.11 Active antiaging ingredients.

Group 1	Group 2	Group 3
Effectiveness proven in double-blind studies	**Effectiveness proven in studies**	**Effectiveness so far without scientific proof**
➢ Vitamin A and its derivatives ➢ Vitamin C ➢ α-Lipoic acid ➢ Peptides, especially pentapeptides	➢ Vitamin E and its derivatives ➢ Vitamin B_3 ➢ Dimethylaminoethanol ➢ Phytohormones such as isoflavones and lignans ➢ Hyaluronic acid	➢ Coenzyme Q10 ➢ Green tea extract ➢ Ginkgo ➢ Resveratrol from grapes ➢ Aloe vera ➢ Kinetin

Source: Data from Ref. [34].

which neither the experimenter nor the probands know who used a cream with the active substance. Group 2 displays other highly effective substances such as vitamins E and B_3. Group 3 contains ingredients such as coenzyme Q10, but their effectiveness has not yet been proven adequately.

First of all, the multi-proven effect of vitamin A has been known for many years: **Retinol** is one of the most effective antiaging substances. It works in deeper skin layers, accelerates cell renewal, reduces the depth of wrinkles, and stimulates the metabolism of the cells. This reduces the premature aging of the skin. The active ingredient smooths the skin and makes it plumper. Because of the great effectiveness, it must be used sparingly to prevent skin irritation (details see Section 6.4). With regular topical application, **vitamin C** ensures a healthy and stable collagen scaffold that keeps the skin wrinkle-free longer. First, it protects the collagen scaffold from free radicals and, secondly, it helps the body to produce new collagen. **α-Lipoic acid** (2–5%) occupies a special status within the antioxidant substances because it exerts an antioxidative effect in both watery and greasy skin structures and additionally regenerates the other antioxidants during the reactions against free radicals [35]. For example, α-lipoic acid ensures that vitamin C, vitamin E, coenzyme Q10, and others are not consumed. According to a study, the Q10 degradation during UV irradiation (sunbeams, solarium) could be reduced by 40%. The UV radiation triggers inflammatory processes in the skin, which damages the skin collagen. These skin damages lead to microwrinkles, which gradually become small wrinkles. Such inflammations lead to faster aging of the skin; in addition to wrinkles and coarse pores, brown spots develop (age spots). α-Lipoic acid not only stops this process but partially reverses it.

Linked amino acids form peptides. The effect of di-, tri-, tetra-, **penta-**, and **hexapeptides** on the skin was deciphered approximately 30 years ago. Since the late 1990s, several variants of peptides are used as antiaging agents in creams. The peptides act as versatile messengers in the skin and trigger the cell metabolism as signal substances. This means increased collagen synthesis, better regeneration, and cell formation of the skin. The results are a firmer skin and firmer contours. On the other hand, other types of peptides can relax the facial muscles, so that mild facial lines are tightened. For example, a cream containing the peptide

Argireline, chemical name acetyl hexapeptide, causes a reduction in the depth of wrinkles by 30–50%. Marketing means that Argireline [36] shall be a "Botox alternative." Palmitoyl pentapeptide-4 and carnosine are other well-researched substances [37]. In the meantime, there are significantly more than 10 interesting candidates.

Hyaluronic acid is an endogenous substance that can store a thousand of times its own volume of water (see also Section 6.3.2). Hyaluronan forms an air-permeable, water-binding film on the skin that reduces the loss of water and moisturizes the skin. At the same time, it triggers a slight tightening effect through the slow drying process. The skin looks smoother and firmer. A temporary padding of the skin from the inside may be possible by using the "low-molecular" 50kDa variant. Small portions of these macromolecules may penetrate the skin through the transfollicular pathway (Section 4.5). Probably, the dissolved polymers diffuse over some hours into the stratum corneum and there expand between the cells, resulting in a tightening effect.

The harmful radicals that act in the skin over a longer period of time lead to a gray, wrinkled skin. A popular mean of combating radicals represents ubiquinol (1–2%). Ubiquinol, also called ubihydroquinone, is a reduced form of the **Coenzyme Q10** (ubiquinone). This compound shows a significantly higher bioavailability than the coenzyme Q10 (CoQ10) mixture, which exists in three redox states [38], namely completely oxidized (ubichinon-10), partially reduced (semiquinone or ubisemiquinone), and completely reduced (ubiquinol). Ubiquinol is a highly effective, fat-soluble antioxidant and is capable of regenerating α-tocopherol. Cellular energy production and antioxidant protection are based on the redox functions of ubiquinol, in which two electrons switch in a redox cycle between ubiquinol (reduced) and the (oxidized) ubiquinone form. The coenzyme stimulates cell renewal, strengthens the connective tissue, and thus ensures a more elastic, more wrinkle-resistant skin. It is a well-known antiaging ingredient, which works on the one hand against premature skin aging by reducing free radicals. On the other hand, Q10 increases the energy metabolism of the cells and thereby increases their resistance and life span. Although the body can produce the coenzyme by itself, production decreases as soon as mid-twenties, while at the same time, the demand increases because of increased oxidative stress.

2-**Dimethylaminoethanol** (DMAE) or deanol stimulates the production of choline. The substance was recently evaluated in a placebo-controlled trial. DMAE [39] is an analog of the B4 vitamin choline and is a precursor of acetylcholine, a well-known neurotransmitter. In a randomized clinical study, 3% DMAE in a facial gel applied daily for 16 weeks has been shown to be safe and efficacious. Forehead lines and periorbital fine wrinkles are alleviated; in addition, lip shape and fullness and the overall appearance of aging skin are improved. Coarse wrinkles, under-eye dark circles, nasolabial folds, sagging neck skin, and neck firmness showed beneficial progress. These effects also remained after a 2-week cessation of application. Unfortunately, DMAE has an amine to ammonia-like odor that needs to be covered.

Phytohormones are hormones found in plants and belong to the secondary phytochemicals. They regulate many biochemical processes in plants and can

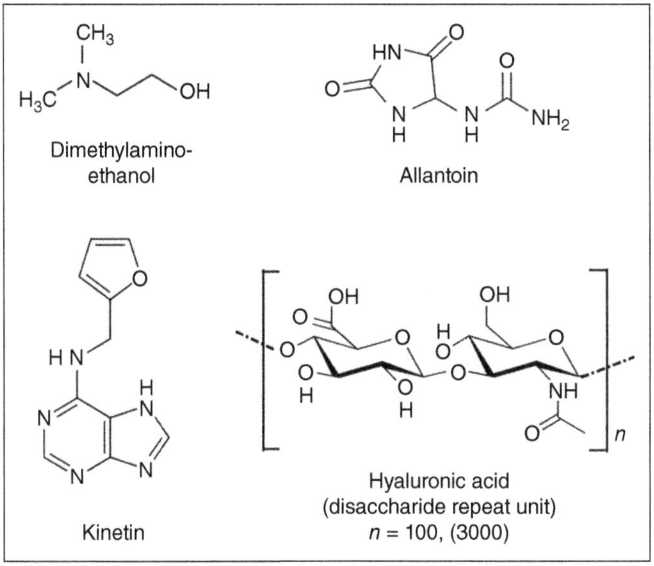

Figure 6.10 Chemical structures of some substances with antiaging effects.

stimulate or inhibit the growth of sprouts, roots, leaves, and seeds. In cosmetics, phytohormones extracted from the plant are used as a mixture of various phytochemicals. A plant with high levels of phytohormones represents soy (glycine soy). Isoflavones belong to a subgroup of phytoestrogens. When used externally on the skin, these antiaging ingredients stimulate collagen synthesis and cell regeneration while slowing collagen degradation. The effects mask the negative impact of estrogen deficiency in the climacteric and thereafter. **Kinetin** belongs to the cytokinins, a class of plant growth substances (phytohormones), which promote cell division or cytokinesis in plant roots and shoots. They are involved primarily in cell growth and differentiation. The plant growth substance (kinetin = N6-furfuryladenine, Figure 6.10) is responsible for the moisture balance and the health of the leaves and delays the aging of the plant cells. Kinetin has a similar effect on human skin cells. The natural changes in the cell aging process are delayed and even reversed. The substance also has strong activities as radical scavenger. In addition, it reduces the TEWL of the skin with high effectiveness. Therefore, kinetin is used successfully in beauty cosmetics for the prevention and reduction of skin aging.

The **polyphenols** contained in many plants are powerful radical scavengers and have proven themselves as antioxidants. **Green tea** contains various amounts of tannins, polyphenols, and caffeine. The polyphenols are of great importance because of their antioxidant, cancer-protective, and immune stimulating effects. A cream containing 2% green tea extracts acts against skin aging and photodamages in the skin. Skin creams with extracts from the **Ginkgo** leaves, the supposedly oldest tree in the world, act as an antiaging agent. The ginkgo improves blood circulation, strengthens the skin, and stimulates cell renewal. This can reduce wrinkling. Grapes and raspberries contain **Resveratrol**, a phytoalexin with antioxidant properties that belongs to the polyphenols and supports collagen synthesis. It develops outstanding performance against free radicals, is anti-inflammatory, and protects against the damage of UV rays. Resveratrol enables a firmer, more even skin structure.

Aloe vera in antiaging creams helps to restructure the structure of the skin, defend against free radicals, and reduce the premature aging of the skin on the face, neck, and décolleté. A comparison of the cosmetic agents with botulinum toxin can be found in the literature [40].

The skin's own **enzymes** also have an antiaging effect. Enzymes are involved in most of the chemical processes in the skin, be it the hydrolysis of vegetable fats and phospholipids by lipases and phospholipases into the important fatty acids or hydrolytic cleavages of various proteins in amino acids to strengthen the NMF. Other enzymes catalyze the synthesis of collagen, which decreases in old age and should be boosted by addition of the vitamins C and E. Except for peeling products, the benefits of externally applied enzymes are not well documented yet, but it might be an interesting approach.

6.7 Essential Oils

Essential oils from plants have been used for thousands of years in various cultures for medicinal and health purposes. They can be applied for personal beauty care, especially for creams, and natural medicine treatments [41]. The low viscous oils and their water-based hydrolats are gained from plant parts such as the seeds, roots, stalks, barks, leaves, fruits, berries, or blossoms and herbs, preferably via steam distillation. In the steam distillation, a two-phase distillate arises. The upper phase forms the essential oil, while some water-soluble ingredients are in the lower water phase. This active ingredient-rich water is called hydrolat. Examples may be

Chamomile water, INCI: Chamaemelum Nobile, Roman Chamomile Water, Organic
Hamamelis water, INCI: Hamamelis Virginia (Witch Hazel) Water, Water, Alcohol
Lavender water, INCI: Lavandula Angustifolia, Lavender Water, Organic
Orange blossom water, INCI: Aqua, Citrus Aurantium, Benzoic Acid, Sorbic Acid
Rockrose water, INCI: Cistus Ladanifer, Rockrose Water, Organic
Rose water, INCI: Rosa Damascena Flower Water, Organic.

For example, the products can be used in a concentration of 100% as facial toner, rarely in amounts of 10–30% in creams. The most interesting ingredients of plants represent the essential oils. The composition of the oil depends on the crop, the extracted parts of the plant, and the process. As a rule, the oils smell strongly of the plant from which they originate. On the skin, many of the oils exhibit a positive health effect. For the application, there are restrictions, depending on the composition of the oil. As essential oils can irritate the skin, the application should be made only in heavily diluted form. In creams, the essential oils are always highly diluted (as opposed to perfume). They benefit from their antioxidant, antimicrobial, and anti-inflammatory properties on and in the skin. These healing oils are increasingly formulated because they act as natural medicine without any side effects. The use of an essential oil in amounts of 0.2–1% instead of a perfume has four key benefits:

- 100% natural base (typically about 30% for perfumes), pure substance preferably with proof of origin
- In many cases, allergy tested
- Selected oils show proven beneficial medical effects in the skin
- Various oils are also approved for food.

All points are important because the essential oils quickly penetrate the skin and enter the bloodstream within 20 minutes ([42]; see also Section 4.3). The proof of the tolerance gives the longtime cream user the certainty that no systemic side effects are to be expected. A classification of some essential oils in fragrance categories is shown in Table 6.12.

The essential oil of each plant shows a unique combination of ingredients and modes of action. Antibacterial, antiviral, and antifungal properties have several oils in common, as highlighted in Table 6.13. Antiseptic oils should preferably be used, selected according to the fragrance. However, the tea tree oil does not fit in any high-quality cosmetics because of the strong smell. Lavender, lemon, and sandalwood should preferably be used if the antimicrobial effect is needed for impure, damaged, or diseased skin such as for the treatment of acne or greasy skin. Because the oil shows only limited effectiveness because of the dilution effect, the concentration should be increased for specialty products.

The essential oils distilled from the plants are suitable for all skin types, but in particular for skin care of impure or dry skin and elderly skin. Many oils

Table 6.12 Fragrance notes of some essential oils after descriptions in the perfumery.

Floral	Citrus	Woody	Green	Fruity	Oriental
Carnation	Bergamot	Cedar	Basil	Apple	Amber
Chrysanthemum	Grapefruit	Fir	Cucumber	Apricot	Musk
Gardenia	Lemon	Hickory	Grass	Black currant	Spice
Hyacinth	Lime	Patchouli	Parsley	Cherry	Woody
Iris	Orange	Pine	Rhubarb	Grape	
Jasmine	Tangerine	Sandalwood	String bean	Melon	
Lavender	Verbena	Vetiver	Watercress	Peach	
Lilac				Pineapple	
Lily				Prune	
Marigold				Raspberry	
Rose				Strawberry	

Source: Data from Ref. [15].

Table 6.13 Antimicrobial properties of essential oils for the skin; effect depends on concentration.

Antibacterial	Antiviral	Antifungal
Cardamom, chamomile, cypress, eucalyptus, ginger, juniper, lavender, lemon, lemongrass, marjoram, mint, orange, pine, rosemary, sage, sandalwood, (tea tree), thyme, and wintergreen	Cinnamon, eucalyptus, lavender, lemon, oregano, sandalwood, (tea tree), and thyme	Eucalyptus, juniper, lavender, lemon, patchouli, sage, sandalwood, (tea tree), and thyme

Source: Data from Ref. [43].

contain an antioxidant and strengthen the skin against wrinkles. Because the oils penetrate quickly into the skin, they can take effect immediately. Depending on the skin type and problem, different oils can be applied (examples):

- *Normal skin*: geranium, jasmine, lavender, and rose.
- *Sensitive skin*: geranium, jasmine, lavender, Roman chamomile, and rose.
- *Oily skin*: bergamot, cedar, cypress, lemon, rose, rosemary, or sandalwood.
- *Impure skin, acne*: chamomile, lavender, manuka, myrrh, myrtle, rosewood, sage, yarrow, and vetiver.
- *Dry skin*: carrot seeds, chamomile, geranium, orange, rose, and Ylang Ylang.
- *Elderly skin*: Carnation, carrot seeds, frankincense, geranium, lavender, myrrh, and rose.
- *Skin allergies*: Roman chamomile.
- *Cellulite*: cypress, geranium, juniper, lemon, rosemary, sandalwood, and blood orange.

The market offers a great selection with more than 500 different essential oils, of which less than 50 are recommended for cosmetic creams. The prices of the expensive oils depend on the qualities. For the addition in high-quality creams, the origin and quality as well as 100% purity of the natural oils should be known. For example, various cosmetic creams contain pure rose oil consisting of more than 300 ingredients. The genuine oil provides a soft and supple skin while acting antiseptic, antiviral, anti-inflammatory, and astringent. It strengthens the skin tension and is ideal for antiaging products. Furthermore, almost all people like the intense scent of the oil. Well-tolerated effective essential oils are listed in Table 6.14. A complete list of essential oils can be found at Wikipedia [44].

The essential oils not only work in the skin but the evaporating parts are also absorbed through the nose and show effects at prolonged exposure, especially in aromatherapy. From the sensory cells of the nose, the fragrance information reaches the brain directly. Every scent has a different effect. They are stimulating or invigorating, soothing, or relaxing. In addition, the fragrances improve well-being, can affect feelings and emotions, and thus make a small contribution to health. Other essential oils increase concentration, relieve pain, and have an anti-inflammatory effect. Some oils, such as original jasmine, show a positive effect on sexuality. Further information can be found in Section 9.6.

Table 6.14 Selection of well-tolerated essential oils from steam distillation.

No	Essential oil	INCI name	Properties
1	Carrot	Daucus Carota Fruit Oil (seed)	Antiwrinkles, cell regenerating, stimulates the lymph channels (reduces swelling), against acne and psoriasis, spicy, peppery scent
2	Chamomile (Roman)	Anthemis Nobilis Flower Oil	Antiseptic, anti-inflammatory, soothing, counteracts blemishes, sweet, fruity scent
3	Geranium	Pelargonium Roseum Leaf Oil	Stimulates cell renewal, antiseptic effect, supports drainage of sebum, astringent, large pores become downsized, stimulates the lymph channels (reduces swelling), scent reminds of roses
4	Jasmine	Jasminum Officinale Oil	Antiseptic, anti-inflammatory, healing, strong, long-lasting aroma, aphrodisiacal
5	Juniper berry	Juniperus Oxycedrus Fruit Oil	Cleansing, anti-inflammatory, firm skin, stimulates the lymph channels, fruity, aromatic fragrance
6	Lavender	Lavandula Angustifolia Oil	Soothing, antibacterial, stimulates cell renewal – known to be effective
7	Lemon	Citrus Limon Seed Oil	Antiseptic, fresh and firm skin, stimulates the lymph channels, fresh smell
8	Orange blossom	Citrus Aurantium Dulcis Flower Oil	Stimulates cell renewal, antiseptic, fresh and firm skin, supports drainage of sebum, popular scent in children
9	Rose	Rosa Damascena Flower Oil	Antiseptic, anti-inflammatory, soothing, strengthening the skin
10	Rosemary	Rosmarinus Officinalis Leaf Oil	Stimulating, circulation-promoting, fresh and firm skin, camphor to forest-like smell
11	Sandalwood	Santalum Album Wood Oil	Antiseptic, moisturizing, soothing, best male fragrance
12	Ylang-Ylang	Cananga Odorata Leaf Oil	Antiseptic, soothing, fresh and firm skin, fragrance: sexuality and sensuality

6.8 Extracts from Plant Parts

The exemplified listed plants can not only be steam distilled but also extracted (Table 6.15). Extracts from the solvent extraction with alcohol and/or water as well as hexane are called tinctures, and the oil-based extracts macerates. The composition of the extract depends on the solvent. In addition to the mentioned historically used solvents, various liquids are common for extraction today. These include propylene glycol, butylene glycol, and glycerol in pure form or in mixtures with water and for the oil-soluble ingredients the sunflower oil or caprylic/capric triglycerides.

Thus, in addition to the vegetable oils, four different liquids can be obtained from many plants, depending on the process, in fact the essential oils, the

Table 6.15 Well-known plant extracts with cosmetic effect.

Common name	INCI name	Extraction solvent	Effect
Arnica	Arnica Montana Flower Extract	Alcohol	Against inflammation
Caffeine	Caffeine	Water, than trichloroethylene (will be removed)	Skin tightening
Centella	Centella Asiatica (Gotu Kola) Leaf Extract	Alcohol/water	Antibacterial, wound healing, reduces scarring, antiaging
Chamomile [45]	Chamomilla Recutita (Matricaria) Flower Extract	Alcohol/water	Soothing, healing effect, anti-inflammation
Green tea	Camellia Sinensis (Green Tea) Leaf Extract	Alcohol/water	Antioxidant, antiaging.
Hamamelis (Witch Hazel)	Hamamelis Virginiana (Witch Hazel) Extract	Alcohol/water, bark/leaf extract	Skin caring, against acne, and inflammation
Hop	Humulus Lupulus (Hops) Extract	Alcohol/water	Application for poorly healing wounds and inflammation, stimulates collagen production
Jiaogulan (Southern Ginseng)	Gynostemma Pentaphyllum Leaf Extract	Alcohol/water	Anti-inflammatory, antioxidant, regulates skin renewal
Lavender	Lavandula Angustifolia Flower Extract	Alcohol/water	Promoting healing and circulation, anti-inflammatory, soothing
Lemon Balm	Melissa Officinalis L.	Alcohol/water; oil	Against herpes, eczema and wounds
Mangosteen	Garcinia Mangostana Fruit Extract	Alcohol/water	Anti-inflammatory, antimicrobial, antifungal, antiviral, allergy-inhibiting; rich in vitamins and minerals
Marigold	Calendula Officinalis L.	Alcohol/water; oil	Against inflammatory and eczemic skin
Rosemary	Rosmarinus Officinalis (Rosemary) Leaf Extract	Alcohol/water; oil	Promotes circulation and skin metabolism
Sage	Salvia Officinalis (Sage) Leaf Extract	Alcohol/water; oil	Anti-inflammatory and antimicrobial properties
Thyme	Thymus Vulgaris Extract	Alcohol/water	Promotes circulation, anti-inflammatory and antimicrobial properties
Saint John's Wort	Hypericum Perforatum Extract	Oil; alcohol/water	Wound treatment, burns, anti-inflammatory

hydrolats, tinctures, and macerates. The products obtained differ significantly in their composition. The essential oils and macerates have lipophilic properties, while the hydrolats and tinctures are largely soluble in water or in a glycerol/water mixture. A particularly gentle method is the extraction with carbon dioxide. With this process, the active ingredients can be obtained in pure form. However, the relatively high costs stand against a broad application. The macerates and tinctures are used in the concentration obtained or concentrated or dried as powder. One method is to take the extracts after removal of the solvent in glycerol/water mixture and sell it in this form or with preservation. Such tinctures require an addition of 1–5% to the cream, calculated on dry substance, in exceptional cases of 10%.

For the recovery of ingredients by extraction of plants, well over a hundred plants can be used. Numerous examples are provided in Table 6.16 and dried products are provided in Table 6.17. It should be noted that the extracts have an

Table 6.16 Cosmetically relevant extracted substances from plants.

Products from plants – Name: INCI name
Aloe vera: Aloe Barbadensis
Andiroba oil: Carapa Guianensis
Apple: Pyrus Malus
Apricot: Prunus Armeniaca Asiatic centella extract: Centella Asiatica
Bamboo extract: Bambusa Arundinacea
Calendula: Calendula Officinalis Capsicum: Capsicum Frutescens
Chamomile: Matricaria Recutita
Chlorella: Chlorella Vulgaris Extract
Citrus: Citrus sp.
Cocoa extract: Theobroma Cacao
Coconut: Cocos Nucifera
Coffee extract: Coffea Arabica Common nettle extract: Urtica Dioica
Cucumber Extract: Cucumis Sativus
Everlasting: Sempervivum Tectorum
Fruit acids AHA: Passiflora Quadrangularis, Ananas Sativus, Vitis Vinifera
Fucus extract: Fucus Vesiculosus
Gentian extract: Gentiana Lutea
Ginger: Zingiber Officinale
Ginkgo: Ginkgo Biloba
Ginseng: Panax Ginseng
Glutange: Hydrolyzed Wheat Protein
Great Burdock: Arctium Majus
Green tea: Camellia Sinensis
Guarana: Paullinia Cupana
Hamamelis extract: Hamamelis Virginiana
Hibiscus: Hibiscus Sabdariffa
Honey extract: Honey
Horsechestnut: Aesculus Hippocastanum
Horsetail extract: Equisetum Arvense
Ivy extract: Hedera Helix
Kelp sea extract: Algae
Kiwi: Actinidia Chinensis
Kola nut extract: Cola Nitida
Licorice: Glycyrrhiza Glabra
Mango: Mangifera Indica
Marshmallow: Althaea officinalis
Oat: Avena Sativa
Oatlange: Hydrolyzed Oat Protein
Olive oil: Olea Europaea
Olive extract: Olea Europaea
Orange: Citrus Sinensis
Papaya: Carica Papaya
Pineapple: Ananas Sativus
Rosehip seed oil: Rosa Moschata
Rosemary: Rosmarinus Officinalis
Royal jelly extract: Royal Jelly
Sage Extract: Salvia Officinalis
Silymarin: Silybum Marianum
Soy extract: Glycine Soja Extract
Soylange: Hydrolyzed Soy Protein
Tea extract: Camellia Sinensis
Thyme: Thymus Vulgaris
Tomato: Solanum Lycopersicum
Turmeric extract: Curcuma Longa
Vanilla extract: Vanilla Planifolia
Vine (red grape): Vinis Vinifera
Wild yam extract: Dioscorea Villosa
Zealange: Hydrolyzed Corn Protein

Source: Data from Ref. [46].

Table 6.17 Dried extracts from plants with relationship to cosmetics.

Name	INCI name	Effect
Acerola (freeze-dried)	Malpighia Punicifolia	Very high in vitamin C
Artichoke	Cynara Scolymus	Metabolic stimulant
Bilberry	Vaccinium Myrtillus	Astringent, against couperose, invigorating, vascular protective
Birch	Betula Alba	Astringent, invigorating, elasticizing, soothing
Bladderwrack	Fucus Vesiculosus	Soothing, softening, stimulating
Butcherbroom	Ruscus Aculeatus	Astringent, invigorating, vascular protective
Chamomile	Chamomilla Matricaria	Soothing, refreshing, anti-inflammatory
Centella	Cedrus Atlantica	Antibiotic, antifungal, and cytostatic
Coneflower and Red Coneflower	Echinacea Purpurea Echinacea Angustifolia	Stimulating, soothing, and firming, against wrinkles
Dandelion	Taraxacum Officinale	Astringent, stimulating
Devil's claw	Harpagophytum Procumbens	Anti-inflammatory and slightly analgesic
Dog rose	Rosa Canina	Anti-inflammatory, vascular protective
Elder	Sambucus Nigra	Anti-inflammatory, antioxidant
Fennel	Foeniculum Vulgare	Fragrant, invigorating, stimulating
Ginseng	Panax Ginseng	Stimulating, firming, elasticizing, invigorating
Green tea	Camellia Sinensis	Astringent, vascular dilator stimulating, antioxidant
Guarana	Paullinia Cupana	Astringent, invigorating
Hawthorn	Crataegus Monogyna	Strengthening, astringent, against couperose, anti-irritant
Hops	Humulus Lupulus	Strengthening, astringent, firming, soothing
Horse chestnut	Aesculus Hippocastanum	Against edema, invigorating, astringent, against couperose
Horse tail	Equisetum Arvense	Elasticizing, astringent, sebostatic
Lady's milk thistle	Silybum Marianum	Antioxidant, anti-inflammatory
Lemon balm	Melissa Officinalis	Soothing, softening, invigorating, anti-inflammatory
Licorice	Glycyrrhiza Glabra	Soothing, refreshing
Mallow	Malva Sylvestris	Softening, anti-irritant, soothing
Olive	Olea Europaea	Emollient, against skin redness, vasodilator
Passion flower	Passiflora Incarnate	Relaxing, calming
Red grape	Vitis Vinifera	Astringent, vascular protective
Willow	Salix Alba	Anti-inflammatory, astringent, cooling

Source: Data from Ref. [46].

intense color and smell typical for the plant. They may not fit the designed cream. In general, there are comparable alternatives among the plant extracts or among the vegetable oils. For the extraction, there is no standard method. Generally, a solvent dissolves ingredients from the crushed parts of plants at room or elevated temperatures. Depending on the production process (solvent, temperature) and on the grinded plant parts, the composition of the extract obtained differs considerably. Caution is advised when buying extracts. In many cases, there is no scientific evidence of the effects in the skin and no specific dosage recommendations. The allergy potential is also not known exactly. It is therefore recommended to use only well-known and well-studied substances with precise origin and scientific description and dosage.

Synthesized **allantoin** corresponds to the substance from the natural extracts of the comfrey plant. The use in cosmetics shows its safe, nontoxic properties, and the high compatibility with other cosmetic raw materials. Allantoin impresses by various beneficial impacts. First, the substance provides a slight moisturizing and significant keratolytic effect, which increase the water content of the horny layer and facilitate the desquamation of upper layers of dead skin cells. Second, allantoin additionally increases the suppleness of the skin, promotes cell proliferation, and wound healing. It has a soothing, anti-irritant, and skin protectant effect. Allantoin is used with 0.1–0.3% for baby and child creams.

6.9 Active Ingredients from the Sea

Seventy percentage of the Earth's surface is covered by the seas. These habitats are home to fascinating, unique plants and organisms, with the highest biodiversity in tropical reefs. However, also marine plants growing in polar region show positive test results for possible use in cosmetics. The cosmetic industry uses already some interesting substances from marine plants. Clinical trials take place with therapeutic effective substances. The potential to discover new drugs is huge. Many scientists are convinced that marine natural products will become very important in cosmetics and pharmaceuticals.

Of industrial importance so far are the unique polysaccharides that occur in various species of algae. These include alginic acid, alginates, and fucoidans from brown algae as well as carrageenans from red algae. Various studies on red and brown algae extracts indicate an antioxidant and antiproliferative action against cancer cells. These are attributed to the fucoidans and polyphenols. Besides, algae include minerals, trace elements, and vitamins. In addition to calcium, magnesium, iron, and potassium, they contain the trace elements iodine, copper, zinc, and selenium, along with manganese, strontium, molybdenum, or germanium. The bioavailable elements are bound to polysaccharides. Furthermore, algae contain the vitamins A, C, and E as well as the wide range of the vitamin B complex, including vitamin B_{12} that is otherwise only found in animal products. These components are important for the immune defense as well as for the vitalization of the skin and the connective tissue (see Table 6.18).

Table 6.18 Active ingredients from the sea [47, 48].

Name	INCI name	Characteristics
Chlorella	Chlorella Vulgaris	Chlorella algae are microalgae that contain valuable active ingredients such as proteins, amino acids, lipids, carbohydrates, vitamins, minerals, and trace elements. The high polysaccharide content ensures skin hydration
Fucus	Fucus Vesiculosus	The bladderwrack is a macroalgae with a high content of trace elements, especially iodine, which stimulates the metabolism. Good anticellulite beauty prescriptions should include bladderwrack extracts. Thanks to the high content of muco-polysaccharides, it gives the skin softness and elasticity
Gumlange	Algae Extract	The polysaccharides of algae extract form a film on the surface of the skin, which protects against dehydration. Gumlange ensures optimal moisturization of the skin
Kelp	Laminaria Digitata	The finger tang is a macro alga with a high content of trace elements, especially iodine. The extract is mainly contained in anticellulite creams and provides moisturizing and soothing properties
Protalgel	Algae Extract, Hydrolyzed Algin, Carrageenan	Protalgel is a mixture of multiple sugars and peptides derived from the enzymatic digestion of microalgae (polysaccharides and proteins). It is extremely moisturizing, nourishing, and soothing
Sealange	Algae Extract, Zinc PCA, Magnesium PCA, Manganese PCA, Calcium PCA	This polysaccharide-rich complex contains an extract of enzymatically hydrolyzed sea algae and a mixture of trace elements. This combination creates a moisturizing, remineralizing, and nourishing complex. It is suitable as an antiaging ingredient and additionally has anti-inflammatory, sebum-regulating and antibacterial properties
Spirulina	Spirulina Platensis Extract	Spirulina is an extract of microalgae. Features: (i) UV-protective; (ii) antioxidant thanks to the carotenoids (beta-carotene) and vitamins (vitamin E) through the reduction of free radicals; (iii) nourishing by high protein content; (iv) moisturizing by the polysaccharides; (v) supports metabolic processes by the content of minerals and trace elements

Various ingredients of algae extracts protect the skin or have a positive effect on certain skin properties [46]:

- UV protection by certain phenols (phlorotannins), carotenoids, the enzyme photolyase, and mycosporine-like amino acids.
- Skin protection by radical scavengers such as tocopherols, polyphenols, β-carotenoids (vitamin A), and other carotenoids that stimulate the immune system.
- Moistening the skin with (sulfated) polysaccharides and mucopolysaccharides (glycosaminoglycans).
- Skin smoothing and regeneration by essential and nonproteinogenic amino acids as well as highly unsaturated fatty acids.

- Brightening by phlorotannins, phloroglucinol, and its oligomers.
- Stimulation of collagen synthesis by low-molecular-weight fucoidans.

Some compounds are unique or in an unusual combination and can synergistically increase the overall effect.

The proven stimulation of collagen synthesis explains the increasing use of fucoidans in cosmetics. The antiaging effect is said to be based on a reduction in skin thickness and an increase in elasticity by applying aqueous fucoidan-containing extracts. High-fucose oligo- and polysaccharides inhibit the matrix metalloproteinases and stimulate collagen and elastin synthesis in the fibroblasts, thereby improving the elasticity of the skin and connective tissue. The result is a younger-looking skin.

6.10 Origin of the Active Ingredients

The extraction of natural oils and other active ingredients from plants requires the use of chemical engineering processes, which are called "unit operations." These include washing, crushing, squeezing and pressing, distilling and steam distilling, deodorizing, solvent extracting, filtering, drying, evaporating, producing powders, grinding, sieving, and others. In addition to the natural substances, nature-identical substances are usually formulated in cosmetic creams (Table 6.19). They can be synthesized by using various chemical pathways or by complicated biotechnological routes. In contrast, purely synthetic substances hardly play a role in cosmetic creams. In order to get to the subsequent products of the plant oils, other chemical reactions must be used. Typical reactions are the hydrolysis of ester compounds (oils), methylation, hydrogenation, esterification, transesterification, and ethoxylation (see also Figure 5.2). For example, the production of cellulose-based raw materials requires a number of chemical processes to obtain the pulp and with their subsequent methylation and ethoxylation.

6.11 Learnings

- ✓ The mainstream cosmetic creams are designed to meet four customer needs. The skin requires substances for moisturizing, revitalizing, smoothing, and antiaging.
- ✓ The permanent use of good cosmetic creams results in improved skin texture, firmer dermal tissue, and diminished signs of aging, all in all to a better look.
- ✓ Proven substances in moisturizers belong to the NMF. These are substances that occur in the cornea and bind moisture.
- ✓ NMF includes in particular short-chain amino acids, PCA and urea, peptides, lactic acid/lactates, citric acid/citrates, and sugar.
- ✓ For the revitalizing of the skin, the vitamins have proven to be effective. Recommended is a mixture of the vitamins B_3 (nicotinamide), C (ascorbic acid), E (tocopherol), and the provitamin B_5 (panthenol) as well as ubiquinol (Coenzyme Q10) and squalene.

Table 6.19 Ingredients from the nature and from chemical modifications of the natural substances, furthermore from nature-identical chemical synthesis.

Basic material	Unit operations	Chemical reaction	Product (examples)
Plant	Squeezing		Oil o Hemp o Borage seed o Soy bean
Plant	Steam distillation		Essential oil o Fennel o Lavender o Rosmarin
Plant	Solvent extraction		Tinctures o Green tea o Hamamelis o Hop Macerates o Lemon Balm o Marigold o Sage
Marine plants	Extraction, drying		o Various algae extracts o Sea salts
Natural materials		Biotechnological synthesis (fermentation)	o Citric acid o Hyaluronic acid o Ascorbic acid (GMO)
Vegetable oils		Hydrolysis (chemical modifying of natural products)	o Mono-, di-glyceride o Glycerol o Fatty acids
Vegetable oils		Methylation (chemical modifying of natural products)	o Fatty acid methyl ester o Glycerol
Fatty acid methyl ester, fatty acids		Hydrogenation (chemical modifying of altered natural products)	o Fatty alcohols
Fatty alcohol, acids, mono-, di-glyceride, sorbit		Ethoxylation (chemical modifying of altered natural products)	Emulsifier o Ethoxylates of fatty alcohols, acids, glycerides o Ethoxylates of sorbitan fatty acid ester
Various substances		Nature identical substances (synthetically produced)	o Vitamins o Urea o PCA o Citral
Various substances		Synthetic products unrelated to nature	o Silicone oils o Butylated hydroxy toluene o EDTA o Sodium hydrogen phosphate

- ✓ To smooth the skin, vegetable oils are useful. For this purpose, a mixture of greasy and less greasy oils should be applied, for example, a mix of shea butter and evening primrose oil.
- ✓ Effective antiaging agents show antioxidant and cell renewal properties. Radical scavengers such as vitamins A (retinol), B_3, C, E, and α-lipoic acid as well as synthetic di-, tri-, tetra-, penta-, and hexapeptides have proven effects.
- ✓ All substances with an antioxidant effect help to reduce the typical signs of aging. These occur especially in people who smoke, drink, often sunbathe, and lead an unhealthy lifestyle.
- ✓ In some cases, it is worthwhile to formulate a plant extract or an extract from marine plants to use their healing properties.
- ✓ The solvents for extraction are alcohol/water mixtures for gaining tinctures or oils for macerates. A new, good but expensive process is the extraction with carbon dioxide.
- ✓ The baby's skin shows many peculiarities that are to be considered in formulating a baby cream. Parabens, dyes, mineral oils, perfume, and potential allergens should not be included.

References

1. Anforderungen an Naturkosmetika, (definition des Bundesministeriums für Gesundheit (BMG), 1992/1993). https://www.fragrancy.de/naturkosmetik.htm (assessed 9 November 2017).
2. Ellsässer, S. (2013). *Körperpflegekunde und Kosmetik*, 3e. Berlin: Springer-Verlag.
3. COSMOS, Trust in organic and natural cosmetics, The COSMOS-standard. https://cosmos-standard.org/the-cosmos-standard/ (assessed 9 November 2017).
4. Draelos, Z.D. (2005). *Cosmeceuticals*, Procedures in Cosmetic Dermatology, 2e. Elsevier Inc.
5. Walters, K.A. and Roberts, M.S. (eds.) (2007). *Dermatologic, Cosmeceutic, and Cosmetic Development: Therapeutic and Novel Approaches*. New York, NY: Taylor & Francis.
6. Draelos, Z.D. (2007). The latest cosmeceutical approaches for anti-aging. *J. Cosmet. Determatol.* 6 (s1): 2.
7. Zuzarte, M., Gonçalves, M.J., Cavaleiro, C. et al. (2011). Chemical composition and antifungal activity of the essential oils of *Lavandula viridis* L'Her. *J. Med. Microbiol.* 60: 612–618.
8. Talakoub, L., Neuhaus, I.M., and Yu, S.S. (2009). Cosmeceuticals. In: *Cosmetic Dermatology* (eds. M. Alam, H.B. Gladstone and R.C. Tung), 7. Philadelphia: Saunders.
9. Lupo, M.P. and Cole, A.L. (2007). Cosmeceutical peptides. *Dermatol. Ther.* 20 (5): 343–349.
10. Rähse, W. (2014). *Industrial Product Design of Solids and Liquids, A Practical Guide*. Weinheim: Wiley-VCH.

11 Medicinal Products Act (Arzneimittelgesetz – AMG), 2006. https://www.gesetze-im-internet.de/englisch_amg/englisch_amg.pdf (assessed 9 November 2017).

12 Lüllmann, H., Mohr, K., and Hein, L. (2006). *Pharmakologie und Toxikologie*, 16e. Georg Thieme Verlag, Stuttgart Chapter 2.6, p. 39 ch. 22, pp. 258–262; 18th edn. 2016.

13 Gabard, B. (2000). *Dermatopharmacology of Topical Preparations: A Product Development-Oriented Approach*. Berlin: Springer.

14 Soler, C. (2005). Protección de la piel frente al frío. *Acófar* (438).

15 Kirk-Othmer (2013). *Chemical Technology of Cosmetics*. Hoboken, NJ: Wiley.

16 BFR Bundesinstitut für Risikobewertung, Vitamin A: Aufnahme über kosmetische Mittel sollte begrenzt werden, Aktualisierte Stellungnahme Nr. 005/2014 des BfR vom 31. Januar 2014. http://www.bfr.bund.de/cm/343/vitamin-a-aufnahme-ueber-kosmetische-mittel-sollte-begrenzt-werden.pdf (assessed 9 November 2017).

17 Gehring, W. (2004). Nicotinic acid/niacin amide and the skin. *J. Cosmet. Dermatol.* 3 (2): 88–93.

18 Ebner, F., Heller, A., Rippke, F., and Tausch, I. (2002). Topical use of dexpanthenol in skin disorders. *Am. J. Clin. Dermatol.* 3 (6): 427–433.

19 Personal care, formulating anti-ageing products with folic acid, 2009. https://www.personalcaremagazine.com/story/5247/formulating-anti-aging-products-with-folic-acid (assessed 9 November 2017).

20 Newbeauty, skin care|active ingredients, this antioxidant is the only one that also stimulates collagen, 2019. https://www.newbeauty.com/blog/dailybeauty/10319-vitamin-c-anti-aging-benefits/ (accessed 7 May 2019).

21 Zielinska, A. and Nowak, I. (2014). Fatty acids in vegetable oils and their importance in cosmetic industry. *CHEMIK* 68 (2): 103–110.

22 Henry Lamotte Oils, Product Portfolio. Cosmetic oils: https://www.lamotte-oils.de/fileadmin/Daten_Oils/Dokumente/Sortiment/Oils_assortment.pdf (assessed 9 November 2017). Organic range: https://www.lamotte-oils.de/fileadmin/Daten_Oils/Dokumente/Sortiment/Organic_assortment.pdf (assessed 9 November 2017).

23 Ecowoman, the green side of life, Wundermittel Sheabutter: Die erstaunliche Wirkung für Haut & Haar. http://www.olinat.co.za/shea-butter.htm#Benefits of shea butter (assessed 9 November 2017).

24 Natural living ideas, 12 surprising benefits of jojoba oil for beautiful skin & hair, 2019. https://www.ecowoman.de/kosmetik/body/wirkung-und-anwendung-von-sheabutter-universale-naturkosmetik-aus-einer-pflanze-5689 (accessed 7 May 2019).

25 Mandeloel.info, Mandelöl – Sanfter Schutz für den Körper und mehr, 2019. http://www.mandeloel.info/ (accessed 7 May 2019).

26 Truth in aging, the honest truth about beauty & personal care products, what is it: evening primrose oil, 2013. https://www.truthinaging.com/review/what-is-it-evening-primrose-oil (accessed 7 May 2019).

27 Provital group, evening primrose oil, 2006. http://www.centerchem.com/Products/DownloadFile.aspx?FileID=6976 (assessed 9 November 2017).

28 Nachtkerzenoel.net, Nachtkerzenöl – eine Wohltat für die Haut und mehr, 2019. http://www.nachtkerzenoel.net/ (assessed 9 November 2017).

29 HanfHaus, general, what is hemp? 2006. https://www.hanfhaus.de/en/general-i-1.html (assessed 9 November 2017).

30 Shikai, borage oil: a little known secret for maintaining healthy skin, 2017. https://www.shikai.com/borage-oil/ (assessed 9 November 2017).

31 Lush, fresh handmade cosmetics, INGREDIENT, Wheatgerm oil, 2019. https://uk.lush.com/ingredients/wheatgerm-oil (assessed 9 November 2017).

32 Good Health Academy (2018). The Surprising Benefits of Safflower Oil for Hair And Skin. http://www.goodhealthacademy.com/beauty-tips/safflower-oil-for-hair-and-skin/ (accessed 7 May 2019).

33 Krist, S. (2013). *Lexikon der pflanzlichen Fette und Öle*, 2e. Wien: Springer-Verlag.

34 Berliner Zeitung, Nur wenige Wirkstoffe aus Anti-Aging-Cremes bremsen nachweislich die Hautalterung, Was wirklich gegen Falten hilft, 2010. https://www.berliner-zeitung.de/nur-wenige-wirkstoffe-aus-anti-aging-cremes-bremsen-nachweislich-die-hautalterung-was-wirklich-gegen-falten-hilft-15064238 (assessed 9 November 2017).

35 Vitaminum Life, Alpha-Liponsäure gegen Hautalterung und Falten, 2016. http://www.alpha-liponsaeure.com/gegen-hautalterung.html (assessed 9 November 2017).

36 Blanes-Mira, C., Clemente, J., Jodas, G. et al. (2002). A synthetic hexapeptide (Argireline) with antiwrinkle activity. *Int. J. Cosmet. Sci.* 24: 303–310.

37 Schagen, S.K. (2017). Topical peptide treatments with effective anti-aging results. *Cosmetics* 4 (2): 16. https://doi.org/10.3390/cosmetics4020016.

38 Coenzyme Q10, from Wikipedia, the free encyclopedia, 2019. https://en.wikipedia.org/wiki/Coenzyme_Q10 (accessed 7 May 2019).

39 Grossman, R. (2005). The role of dimethylaminoethanol in cosmetic dermatology. *Am. J. Clin. Dermatol.* 6 (1): 39–47. http://www.ncbi.nlm.nih.gov/pubmed/15675889 (assessed 9 November 2017).

40 Zoe Diana Draelos, MD, Dermatology Consulting Services, Cosmeceuticals, Topical Cosmeceuticals Attempting to Mimic Injectable Botulinum Toxin: Is It Possible? 2019. http://www.zoedraelos.com/articles/cosmeceuticals/ (accessed 7 May 2019).

41 Higley, C. and Higley, A. (2005). *Reference Guide for Essential Oils*, 9e. Abundant Health.

42 Tisserand, R.B. and Young, R. (2013). *Essential Oil Safety*, 2e. Churchill Livingstone.

43 smarticular, Das richtige ätherische Öl für jeden Zweck, 2019. https://www.smarticular.net/das-richtige-aetherische-oel-fuer-jeden-zweck/ (accessed 7 May 2019).

44 List of essential oils, From Wikipedia, the free encyclopedia, 2019. https://en.wikipedia.org/wiki/List_of_essential_oils (accessed 7 May 2019).

45 Srivastava, J.K., Shankar, E., and Gupta, S. (2010). Chamomile: a herbal medicine of the past with bright future. *Mol. Med. Report* 3 (6): 895–901. https://doi.org/10.3892/mmr.2010.377.

46 http://www.mani-gmbh.com/web/actives.398.html (assessed 4 February 2018).

47 Deutsche Apotheker Zeitung, NATURSTOFFE, Makroalgen – Potenzial für Pharmazie und Kosmetik, 2010. https://www.deutsche-apotheker-zeitung.de/daz-az/2010/daz-50-2010/makroalgen-potenzial-fuer-pharmazie-und-kosmetik (assessed 5 February 2018).

48 woojin-korea, main item, 2005. http://woojin-korea.com/products/16LAN/LAN1800.htm (assessed 6 February 2018).

7

Active Ingredients for Special Products

7.1 Definition of Special Creams

This chapter deals with active ingredients for cosmeceuticals [1, 2]. Cosmeceuticals, another name for special cosmetic creams, aim at the area between the typical cosmetic and the therapeutic agents [3]. They are characterized by clearly detectable effects. These specialties, which have become indispensable in our lives, fulfill important tasks, as described below for creams with proven active ingredients.

Every formulator of recipes needs to check whether the proposed ingredients are suitable for their special cream or whether better alternatives exist. For most substances, there are 5–100 alternatives, but they must fit into the formulation. The statements and suggestions made here are essentially based on own experience and tested substances, approved by the European Cosmetics Regulation. The special creams and lotions include acne, after sun, babies, bedsore (decubitus), cellulite, depilatory, dry skin sites and eczema, foot and athlete's foot, hands, herpes, horny layers, lotions, tanners, and water loss (shin) as well as ultraviolet (UV) protection. Surprisingly, only a few organic products exist in the group of specialties that really support the protection or healing, even if the terms "cure" and "alleviation" are not allowed on the packaging of a cosmetic product. Table 7.1 demonstrates the delimitation of the special products against cosmetics and therapeutics. All creams discussed here can be assigned to the group of cosmeceuticals. The skin disorders and their control are extensively described in the literature [4]. In many cases, cosmetic creams can promote healing.

7.2 Antiacne Creams for Blemished Skin

Acne [5] is a common skin disease that has affected 85% of the population for a period of time. Because of the hormonal changes during puberty, acne occurs mainly in adolescents over the age of 11. Almost every teenager has sometime discovered the characteristic blackheads and pimples in his face. About 60% of teenagers suffer from harmless, physiological acne for one to two years, characterized by the appearance of some blackheads and pimples. This pre-acne can be treated by cosmetic and over-the-counter (OTC) remedies from the pharmacy. The situation is totally different with regard to clinical acne, which affects

Cosmetic Creams: Development, Manufacture and Marketing of Effective Skin Care Products,
First Edition. Wilfried Rähse.

Table 7.1 Delimitation of cosmeceuticals from simple cosmetic creams and from therapeutic skin creams.

Subject **Manufacturer**	**Cosmetics (skin care)** **Cosmetic industry**	**Cosmeceuticals (medical skin care)** **Cosmetic industry**	**Dermatics (medicament)** **Pharmaceutical industry**
Tasks	Beauty ➢ Fragrancy ➢ Care ➢ Moisture	Health ➢ Care/protection/ prevention/concomitant therapy ➢ Delay skin aging ➢ UV protection ➢ Elimination of dry skin ➢ Acne, cellulite, herpes ➢ Medical treatment of skin areas	Healing of skin diseases ➢ Cure, alleviation ➢ Eczema ➢ Dermatitis/ psoriasis ➢ Seborrheic keratosis ➢ Microbial infestation, inflammation ➢ Rashes ➢ Wounds
Active ingredients			
• Number	1 (Advertised)	>2	Mostly 1
• Concentrations	Low	Medium to high	Effectiveness and age dependent (children, adults)
Additives			
• Perfume	Yes	No	No
• Dye	In some cases, yes	No	No
Emulsion type (preferred)	o/w	o/w	w/o
Water (%)	70 to >80	45–70	60–80?
Container	(Pump) bottles; crucibles (glass or plastic)	Dispensers; airless dispenser	tubes

? – Value is unknown or not safe.

about 40% of the population and can last for between 5 and 20 years and requires specialist treatment. In severe cases, the disease does not cure completely. The release of male hormones can also trigger the formation of pimples and even of acne in elderly years.

Acne is common in blemished skin, which is characterized by a moist-greasy condition. During the acne, blackheads, the so-called comedones, develop, sometimes forming small red pustules. The skin condition can be referred to as physiological or pre-acne. Acne is a condition triggered by the male hormones (testosterone), which is downsized by estrogens. Therefore, acne is a typical side effect of puberty. It is also caused by stress, fatigue, sugar, cigarettes, alcohol, and others.

It occurs in the face during the teenage years and ends some years later. The external signs are found on the nose, forehead, and chin, and sometimes in the case of oily skin on the whole face. In girls, the effects of acne are caused by the female cycle. In the second half of the female cycle, the production of testosterone is increased; this is also observed during pregnancy.

In adolescents during puberty, the processes in the skin change. Under the influence of male sex hormones (androgens), the sebaceous glands enlarge and produce in a hyperfunction more sebum. In healthy skin, the sebum freely penetrates through the sebaceous duct to the outside of the skin. In the case of acne, because of the androgen stimulation, the formation of horn material also increases in duct lining cells. The excess horn material eventually clogs the excretory duct. In the first phase, the nonvisible microcomedones emerge, resulting in a second phase to the acne typical skin lesions including open and closed blackheads. If the sebum plug lies below the closed skin surface, the comedo shimmers through the skin as a hard, bright point. The position of the plug in the surface creates an open comedo, known as a blackhead. The described impurities can inflame. The metabolic processes of bacteria and other pathogens, settling on the inflamed sites, lead to infections and redness of the skin. At the top of the formed pimple, clearly visible white pus often arises (Figure 7.1).

Explained in more detail, a follicular canal can react sensitive to the enzymatic degradation products of the sebum lipids. Then, the horny layer grows increasingly (hyperkeratosis) in the duct or in the sebaceous gland, which enlarges. In both, a mixture of sebum, horn cells, bacteria, and hair parts seals the canal, a blackhead emerges at the exit of the sebaceous duct. The typical appearance of the blackheads arises from the reaction of the embedded melanin with the oxygen of the air, so that the plug is dark colored. The sebaceous glands produce increased skin fat. The plugs prevent the thicker skin fat from flowing away, causing the sebum to accumulate more and more in the skin. Therefore, the sebaceous gland and the duct inflate, and in certain circumstances the walls, can tear. Then, the content of the comedo pours into the tissue and triggers inflammatory reactions with pus formation. In less dramatic cases, bacterial enzymes and sebaceous degraded products can lead to irritation of the walls. That means, inflammation with pus formation occurs.

Because of the overactivity of the sebaceous gland, the bacterium "propionibacterium acnes" finds improved growth conditions in the outlet of the sebaceous duct and proliferates. The bacteria have enzymes that degrade the sebum and keep the inflammation going. Special immune cells migrate into the region and loosen up the wall of the duct. Therefore, they facilitate and intensify the back and forth of messengers of inflammation, causing pustules, papules, nodules, and scars on the skin. Encapsulation of the bacteria in the follicular duct results in antibiotic resistance and poor response to some acne remedies. External factors that promote obstruction of the sebaceous ducts must be avoided.

Clinical acne should be unconditionally treated by a dermatologist. Only the precursors can be limited in the effects by cosmetic means. The clinical form of acne is characterized by a greasy skin with a chronic inflammation of the hair follicles. As a result, numerous comedones, papules, pustules, cysts, and scars

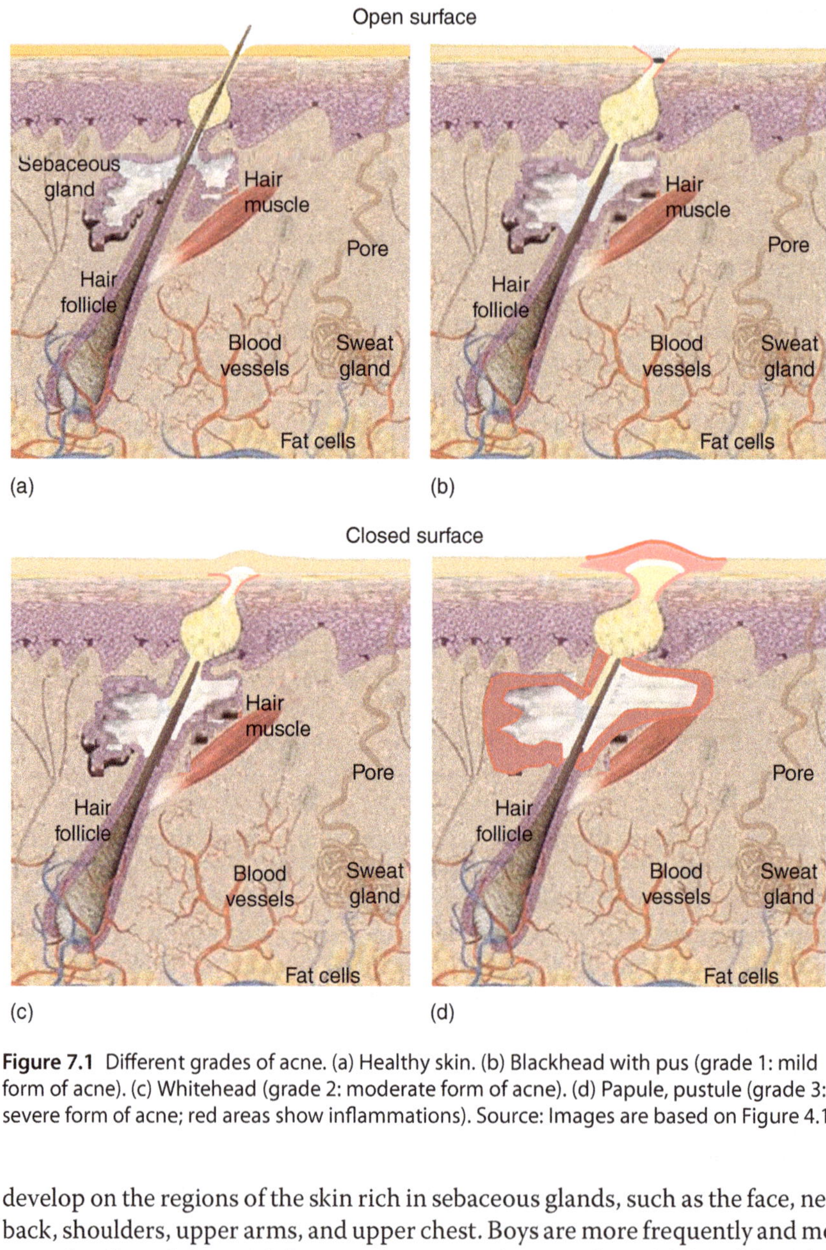

Figure 7.1 Different grades of acne. (a) Healthy skin. (b) Blackhead with pus (grade 1: mild form of acne). (c) Whitehead (grade 2: moderate form of acne). (d) Papule, pustule (grade 3: severe form of acne; red areas show inflammations). Source: Images are based on Figure 4.1.

develop on the regions of the skin rich in sebaceous glands, such as the face, neck, back, shoulders, upper arms, and upper chest. Boys are more frequently and more severely affected than girls because in acne, an increased sensitivity of the sebum glands to androgens (male sex hormones) can be detected. In individual cases, it also occurs in baby and toddler as well as in adulthood to acne. The following conditions are involved in the development of acne:

- Increased production of sebum and sweat due to seborrhoea (fat-moist condition of the skin).
- Sebum gland hyperfunction.

- ➢ Cornification disorder in the sebaceous ducts.
- ➢ Genetic predisposition.
- ➢ Colonization of the follicular exit canal with propioni bacteria.
- ➢ Tendency to inflammatory mediators in existing comedones (inflammatory response).

In acne vulgaris, there are several manifestations, characterized by various skin changes:

- *Acne comedonica*: Open and closed blackheads that are not or hardly inflamed characterize this form of acne. There are more closed than open blackheads on the face, especially in the lateral nasal, chin, and forehead area.
- *Acne papulopustulosa*: In addition to blackheads, inflammatory pustules (with pus-filled protuberances of the excretory duct) and painful papules (up to 5 mm in size) and small nodules (5–10 mm in size) appear. Mostly, the face is affected, but this form of acne can also occur on the chest, back, and upper arms. Scarring arises from the inflammation.
- *Acne conglobata* is a severe form of acne that affects especially men. It forms 1–2 cm large, highly inflammatory and painful nodes that merge with each other in the progress of the disease or encapsulate as deep abscesses. Once the inflammation has subsided, the characteristic acne scars become visible.

In addition, there are other special forms of acne, which will not be discussed here.

Washing the affected areas should be done only once a day with pure water or with a synthetic pH 5 liquid soap based on ether sulfates, sulfosuccinates, or betaines. Alkaline soap and alkyl sulfates have a comedogenic effect and must be avoided. The recommended surfactants have low irritation potentials and mild skin-drying effects. A peeling of the affected skin sites should be done several times a week to open the blackhead and to facilitate the sebum flow out of the duct.

For tonifying the skin, alcoholic facial waters (30%) are suitable that act antimicrobial, keratolytic, anti-inflammatory, and astringent (see Table A.4). The same properties should include light, moisturizing care creams. Active ingredients for such products are listed in Table 7.2. A recipe for a cream to reduce the effects of acne can be found in Section 8.6.

As therapeutic agent, benzoyl peroxide has proven to be the only effective substance for the treatment of acne. However, the use of this skin-irritating substance is banned in cosmetic products but is allowed up to 10% in OTC products. The highest dose with 10 % of the active ingredient should only be used on the doctor's advice for severe acne. For mild to moderate acne, the less concentrated agents are suitable with 3% and 5 % active ingredient. Benzoyl peroxide represents a chemical peeling agent for the skin that removes superficial dandruff and dries out the oily skin. The bacteria present are killed by oxidation reactions, so that the inflammations decay. The positive effect should be clearly visible after four to six weeks. Because of the severe skin irritation, a suitable cosmetic acne cream should always be used in between to support the medicine and provide an intensified skin care.

Table 7.2 Cosmetic active ingredients to combat acne.

No.	Substance group	Possible ingredients	Amount (%)
1	Vitamins/pro-vitamins (Table 6.6)	A, retinol	0.3
		B_3, nicotinamide	3–6
		B_6, pyridoxine-HCl	2
		C, e.g. magnesium ascorbyl phosphate (in combination with vitamin E)	5–6 0.5–1
2	Keratolytic substances (Figure 6.5)	Urea	10
		Salicylic acid	<2
		AHAs and their salts (glycolic, lactic, tartaric, malic, citric, and gluconic acid)	5
		Allantoin	0.1–0.5
3	Vegetable oils rich in linoleic acid (Table 6.10);	Safflower, borage, evening primrose, hemp oil	5–20
	Antiacne effect	Shea butter, jojoba oil	3–12
4	Essential oils (Table 6.14)	Carrot, hamamelis, lavender, lemon and sandalwood and other	0.5–5
5	Minerals	(Zinc oxide), zinc stearate, Zn-PCA (0.3%), zinc sulfate	<5

7.3 After-Sun Creams and Lotions for Reddened Skin

Heat, light, and sun have a positive effect on the body and contribute to a physical and mental well-being. However, too much sun is harmful to the skin and can cause burns. Signs of excessive exposure to the skin are redness, a feeling of tension in the affected areas of the skin, slight swelling or even blistering, and pain when touching the skin. For regeneration, the skin needs about four to eight weeks. The more sunburn the skin has to endure, the higher the risk of skin cancer. Many people underestimate the effects of excessive sunbathing. Severe sunburn causes irreparable damage to the skin. Many years later, the consequences become visible in the form of premature aging and wrinkles. Also, the risk of a skin cancer increases. Before sunbathing or a longer stay in the sun, the use of a sunscreen is required. However, in the case of sunburn, it is advisable to treat affected areas with an after-sun cream/lotion. This cools and soothes the skin and reduces the skin tensions. In the case of severe blistering sunburn, a doctor should be consulted.

A number of agents have been proven to treat sunburn. These include the moisturizers and antioxidants, as well as the anti-inflammatory, soothing, and nourishing substances. Suggestions are shown in Table 7.3, formulations in Section 8.7. The most effective is a combination of active ingredients from all areas.

Table 7.3 Cosmetic ingredients for reducing the consequences of sunburn.

No.	Substance group	Possible ingredients	Amount (%)
1	Moisturizing substances	*Aloe vera* powder (1 : 200)	0.1–0.3
		Hyaluronic acid (high molecular weight)	0.1–0.5
		NMF (Table 6.4)	5–20
		Protein or hydrolysates (Collagen, elastin, soy, wheat)	3–8
2	Anti-inflammatory substances	Panthenol	2–5
		Allantoin	0.1–0.25
		Bisabolol (from chamomiles)	0.6–1
		Hamamelis extract	0.5–10
3	Antioxidants	Magnesium ascorbyl phosphate	1–5
		Tocopherol	0.5–5
		α-Lipoic acid	2–5
		Ubiquinol	1–2
		Squalene	1–5
		Further substances: see Table 5.11	
4	Vegetable oils rich in linoleic acid and healing oils	Safflower, borage, evening primrose, hemp oil,	5–20
		Shea butter, jojoba oil	3–10
5	Skin-soothing substances	*Essential oils*: Chamomile (Roman), jasmine, juniper berry, lavender, rose.	1–10
		Plant extracts: Chamomile, hamamelis, hop, lavender, thyme, Saint John's Wort	3–15

7.4 Creams for Baby Skin

Baby's skin is very sensitive and considerably different from the skin of adults. The following differences exist [6] and must be noted during the formulation of a cream:

- The skin is much thinner.
- Baby skin is less resistant to external influences and sensitive to physical, chemical, and microbiological stimuli.
- Baby skin tends to dry out.
- The horny layer is less compact and thus more permeable.
- The skin contains relatively more saturated fatty acids.
- The collagen layer is not fully developed.
- Sebum and sweat glands are not yet fully formed.
- The hypolipid film (acid mantle) is not yet intact.

- The pH of the skin regulates from 6.9 to 5.5 in the first few weeks. During this time, the skin is thus more susceptible to germs.
- Melanocytes already exist but are not yet fully functional.
- Hyperkeratosis to protect against sunlight is not there yet. Beware of UV rays.

Body temperature regulation is difficult for babies and comparatively slow. As the ratio of the skin surface to the volume of the body is high in babies, they cool down quickly. The sweat glands are not working as in adults, so the compensation of high temperatures through sweating is restricted. The not yet fully developed blood vessels expand and contract less with temperature changes.

These physiological peculiarities result in some boundary conditions in the formulation of creams for babies:

- Do not use substances with allergic potential such as plant extracts or perfumes or dyes. This general rule for sensitive persons should be strictly observed for babies.
- Do not use potentially problematic substances or preservatives because skin barrier is yet weak and permeable.
- Do not use creams with an alkaline pH or corresponding substances. Because baby's skin has problems to adjust the right pH, formulate the cream with a buffer in pH range 5–5.5.
- Use vegetable oil to grease the skin. The self-lubrication of baby's skin is slow.
- Use proven moisturizing substances.

Baby's skin is sensitive against cold, UV rays, and other environmental influences. After bathing the baby, the cream should refat with vegetable oils and moisturize the skin. For the face and body, an oil in water (o/w) cream is recommended. On the other hand, a water in oil (w/o) cream better protects the diaper area because of the effect of 50% more oil that is in the continuous phase. This oil-rich cream remains considerably longer on the skin and has a higher protective and water-repellent effect, especially when using some silicone oils. This cream should contain zinc, such as the effective zinc stearate, to combat inflammation. Thus, the use of two different products for optimal skin care is required.

As children grow older, skin thickness and resistance continue to increase. At the age of about six years, the skin and its appendages are fully mature and conform to the structure of adults. At the age of 11–12 years, sweat gland activity increases. Only when the hormonal changes begin with puberty, differences in the structure and function of the skin of girls and boys are formed. With careful selection of the modules, no different basic creams are required for children and adults or for women and men.

For care and refatting, the discussed vegetable oils are best suited. Baby's milk crust areas can be treated well with an o/w cream containing evening primrose oil. Other caring substances suitable for baby skin include allantoin and panthenol as well as glycerol and sorbitol as moisturizing substances. All ingredients mentioned are already included in the modules. The best UV protection for babies is to prevent contact with the sun rays. In other cases, for sun protection of babies older than 12 months, a product with a broad-spectrum sun protection factor (SPF) of 20 made with naturally occurring minerals such as titanium dioxide and zinc oxide (in coated form) works best and is well tolerated.

The commercially available baby creams focus on a few ingredients. They often contain lanolin (sheep's wool wax) and beeswax, allantoin, bisabolol, panthenol, chamomile, and marigold as well as zinc oxide. Many of them, although natural substances, show a not to be underestimated allergic potential and should be avoided in particularly sensitive baby skin.

7.5 Prophylaxis at Risk of Pressure Ulcers (Bedsore)

The disease occurs especially in the elderly. Pressure ulcers are local damage to the skin and underlying tissue due to prolonged pressure loading. This strain disturbs or prevents the required blood supply of the skin. Bedsore occurs in patients with reduced mobility, for example, if they are bedridden for a long time or can barely move in a wheelchair. The most common sites are the skin overlying the sacrum, coccyx, heels, or the hips, but other sites such as the elbows, knees, ankles, back of shoulders, or the back of the cranium can be affected. At these sites, the pressure cannot be sufficiently distributed because there is no subcutaneous fatty tissue. The occurrence of bedsores indicates mistakes in the care. The number of cases in a nursing home is therefore considered a measure of care quality.

If a local pressure load lasts longer, the cells are undersupplied with oxygen (hypoxia) and nutrients. The oxygen partial pressure decreases (ischemia) and toxic acidic metabolites arise. This leads to necrotization of the tissue and irreversible damage to the nerve cells. In healthy people, the increase in acidic metabolites triggers a reflex to the rearrangement and thus relief of endangered skin. In older and sick people, these reflexes are often limited or even absent, so that the required relief of the tissue is omitted. As a result of the acidification of the tissue, the body expands the vessels (vascular dilatation) after the rearrangement of the person and supplies these skin areas with blood again. The result is a permanent reddening of the skin, pressure ulcer grade I. The other stages of the disease should not be described here.

Used prophylactically, a cosmetic cream can help to prevent or reduce the onset of pressure sores. Therefore, a corresponding cream must be formulated and applied to the endangered skin areas at regular intervals. The prophylactic care takes place in particular with a mixture of oils and vitamins. For the protection of the skin, gas-permeable silicone oil is indispensable, for example, the quality silicone oil M 500.

7.6 Improving the Appearance of the Skin in the Case of Cellulite

Cellulite is a constitutionally induced, noninflammatory change in the subcutaneous adipose tissue of the thigh and buttocks of women. The dellen-shaped skin surface reminiscent of the orange surface is perceived as unpleasant. The development of cellulite is decided by the following factors:

- genes (predisposition)
- strength of the connective tissue
- body fat percentage
- female hormones, especially estrogen
- nutrition and movements (sports).

The image of cellulite is getting worse at:

- one-sided nutrition
- lack of muscle mass – even in normal weight or slim people
- dehydration
- heavy smoking and alcohol
- increasing weight or after crash diets.

In overweight or weak connective tissue, the skin modification occurs at a young age. With age, more than 80% of women get cellulite, sometimes weaker, sometimes more pronounced. As men have a different structure of the connective tissue because of their hormones, cellulite does not occur in men. Cellulite develops in the subcutaneous fatty tissue as pads (enlarged fat cells) with slight congestion of the lymph. In women, the fatty tissue is permeated by connective tissue, latticed collagen strands. Because of the hormonal changes in the menstrual cycle, the structures swell to a greater or lesser extent, which makes the collagen ligaments visible. As cellulite affects deep skin structures, creams cannot eliminate them, but only cause superficial improvements. To achieve an effect, slightly penetrating substances should be applied. Suitable for anticellulite creams/oils are vegetable oils, especially essential oils or extracts from plants or from the sea, so that they remain in the skin for some time, which make sense to thicken the oils with a known viscosity regulator. Some essential oils that should act strongly against cellulite are listed below. If possible, it is recommended to formulate mixtures of the oils in higher concentrations in the cream. Preference is given to a mixture or partial mixture of

- Juniper, fennel, geranium, rosemary, lemon, and rosewood.
- Further suitable essential oils from plants are anise, cedar, clementine, cypress, frankincense, grapefruit, immortelle, lavender, lemongrass, mint, myrtle, orange, bitter orange, oregano, rose, sage, and ylang ylang.
- Effective plant extracts: Mangosteen.

In one study, it is proven that mangosteen reduces the accumulation of fat cells and promotes the drainage effect by stimulating the microcirculation as well as firms the skin by stimulating hyaluronic acid synthesis. Some anticellulite oils contain birch extract (International Nomenclature of Cosmetic Ingredients [INCI]: *Betula alba* bud extract) for its astringent effect. Phytosterols from soy should be useful.

- Seaweed extracts: *Fucus vesiculosus*, *Haematococcus pluvialis*, and *Laminaria digitata*.
- Other effective substances: Caffeine, L-carnitine, and panthenol.

The salt from the sea, especially from the Dead Sea, is valuable because of the high concentrations of magnesium, calcium, potassium, and bromide ions.

7.7 Chemical Removal of Unwanted Hair (Depilatory Cream)

Even if the depilatory cream is not one of the skin care products, it should be briefly addressed here. Please note that the regulations for the approval of depilatory creams vary significantly from country to country. In appearance, on the one hand, the effect on other people is determined by the condition of the skin, on the other hand by the hairstyle, the hair fullness, and their color. To maintain a healthy skin condition, skin care creams exist. Agents for strengthening hair growth and hair color are not dealt with in this book. In addition to the hair on the head, unwanted hair grows on many parts of the body that many people want to remove. These are hairs in the armpits, on the chest (men), at the genitals, arms, and legs. Especially unpopular with women are hair on the face (lady beard), which fall in eye. The most women remove all these hairs. Men often do not like breast, abdominal, and back hairs. Both men and women want a hair-free intimate area for reasons of hygiene. For many people, the prerequisite for a well-groomed skin is a smooth, hairless skin, which looks healthy and beautiful.

The unwanted hair can be removed by various methods. This does not concern beard growth in men. The fast-growing beard needs to be shaved off. On the one hand, during epilation, the gentle growing hair is torn out of the skin with wax or rotating epilators. The skin stays hairless for about one to two months. After the painful electroepilation, the hair disappears for months, sometimes even forever. The depilation procedures, on the other hand, are painless. A shave can solve the problem, or the hair can be cut by chemical agents near the root. After applying the depilatory cream in thick layer, the reactants cleave the sulfur bridges (–S–S–) in the amino acid sequence near the hair root, where the keratin is not fully hardened and horny. This process is carried out with the help of thioglycol salts in a strong alkaline medium. The limitations of their use are described in the Cosmetic Regulation (see Figure 1.6). The reaction lasts at least 5 and a maximum of 15 minutes, usually 10 minutes. The time depends on the salt type and its concentration as well as the alkalinity. Because of the high pH of about 12, the product should be removed after 15 minutes at the latest with a paper tissue or a household paper. Then, it is advisable to wash the appropriate places with plenty of water. After drying, the skin needs a weakly acidic, buffered skin care cream (pH 5), which restores the skin pH and calms the irritated skin. Depilation is a disadvantage because the first stubble appears already after about one to two weeks. Therefore, the process must be repeated every two to four weeks.

Many products in the market contain mineral oil. The use cannot be recommended. Better is the usage of a natural vegetable oil, such as almond or jojoba oil and shea butter. Potassium thioglycolate, which ensures odorless hair removal when freshly used, has proven to be a suitable hair-splitting salt. To adjust the alkalinity, the amounts of calcium and potassium hydroxide and their ratio are crucial. A recipe suggestion can be found in Appendix A8. The label must contain the following warnings: Contains thioglycolate and alkali. Avoid contact with eyes. In the event of contact with eyes, rinse immediately with plenty of water and seek medical advice. Follow the instructions. Keep out of reach of children.

7.8 Treatment of Eczema

The term “skin eczema” refers to the widespread inflammatory, but noninfectious skin diseases. About 40 % of all skin diseases are diagnosed as skin eczema, which are noticeable by itchy, reddened skin, and are very uncomfortable for those affected. First, they appear as an overreaction of the skin to external influences, i.e. contact with mechanical, chemical, or biological noxae. This eczema is called exogenous skin dermatitis. The eczema-causing substances may be chemicals, metals, cosmetics, hair dyes, acids, resins, preservatives, or mineral oils, as well as latex gloves. An increased risk of developing exogenous skin eczema affects people who frequently come into contact with various substances. These are water, soaps, and high humidity as well as some cleaning agents and metals or metal compounds, furthermore allergy triggering plants. Regular contact leads to irritation and then damage to the natural skin barrier. Skin defects represent the onset of dermatitis. Hairdressers and beauticians, painters, bakers, cooks and confectioners as well as cleaners, nurses, or gardeners and people who process metals or metal compounds are particularly at risk. Even strong UV radiation during intensive sunbathing can lead to exogenous dermatitis. Above average, contact dermatitis affects those people who have a dry and slightly scaly, very sensitive skin.

Second, internal influences in the body can lead to endogenous dermatitis. In most cases, there is a hereditary predisposition. The most common form of endogenous eczema represents atopic dermatitis. Children and adults who suffer from allergies such as food intolerance, asthma, animal dander, or hay fever are particularly often affected. The diseases occur acutely or chronically in different places such as hands, feet, arms, head, and anal region. Mild diseases are manifested in local reddening of the skin, which usually heals by itself a few days after the onset of symptoms. Serious cases can persist for months. In the case of severe skin irritation, skin redness may be accompanied by the formation of tiny fluid-filled blisters. The strong itching caused a considerable burden. After bursting, the bubbles wet the affected areas and form thin, visually striking crusts on the surface.

Repeated acute skin eczema occurring at the same site can become chronic. These too are characterized by a strong itching. The skin permanently forms dander and the resulting cornification appears cracked and rough. Thickening and lichenification can arise in the course of the chronic disease. In addition, the inflammatory skin sites represent a gate of entry for various germs such as fungi, bacteria, or viruses that live on the surface of the skin and hinder healing.

The most well-known skin disease is the hereditary transferred atopic dermatitis. About 10% of the children are affected. In most cases, it disappears during puberty, so that only 1–3% of adults suffer from atopic dermatitis with an upward trend. Because of the lack of lipids, the ceramides of the horny layer are produced quantitatively and qualitatively incorrect at the affected sites. Therefore, the lipid bilayer (kit) is insufficiently composed. The other abnormalities of the diseased skin concern the levels of γ-linolenic acid and urea. Today, it is assumed that a deficiency in lipid metabolism does not produce enough γ-linolenic acid from linoleic acid. The γ-linolenic acid is essential for the synthesis of the prostaglandin PGE1. The absence of PGE1 in the skin explains some pathological manifestations of atopy, such as the weakened immune system, the tendency to inflammation, and the spontaneous response to allergens of natural origin. The allergies are triggered by contact with pollen, certain foods, and animal hair.

In neurodermitic patients, the amount of urea in the skin is only about 15–30% of healthy skin. This increases water loss, reduces sebum production, and disturbs perspiration. The sweat does not flow outward but migrates into the surrounding skin areas and exerts strong itching. An infestation with microorganisms occurs by scratching, which further worsens the skin condition there. All this leads to a very dry, irritated, scaly, and itchy skin. The complex therapy should not be discussed here. Usually, the application of a cortisone-containing prescription cream, especially in bouts, takes place. Subsequently, the dry, chapped skin can be smoothed with an effective cosmetic cream. As described, it must contain lipids containing both linoleic and linolenic acid. Furthermore, the addition of plenty urea is important. Third, the affected areas must be freed from microorganisms. This can be achieved simply by a high concentration of sodium chloride. Such a cosmetic cream has proven its effectiveness in more than a hundred cases. A similar formulation, also suitable for dry skin areas, is shown in Table 8.6.

7.9 Cream for the Feet and Against Athlete's Foot

With damaged skin and/or a weakened immune system, the fungal spores overcome the skin barrier and penetrate into the horny layer with their cell threads, the so-called hyphae. There the fungi multiply. The growth leads to symptoms such as itching and inflammation. Fungal infections accelerate the renewal of the horny layer and show an increased scaling. The shedding dander transfers the infection to others because the fungal spores survive for days or weeks in them and other people can become infected by contact. The risk of infection exists in people who walk barefoot in a humid environment. In particular, infestation occurs in public swimming pools and saunas, on carpeted floors of hotel rooms, and in dressing rooms. Fungi love it moist. Therefore, socks and stockings made of synthetic fibers, which impede air circulation and do not absorb sweat, should be avoided. There, a humid "greenhouse climate" can arise, which creates the best conditions for the proliferation of fungi. Moisture also likes to stay in the very tight 4th toe space, which is why the athlete's foot is most common there. In these sites, the heavily soaked skin barrier does not work anymore.

The athlete's foot (tinea pedis) is a filamentous fungus that affects the horn substances, i.e. skin, hair, and nails in the foot area. It prefers to settle between the toes and draws attention to itself by a strong itching, redness, and scaling as well as wetting. Another form of athlete's foot favors the infection with shoots and filamentous fungi (tinea unguium), which settle under the toenail and cause nail deformation and discoloration. About 5–12% of Europeans are aware of these symptoms. Nail fungus is very difficult to treat because access is difficult. The fungi are transferred from person to person, love a humid environment, and can therefore also be found in the swimming pool, in the sauna, and in the sweat of the feet. Sports and work shoes should be as permeable to air as possible, always dry well, and be disinfected from time to time.

A foot cream should provide the skin with lipids, moisturizing factors, and vitamins. It is possible to formulate a nourishing cream, which also gradually kills the fungi softly in a natural way. The effectiveness against fungi takes place over several measures/active substances:

- The pH of the skin on the foot is set to 4.5 (the fungi need a weak alkaline environment).
- An acid/base buffer ensures the constancy of the pH over a longer period of time.
- Sorbic acid stops and reduces existing microorganisms.
- Sea salts support the reduction of fungi.
- Lavender oil destroys the cell walls and kills the fungi (evidenced by a study of Portuguese scientists [7]).
- High-quality lavender oil, judged by the proportions of linalool and linalyl acetate, should be present in effective amounts.

A recipe suggestion can be found in Table 8.18. A similar formulation has already proven itself.

7.10 Cream for Hands

The composition of a hand cream is tailored to the hand care. The skin of the back of the hand is thin, almost comparable to the face skin, has only few sebaceous glands and subcutaneous fatty tissue and is therefore particularly sensitive. The withdrawal of water causes a slow drying out of the skin. This happens in autumn and winter by cold air and by heating air or by frequent, intensive washing with surfactant-containing agents, so that the skin appears brittle and dry. Even small cracks can occur in the hand and germs can penetrate. Therefore, there are many hand cream specialists. Examples are the Rosacea cream for people with this skin disease, creams for extremely dry hands or for sun protection, with antiaging effect against age spots, or hand cream for sweaty hands. Hand creams should not be greasy, give the skin the necessary moisture, move in quickly, distribute well, and be economical.

Hand creams should be suitable for all skin types, i.e. for normal, dry, or very dry skin. The skin condition decides the frequency of use. A good hand cream must not be sticky. A cream does not appear suitable for everyday use, if a greasy film remains on the hands for a longer time. When developing a hand cream, there always occurs the conflict between the effect and the stickiness because an effective cream must have fatty components. The only solution cannot be to dilute the cream with water but requires a careful selection of suitable vegetable oils. Here, for example, the content of greasy oils could be reduced or replaced by medium dry oils or fatty alcohols. In any case, a thin application on the skin is recommended, supported by a small discharge hole on the dispenser.

The main purpose of the hand cream is to moisturize the skin. The substances of the natural moisturizing factor (NMF) are suitable for this. Furthermore, some lipids are needed for care and protection. Useful and frequently used vegetable oils are avocado, jojoba, shea butter, almond oil, and others. To minimize the washing-out effect of lipophilic care and protective components and to keep the skin supple, the addition of silicone oil is recommended. It has a protective and water-repellent effect. Also, long-chain fatty alcohols such as octyldodecanol or low-melting waxes (cetyl palmitate, myristyl myristate) can also take over protective functions.

Many consumers want a cream without perfume (mixture with high synthetic content). As an alternative, the fragrance can then arise from a single vegetable essential oil, as described in previous chapters as the best option. The recipe suggestion, provided in Table 8.19, includes an essential oil of geranium for scenting, which creates a delicate rose scent. Basically, all face creams discussed here are also suitable for the hands. The formulation adapted to the hands is more water-rich. Compared to face creams, the raw material costs are significantly lower.

7.11 Antiherpes Cream

Viral infections caused by herpes simplex viruses (HSVs), which in most cases occur at the transitional area of facial skin to lip red and rarely in the genital area, are colloquially referred to as "herpes" [8]. The pathogens are, on the one hand, the most common herpes simplex virus 1 (HSV-1) and, on the other hand, the herpes simplex virus 2 (HSV-2). After a first infection, the virus remains lifelong at rest (latency) in the organism, which is called a persistent infection. For the HSV, which is widespread worldwide, humans are the natural host. Already in infancy, HSV-1 can be acquired through salivary contact and smear infection in the family. Frequently, the virus occurs at the end of puberty and less often afterward. In Germany, antibodies to HSV-1 could be detected in more than 84% of the cases during random sampling. Preferably, the areas become infected at the transition from mucous membrane to normal skin. Here, the viruses multiply in the epithelial cells. The spread of the HSV within the epithelium is done by destruction of the host cells, which manifests itself in an inflammatory reaction characterized by skin blisters. In the blister fluid (exudate), the viruses accumulate and are present there in high concentration (>100 000 PFU/μl).

In herpes labialis, sore spots with blisters ("fever blisters") occur on the mouth. The infection is usually triggered by reactivation of an already existing HSV infection. About 40% of the population experiences a reactivation visible as blisters at least once in life. A trigger of herpes recurrence can be an acute emotional stress. A herpes infection begins with mild tingling and itching, followed by the first blisters around the lips. Untreated, it usually takes about 7–10 days. With the right remedies, wound healing can be accelerated. In the pharmacy, there are several cold sore creams. They contain antiviral agents, such as penciclovir, acyclovir, or Zovirax, which inhibit the multiplication of viruses and thus prevent the spread of herpes. They shorten the duration of the infection by about two days. The ointments work only when applied at the first tingling. At this point, the viruses have not multiplied yet.

A suitable cosmetic cream can do at least the same without side effects. Applied at a very early stage, visible inflammations are prevented. Used a little later, there is an almost immediate cessation of itching and a reduction of the rash. Bubbles do not occur then. If used too late, a nonitchy rash develops. Suitable creams have the following characteristics:

- The pH is at or below 5 and does not change (buffer).
- Sorbic acid supports the effect.
- Salts may be antiviral when properly formulated.

For a convenient application, but also to increase the economy, it is recommended to fill the cream in a 5 or 10 ml tube. The formulation in Table 8.6, which also works in atopic dermatitis, is particularly suitable for an antiherpes cream. There are positive experiences with similar recipes in many cases.

7.12 Cream for Removing Thick Horny Layers (Callus)

Some people suffer from the formation of thick horny layers under their feet that can interfere with walking and therefore need to be eliminated. Dry skin and the formation of calluses by a disturbed skin barrier is the most common foot problem, especially for women and the elderly. A healthy skin is supple and resistant. Above all, the horny layer protects the deeper tissue layers from external influences, pressure, and friction. Because of the high load on the feet, the horny layer is thickened just at the soles of the feet in the ball and heel area. In the case of moisture loss, it can become dry and cracked. In particular, the so-called callus by pressure and friction over time occurs by wearing poorly fitting shoes. Furthermore, people who have foot deformities such as the lower and splayed feet or who have disposition to hammertoes, heel spurs, or hallux valgus are particularly susceptible to corneal development. The epidermis, the outer skin layer, responds to these stresses with increased corneal formation (hyperkeratosis), so that ugly calluses form. The corneal formation is thus a normal protective reaction of the skin. Cornea usually arises on the soles of the feet. Heavy, continuous stress can also affect the palms and fingers, which triggers thickening of the skin. Callus is usually not painful, but a constant, excessive stress can lead to the formation of cracks, which cause pain.

To remove the thick cornea, there are two options, mechanical or chemical removal. The mechanical removal of the layers, often performed following a foot bath, relies on special paring knives and pumice. The chemical method is carried out with keratolytic creams, containing substances that lead to softening of the cornea and in higher concentrations to dissolution. The chemical substances can be used in a professional pedicure or privately. Cosmetic creams discussed here are designed for private use. The approval for cosmetics or as an OTC cream must be checked on a case-by-case basis.

First, the most suitable substance for removing callus should be urea. The use of urea is not allowed in natural cosmetics because of its synthetic production. Urea has a softening and antipruritic effect in the concentration range above 10%. In amounts of more than 20%, the urea–water solution dissolves the callus. In addition, it spends intensive moisture for the dry skin. Second, another possibility represents the use of α-hydroxy acids (AHAs) (see Section 6.3.2), which have a peeling effect at high concentrations (>40%) and low pH (<4). Beauticians work with these agents. For private use, concentrations of about 10% at pH 4.5 are sufficient to produce a slight peeling effect and remove dander. Also, AHAs are moisturizing. Third, allantoin shows keratolytic properties and can act alone or better in combination with others. Concentrations in the range of 0.3–0.5% are sufficient. The amount of 0.5% represents the solubility limit in water at 20 °C. There is one recipe example for a rich cream in Chapter 8, Table 8.20. The use of salicylic acid is another alternative. It has a keratolytic effect and is also suitable as a preservative. The maximum amount is 0.5% according to the European Cosmetic Regulation.

7.13 Lotions for Body Care

A body lotion is a moisturizing care cream intended for the whole body. Thanks to its fairly fluid consistency, it is easy to disperse and the low concentration of ingredients ensures fast absorption. The lower fat content compared to the creams allows immediate dressing after application. A well-balanced body lotion should be suitable for all skin types. For the care of the face, neck, and décolleté, the body lotion can be used without any problems, but there are significantly better creams for these body areas, presented in Sections 8.2–8.5.

A body lotion is used to care for the entire body, i.e. the chest, the lower body, back and buttocks, and especially the legs. A regular application every day or week is useful, depending on the skin condition. The body lotion should prevent skin dryness and keep the skin supple. Applying the lotion after the daily shower brings a cool, refreshing feeling, in particular after sport activities. The skin is smoother and does not dry out immediately. For women, moisturizing and lipid delivery is more important than for men because their skin is thinner and tender. Furthermore, the use of a body lotion is useful on the one hand in autumn/winter, when the body was exposed to cold, dry air for a long time, and the dry skin should be moisturized. On the other hand, after a long bath in the sea or lake, it is always recommended to support the regeneration of stressed skin with a body lotion.

The composition of a body lotion should consist of moisturizing and lipid-supplying substances and vitalizing vitamins. A proposal for the formulation is presented in Table 8.21. Lotions may also be produced about diluting existing formulations with about 10% to maximum 25% water (Tables 8.2–8.5). The preservation is already designed accordingly. Possibly, the amount of the emulsifier or coemulsifier (thickener) must be adjusted.

7.14 Cream for Itchy Shins

In autumn and winter, many people tend to too dry skin. Especially, they suffer from dry, itchy shins, in particular older women. Itchy legs are very annoying for those affected. The itching can be so bad that sufferers scratch the skin bloody. As there are no fat pads that also store heat and moisture, the shins dry out very quickly. A cause of the sensitive skin on the tibia can be a genetic predisposition. Further, the itching appears as a symptom of a skin disease such as psoriasis or neurodermatitis or an allergy. In addition, it can be a "dried out" skin area, which is caused by very dry, especially cold air. The cold air also makes the skin dry and scaly at this point. Blood vessels contract in the cold and restrict blood flow, which reduces the transport of nutrients and oxygen. As a result, the sebaceous glands, responsible for the formation of the protective fat film on the skin, are no longer sufficiently supplied with nutrients. They work more slowly, which means that the hydrolipid film is no longer created everywhere. The acid mantle has "holes," which causes the skin to dry out quickly. As the shins have hardly any fat deposits, the itching develops at these points. The incomplete protective film makes the skin more vulnerable and causes a rapid dehydration, detectable over a high transepidermal water loss (TEWL).

The cosmetic can help with an existing product or with a special product, which should be applied daily, and in severe cases twice a day. To limit the TEWL, the cream should be high in fat and moisturizing. In particular, the formulations in Table 8.2 (w/o) and in Tables 8.5 and 8.6 (o/w) are suitable for this purpose.

7.15 Self-tanning Cream

Tanning of the skin without harmful UV light can be achieved with a self-tanner [9]. This is a cosmetic product that comes mostly in the form of a cream on the market, but also as milk, lotion, gel, foam, or spray. People, who want to stand out even in the less sunny seasons by a healthy, slightly brown-tinted skin without going to the solarium, use self-tanners. In many cases, they apply the cream only on the face, neck, and décolleté, less often on the whole body. As active ingredients, two substances come into question, namely the most commonly used dihydroxyacetone (DHA) and/or the rarely used erythrulose (Figure 7.2).

DHA penetrates the epidermis and reacts within two to four hours with keratin (protein) to the melanoids. These are the melanin-like brown dyes. Sometime

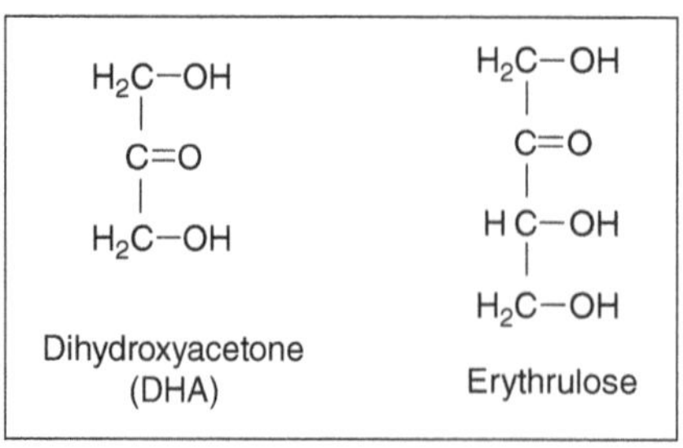

Figure 7.2 Active ingredients for self-tanning.

after the application, the cream is completely absorbed. As the reaction takes place in the skin, it cannot be washed off or rubbed off. Shaving or sweating may negatively affect the tanning result in the first two hours (chipmunks). Problem areas with self-tanner are the knees and elbows because there the cornea is thicker and the self-tanner can react more strongly. Here, a peeling helps before tanning. The upper layers of the skin lose countless dander every day, thereby daily color pigments are also lost with the scales. The tint fades after two to five days; after two weeks, the color completely disappears. Therefore, the cream should be applied regularly every day. The color intensity depends on the application concentration, which is usually between 2% and 5%. A self-tanner is not a sunscreen and thus does not protect against harmful UV radiation. However, the product can be additionally equipped with UV filters.

DHA is an intermediate of the carbohydrate metabolism in the skin. The substance is harmless and does not act systemically. The skin tanning method is considered harmless. The cream should be bottled in an opaque dispenser and stored cool or below 20 °C. At high temperatures, formaldehyde can be split off. Chemically and thermally more stable is erythrulose, a sugar found in plants and lichens. The tanning effect does not depend so much on the concentration and skin structure as for DHA. The tan lasts longer and is less intense. Therefore, it makes sense to use the two substances at the same time. It is recommended to supplement the body lotion (Table 8.21) with these active ingredients. The concentrations determine the intended effect. A slight tan looks good and does not immediately reveal that a cream is used.

7.16 Sunscreens (UV Protection)

7.16.1 Solar Radiation

The spectrum of solar radiation ranges from short-wave ionic cosmic rays, consisting of γ- and X-rays, to UV and visible (VIS) rays to infrared (IR) and microwaves (Figure 7.3). The shorter the wavelength, the higher the energy of the radiation is. Therefore, short waves of the UV radiation can trigger chemical reactions. The intensity of solar radiation is lower on Earth than outside the atmosphere. The atmosphere weakens solar radiation through absorption and scattering. Both physical processes are wavelength dependent. From the visible light, the short-wave blue light is scattered more strongly by the air molecules than the long-wave red (Rayleigh scattering). As the predominantly blue light is reflected, the cloudless, clear sky appears blue. Almost half of the solar radiation is the visible light. At sunshine, the rays reach the Earth's surface only slightly attenuated by the atmosphere. Otherwise, the clouds additionally weaken the light. The invisible component predominantly consists of near-infrared radiation (NIR). Of this, about one-quarter is absorbed in the atmosphere by water molecules. The UV radiation accounts for less than 10% of the radiation. The incoming UV radiation can be divided into UVA and particularly high-energy UVB components. Fortunately, the harmful UVB is reduced by the ozone layer so that only small portions reach the earth. The UVA rays are hardly scattered

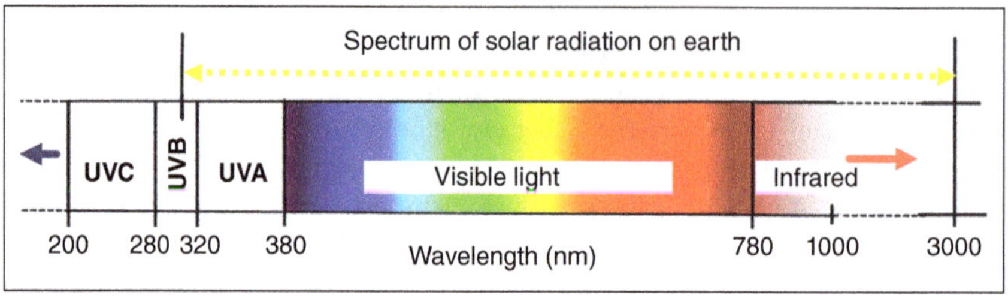

Figure 7.3 Spectrum of sunlight between 300 and 3000 nm.

but reduced by clouds and pollution. At these, a scattering of UVB rays takes place, so that this radiation can be high even in cloudy skies and cause sunburn.

7.16.2 Character of UV Rays

The shorter the wave, the more energetic and dangerous are the rays. The most energetic ones can trigger chemical reactions by splitting a connection. This leads to unwanted changes of the biological processes in the skin. The relatively long UVA waves (315–380 nm) can penetrate deep into the biological tissue to the dermis. They cause a direct pigmentation (conformational change of melanin), a weak tanning lasting only for hours. UVA radiation can damage collagens in the skin, resulting in loss of elasticity and premature aging. UVA rays produce only slight sunburn. The UVB rays (280–315 nm) have shorter wavelengths and are more energetic. They penetrate less deeply into the biological tissue than the UVA rays and reach at most the basal layer. In the skin, the rays scatter, which cause sunburn (erythema effect, red skin) and form with a delay of 72 hours **melanin** and vitamin D_3. This indirect pigmentation results in a delayed, long-term browning with true **photoprotection** (Figure 7.4) through thickening the top skin layer (light callosity = hyperkeratosis). Note the risk of melanoma due to the formation of free radicals.

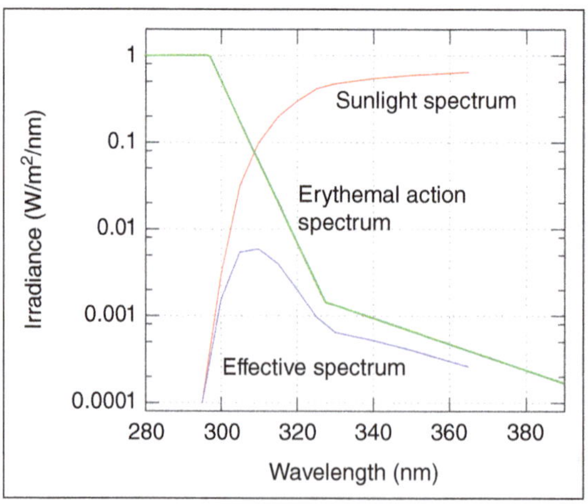

Figure 7.4 Erythemal action spectrum. Source: Wikipedia [10]. https://en.wikipedia.org/wiki/Ultraviolet. Licensed Under CC BY 3.0.

Throughout the lifetime of humans, defects generated by UV radiation accumulate in the skin. Radiation damage in childhood and adolescence must therefore be avoided at all costs. UV can cause inflammation, immune changes, physical changes, impaired wound healing, and DNA damage, which in the worst case promotes cellular carcinogenesis (melanoma and other). The consequences of intense, frequent sunbathing are later visible by the premature aging of the skin, in particular the skin of the face.

UV rays are reflected from light areas such as water, snow, ice, sand, and smooth surfaces on buildings (concrete and plastic), which significantly amplifies the radiation. In addition, the UV radiation is height dependent. In the mountains, the direct UVB exposure increases by 15% per 1000 m altitude, so there is an increase of 45% at 3000 m. The UVA radiation increases by about 10% per 1000 m and is at 3000 m over 30% higher than at the sea level. This is the reason why a sunscreen with a particularly high protection factor should be chosen when skiing. Water offers only a weak protection against the UV rays. During swimming, about 75–90% of the rays reach the body, with greater attenuation of UVB. At a depth of 1.50 m under water, 40% UVB and 70% UVA can still be detected. On the water, the reflection leads to a strong radiation of the harmful UV rays. People who are in these places need strong protection against the UV radiation through a regularly applied waterproof sunscreen and care through an after-sun lotion. As Table 7.4 shows, the UV rays also have a number of vital tasks. Therefore, a balance is crucial: sunlight is important for our body, too much is harmful. Unfortunately, the limit is often exceeded by many people.

UVA rays cause only a short-lasting instant pigmentation, which provides virtually no protection. The tanning results from an oxidation of melanin precursors, whereby they darken. As the reaction is reversible, the tan fades after a few days. In contrast, the UVB rays stimulate the **melanin synthesis** (dye), which takes place through the melanosomes (organelles) in the melanocytes (cells). Melanin and the melanosomes migrate into the surrounding keratinocytes. There, the melanin settles around the cell nuclei to protect the DNA. The protection is done by reflection of the UV rays, which are converted into heat. This process depends

Table 7.4 Risks and possible benefits for the skin through frequent, intense sunlight [6].

Radiation	Risks	Benefits
UVA	o Photo-skin aging (wrinkles) o Elastosis o Age spots o Sun allergy o Mallorca acne (photodermatosis) o Photoallergic and phototoxic reactions	• Immediate tanning (unstable) • Immunomodulation • Improved sentiment
UVB	o Sunburn o Immunosuppression o Eye damage o Skin lesions o Skin cancer	• Tanning (resistant) • Germicidal • Vitamin D synthesis • Immunomodulation • Light callosity

Source: Ellsässer 2013 [6]. Reproduced with permission of Springer Nature.

on the strength of melanin production that in turn is based on the skin type. The darker the skin, the better the protection is.

In addition, the UVB rays stimulate the development of the **light callosity**. The self-protection of the skin against sun rays, consisting of tanning and light callosity, is achieved with careful tanning without sunburn after about two to three weeks. From this period, self-protection is fully effective. In people with bright skin, self-protection is low with a SPF of <3. Very dark-skinned people can reach an SPF of 10. For many people, the protection through the light callosity lies at a value of SPF <4. To eliminate DNA and cell damage and to neutralize radicals, the skin has repair systems, thereby a variety of suitable chemical compounds. These include vitamins A, C, and E; many radical scavengers; and antiaging substances (see Chapter 6), glutathione, zinc, selenium ions, and others. These substances are often also components of effective creams to strengthen the repair system.

7.16.3 Radiation-Induced Damage in the Skin

Fortunately, the healthy skin is very well protected against exogenous noxae. In the case of sunlight, the self-protection of the skin is far from adequate for today's recreational habits and the beauty ideal of bronze-colored skin. Problems occur when the natural protection mechanism is overwhelmed during long stays [9, 11]. In light-skinned people, this limit is quickly reached or exceeded. Therefore, all who move in the sun should protect their skin with sunscreen. It is easily possible to provide the formulation of a face cream with chemical or physical sunscreen ingredients. At the moment (III/2018), these products are very popular. The suitability is only given if the other ingredients are resistant to oxidation. This requirement is usually not met by vegetable and essential oils or perfumes and other. A lot of face creams with sunscreen are not stable enough against oxidation for a longer stays in the sun and therefore unsuitable. Therefore, be careful when formulating or buying.

Sunburn (**sun erythema**) develops when the skin's own protective mechanisms against UVB are overloaded by intensive radiation. The absorption of UVB rays damages or even kills some cells in the epidermis. Furthermore, the skin begins to redden and burn. Upon further irradiation, blisters may develop, as well as a generalized malaise that is identical to a burn. Some hours after the excessive sun exposure, the erythema develops. If the redness already occurs during sunbathing, a strong burn must be expected. UVB radiation causes so much damage to the cells and the DNA that the repair systems in the skin are overloaded. There is an increase in radicals, damage to cells and DNA, and foreign antigens that can potentially lead to cancer in the long term. The damaged cells send messengers into the body via the blood, leading to systemic **immunosuppression**.

An excess of **UVA** radiation can cause skin aging. The rays reach deep into the dermis up to the fibroblasts, cause damages there, and trigger increasingly the degradation of collagen, enhancing the synthesis of elastic fibers. However, these fibers degenerate after a short time and clump together. This process known as **elastosis** reduces the elasticity of the skin and causes deep wrinkles and unnatural, tanned-looking skin. The wrinkles formed by photoaging look different (more pronounced) than the wrinkles of the normal aging process defined by the genetic code. The UVA radiation also reaches the blood vessels in the dermis. If photosensible molecules (drugs, endogenous substances) are present in the bloodstream, under the influence of the radiation, they can change and trigger **phototoxic** reactions. In a similar way, antigens can arise that lead to **photoallergic** reactions. The effects manifest two to three days after the irradiation.

The different forms of light dermatoses are called "sun allergy." **Polymorphic light eruption** (PALE or polymorphic photodermatosis) is most common and characterized by wheals, redness, and itching. About 10–15% of the light-skinned population, preferably women, experience a PALE that arises almost entirely because of the influence of **UVA** rays. Again, the sunbathing must be adequately protected by a sunscreen. In addition, the cream should strengthen the repair system of the skin with the appropriate active ingredients. Useful is a prophylaxis by β-carotene and calcium compounds, which should begin four to six weeks before sunbathing.

Triggered by the UVA rays, organic cream ingredients on the skin may be possibly oxidized through radicals to lipid peroxides. The oil components of the cream must be selected on the basis of this problem; this means, in sunscreens, only oxidation-resistant oils should be used (but nevertheless no mineral oils). The lipid peroxides can cause inflammation in the area of the hair follicles, which triggers redness, itching, pustules, and papules, similar in appearance to acne. Therefore, these pathological skin reactions are falsely referred to as **Mallorca Acne**. Particularly affected are people who are prone to acne. This example demonstrates the importance of fine-tuning the cream composition.

According to German statistics, approximately 207 000 people are newly diagnosed with white skin cancer (basal cell cancer 137 000, squamous cell carcinoma 70 000). Malignant melanoma affected about 20 000 people and nearly 3000 people died of it (2010). This melanoma, called **black skin cancer**, is the most aggressive form. The melanocytes degenerate and form a very aggressive cancer. Some melanomas become rapidly larger and metastasize in lymph nodes and internal organs that can lead to death. **Spinalioma**, also known as squamous cell carcinoma, is caused by a light-related cornification disorder (actinic keratosis). This form leads to scaling and encrustation, resulting in a sandpaper-like appearance of the skin. On the one hand, it can be easily eliminated by drugs as well as by icing or laser therapy; on the other hand, it can break into the dermis when left untreated and metastasize via the lymphatic system. **Basal cell carcinoma** is the most common type of skin cancer. It does not metastasize but can penetrate deeper into the skin to bone and cartilage. Again, it can be fatal in severe cases. The development of skin cancer is primarily triggered by UV rays.

Short-wave **IR rays** (780–1400 nm) reach the blood vessels in the subcutis and causes there an increase in Brownian motion. The energy absorbed by the blood initially leads to a warm sensation. This starts a process of drying the skin. In the case of too intense sunshine, possibly combined with extensive sports activities, the body loses a lot of water and salts through sweat secretion. The blood in the skin can no longer be sufficiently cooled, heats up continuously, and promotes the heat inside the body. The body overheats and responds by a **heat collapse**, sunstroke with muscle cramps, dehydration, fever, or cerebral edema. Long-lasting, regular IR light irradiation may, according to studies, induce increased expression of various enzymes. An example is the matrix metalloproteinase-1 (MMP-1), which degrades collagen and elastin [6]. This leads to a degeneration of the connective tissue fibers and consequently to a loss of elasticity and water retention. The skin ages faster, especially in combination with the negative effects from long-lasting UV radiation.

7.16.4 Sunscreen Substances According to the Cosmetics Regulations Worldwide

Sufficient sun protection through textiles is no longer available today. Fashion trends and fabrics have changed significantly over time. There are almost no hats left, no long, high-necked dresses, jackets, and pants in the summertime. In addition, there is the beauty ideal of a tanned or brown skin, which seduced to excessive sunbathing of the entire body. For these reasons, the skin must be sustainably protected for longer periods of time before negative side effects of the sun stay occur. Substances used for sun protection can be subdivided into groups according to Table 7.5.

Under European law, UV filters are defined by their purpose as "substances that are exclusively or predominantly intended to protect the skin by absorption, reflection, or scattering of certain UV radiation against certain UV radiation." In Europe, sunscreen products (SPF > 6) are considered cosmetics. Only UV filters listed in the positive list of the Cosmetics Regulation may be used. To protect against UVB and UVA rays, the substances alone or in combination must act as a broadband filter over the entire UV range.

When formulating the sunscreen, there are two different ways to protect the skin from UV rays. On the one hand, inorganic pigments are incorporated as nanoparticles with preferred particle diameters between 20 and 60 nm (up to 500 nm), which act in combination as a broad-spectrum filter. They scatter and reflect the rays; the smaller the particle, the better the result is. On the other hand, suitable organic substances protect wavelength dependent from UV radiation by absorption and conversion into heat. Natural cosmetics only accept the inorganic particles, namely titanium dioxide and zinc oxide, with diameters above 100 nm. These products, when applied to the skin, may produce thin white covers that

Table 7.5 Substance groups for the sunscreen.

No.	Sun protection	UV band
1	Organic chemical substances	UVB-filter (280–320 nm)
2		UVA filter (320–400 nm)
3		Broad-spectrum filter (280–400 nm)
4	Inorganic chemical substances	Titanium dioxide (250–340 nm)
5		Zinc oxide (250–380 nm)
6	Biological prophylaxis for UV and for IRA rays	Calcium, folic acid, nicotinamide, and other
7		Antioxidants like Vitamins E and C, Coenzyme Q 10, flavonoids, and others

disappear after some time. Incorporation of nanoparticles into creams requires an energy-intensive suspending, preferably in an oil phase, or the purchase of concentrated TiO_2 suspension. At high pigment concentrations, such a cream can be less well spread on the skin in comparison to creams with organic filters.

Using micro- instead of nanoparticles, the whole skin would become chalky white. Therefore, only largely invisible nanoparticles are used today. Nanoparticles are not resorbed in the skin, they do not trigger reactions in the skin, and they do not have a systemic effect. Therefore, these products are suitable for children's skin; however, the UVA protection does not meet the latest EC recommendations completely. Several years ago, it was assumed that the extreme fine invisible nanoparticles in the range of 10–60 nm should be able to penetrate the skin via the follicular channels. Recent studies demonstrate that nanoparticles (>40 nm) can only enter the stratum corneum. As there are no studies to the contrary, nanoparticles in external application creams are now considered safe. For particle diameters below 100 nm, however, they must be marked in the INCI list with the addition of "nano." Nano zinc oxide particles agglomerate during preparation so that the effective diameter in the cream is between 250 and 500 nm.

To reduce the photocatalytic reactions and to increase the water-repellent properties, the nanoparticles are coated with hydrophobic liquids, whereby the diameters might increase to values of more than some hundred nanometers. The coating is carried out either with inorganic nanoparticles such as alumina and silica or especially with organic compounds such as fatty acids or silicones (INCI: Simethicone), with silicones being preferred for cosmetic applications (and this in the natural cosmetics!). Nevertheless, nanoparticles should not be inhaled. This means a waiver of nanoparticle sun sprays.

The photocatalytic activity of nanoparticles is reduced by the coating layer, but not to zero. However, the reactions appear to have no negative effects in the skin, as they take place in the horny layer and not in living cells. Titanium dioxide acts significantly stronger against UVB radiation as zinc oxide. However, the bandwidth of zinc oxide is superior in the near UVA range, but not quite sufficient. Therefore, a mixture of the two pigments with an organic UVA filter should be formulated that also covers the bandwidth in the near UVA range. Because uncoated nanoparticles promote oxidations, these sunscreens should necessarily contain oxidation-stable ingredients and radical scavengers to protect the skin. Thus, the use of uncoated nanoparticles is less recommended. Largely safe particles require a careful and effective silicone coating. An example represents the SoFiTix broad-spectrum powder [12], consisting of 60% silicon-coated titanium dioxide and 40% zinc oxide (INCI: Titanium Dioxide, Simethicone, Zinc Oxide; in the cream are all particles >100 nm). One percentage of this powder gives an SPF of 2, and 10% gives an SPF of 20. The SPF of 20 means 95% protection against UVB rays, which should be sufficient for most people with repeated use during the day under normal conditions.

Organic substances with alternating double bonds can absorb UV rays and convert them into heat. Depending on the molecular arrangement, UVB and UVA or broad-spectrum filters result, which were selected for their potency and stability (Table 7.6). Each substance has a characteristic peak in the absorption spectrum like a fingerprint. To cover the entire UV spectrum between 280 and 400 nm,

Table 7.6 UV filters approved in accordance with the EC Cosmetics Directive [13–15]; their chemical formulas are described in the literature [6].

No.; name abbreviation	INCI name/*trade name*	EC number	CAS number	Quantity approved US	Maximum concentration EC regulation	Maximum concentration Australia	Filter type, band (nm)
1 PABA	PABA (4-Aminobenzoic Acid)/*PABA (Merck)*	205-753-0	150-13-0	5–15%	5%		UVB 283 max.
2 CBM	Camphor Benzalkonium Methosulfate/*Mexoryl SK*	258-190-8	52793-97-2		6%	6%	UVB 280–320
3 HMS	Homosalate/*Eusolex HMS*	204-260-8	118-56-9	4–15%	10%	15%	UVB 295–315, 308 max.
4 BP3, B-3	Benzophenone-3/*Eusolex 4360, Neo Heliopan BB, Escalol 567 Spectra-Sorb UV-9, Uvinul M-40*	205-031-5	131-57-7	2–6% Oxybenzone	10%	10%	Broad-spectrum 280–400, 288 and 329 max.
5	Deleted						
6 PBSA	Phenylbenzimidazole Sulfonic Acid[a),b)]/*Eusolex 232, Parsol HS*	248-502-0	27503-81-7	1–4% Ensulizole	8% as acid	4%	UVB 280–320, 310 max.
7 TDSA	Terephthalylidene Dicamphor Sulfonic Acid[a)]/*Mexoryl SX*	410-960-6	92761-26-7, 90457-82-2	<10% Canada; Ecamsule	10% as acid	10%	UVA 320–400, 345 max.
8 BMDM	Butyl Methoxydibenzoyl Methane[a)]/*Eusolex 9020, Parsol 1789, Neo Heliopan 357*	274-581-6	70356-09-1	3% Avobenzone	5%	5%	UVA 315–380, 358 max.
9 BCSA	Benzylidene Camphor Sulfonic Acid/*Mexoryl SL*		56039-58-8		6% as acid	6%	UVB 293 max.
10 OC, OCR	Octocrylene[b)]/*Neo Heliopan 303, Uvinul N-539, Escalol 597, Parsol 340, Eusolex OCR*	228-250-8	6197-30-4	7–10%	10% as acid		UVB 303 max.

11 PBC	Polyacrylamidomethyl Benzylidene Camphor/*Mexoryl SW*		113783-61-2		6%		UVB 295 max.
12 EHMC	Ethylhexyl Methoxycinnamate/*Eusolex 2292, Parsol MCX, Escalol 557, Neo Heliopan AV, Uvinul MC 80*	226-775-7	5466-77-3	7.5% Octinoxate	10%		UVB 280–320, 308 max.
13	PEG-25 PABA/*Uvinul P 25*		116242-27-4		10%	10%	UVB 280–320, 308 max.
14 IMC	Isoamyl *p*-Methoxycinnamate/*Neo Heliopan E 1000*	275-702-5	71617-10-2		10%	10%	UVB 280–320, 308 max.
15 EHT	Octyl Triazone, Ethylhexyl Triazone[a),b)]/*Uvinul T 150*	402-070-1	88122-99-0		5%		UVB 280–320, 312 max.
16 DTS	Drometrizole Trisiloxane[a),b)]/*Mexoryl XL, Silatrizole*		155633-54-8		15%	15%	Broad-spectrum, 280–400; 303 u. 344 max.
17 DEBT	Diethylhexyl Butamido Triazone[a),b)]/*Uvasorb HEB*		154702-15-5		10%		UVB 312 max.
18 MBC	4-Methylbenzylidene Camphor/*Eusolex 6300, Parsol 5000*	253-242-6	38102-62-4, 36861-47-9		4%	4%	UVB 295 max.
19 BC	3-Benzylidene Camphor/*Mexoryl SDS-20*	239-139-9	15087-24-8		2%		UVB 295 max.
20 EHS	Ethylhexyl Salicylate[b)]/*Neo Heliopan OS*	118-60-5	204-263-4	5%	5%		UVB 310 max.
21 EDP?	Ethylhexyl Dimethyl PABA/*Eusolex 6007*	244-289-3	21245-02-3	1.4–8%	8%		UVB

(Continued)

Table 7.6 (Continued)

No.; name abbreviation	INCI name/*trade name*	EC number	CAS number	Quantity approved US	Maximum concentration EC regulation	Maximum concentration Australia	Filter type, band (nm)
22 BP4, BP5, B-4, B-5	Benzophenone-4, Benzophenone-5/*Uvinul MS 40, Uvasorb S 5*	223-772-2	4065-45-6, 6628-37-1	5–10% Sulisobenzone	5% as acid	10%	UVB, short waves UVA; 286 and 324
23 MBBT	Methylene Bis-Benzotriazolyl Tetramethylbutylphenol[b)]/*Tinosorb M*	403-800-1	103597-45-1		10%		Broad-spectrum; 306 and 348 max.
24 DPDT	Disodium Phenyl Dibenzimidazole Tetrasulfonate/*Neo Heliopan AP*	429-750-0	180898-37-7		10% as acid	10%	UVA
25 BEMT	Bis-Ethylhexyloxyphenol Methoxyphenyl Triazine[a),b)]/*Tinosorb S*		187393-00-6		10%	10%	Broad-spectrum, 310 and 340 max.
26	Polysilicone-15/*Parsol SLX*	426-000-4	207574-74-1		10%	10%	UVB, 310 max.
27	Titanium Dioxide	236-675-5, 205-280-1, 215-282-2	13463-67-7, 1317-70-0, 1317-80-2	2–25%	25%	25%	Broad-spectrum. 250–340
28 DHHB	Diethylamino Hydroxybenzoyl Hexyl Benzoate[a)]/*Uvinul A*[b)]	443-860-6	302776-68-7		10% in sun-screen	10%	UVA
	Zinc oxide from CosIng, temporarily approved as a UV filter						
30a	Zinc Oxide (nano)	215-222-5	1314-13-2	25%	25%	No limit	Broad-spectrum, 250–380

? – Value is unknown or not safe.

a) Recommended due to excellent photostability [16], alone or in combination; in addition, 23, 27, and 30a.

b) Recommended due to excellent photostability [17], measured individually.

sunscreen with organic filters should contain several substances for complete UVA and UVB protection. As the substances in the skin probably act predominantly in the skin, the cream needs about 15 minutes after application before the UV protection has formed. In contrast, particle-containing creams act immediately after application. Under strong UV light, the organic compounds may undergo decomposition reactions and, in some cases, reach the bloodstream. Particularly, UVA filters are photoinstable and must be stabilized by suitable UVB filters [18], or they lose their effectiveness during the sunbath. For instance, the UVA filter, butyl methoxydibenzoyl methane, undergoes an internal molecular hydrogen rearrangement through UVB radiation, which reduces the absorption strength considerably. Suitable UVB filters prevent this reaction. Some filters, such as ethylhexyl methoxycinnamate (No. 12), disintegrate under strong sun light. The compound can be detected in the urine. That is why it is important to combine the organic filters properly to stabilize them (see Table 7.7). Benzophenones are discussed as a trigger for cancer. All substances that can change under UV light should be avoided. Table 7.6 shows the recommendable stable filters, marked by an asterisk. The organic filters have fallen into disrepute because the internet articles do not distinguish between the stable and unstable filters. There are enough photostable substances for the formulation of effective sunscreens with high UV protection. In addition, the nanoscale organic powder (No. 23; Annex VI) and the inorganic pigments such as titanium dioxide and zinc oxide are included, a total of 11 photostable substances.

Table 7.7 Examples for proved stable combinations of UV filters [18]; the numbers refer to Table 7.6; the combination with No. 12 leads to instabilities.

Numbers	Organic UV filter	Pigment
8	UVA	
10	UVB	
7	UVA	
8	UVA	
10	UVB	
27		Broad spectrum
7	UVA	
8	UVA	
18	UVB	
27		Broad spectrum
7	UVA	
8	UVA	
10	UVB	
18	UVB	
27		Broad spectrum

Source: Data from Ref. [13].

Less stable organic UV filter may react in UV light and penetrate through the skin, disrupt metabolic processes, and can cause allergies. The risk of serious side effects in the skin is low because the substances listed in the Cosmetics EC-Regulation are for the most part widely investigated in the approved concentration range. However, beware that traces and cleavage products of some organic filters were detected in breast milk, urine, blood plasma, and sperm, comparable with ingredients of perfumes. Therefore, the use depends on the stability of the filter. For ethical reasons, manufacturers should only launch photostable formulations on the market. Please note the toxicological data of the manufacturer/supplier.

On the one hand, some organic UV filters are not undisputed because unwanted reactions in the skin cannot be ruled out. On the other hand, all filters can be washed off during swimming in water; in particular, there is a high risk with non-water-repellent sunscreen. The first problem is that there is no UV protection in and just after the water. Second, the organic compounds unnecessarily disturb the ecological balance in the sea, more than the coated nanoparticles, for example, near corals. Although water-repellent creams can help to reduce the effect, an influence is not excluded.

An interesting compound among the legal substances represents the methylene bis-benzotriazolyl tetramethylbutylphenol (No. 23 in the Annex VI of the Cosmetics Regulation). It is used in the form of organic nanoparticles with approximately 100–200 nm in diameter, insoluble in water and in lipids, and absorbs as a broad-spectrum filter in the range of 290–385 nm with maxima at 305 and 360 nm. These nanoparticles combine the protective effect of the inorganic pigments through scattering and reflection with the absorption properties of chemical filters. Because of their particle sizes and poor solubility, the pigments cannot penetrate the skin and cause photoallergic or toxic reactions. The substance with the trade name Tinosorb® M is also very well tolerated by the skin.

The 28 chemical substances in the positive list of the EC Cosmetics Regulation ([13], Annex VI) must meet many requirements. This list also exhibits the maximum use concentration. The requirements are not met by all substances to 100%:

- Skin-friendly without allergic potency
- Do not trigger a phototoxic or allergic reaction
- Do not get into the blood
- Show wide filter spectrum (UVA, UVB, or broad-spectrum)
- High efficiency in already low concentration
- Long duration of action
- Sweat and water resistance (desired)
- Good adhesion on the skin surface
- Stability against UV light (photostable), air (oxidation), and heat (temperature)
- No intra- and intermolecular reactions (inert)
- Good solubility in the cream base
- Unobtrusive in smell and color

7.16.5 Application and Warning Notices

According to EC Commission recommendations, the following information about the effect of sunscreen must be given:

- SPF, as determined by the International Sun Protection Factor Test Method (in use since 2006);
- UVA factor, according to the "persistent pigment darkening method";
- Water resistance, if possible with % indication or test method.

UVA protection should be at least one-third of the SPF (UVB). Furthermore, all sunscreen packages (SPF > 6) or labels should contain the following information [15]:

Application and warning notices:

- Apply before sunbathing.
- Apply several times to maintain the sunscreen, especially after staying in the water.
- Apply sunscreen generously; low application quantities reduce the protective performance.
- Protect babies and toddlers from direct sunlight.
- Use protective clothing and sunscreen with a high SPF (greater than 25) for infants and young children.
- Even sunscreens with high SPFs do not provide complete protection against UV rays.
- Excessive sun exposure poses a serious health risk.
- Do not spend too long in the sun despite using a sunscreen.
- Avoid intense midday sun.

In addition, the use of the terms "sun blocker" or "complete protection" or "protection for the whole day" is prohibited. This rule applies almost everywhere in the world.

In Europe, sunscreens are considered as cosmetics because the protection of skin from damage due to sun exposure is referred to as a cosmetic action. The rules will be constantly updated as scientific progress occurs, based on the Scientific Committee of the Consumer Safety's (SCCS) recommendations on safety of ingredients (see Section 1.7). Many of well-proven UV filters, widespread in Europe (see Table 7.6), are not yet approved in the United States. This is one reason why US dermatologists recommend the use of zinc oxide and titanium dioxide. In the United States, sunscreens are covered by the Medicines Act and marketed as OTC drugs. The approval of a new UV filter at the Food and Drug Administration (FDA) is a costly and lengthy process. Therefore, the US cosmetics companies shy away from the costs of approval. Using the New Drug Application (NDA) process for OTC drug approval, a company submitted data to the FDA on the safety and efficacy of OTC sunscreen products. The FDA approved OTC sunscreen drug products that may also contain further active ingredients. As a result, positive tested UV filters (Table 7.8) can sometimes only be used in

Table 7.8 Approved sunscreens in the United States and their maximum concentration [19].

No.	Name	Maximum concentration (%)	No.	Name	Maximum concentration (%)
1	Aminobenzoic acid (PABA)	15	12	Phenylbenzimidazole sulfonic acid	4
2	Avobenzone	3	13	Sulisobenzone	10
3	Cinoxate	3	14	Titanium dioxide	25
4	Dioxybenzone	3	15	Trolamine salicylate	12
5	Homosalate	15	16	Zinc oxide	25
6	Menthyl anthranilate	5	17	Ensulizole	4
7	Octocrylene	10	18	Meradimate	5
8	Octyl methoxycinnamate	7.5	19	Octinoxate	7.5
9	Octyl salicylate	5	20	Octisalate	5
10	Oxybenzone	6	21	Octocrylene	10
11	Padimate O	8	22	Oxybenzone	6

Source: Data from Ref. [19].

combination with other filters. Therefore, the list of European regulation is more modern and adapted to the state of knowledge. Conversely, substances commonly used in the United States [20] are not authorized in Europe. For the same substances, the maximum permitted levels between the United States and Europe may vary significantly.

The FDA has not approved any new filters in the past 10 years. The Sunscreen Innovation Act (SIA 2014), created to take over the substances authorized in Europe, have not really change the situation until now. The FDA approval of the well-examined substances such as miloxate, bemotrizinol, bisoctrizole, drometrizole trisiloxane, ecamsule, enzacamene, iscotrizinol, and octyl triazone is still far away [19]. The "final rule" (Table 7.9) deals mostly with the labeling. Wording is strictly regulated with regard to relevant information like water resistance or sunscreen with a broad spectrum (SPF > 15). Namely, these products may bear the claim: "if used as directed with other sun protection measures, (the product) decreases the risk of skin cancer and early skin ageing caused by the sun." Single states such as California have specific state rules. They demand to print the following warning on labels of products, e.g. if it contains the cancer-causing chemical benzophenone: "WARNING: This product contains benzophenone, a chemical known to the State of California to cause cancer" [19]. For each recipe, therefore, it must be carefully checked whether the substances and their maximum quantities comply with the regulations in the respective country or region. This applies worldwide.

Canada's directive represents a blend of the EU and US regulations. The regulation of cosmetics takes place under similar conditions. However, sunscreens are classified in different ways according to the single ingredient and may fall either

Table 7.9 Final rule for the labeling of sunscreens in the United States [19]; partially does not comply with the EC Cosmetics Regulation.

Information	Final rule
Active ingredients/ purpose	**Name of ingredients, amounts, purpose (sunscreen)**
Uses	"Helps to prevent sunburn," optional "if used as directed with other sun protection measures (see Directions), decreases the risk of skin cancer and early skin ageing caused by the sun"
Warnings	"For sunscreen products that are not broad spectrum or for products that are broad spectrum with an SPF value less than 15, **Skin Cancer/Skin Aging Alert** (in bold font). Spending time in the sun increases your risk of skin cancer and early skin ageing. This product has been shown only to help prevent sunburn, **not** (in bold font) skin cancer or skin aging." "For external use only." "Do not use on damaged or broken skin." "Stop use and ask a doctor if skin rash occurs." "When using this product, keep out of eyes. Rinse with water to remove." "Keep out of reach of children. If swallowed, get medical help or contact a Poison Control Center right away."
Directions	Non-water-resistant product: o Apply liberally 15 min before sun exposure. o Use a water-resistant sunscreen if swimming or sweating. o Reapply at least every 2 h. o Children under 6 mo. Ask a doctor. Water-resistant product: o Apply liberally 15 min before sun exposure. o Reapply after 40 (or 80) min of swimming or sweating. o Immediately after towel drying. o At least every 2 h. o Children under 6 mo. Ask a doctor. Water-resistant and non-water-resistant products, For sunscreens with broad-spectrum SPF values of 15 and higher: o **Sun protection measures** (in bold font). Spending time in the sun increases your risk of skin cancer and early skin aging. To decrease this risk, regularly use a sunscreen with a broad-spectrum SPF value of 15 or higher and sun protection measures including ✓ Limit time in the sun, especially from 10 a.m. to 2 p.m. ✓ Wear long-sleeved shirt, pants, hats, and sunglasses.
Inactive ingredients	List of all inactive ingredients in alphabetical order.
Other information	"Protect this product from excessive heat and direct sun."
Questions	No statement required.

Source: Data from Ref. [19].

under the Natural Health Product or, in most cases, the Drug Products regulations. In New Zealand, some of the Middle East/Arabic countries, Turkey, and ASEAN countries, the EC Cosmetics Regulation is accepted and applied as it is or with minor changes. The main point is to check the actual local legislative status of the ingredients and the rules for labeling. The warnings on the label of ASEAN products are less stringent than from those of the United States. Examples of recommended warnings:

- Do not stay too long in the sun while using a sunscreen.
- Reapply frequently to maintain protection, especially after sweating, swimming, or toweling.
- The use of sunscreens is one way to reduce the dangers from sun exposure.
- Instructions for use to ensure that sufficient quantity is applied, e.g. pictogram and illustration.
- Overexposure to the sun is a serious health threat.

In China, sunscreens are classified as "special use cosmetics." This means that the products must be registered with the State Food and Drug Administration (SFDA). The authority will check the safety and efficacy of the product, along with the SPF and other parameters. China has aligned itself with the EC Cosmetics Regulation. The regulation and approved substances are largely harmonized between China and Europe. However, China requires a mandatory in vivo UVA test.

Because of the geographical and meteorological position of Australia, the sun protection is given high attention. A cream containing an ingredient with sunscreen properties is a cosmetic sunscreen there when the primary purpose of the product is not the sun protection. The rules for the approval differ, but the maximum concentrations of the substances are largely consistent with the EC Directive. Cosmetics are regulated by the National Industrial Chemicals Notification & Assessment Scheme (NICNAS) and real sunscreens as therapeutic products by the Therapeutic Goods Administration (TGA). Sunscreens in Australia often contain nano titanium dioxide and zinc oxide. A declaration of the nanosize is not necessary. Therapeutic sunscreens with broad-spectrum protection (SPF of 30 or higher) may have indications on the label like:

- May assist in preventing some skin cancers.
- May reduce the risk of some skin cancers.
- Can aid in the prevention of solar keratoses.
- Can aid in the prevention of sunspots.

In the Single Customs Union, consisting of Russia, Belarus, and Kazakhstan, sunscreens are classified and registered as cosmetics according to the Technical Regulation for the Safety of Cosmetic Products. The rules and substances from a positive list resemble the EU regulation. Also, the labeling is similar to the EU regulations for sunscreens with a maximum SPF of 50+, PA+++ marking (see Section 7.16.6), and the broad spectrum. Because of the climate conditions, products need to be evaluated for thermal stability.

The Mercosur Common Market includes Argentina, Brazil, Paraguay, and Uruguay. The group has a technical regulation establishing labeling and safety

Table 7.10 Worldwide classification and labeling of sunscreens [19].

Area	Classification	SPF maximum value	UVA label
Europe	Cosmetics	50+	UVA, PA+
USA	OTC	50	Broad spectrum
Canada	Drugs	50+	Broad spectrum
Australia	Therapeutics/cosmetics	50+	PA+
New Zealand	Cosmetics	50+	
India	Cosmetics	50	UVA protection, PA+
China	Special cosmetics	50+	PA+
ASEAN	Cosmetics	? (100)	UVA
Mercosur	Cosmetics	50+ (100)	UVA
Mexico	Cosmetics	50+ (60)	UVA
Russia	Cosmetics	50+	UVA
Protection grade for UVA: PA+ stands for PA+, ++, +++, (++++)			

? – Value is unknown or not safe.
Source: Data from Ref. [19].

requirements for sunscreens. Brazil's regulations differ in some points. The technical framework follows the EU recommendation of 2006. Among other things, the label requires the indication of the SPF, the water resistance, and the UVA protection, which must be at least one-third of the UVB. Products must be registered. An overview of the product labeling in different countries is presented in Table 7.10.

7.16.6 Measurements for the Determination of Sun Protection

The skin must be protected from the negative effects of UV radiation. In addition to the desired tan, staying in the sun triggers some unpleasant side effects such as sunburn and premature aging of the skin by free radicals, as well as in extreme cases, cancer. Skin aging is initially not noticeable and makes itself felt only after decades, preferably in the face. It is proportional to the number of extreme solar radiation that led to sunburn. For skin aging, too long stays in the sun are responsible, especially in the childhood and adolescence. The skin type of persons determines the sensitivity to sunbeams. Sensitivity depends on the color of the skin, hair, and eyes. People with fair complexion (Nordic type) tolerate significantly less sun than people with darker skin (Mediterranean type). The effectiveness of the self-protection in minutes multiplied by the SPF of the cream gives the time of sun protection by the cream. If the self-protection for the Nordic type takes 5 minutes, then a cream with the SPF of 30 acts over a total protection time of 150 minutes. The other way around SPF 30 means that only 1/30th of the burning radiation reaches the skin through the recommended thickness of sunscreen, which is slightly thicker than what people think. On average, 2 mg/cm^2 should be used, i.e. about 30 g from head to toe. The sun protection drops considerably when less cream is applied. Using half the required amount of sunscreen

Table 7.11 SPF classes.

SPF class	SPF	% UVB blocked
Low protection	6	83.3
	10	90
Medium protection	15	93.3
	20	95
	25	96
High protection	30	96.7
	50	98
Very high protection	50+ (60 measured)	

only provides the square root of the SPF. Therefore, a half application of an SPF 30 sunscreen only provides an effective SPF of 5.5.

According to the EC Directive, there are only four SPF classes: base (low), medium, high, and very high with eight SPFs (Table 7.11). The value 6 means 6.0–9.9 and so on. This assignment applies today throughout the world. Products with SPFs below 6 are no longer part of the sunscreens because of the low protective effect. The predominant purpose of the products (sun protection) is not met. An SPF of 10 absorbs 90% of the UVB rays. A doubling of the active ingredients for increasing the SPF from 10 to 20 only causes a rise of the absorption by 5–95%. Another doubling gives only 2–3% more absorption. Higher SPFs than 30 are generally not worthwhile. The standard method does not allow to identify reproducible differences in the range of SPF 60–100, which are also not experienced by persons. Because of the greatly increased amount of organic active substances, the burden on the skin can increase due to the cleavage products with negative consequences. Therefore, it is important to use only photostable organic compounds. Regardless of the SPF in the range of 50+, a residual radiation of 2% always remains.

The principle for the in vivo measurement of the SPF is simple. Sun-weaned skin areas are exposed to sunlight with and without the sunscreen and the time to erythema is measured (ISO 24444). The time is called minimal erythema dose (MED) and depends on the skin type of the person, who can stay in the sun without protection between 5–10 minutes (toddlers, and Celtic type I) and 30–45 minutes (Mediterranean type IV; s. Table 7.12). The SPF is then defined as

$$\text{SPF} = \text{MED with sunscreen}/\text{MED without sunscreen}$$

and the total time of protection

$$\text{MED with sunscreen} = \text{MED without sunscreen} \times \text{SPF of the sunscreen.}$$

The effect of the UVA rays in the skin is determined by a globally used in vivo measurement method, developed in 1996 in Japan. The skin on the back of 10 subjects, just a small untreated and a treated area with 2 mg/cm^2 cream, is irradiated for a set time with a solar-simulated UVA light (special

Table 7.12 Minimal erythema dose for different skin types.

Type	Skin/hair color	Description	MED (J/m^2)	Self-protection per day (min)
Toddler, bright, European	White and very pale	Fast sunburn, never tans	About 100	5
I Celtic	Very light skin, many freckles, blonde, or red hair	Immediate sunburn, never tans	200	5–10
II Bright European	Light skin, often freckles, blonde to dark blonde hair	Fast sunburn, slight tans	250	10–20
III European dark	Slightly tinted skin, dark blond to brown hair	Lasting tan, sometimes sunburn	300	20–30
IV Mediterranean	Tanned skin, dark or black hair	Fast, intense tan, rarely sunburn	450	30–45
V Middle Eastern, Asian, Latin American	Dark, olive-brown skin, black hair	Always brown, very rarely sunburn	600	—
VI African black	Black skin, frizzy black hair	Always black, very rarely sunburn	1000	—

mercury vapor or xenon lamp) in the wave range of 320–400 nm. The duration amounts to 15 minutes for determination of immediate pigment darkening (IPD) and 2 hours to measure the persistent pigment darkening (PPD) value by a colorimetric and visual technique. Measured is the tanning of the skin (pigmentation). The increase in pigmentation during 2 hours of solar-simulated UVA irradiation ensues certainly not because of melanin protein synthesis but because the synthesis requires at least 18–24 hours. The ratio of the two PPD values (treated/untreated) is a measure of the protective effect of the cream against UVA radiation. PPD of 10 and more means that more than 90% of the UVA rays are absorbed. The EC Recommendation of 2006 suggests that these values should reach at least one-third of the SPF value (SPF = 30, PPD = 10). To visualize this positive property, the products may carry a characteristic UVA circle (Figure 7.5). Many creams do not provide this desirable UVA protection, especially those containing only pigments, e.g. titanium dioxide or titanium dioxide/zinc oxide. The UVA protection on high level according to the recommendations of the European Commission might require the addition of an organic filter.

Figure 7.5 Symbol for compliance with the recommended EC-UVA protection (PPD > 1/3 SPF).

In the long term, the indication of the SPF on sunscreen should be waived. It is replaced by the protection grades PA+ to PA++++ against UV rays. The target is that customers can classify the total UV protection at a glance. The internationally used factor marks the effect strength with different numbers of + signs. Initially, the factors have been defined by the Japan Cosmetic Industry Association (JCIA) for the UVA protection. Today, the meaning is broadened by the correlation (SPF $\leq$ 3 PPD) that combines the UVB with the UVA protection. Thus, this key figure provides information about the entire UV protection:

- PA+ means the sunscreen can provide UVA protection with a factor of PPD between 2 and 4. Then, the SPF should be 6–12.
- PA++ provides moderate protection against UVA rays with a factor of PPD between 4 and 8 and with SPF values between 12 and 24. It is ideal for normal skin individual exposed to medium UV radiation.
- PA+++ is designed for normal skin that exposes to very strong UV radiation with an effective UVA protection, characterized by a PPD value between 8 and 16 and an SPF of 24–48.
- PA++++ provides a high level of protection for toddlers and persons with sensitive skin; PPD > 16 (SPF > 48). This class is new. In some countries, PA+++ is still the highest level with a PPD > 8.

The determination of sunscreen water resistance takes place by a method of the European Cosmetic and Perfumery Association (Colipa). Unfortunately, the method is not standard and not internationally recognized. The SPF measurements are performed on probands who have had the product applied on their backs. The cream in the right amount (2 mg/cm^2) absorbs for 20–30 minutes. This is followed by 20 minutes of water contact (T = 29 $\pm$ 2 °C) in a bathtub with a water circulation system or a similar system. Then, the water dries on the skin for 15 minutes in the air. The SPF on the back is determined immediately afterward and compared to the baseline before water contact. If the SPF is still >50%, the product is called water resistant. A value over 80% refers to a very water-resistant product. The developer of the method would have to set exact standards for all parameters (type of water, water movement, pressure, etc.). Then, maybe the method would be internationally recognized. Some manufacturers make their own tests, which then lead to noncomparable ratings such as waterproof, seawater-resistant, water-repellent, or sweat-proof. A real water-resistant sunscreen means that the organic filters must be fixed in the skin, so they cannot be washed off. This is achievable by incorporating the active substances into liposomes (see Section 5.5) or better in nanoparticles, wherein the core contains the organic filter (such as 23 in Table 7.6) and the shell consists of nonionic surfactants or polymers (Lipopearls®, Nanopearls®).

7.16.7 Recommended Active Ingredients

On the market, there are typical sunscreens in the form of lotions, but also as creams, gels, and sprays to buy. The attractive-looking gels are not recommended because the required water-soluble UV filters can cleave under sunlight. The sprays provide a practical solution of application. Two points speak against the use. For spraying, the lotions must be relatively thin and the small droplets can be inhaled (may contain solid nanoparticles) and, on the other hand, the sprayed-on film is often too thin to ensure the necessary protection. Therefore, the use of creams and lotions seem to be the best method for the prescribed uniform, relatively thick application in which creams are mostly used only for the face and lotions for the whole body. For sun protection on holiday at the waterside or in the mountains, the lotions present in the form of o/w and w/o emulsions are particularly suitable.

In Europe, today Tinosorb S (25 in Table 7.6) is widely used as an organic UV broadband filter in sunscreens. A further development represents the nanoparticulate Tinosorb M (23 in Table 7.6). Relatively new is the Mexoplex® filter system, a combination of Mexoryl® XL (16 in Table 7.5), Mexoryl® SX (7), and Tinosorb S (25), an outstanding combination of photostable filters with broadband action and high additional UVA protection [21], higher than demanded. To achieve high protection factors, fewer chemicals are needed overall (about 15%).

Everywhere on the Internet is to be read that pigmented sunscreens should be taken for children. It makes sense to avoid photounstable organic filters. However, as there are enough stable compounds, this limited recommendation is incomprehensible. When using only the inorganic oxide mixture, it must be ensured that the important UVA protection (1/3 SPF) is often not achieved at high SPF values. This means a reduction in UVA protection with prolonged sun exposure, while the complete protection against the UVB rays still exists. The lack of complete protection requires the addition of an organic UVA filter. Especially for SPF values in the range of 20–30 and higher, the formulation with inorganic nanoparticles (TiO_2, coated with silicone oil, ZnO) should be expanded by adding an organic UVA filter such as Mexoryl SX (7) for a safe UVA protection. In addition, ZnO exhibits useful antiseptic properties. The pH value of the cream should be above 5 because zinc oxide can dissolve in acidic water.

All components in the sunscreen lotion/cream must be photo- and oxidation-stable. Suitable vegetable oils include jojoba, marula, squalane, coconut, and the vegetable butters such as cocoa, mango, and shea butter. Particularly interesting first is (phyto-)squalane, which is obtained by the hydrogenation of squalene or by direct isolation from plants, and second the MCT-oil (Caprylic/Capric Triglyceride, C8/C10). In addition, sunscreen should contain antioxidants such as vitamins C and E, phytosterols, and many more (Section 5.3.6). The use of soothing substances such as panthenol and Ectoin® is also sensible. As stated earlier, each cream should contain moisturizing components (Section 6.3). Even in a sunscreen, this effect is essential.

7.16.8 Care Creams with Sun Protection

Various face creams are on the market without and with UV filters:

(1) Usual day creams (moisturizer, vitalizer, smoother, and antiaging; Chapter 6) without filters;
(2) The same cosmetic day creams with an SPF below 6, as moisturizer with a low sun protection;
(3) Various day creams with SPF values of 15 to 50+ belong to the sunscreens and are subject to their regulations.

The use of a face care cream is age- and sex-dependent. The thinner skin of women requires more attention, care, and protection than men's skin. The care should start at a young age with a moisturizer and be continued from the age of 35 years with a vitalizer and later with an antiaging cream (Figure 7.6). Well-formulated creams meet the needs of most people. If a fair-skinned, sensitive face is repeatedly exposed to sunlight for short moments while working, it is probably

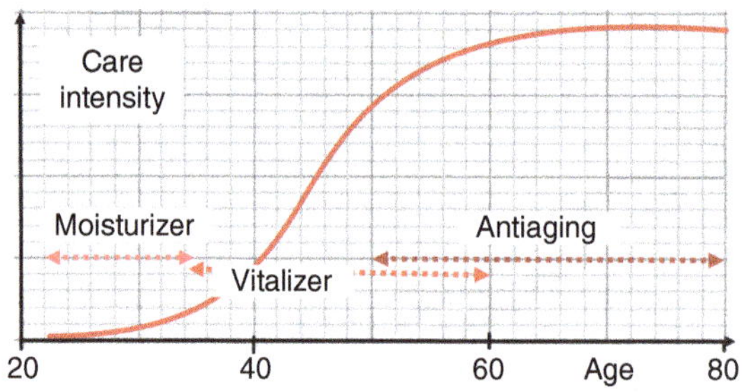

Figure 7.6 Intensity of skin care with age.

Figure 7.7 Examples of face creams with sunscreen. Source: Courtesy of Douglas.

sufficient to use a day cream with a low SPF. It should be noted that because of the smaller thickness of the application, a protection factor of 2–3 results. In the case of longer stays or in countries with strong sunlight, it is necessary to choose a cream with full sun protection, which means SPF of 30 or more. These UV filter-containing products are coupled with moisturizer, vitalizer, antiaging, self-tanner, or makeup tint, as shown in Figure 7.7. In the (very expensive) products shown in Figure 7.7, it is noticeable that only two creams carry the new, recommended labeling, namely the combination of SPF with UVA/PA++. In the leisure, a high SPF is recommended, as usually the times in the sun are not so carefully considered. A shower and the after-sun care should always follow a stay in the sun (Section 7.3).

The skin around the eyes is characterized by a thin epidermis, a loose connective tissue without fat cells, and only a few sebum glands. As a result, the under-eye area tends to become dry and form small wrinkles or, on the other hand, to store water in the loose connective tissue. This temporary swelling, in some cases associated with dark shadows, is usually pent-up lymphatic fluid, which disturb the appearance. A healthy lifestyle, cool compresses, and creams can help. Eye creams should be low in fat and contain, for example, hyaluronic acid, antioxidants such as Coenzyme Q10 and retinol, as well as aloe, chamomile, protein hydrolysates, vitamins A, C, E, and panthenol (see Table 8.22). The eye cream may contain UV filters for the seasons with strong sunlight, preferably Mexoplex. How a water-resistant sunscreen lotion could be formulated is shown in Table 8.23.

7.17 Comment on Cosmeceuticals

In addition to the applications of special cosmetics described here, there are many more areas for cosmetic creams that should be mentioned: Peeling creams, vegan cosmetics, skin bleaching, face cream with makeup, lip care, nose ointment, after-laser treatment, nail cream, hair color cream, creams to increase or decrease hair growth, antipimple cream, antiperspirant cream, cream for the genital area, wound cream (OTC), and others.

7.18 Learnings

- ✓ Specialized cosmetic creams formulated to have special effects are called cosmeceuticals.
- ✓ Cosmeceuticals fill the gap between the usual cosmetic and therapeutic creams.
- ✓ They are suitable for the protection of the skin (UV protection), for prophylaxis (bedsore), and for alleviating the effects of stress (after sun) or diseases (acne) as well as for the removal of skin problems (scaly skin).
- ✓ Further examples are baby creams, cellulite, eczema, hand and foot creams, herpes, horny layers, nails, peeling, self-tanner, and others.

✓ Proven active ingredients for each application are discussed and in Chapter 8 formulated.

✓ Cosmeceuticals show that cosmetics can do more than just moisturize and perfume the skin.

References

1 Barel, A.O., Paye, M., and Maibach, H.I. (2014). *Handbook of Cosmetic Science and Technology*, 4e. CRC Press, Taylor & Francis Group.

2 Draelos, Z.D. (2015). *Cosmeceuticals*, Procedures in Cosmetic Dermatology Series, 3e. Elsevier.

3 Elsner, P. and Maibach, H.I. (eds.) (2000). *Cosmeceuticals: Drugs vs. Cosmetics*, Cosmetic Science and Technology Series, vol. 23. New York, NY: Marcel Dekker.

4 Lees, M. (2012). *Skin Care: Beyond the Basics*, 4e. Cengage Learning.

5 Penzer, R. and Ersser, S. (2010). *Principles of Skin Care: A Guide for Nurses and Health Care Practitioners*. Wiley Blackwell.

6 Ellsässer, S. (2013). *Körperpflegekunde und Kosmetik*, 3e. Berlin: Springer-Verlag.

7 Zuzarte, M., Gonçalves, M.J., Cavaleiro, C. et al. (2011). Chemical composition and antifungal activity of the essential oils of *Lavandula viridis* L'H'er. *J. Med. Microbiol.* 60: 612–618.

8 Stanberry, L.R. (2006). *Understanding Herpes: Revised Second Edition*. Jackson, MS: University Press of Mississippi.

9 Draelos, Z.D. (2016). *Cosmetic Dermatology: Products and Procedures*, 2e. Wiley Blackwell.

10 Ultraviolet, from Wikipedia, the free encyclopedia, 2019. https://en.wikipedia.org/wiki/Ultraviolet#/media/File:Erythemal_action_spectrum.svg (accessed 7 May 2019).

11 Lowe, N.J., Shaath, N.A., and Pathak, M.A. (eds.) (1997). *Sunscreens: Development: Evaluation, and Regulatory Aspects*, 2e. Marcel Dekker.

12 Käser, H. (2016). *Naturkosmetische Rohstoffe, Wirkung, Verarbeitung, kosmetischer Einsatz*, 5e. Linz: Verlag Freya.

13 REGULATION (EC) No 1223/2009 OF THE EUROPEAN PARLIAMENT AND OF THE COUNCIL of 30 November 2009 on cosmetic products. https://eur-lex.europa.eu/legal-content/EN/TXT/PDF/?uri=CELEX:32009R1223&from=EN (assessed 26 April 2018).

14 DocPlayer, Eidgenössisches Departement des Innern, EDI Bundesamt für Lebensmittelsicherheit und Veterinärwesen, Liste der zugelassenen UV-Filter in der Verordnung über kosmetische Mittel, 2016. http://docplayer.org/25599710-Liste-der-zugelassenen-uv-filter-in-der-verordnung-ueber-kosmetische-mittel-vkos-sr-1.html (assessed 26 April 2018).

15 PZ Pharmazeutische Zeitung, Sonnenschutzmittel, Bessere Deklaration schützt Verbraucher, 2007. https://www.pharmazeutische-zeitung.de/index.php?id=3124 (assessed 26 April 2018).

16 Beyer, N., Fa. Beyer & Söhne, UV-Filter in Sonnencremes: Welche sind wirklich gut?. https://www.beyer-soehne.de/uv-filter-in-sonnencremes-welche-sind-wirklich-gut (assessed 26 April 2018).
17 Hauri, U., Lutolf, B., Schlegel, U., and Hohl, C. (2004). Determination of photodegradation of UV filters in sunscreens by HPLC/DAD and HPLC/MS. *Mitt. Lebensmittelunters. Hyg.* 95: 147–161.
18 PZ Pharmazeutische Zeitung, Sonnenschutz, Manchmal trügt der Schein, 2006. https://www.pharmazeutische-zeitung.de/index.php?id=1350 (assessed 26 April 2018).
19 Pirotta, G. (2015). An overview of sunscreen regulations in the world, Monographic special issue: Sun care. *Household Pers. Care Today* 10 (4). https://www.researchgate.net/publication/283515177_An_overview_of_sunscreen_regulations_in_the_world (assessed 26 April 2018).
20 Kirk-Othmer (2013). *Chemical Technology of Cosmetics.* Hoboken, NJ: Wiley.
21 Wolf, E., Avoxa, Pharmazeutische Zeitung, pta Forum, Sonnenschutzpräparate Farbe bekennen, Ausgabe 12/2015. https://ptaforum.pharmazeutische-zeitung.de/index.php?id=6904 (assessed 26 April 2018).

8

Proposals for the Formulation of Creams

8.1 General Remarks

In general, o/w-emulsions absorb faster compared to w/o-emulsions. They are easier to apply, more pleasant, and significantly frequently more effective. W/o-emulsions leave an oily, sometimes sticky, feeling on the skin. The film on the skin causes an occlusion, i.e. reduction of water evaporation and oxygen exchange. The advantage lies in the long-lasting effect of the formulation that offers clear advantages in some cases. This is the reason why therapeutic creams are often formulated with oil as continuous phase. Cosmetic creams should not leave a visible and tactile film on the skin. Therefore, the velvety creams with the oil droplets in the water phase are rightly preferred by consumers. In any case, each cream should contain some substances with a moisturizing, vitalizing, smoothing, and antiaging effect because every adult needs this combination. To underline the main task such as extreme skin smoothing, corresponding special ingredients can still be added. Based on examples, this chapter explains how to extend the existing modules (see Section 5.2) with an additional "active ingredient module" to generate the desired product, specialist for skin care. On the other hand, the recipes should serve as a suggestion for the own formulations. Depending on the type of cream and own experiences/ideas, everyone can change the formulation by adding or omitting excipients and active substances. It may also be advisable to optimize the formulation using the own cost numbers. Changes might apply in particular the fragrance, thickener, and emulsifier system.

Most creams can be applied to the face and the whole body without restrictions. In this chapter, a lot of proposals for rich formulations are written down. If necessary, rich formulations can be reduced and/or diluted with water (5–20%) to obtain the desired consistency of the cream or body lotion. With a 20% water addition, all components in the final recipe must be divided by 1.2, so that 120% yields 100% again. In this way, all cream recipes can be transferred into lotions. The preservation and the thickeners are already sized accordingly.

The proposed formulations provide a framework that may be otherwise filled out. An exact copying also presupposes the production method described.

Cosmetic Creams: Development, Manufacture and Marketing of Effective Skin Care Products,
First Edition. Wilfried Rähse.

In order to realize one's own ideas, the modified proposal of the specified formulations requires a subsequent optimization. Either the consistency of the formulated cream conforms to the requirements or it appears to be too fluid and must be thickened by addition of further substances or it is converted into a body lotion. An almost firm consistency is not recommended for a cream because the important ingredients do not diffuse into the skin sufficiently. Even the fragrance is in kind and quantity only, a suggestion that can be changed according to your own ideas in another essential oil or even in a perfume.

Personal Note 2: All formulas were compiled from experience, but not verified in experiments, also not the interesting combination of ceteareth-30 with glycerol monostearate. The percentages for the ingredients always refer to pure substance (about 100%), even if they are only available in solution. The oil content may be reduced, if the creams appear too rich when used sparingly. Check the recipes taken over whether they comply with the legal regulations and safety requirements.

8.2 Moisturizers

Various causes are responsible for the formation of dry skin. On the one hand, a tendency can be inherited, but also external influences, habits and nutrition, and age play a role. With mature skin, the elasticity and tension gradually diminish because they store less fat and moisture. On the other hand, any washing with soap completely removes the protective mantle. Usually, the skin is able to regenerate itself. The older the person, the hotter the water and the more surfactants are present in alkaline environment, the more the skin suffers and needs to regenerate. Another burden can occur through warm moist air in the bathroom that leads at insufficiently preserved creams to an increase of unwanted germs, especially in open pots. This deficiency should be avoided by the preservative in good creams and by appropriate containers in the form of airless dispensers.

In many cases, the cream's job is to bring the moisture content in the skin largely into the condition of young, healthy skin. Generally, the upper layers of skin (epidermis) need a water balance typical of healthy skin. Above all, the horny layer must have the right moisture content of 15–20% to look healthy and beautiful. When it drops below this range, the skin feels dry. To avoid such conditions, characterized by skin slackening, each cream should at least make a small contribution to regulating the water balance. An effective moisturizer significantly increases the firm appearance of the skin, especially in the elderly. Substances that provide more and longer lasting moisture are described in Section 6.3.

Usually, each cream includes several moisturizing substances. In these examples, the base module already contains various moisturizing ingredients, in particular glycerol, sorbitol, and hyaluronic acid (with the high molecular weight of 1–1.5 MDa), all known as effective agents for this application. Because of the large selection of ingredients, there are several concerted alternatives available to increase the share of humectants to 10% and fill the gap of the modules.

After own experiences, everyone can change or supplement the suggested ingredients. In this way, it is possible to produce very effective moisturizers. All known selected substances are well tolerated by the skin. Some of the varieties of alternatives are shown in Table 8.1. The discussion of the proposed recipes ensues without considering the cost, the desired markets (region), and market segments.

For a pleasant feeling on the skin, aloe represents an appropriate and effective ingredient. It is particularly easy to use the gently dried powder (1 : 200) instead of the gel. During the preparation, the gel is formed that highlights the effect of the moisturizer. Furthermore, urea and lactate as well as amino acids as indispensable components of the skin should not be missing. The high-molecular-weight hyaluronic acid ensures a lasting effect. To mask the creamy inherent smell, which is reminiscent of therapeutic creams, it is advisable to add an essential oil (see Section 6.7) in quantities of 0.1–0.25%. An alternative offers the addition of cyclodextrins to extinguish the smell. The third possibility concerns the conscious use of the positive effects of an essential oil and at the same time the smell. A large number of certified oils exist, covering a wide range of fragrances and medical effects. Many odors of the oils are perceived as pleasant. Add about 0.4–0.5%.

The alternatives described deal with some of many possible variants. Table 8.2 shows the complete formulation of a day cream, based on the modules and the alternative M2 in Table 8.1. The recipe for a night cream (see Table A.1) contains the substances of alternative M1 and another emulsifier system.

8.3 Vitalizing Creams

To change a pale facial skin in a healthy color state, it is recommended to use a vitalizing cream. This should stimulate not only the cell renewal but also the blood circulation. Of course, a moisturizing effect is included, expected by the consumer. For skin vitalization, the vitamins represent the first choice. With regular use, the interaction of several vitamins has a sustained invigorating effect on the skin structure. In addition, the antioxidant action of vitamins results in an antiaging effect, so that vitalization and antiaging take place simultaneously with an emphasis on vitalization, depending on the quantity ratios. In the case of antiaging products, the amounts are exactly the opposite. Table 8.3 shows some alternatives for this type of creams. Retinaldehyde, magnesium ascorbyl phosphate, α-hydroxy acids (AHAs), and α-lipoic acid can be used as particularly suitable for the vitalizing effect. Of the essential oils, rosemary should be the most effective. These substances can be found in the three alternatives. The recipe, based on the alternative V1 from Table 8.3, is depicted in Table 8.4. To demonstrate the diversity of the formulations, the emulsifier systems and the lipid module are varied, as described in Chapter 5. The alternative V3 with the tested modules can be seen in Table A.2. It should be noted that a high vitamin C concentration can cause whitening of the skin. If this is undesirable, the concentration should remain below 3%.

Table 8.1 Compositions of the "active ingredient module" for moisturizer in the alternatives 1–3.

Module	Ingredient	INCI name	Amount (%)
Humectants, which are already part of the modules			
Moisturizing substances	Glycerol	Glycerin	3
	Sorbitol	Sorbitol	3
	Hyaluronic acid	Hyaluronic Acid	0.1
	Allantoin	Allantoin	0.1
Substances with moisturizing effects in other modules	Trisodium citrate dihydrate	Sodium Citrate	0.64
	Citric acid monohydrate	Citric Acid	0.16
	Silicone oil	Dimethicone	1.5
	Xanthan	Xanthan Gum	0.3
	D-Panthenol	Panthenol	2.5
Module "Active ingredients" (moisturizer)			
Alternative M1	Urea	Urea	5
	Glycine	Glycine	2
	Hyaluronic acid	Hyaluronic Acid	0.1
	Lavender oil	Lavandula Angustifolia Oil	0.5
	Water	Aqua	2.4
Alternative M2	Urea	Urea	3
	Aloe vera powder 1 : 200	Aloe Barbadensis Leaf Juice Powder	0.15[a)]
	Hyaluronic acid	Hyaluronic Acid	0.15
	Lemon	Citrus Limon Seed Oil	0.2
	Water	Aqua	6.5
Alternative M3	Urea	Urea	4
	Sodium pyrrolidone carboxylic acid (PCA)	Sodium Pyrrolidone Carboxylic Acid	3
	Sodium lactate	Sodium Lactate	1.8
	Lactic acid	Lactic Acid	0.9
	Chamomile	Anthemis Nobilis Flower Oil	0.3

a) Corresponds to 30% Aloe vera gel.

Table 8.2 Suggestion for a daytime moisturizer (based on alternative M2, Table 8.1).

No	Ingredient	Amount (%)	INCI name	EC number
1	Water (cosm. – cosmetic quality)	61.45	Aqua	231-791-2
2	Sweet almond oil	4	Prunus Amygdalus Sativa Kernel Oil	291-063-5
3	Cetyl stearyl alcohol	3.5	Cetearyl Alcohol	267-008-6
4	Evening primrose oil	3	Oenothera Biennis Oil	289-859-2
5	Hemp oil	3	Cannabis Sativa Seed Oil	289-644-3
6	Shea butter	3	Butyrospermum Parkii Butter	293-515-7
7	Urea	3	Urea	200-315-5
8	Glycerol	3	Glycerin	200-289-5
9	Sorbitol	3	Sorbitol	200-061-5
10	D-Panthenol	2.5	Panthenol	201-327-3/ 240-540-6
11	Nicotinamide	2	Niacinamide	202-713-4
12	Ethanol	2	Alcohol	200-578-6
13	Cetyl stearyl ether 30EO	1.5	Ceteareth-30	68439-49-6 CAS
14	Silicone oil M350	1.5	Dimethicone	63148-62-9 CAS/9006-65-9 CAS/9016-00-6 CAS
15	Vitamin E (natural, D-α; 0.7)/sunflower oil (0.3)	1	Tocopherol/ Helianthus Annuus	59-02-9 (D) CAS/232-273-9
16	Trisodium citrate dihydrate	0.64	Sodium Citrate	200-675-3
17	Glycerol monostearates	0.5	Glyceryl Stearate	250-705-4/ 286-490-9
18	Xanthan	0.3	Xanthan Gum	234-394-2
19	Potassium sorbate	0.25	Potassium Sorbate	246-376-1
20	Hyaluronic acid	0.25	Hyaluronic Acid	232-678-0
21	Lemon	0.2	Citrus Limon Seed Oil	296-174-2/ 284-515-8/ 285-359-3
22	Citric acid monohydrate	0.16	Citric Acid	201-069-1
23	Aloe vera powder (1 : 200)	0.15	Aloe Barbadensis Leaf Juice Powder	287-390-8/ 305-181-2
24	Allantoin	0.1	Allantoin	202-592-8
	Sum (without water):	38.55%		

Table 8.3 Compositions of the "active ingredient module" for vitalizers in the alternatives 1–3.

Module	Ingredient	INCI name	Amount (%)
Vitalizing ingredients, which are already part of the modules			
Vitamins	D-Panthenol	Panthenol	2.5
	Nicotinamide	Niacinamide	2
	Vitamin E (natural)/sunflower oil (70 : 30)	Tocopherol/ Helianthus Annuus	1
Buffer (AHAs)	Trisodium citrate dihydrate	Sodium Citrate	0.64
	Citric acid monohydrate	Citric Acid	0.16
Module "Active ingredients" (vitalizer)			
Alternative V1	Urea	Urea	4.5
	Hyaluronic acid	Hyaluronic Acid	0.1
	Retin aldehyde	Retinal	0.25
	Magnesium ascorbyl phosphate	Magnesium Ascorbyl Phosphate	2.5
	D-Panthenol	Panthenol	2.15
	Geranium	Pelargonium Roseum Leaf Oil	0.5
Alternative V2	Hyaluronic acid	Hyaluronic Acid	0.2
	Sodium lactate	Sodium Lactate	2
	Magnesium ascorbyl phosphate	Magnesium Ascorbyl Phosphate	3.5
	Vitamin E (natural)/sunflower oil (70 : 30)	Tocopherol/ Helianthus Annuus	1.5
	D-Panthenol	Panthenol	2.5
	Orange blossom	Citrus Aurantium Dulcis Flower Oil	0.3
Alternative V3	Sodium lactate	Sodium Lactate	1.6
	Lactic acid	Lactic Acid	0.9
	Magnesium ascorbyl phosphate	Magnesium Ascorbyl Phosphate	3
	α-Lipoic acid	Thioctic Acid	2
	Rosemary	Rosmarinus Officinalis Leaf Oil	0.4
	Water	Aqua	2.1

Table 8.4 Formulation of a vitalizing cream (according to V1; with alternative vegetable oils and emulsifiers).

No	Ingredient	Amount (%)	INCI name	EC number
1	Water	53.95	Aqua	231-791-2
2	D-Panthenol	4.65	Panthenol	201-327-3/240-540-6
3	Urea	4.5	Urea	200-315-5
4	Jojoba oil	4	Simmondsia Chinensis Seed Oil	289-964-3
5	Myristyl myristat	4	Myristyl Myristat	221-787-9
6	Polyglyceryl-3 dicitrate/stearate	4	Polyglyceryl-3 Dicitrate/Stearate	—
7	Wheat germ oil	3	Triticum Vulgare Oil	281-689-7
8	Safflower oil	3	Carthamus Tinctorius Seed Oil	232-276-5
9	Borage seed oil	3	Borago officinalis Seed Oil	225234-12-8 CAS
10	Glycerol	3	Glycerin	200-289-5
11	Sorbitol	3	Sorbitol	200-061-5
12	Magnesium ascorbyl phosphate	2.5	Magnesium Ascorbyl Phosphate	114040-31-2/ 113170-55-1 CAS
13	Nicotinamide	2	Niacinamide	202-713-4
14	Ethanol	2	Alcohol	200-578-6
15	Vitamin E (natural, D-α; 0.7)/sunflower oil (0.3)	1	Tocopherol/ Helianthus Annuus	59-02-9 (D) CAS/232-273-9
16	Trisodium citrate dihydrate	0.64	Sodium Citrate	200-675-3
17	Geranium	0.5	Pelargonium Roseum Leaf Oil	290-144-2
18	Xanthan	0.3	Xanthan Gum	234-394-2
19	Retinaldehyde	0.25	Retinal	204-135-8
20	Potassium sorbate	0.25	Potassium Sorbate	246-376-1
21	Hyaluronic acid	0.2	Hyaluronic Acid	232-678-0
22	Citric acid monohydrate	0.16	Citric Acid	201-069-1
23	Allantoin	0.1	Allantoin	202-592-8
	Sum (without water):	46.05		

8.4 Creams with Smoothing Properties

Many people suffer from dry skin or, more commonly, dry skin sites. These sites are often characterized by a strong deficit of urea, in some cases also of γ-linolenic acid, and by a colonization of microorganisms. With this clarification of the recognized medical causes, a cream can be specifically formulated for the removal of the dry spots and dry skin areas by high additions of urea, evening primrose or borage oil, and sea salt to kill the bacteria. An effective preservation system helps to eliminate the microorganism. For this, sorbic acid at a working pH around 5 is the best choice. Such a cream also works very well on neurodermitic skin, which becomes completely smooth after few days. It complements the cortisone-based prescription cream and provides an example of the therapeutic effect of cosmetic creams (cosmeceuticals).

The described modules (Chapter 5) complement each other to 90%. The formulation, described in Table 8.5, takes 110% (90 + 20). This fact results in no

Table 8.5 Composition of the "active ingredient module" for skin-smoothing creams.

Module	Ingredient	INCI name	Amount (%)
Smoothing ingredients, which are already part of the modules			
Lipids	Sweet almond oil	Prunus Amygdalus Dulcis Oil	4
	Evening primrose oil	Oenothera Biennis Oil	3
	Hemp oil	Cannabis Sativa (Hemp) Seed Oil	3
	Shea butter	Butyrospermum Parkii Butter	3
Vitamins	D-Panthenol	Panthenol	2.5
	Nicotinamide	Niacinamide	2
	Vitamin E (natural)/sunflower oil (70 : 30)	Tocopherol/ Helianthus Annuus	1
Module "Active ingredients" (smoother)			
Alternative S1	Urea	Urea	6
	Magnesium ascorbyl phosphate	Magnesium Ascorbyl Phosphate	2.5
	D-Panthenol	Panthenol	2
	Salt of the Dead Sea	Maris Sal	1
	Sea salt	Sea Salt	3
	Jasmine	Jasminum Officinale Oil	0.3
	Cetyl stearyl alcohol	Cetearyl Alcohol	0.5
	Water	Aqua	1.7

changes during manufacturing. However, finally, after production, the recipe must be corrected by adjusting the components by a factor of 0.909. In this case, all ingredients have a 10% reduced amount. Another possibility without reduced amounts, taken here, is based on the 80% modules recipe, produced with 10% less water (Table 8.6). For emulsification, the system cetyl stearyl ether-30 EO/cetyl stearyl alcohol is suggested because it is capable of processing the large amount of salts (>5%). Other systems have difficulties in emulsification; this means that in the presence of many salts, no stable emulsion arises.

Table 8.6 Formulation of a strong smoothing cream against dry skin sites, in particular for areas with atopic eczema; the formula is useful for all kinds of problems in adults.

No	Ingredient	Amount (%)	INCI name	EC number
1	Water (cosm.)	47.65	Aqua	231-791-2
2	Urea	6	Urea	200-315-5
3	D-Panthenol	4.5	Panthenol	201-327-3/ 240-540-6
4	Sweet almond oil	4	Prunus Amygdalus Sativa Kernel Oil	291-063-5
5	Cetyl stearyl alcohol	4	Cetearyl Alcohol	267-008-6
6	Evening primrose oil	3	Oenothera Biennis Oil	289-859-2
7	Hemp oil	3	Cannabis Sativa Seed Oil	289-644-3
8	Shea butter	3	Butyrospermum Parkii Butter	293-515-7
9	Glycerol	3	Glycerin	200-289-5
10	Sorbitol	3	Sorbitol	200-061-5
11	Sea salt	3	Sea Salt	231-598-3
12	Magnesium ascorbyl phosphate	2.5	Magnesium Ascorbyl Phosphate	114040-31-2/ 113170-55-1 CAS
13	Nicotinamide	2	Niacinamide	202-713-4
14	Sodium lactate	2	Sodium Lactate	200-772-0/ 212-762-3
15	Ethanol	2	Alcohol	200-578-6
16	Cetyl stearyl ether 30EO	1.5	Ceteareth-30	68439-49-6 CAS
17	Silicone oil M350	1.5	Dimethicone	63148-62-9 CAS/9006-65-9 CAS/9016-00-6 CAS
18	Dead Sea salt	1	Sea Salt (Dead Sea Salt)	231-598-3
19	Vitamin E (natural, D-α; 0.7)/sunflower oil (0.3)	1	Tocopherol/ Helianthus Annuus	59-02-9 (D) CAS/232-273-9

(Continued)

Table 8.6 (Continued)

No	Ingredient	Amount (%)	INCI name	EC number
20	Trisodium citrate dihydrate	0.64	Sodium Citrate	200-675-3
21	Glycerol monostearates	0.5	Glyceryl Stearate	250-705-4/ 286-490-9
22	Jasmine	0.3	Jasminum Officinale Oil	289-960-1
23	Xanthan	0.3	Xanthan Gum	234-394-2
24	Potassium sorbate	0.25	Potassium Sorbate	246-376-1
25	Citric acid monohydrate	0.16	Citric Acid	201-069-1
26	Hyaluronic acid	0.1	Hyaluronic Acid	232-678-0
27	Allantoin	0.1	Allantoin	202-592-8
	Sum (without water):	52.35		

8.5 Antiaging Creams

In the correct concentration, vitamins A, B_3, C, and E have a strong wrinkle-smoothing effect, as in Section 6.6 described. In particular, the compounds of vitamin A should be mentioned here. The α-lipoic acid and kinetin represent high effective ingredients for antiaging creams with visible results in the face after prolonged use and should preferably be formulated. The same applies for 2-dimethylaminoethanol (DMAE), but because of the rather unpleasant smell, an odor reduction via cyclodextrins is required. The determination of the correct quantities for this ensues in laboratory tests. Also, ubiquinol (reduced form of coenzyme Q10) can provide a positive contribution. Some customized peptides, which are known for their wrinkle smoothing after six weeks of application, are less frequently used because of their high prices. Expensive antiaging creams should contain these in effective amounts and additionally α-lipoic acid and kinetin for the best wrinkle-smoothing results. Naturally, vitalizing and moisturizing substances are indispensable for a convincing cream. By formulating reasonable levels of effective antiaging ingredients, any company can put beneficial antiaging creams on the market.

Aloe vera also shows, in addition to the moisturizing, a firming effect. In particular, the degraded hyaluronic acid having a molecular weight of 20–50 kDa acts as a wrinkle-smoothing agent and is preferred for cost-effective reasons (0.05–0.2%; small amounts, high impact). Table 8.7 displays three alternatives for antiaging complexes and Table 8.8 an example for the composition of a potent cream. In Annex A3, the alternative AA3 has been incorporated into the module-based standard formulation.

Figure 8.1 shows practical examples of the base creams (moisturizer, vitalizer, and antiaging). The examples are selected by the different design of the storage

Table 8.7 Composition of the "active ingredient module" for antiaging creams in the alternatives 1–3.

Module	Ingredient	INCI name	Amount (%)
Ingredients that can show antiaging effects (part of the modules)			
Lipids	Sweet almond oil	Prunus Amygdalus Dulcis oil	4
	Evening primrose oil	Oenothera Biennis oil	3
	Hemp oil	Cannabis Sativa (Hemp) Seed Oil	3
	Shea butter	Butyrospermum Parkii Butter	3
Vitamins	D-Panthenol	Panthenol	2.5
	Nicotinamide	Niacinamide	2
	Vitamin E (natural)/sunflower oil (70 : 30)	Tocopherol/Helianthus Annuus	1
Module "Active ingredients" (antiaging)			
Alternative AA1	Urea	Urea	3
	Magnesium ascorbyl phosphate	Magnesium Ascorbyl Phosphate	1.4
	D-Panthenol	Panthenol	2
	Retinaldehyde	Retinal	0.2
	α-Lipoic acid	Thioctic Acid	3
	Aloe vera powder (1 : 200)	Aloe Barbadensis Leaf Juice Powder	0.1
	Jasmine	Jasminum Officinale Oil	0.3
Alternative AA2	Magnesium ascorbyl phosphate	Magnesium Ascorbyl Phosphate	3
	D-Panthenol	Panthenol	2
	Hyaluronic acid (degraded)	Hyaluronic Acid	0.1
	Ubiquinol	Ubiquinol	2
	α-Lipoic acid	Thioctic Acid	2
	Water	Aqua	0.9
Alternative AA3	Glycine	Glycine	2
	Magnesium ascorbyl phosphate	Magnesium Ascorbyl Phosphate	2.5
	D-Panthenol	Panthenol	1
	Retinaldehyde	Retinal	0.2
	Kinetin	Kinetin	4.1
	Rose	Rosa Damascena Flower Oil	0.2

Table 8.8 Formulation of a potent antiaging cream (based on alternative AA1; with alternative vegetable oils and emulsifiers, supplemented by glycine and lysolecithin).

No	Ingredient	Amount (%)	INCI name	EC number
1	Water	51.45	Aqua	231-791-2
2	Panthenol	4.5	Panthenol	201-327-3/ 240-540-6
3	Jojoba oil	4	Simmondsia Chinensis Seed Oil	289-964-3
4	Polyglyceryl-3 dicitrate/stearate	4	Polyglyceryl-3 Dicitrate/Stearate	—
5	Myristyl myristat	4	Myristyl Myristat	221-787-9
6	α-Lipoic acid	3	Thioctic Acid	214-071-2
7	Wheat germ oil	3	Triticum Vulgare Oil	281-689-7
8	Safflower oil	3	Carthamus Tinctorius Seed Oil	232-276-5
9	Borage seed oil	3	Borago officinalis Seed Oil	225234-12-8 CAS
10	Urea	3	Urea	200-315-5
11	Glycerol	3	Glycerin	200-289-5
12	Sorbitol	3	Sorbitol	200-061-5
13	Glycine	2	Glycine	200-272-2
14	Nicotinamide	2	Niacinamide	202-713-4
15	Ethanol	2	Alcohol	200-578-6
16	Magnesium ascorbyl phosphate	1.4	Magnesium Ascorbyl Phosphate	114040-31-2/ 113170-55-1 CAS
17	Vitamin E (natural, D-α; 0.7)/sunflower oil (0.3)	1	Tocopherol/ Helianthus Annuus	59-02-9 (D) CAS/232-273-9
18	Trisodium citrate dihydrate	0.64	Sodium Citrate	200-675-3
19	Lysolecithin, de-oiled (powder)	0.5	Lysolecithin	288-318-8
20	Jasmine	0.3	Jasminum Officinale Oil	289-960-1
21	Xanthan	0.3	Xanthan Gum	234-394-2
22	Potassium sorbate	0.25	Potassium Sorbate	246-376-1
23	Retinaldehyde	0.2	Retinal	204-135-8
24	Citric acid monohydrate	0.16	Citric Acid	201-069-1
25	Hyaluronic acid	0.1	Hyaluronic Acid	232-678-0
26	Allantoin	0.1	Allantoin	202-592-8
27	Aloe vera powder (1 : 200)	0.1	Aloe Barbadensis Leaf Juice Powder	
	Sum (without water):	48.55		

Figure 8.1 Design of basic cosmetic creams. Source: Courtesy of Douglas.

containers in shape and color. Some examples of special creams are displayed in Figure 8.2. The descriptions are provided in the next chapters.

8.6 Acne Creams

The best recommendation for an effective acne medicine represents benzoyl peroxide-containing therapeutic creams with 3%, 5%, or 10% active substance, depending on the severity of the disease. These products should always be combined with a nourishing cosmetic acne cream for the alleviation of skin irritation and drying out of the skin. Most creams, especially too fat creams (such as night creams), are unsuitable because they can clog the skin, make comedones, and aggravate the acne. Acne creams should contain oils rich in triunsaturated acids. Because the evening primrose and hemp oil have high linoleic/linolenic acid contents, which act against acne and care for the skin, the proposed acne cream may be based on the modules. Shea butter is also effective against acne, and almond oil nourishes the skin and generates no comedones. Allantoin with its keratolytic properties acts significantly stronger than urea and is an important component for an acne cream. Pyridoxine, part of the vitamin B_6 complex, regulates excessive sebum production and is therefore indispensable for moderate acne. Under use of Table 7.1, the two alternatives shown in Table 8.9 were generated. The proposed formulation of the alternative AC1 (Table 8.10) acts strongly against acne. An additional moisturizing effect provides the alternative AC2. The essential oils used reduce the bacterial activities as well as the preservative system and the acidic pH of 5. During the

Figure 8.2 Examples of special creams. Source: Courtesy of Douglas.

application of acne creams, it is advisable to use the smoothing cream in between (Table 8.6), which nourishes the skin and kills bacteria totally for a long time. In Table A.4, antiacne tonic water is displayed.

8.7 After Sun Creams/Lotions

The use of a special product helps to cool, soothe, and regenerate burned skin. Selected ingredients such as the aloe vera gel contribute to this because the gel cools and soothes the skin in higher concentrations. The gently dried aloe powder is a pleasant form of application because in the water of the emulsion, it produces a gel that corresponds to 200 times the amount of the powder. This is how 0.3% powder turns into a gel that covers the entire water phase. As further moisturizing substances, urea, amino acids, and AHAs prove themselves suitable for the dry skin. Panthenol is highly effective to accelerate the healing process, also the hamamelis extract helps. Very important to minimize long-term damage and speed up healing is the use of antioxidants. For this, the combination of vitamins C and E with α-lipoic acid should provide the best effect. For maintenance, the proven mixture of vegetable oils should be used. To soothe the skin, chamomile (Roman) and jasmine are the most suitable essential oils. Examples for possible

Table 8.9 Composition of the "active ingredient module" for antiacne creams in the alternatives 1 and 2.

Module	Ingredient	INCI name	Amount (%)
Ingredients that show antiacne effects (part of the modules)			
Lipids	Evening primrose oil	Oenothera Biennis Oil	3
	Hemp oil	Cannabis Sativa (Hemp) Seed Oil	3
	Shea butter	Butyrospermum Parkii Butter	3
Vitamins	D-Panthenol	Panthenol	2.5
	Nicotinamide	Niacinamide	2
	Vitamin E (natural)/sunflower oil (70 : 30)	Tocopherol/ Helianthus Annuus	1
Module "Active ingredients" (antiacne)			
Alternative AC1	Lavender	Lavandula Angustifolia Oil	2.6
	Magnesium ascorbyl phosphate	Magnesium Ascorbyl Phosphate	3
	Nicotinamide	Niacinamide	2
	Retinaldehyde	Retinal	0.2
	Pyridoxine-HCl	Pyridoxine-HCl	2
	Allantoin	Allantoin	0.2
Alternative AC2	Magnesium ascorbyl phosphate	Magnesium Ascorbyl Phosphate	2.5
	Allantoin	Allantoin	0.2
	Lactic acid	Lactic Acid	1.6
	Sodium lactate	Sodium Lactate	2.5
	Hamamelis	Hamamelis Virginiana	3.2

compositions of the active ingredient module can be found in Table 8.11, the complete formulations in Table 8.12 and in the Table A.5.

8.8 Baby Cream

The warm, soft, and fragrant baby skin is particularly sensitive. It is five times thinner than the skin of adults and can barely counteract germs and environmental influences. That is why proper care is important in the first few months. Regular application of a cream does not affect the baby's ability to produce its own fat and moisture. Weekly skin care after bathing is useful because babies and young children hardly produce sebum and their barrier function is weaker compared with adults. The moisturizing cream can prevent dehydration. Creams

Table 8.10 Formulation of caring cream against acne (AC1 and the modules).

No	Ingredient	Amount (%)	INCI name	EC number
1	Water (cosm.)	54.95	Aqua	231-791-2
2	Nicotinamide	4	Niacinamide	202-713-4
3	Sweet almond oil	4	Prunus Amygdalus Sativa Kernel Oil	291-063-5
4	Cetyl stearyl alcohol	3.5	Cetearyl Alcohol	267-008-6
5	Magnesium ascorbyl phosphate	3	Magnesium Ascorbyl Phosphate	114040-31-2/ 113170-55-1 CAS
6	Evening primrose oil	3	Oenothera Biennis Oil	289-859-2
7	Hemp oil	3	Cannabis Sativa Seed Oil	289-644-3
8	Shea butter	3	Butyrospermum Parkii Butter	293-515-7
9	Glycerol	3	Glycerin	200-289-5
10	Sorbitol	3	Sorbitol	200-061-5
11	Lavender	2.6	Lavandula Angustifolia Oil	289-995-2
12	D-Panthenol	2.5	Panthenol	201-327-3/ 240-540-6
13	Pyridoxine-HCl	2	Pyridoxine-HCl	200-386-2
14	Ethanol	2	Alcohol	200-578-6
15	Cetyl stearyl ether 30EO	1.5	Ceteareth-30	68439-49-6 CAS
16	Silicone oil M350	1.5	Dimethicone	63148-62-9 CAS/9006-65-9 CAS/9016-00-6 CAS
17	Vitamin E (natural, D-α; 0.7)/sunflower oil (0.3)	1	Tocopherol/ Helianthus Annuus	59-02-9 (D) CAS/232-273-9
18	Trisodium citrate dihydrate	0.64	Sodium Citrate	200-675-3
19	Glycerol monostearates	0.5	Glyceryl Stearate	250-705-4/ 286-490-9
20	Allantoin	0.3	Allantoin	202-592-8
21	Xanthan	0.3	Xanthan Gum	234-394-2
22	Potassium sorbate	0.25	Potassium Sorbate	246-376-1
23	Jasmine	0.2	Jasminum Officinale Oil	289-960-1
24	Citric acid monohydrate	0.16	Citric Acid	201-069-1
25	Hyaluronic acid	0.1	Hyaluronic acid	232-678-0
	Sum (without water):	45.05		

Table 8.11 Composition of the "active ingredient module" for after sun creams in the alternatives 1 and 2.

Module	Ingredient	INCI name	Amount (%)
Ingredients that helps with sunburn (part of the modules)			
Lipids	Sweet almond oil	Prunus Amygdalus Dulcis Oil	4
	Evening primrose oil	Oenothera Biennis Oil	3
	Hemp oil	Cannabis Sativa (Hemp) Seed Oil	3
	Shea butter	Butyrospermum Parkii Butter	3
Vitamins	D-Panthenol	Panthenol	2.5
	Nicotinamide	Niacinamide	2
	Vitamin E (natural)/sunflower oil (70 : 30)	Tocopherol/ Helianthus Annuus	1
Moisturizing substances	Glycerol	Glycerin	3
	Sorbitol	Sorbitol	3
	Hyaluronic acid	Hyaluronic Acid	0.1
	Allantoin	Allantoin	0.1
Module "Active ingredients" (after sun)			
Alternative AS1	D-Panthenol	Panthenol	2
	Hydrolyzed soy protein	Hydrolyzed Soy Protein	2.5
	Magnesium ascorbyl phosphate	Magnesium Ascorbyl Phosphate	2
	α-Lipoic acid	Thioctic Acid	1
	Jojoba oil	Simmondsia Chinensis Seed Oil	2
	Hyaluronic acid	Hyaluronic Acid	0.1
	Aloe vera powder (1 : 200)	Aloe Barbadensis Leaf Juice Powder	0.2
	Jasmine	Jasminum Officinale Oil	0.2
Alternative AS2	D-Panthenol	Panthenol	2.5
	Magnesium ascorbyl phosphate	Magnesium Ascorbyl Phosphate	1
	Shea butter	Butyrospermum Parkii Butter	3
	Urea	Urea	2.9
	Aloe vera powder (1 : 200)	Aloe Barbadensis Leaf Juice Powder	0.25
	Chamomile (Roman)	Anthemis Nobilis Flower Oil	0.35

Table 8.12 Formulation of an effective after sun cream (analogous to AS1; alternative vegetable oils and emulsifiers).

No	Ingredient	Amount (%)	INCI name	EC number
1	Water	60.15	Aqua	231-791-2
2	Jojoba oil	6	Simmondsia Chinensis Seed Oil	289-964-3
3	D-Panthenol	4.5	Panthenol	201-327-3/ 240-540-6
4	Polyglyceryl-3 dicitrate/stearate	4	Polyglyceryl-3 Dicitrate/Stearate	—
5	Myristyl myristat	4	Myristyl Myristat	221-787-9
6	Borage seed oil	3	Borago officinalis Seed Oil	225234-12-8 CAS
7	Glycerol	3	Glycerin	200-289-5
8	Sorbitol	3	Sorbitol	200-061-5
9	Hydrolyzed soy protein	2.5	Hydrolyzed Soy Protein	271-770-5
10	Magnesium ascorbyl phosphate	2	Magnesium Ascorbyl Phosphate	114040-31-2/ 113170-55-1 CAS
11	Nicotinamide	2	Niacinamide	202-713-4
12	Ethanol	2	Alcohol	200-578-6
13	α-Lipoic acid	1	Thioctic Acid	214-071-2
14	Vitamin E (natural, D-α; 0.7)/sunflower oil (0.3)	1	Tocopherol/ Helianthus Annuus	59-02-9 (D) CAS/232-273-9
15	Trisodium citrate dihydrate	0.64	Sodium Citrate	200-675-3
16	Xanthan	0.3	Xanthan Gum	234-394-2
17	Potassium sorbate	0.25	Potassium Sorbate	246-376-1
18	Aloe vera powder (1 : 200)	0.2	Aloe Barbadensis Leaf Juice Powder	287-390-8/ 305-181-2
19	Hyaluronic acid	0.2	Hyaluronic Acid	232-678-0
20	Citric acid monohydrate	0.16	Citric Acid	201-069-1
21	Allantoin	0.1	Allantoin	202-592-8
	Sum (without water):	39.85		

for the diaper area should contain more oils and zinc salts and be applied so that a very thin protective film is formed. The restriction to w/o-emulsions is outdated by modern care ingredients. The o/w-emulsions have advantages because they are easier to spread and easier to absorb. In winter, a high-fat cream serves as a cold protection for the face.

The ingredients for a baby cream are usually chosen because of their low allergy potential. The known humectants and moisturizer have no side effects. They reduce the dehydration of the skin by increasing its water binding capacity in addition to the skin's own moisturizing substances (natural moisturizing factor, NMF). Suitable for this are the classical moisturizing substances such as glycerol, sorbitol, and all other components of the NMF. The excipients required in creams, such as thickeners and emulsifiers, are largely neutral on the skin. After natural cosmetics, EO-containing emulsifiers should promote penetration and emulsify skin fats. However, there are no scientifically verifiable works for the emulsification of skin fats with creams or for the infiltration of harmful substances (from where should they come?). The Cosmetic Ingredients Review (CIR)-expert panel notes that there is enough data to rate Ceteareths as "safe as used" for cosmetics. Decisive for the evaluation of a product is its compatibility in dermatological tests and the calculation of the Margin of Safety (Sections 13.4 and 13.5). Also, for the preservative sorbic acid, there are no studies that make a use in baby creams questionable. On the contrary, in the Cosmetics Regulation, there is a maximum limit of 0.6% but no restrictions for babies or toddlers. As an alternative emulsifier, polyglyceryl-3 dicitrate/stearate can be used in an amount of 3%. Unperfumed skin creams are often rejected because of the greasy medical odor. The addition of perfume, however, is controversial because of the allergy potential of various components. A compromise lies in the fragrance, generated with a single essential oil. The oil should be selected based on a low allergic potential.

For the lipid supply, a wide range of appropriate vegetable oils exist. Mineral oil and Vaseline, which are not components of the skin lipids, should be avoided if possible, although they pose no risk of allergies. Silicone oils (International Nomenclature of Cosmetic Ingredients, INCI: Dimethicone) have molecular weights between 162 and more than 100 000 Da. For cosmetics, linear products between 6000 and 12 000 Da or with a viscosity in the middle range of 200 to maximum 1000 mPas should be chosen (for example, silicone oil M 350 or M 500). These oils do not penetrate the skin, have a high level of tolerability without any potential for allergies, are permeable to gas, and easily distributed (Section 5.3.1). In the diaper zone, they protect the baby reliably against wetness. This makes them ideal for use in baby creams. Another recommendation represents the rapeseed phytosterol oil (INCI name: Brassica Campestris Sterols) for the use in baby creams because it not only supports the barrier function but also inhibits inflammatory effects in concentrations of 0.5–3%. The pH of the care cream exerts a strong influence on the cream effect. At a physiological pH of 5.5, a more hydrating effect may be observed than with the analogous mixture at neutral pH 7. Therefore, to minimize the microorganisms, the pH should be adjusted via a buffer system.

Many of the ingredients commonly found in baby creams should be critically scrutinized. The use of lanolin (wool wax) and beeswax, both of which are of natural origin, can trigger allergies. This also applies to chamomile as well as to bisabolol and calendula extract, often used as calming ingredients. Vegetable lipids such as sweet almond oil, shea butter, jojoba oil, and phytosterols are very popular and unproblematic because of their care and protective effect. Jasmine oil, which smells pleasant and has a calming effect, is ideal for use in baby creams. A proposal for the formulation of a baby cream outside the module system is described in Table 8.13.

Table 8.13 Formulation of a novel baby cream for the diaper area (for the body cream, it is possible to remove the silicone oil from the formulation).

No	Ingredients	Amount (%)	INCI name	EC number
1	Water (cosm.)	68.35	Aqua	231-791-2
2	Shea butter	4	Butyrospermum Parkii Butter	293-515-7
3	Evening primrose oil	4	Oenothera Biennis Oil	289-859-2
4	Sweet almond oil	3	Prunus Amygdalus Sativa Kernel Oil	291-063-5
5	Glycerol	3	Glycerin	200-289-5
6	Sorbitol	3	Sorbitol	200-061-5
7	Cetyl stearyl alcohol	3	Cetearyl Alcohol	267-008-6
8	Silicone oil M350	2.5	Dimethicone	63148-62-9 CAS/9006-65-9 CAS/9016-00-6 CAS
9	D-Panthenol	1.5	Panthenol	201-327-3/ 240-540-6
10	Hydrolyzed soy protein	1.5	Hydrolyzed Soy Protein	271-770-5
11	Rapeseed sterols	1	Brassica Campestris (Rapeseed) Sterols	292-737-1
12	Magnesium ascorbyl phosphate	1	Magnesium Ascorbyl Phosphate	114040-31-2/ 113170-55-1 CAS
13	Vitamin E (natural, D-α; 0.7)/sunflower oil (0.3)	1	Tocopherol/ Helianthus Annuus	59-02-9 (D) CAS/232-273-9
14	Cetyl stearyl ether 30EO	0.9	Ceteareth-30	68439-49-6 CAS
15	Nicotinamide	0.5	Niacinamide	202-713-4
16	Glycerol monostearates	0.5	Glyceryl Stearate	250-705-4/ 286-490-9
17	Trisodium citrate dihydrate	0.32	Sodium Citrate	200-675-3
18	Xanthan	0.3	Xanthan Gum	234-394-2
19	Jasmine	0.2	Jasminum Officinale Oil	289-960-1
20	Potassium sorbate	0.15	Potassium Sorbate	246-376-1
21	Hyaluronic acid	0.1	Hyaluronic acid	232-678-0
22	Allantoin	0.1	Allantoin	202-592-8
23	Citric acid monohydrate	0.08	Citric Acid	201-069-1
	Sum (without water):	31.65		

8.9 Bedsore Cream

For the prevention and reduction of pressure ulcers in bedridden people, the following applications are recommended:

- Optimal personal care by trained personnel
- Suitable mattress
- Regular, intensive skin care with a suitable cream of the endangered areas.

There is almost no convincing cream in the field of prophylaxis. This gap can definitely close the cosmetics. Important ingredients of the cream should be vegetable oils. For this purpose, the oils already described in the modules come into question, such as the shea butter, the jojoba oil, and the linoleic acid-rich oils. Furthermore, silicone oils must be applied to protect the skin in reasonable quantities. This oil offers protection against moisture as well as mechanical and chemical influences and is in the INCI list just behind the water. Vitamins are also indispensable, especially B_3, C, and E. As in all applications, effective moisturizing substances are required. The cream has to make the skin soft but resistant; they protect without producing an occlusive effect. Furthermore, this includes a full supply of the older skin with the necessary ingredients. Formulation instructions are given in Table 8.14. A complete formula, based on the modules, can be found on the one hand in Table 8.15 and on the other hand as an intensive cream in Table A.6.

8.10 Cellulite Cream

Improving the appearance of the thighs of a woman with cellulite is a difficult task. By means of a cream, a temporary or final elimination of cellulite is not possible. With regular use, the cream can strengthen the tissues and moisturize and firm the skin, leaving the skin looking healthier. To increase the blood flow and to improve the effect, the special cream should be massaged into the skin. Furthermore, it makes sense to relax the skin structure beforehand via a hot bath or shower or sauna so that the active ingredients can penetrate more easily. For a visible, lasting effect, a daily intensive application must be made with a carefully formulated cream. This special cream should contain ingredients with a broad spectrum of effects. First, they must excite the lymph channels and reduce swelling. In addition, the ingredients should stimulate cell growth, act astringent, promote the circulation, and strengthen the tissue. The effects of all the substances together give the improved appearance. Proven substances are characterized in Section 7.9.

The recipes proposed (Table 8.16) emerge from the modules described, whereby the 80% variant comes into consideration. The oil-rich formula (alternative C1; Table 8.17) benefits from the action of the mangosteen and essential oils, especially the lemon oil. The lemon oil is known for its pronounced anticellulite effect. The second alternative C2, shown in Table A.7, relies on the active substances from the sea and considers algae components for hydration, strengthening of tissue, and acne control.

Table 8.14 Composition of the "active ingredient module" for bedsore creams in the alternatives 1 and 2.

Module	Ingredient	INCI name	Amount (%)
Ingredients that helps to prevent pressure ulcer (part of the modules)			
Lipids	Sweet almond oil	Prunus Amygdalus Dulcis Oil	4
	Evening primrose oil	Oenothera Biennis Oil	3
	Hemp oil	Cannabis Sativa (Hemp) Seed Oil	3
	Shea butter	Butyrospermum Parkii Butter	3
Vitamins	D-Panthenol	Panthenol	2.5
	Nicotinamide	Niacinamide	2
	Vitamin E (natural)/sunflower oil (70 : 30)	Tocopherol/ Helianthus Annuus	1
Moisturizing substances	Glycerol	Glycerin	3
	Sorbitol	Sorbitol	3
	Hyaluronic acid	Hyaluronic Acid	0.1
	Allantoin	Allantoin	0.1
Co-emulsifier	Cetyl stearyl alcohol	Cetearyl Alcohol	3.5
	Silicone oil	Dimethicone	1.5
Module "Active ingredients" (pressure ulcer)			
Alternative PU1	D-Panthenol	Panthenol	1.5
	Magnesium ascorbyl phosphate	Magnesium Ascorbyl Phosphate	1
	Shea butter	Butyrospermum Parkii Butter	1
	Hyaluronic acid	Hyaluronic Acid	0.1
	Urea	Urea	2.5
	Silicone oil	Dimethicone	3
	Rosemary	Rosmarinus Officinalis Leaf Oil	0.9
Alternative PU2 "intensive cream"	D-Panthenol	Panthenol	2
	Magnesium ascorbyl phosphate	Magnesium Ascorbyl Phosphate	1
	Jojoba oil	Simmondsia Chinensis Seed Oil	4
	Hyaluronic acid	Hyaluronic Acid	0.1
	Urea	Urea	4
	Silicone oil	Dimethicone	4.5
	Sea salt	Sea Salt	3
	Ylang-Ylang	Cananga Odorata Leaf Oil	1.4

Table 8.15 Formulation of a decubitus cream for bedridden people for prophylaxis (according to PU1).

No	Ingredient	Amount (%)	INCI name	EC number
1	Water (cosm.)	55.95	Aqua	231-791-2
2	Silicone oil M500	4.5	Dimethicone	63148-62-9 CAS/ 9006-65-9 CAS/9016-00-6 CAS
3	D-Panthenol	4	Panthenol	201-327-3/ 240-540-6
4	Shea butter	4	Butyrospermum Parkii Butter	293-515-7
5	Sweet almond oil	4	Prunus Amygdalus Sativa Kernel Oil	291-063-5
6	Cetyl stearyl alcohol	3.5	Cetearyl Alcohol	267-008-6
7	Evening primrose oil	3	Oenothera Biennis Oil	289-859-2
8	Hemp oil	3	Cannabis Sativa Seed Oil	289-644-3
9	Glycerol	3	Glycerin	200-289-5
10	Sorbitol	3	Sorbitol	200-061-5
11	Urea	2.5	Urea	200-315-5
12	Ethanol	2	Alcohol	200-578-6
13	Cetyl stearyl ether 30EO	1.5	Ceteareth-30	68439-49-6 CAS
14	Nicotinamide	1	Niacinamide	202-713-4
15	Vitamin E (natural, D-α; 0.7)/sunflower oil (0.3)	1	Tocopherol/ Helianthus Annuus	59-02-9 (D) CAS/232-273-9
16	Magnesium ascorbyl phosphate	1	Magnesium Ascorbyl Phosphate	114040-31-2/ 113170-55-1 CAS
17	Rosemary	0.9	Rosmarinus Officinalis Leaf Oil	283-291-9
18	Trisodium citrate dihydrate	0.64	Sodium Citrate	200-675-3
19	Glycerol monostearates	0.5	Glyceryl Stearate	250-705-4/ 286-490-9
20	Xanthan	0.3	Xanthan Gum	234-394-2
21	Potassium sorbate	0.25	Potassium Sorbate	246-376-1
22	Hyaluronic acid	0.2	Hyaluronic Acid	232-678-0
23	Citric acid monohydrate	0.16	Citric Acid	201-069-1
24	Allantoin	0.1	Allantoin	202-592-8
	Sum (without water):	44.05		

Table 8.16 Composition of the "active ingredient module" for an anticellulite cream.

Module	Ingredient	INCI name	Amount (%)
Ingredients that helps to improve the appearance of women's thighs (part of the modules)			
Lipids	Sweet almond oil	Prunus Amygdalus Dulcis Oil	4
	Evening primrose oil	Oenothera Biennis Oil	3
	Hemp oil	Cannabis Sativa (Hemp) Seed Oil	3
	Shea butter	Butyrospermum Parkii Butter	3
Vitamins	D-Panthenol	Panthenol	2.5
	Nicotinamide	Niacinamide	2
	Vitamin E (natural)/ sunflower oil (70 : 30)	Tocopherol/ Helianthus Annuus	1
Moisturizing substances	Glycerol	Glycerin	3
	Sorbitol	Sorbitol	3
	Hyaluronic acid	Hyaluronic Acid	0.1
	Allantoin	Allantoin	0.1
Module "Active ingredients" (cellulite)			
Alternative C1	D-Panthenol	Panthenol	1.8
	Magnesium ascorbyl phosphate	Magnesium Ascorbyl Phosphate	1
	Shea butter	Butyrospermum Parkii Butter	1
	Hyaluronic (degraded)	Hyaluronic Acid	0.2
	Urea	Urea	4
	Mangosteen	Garcinia Mangostana Fruit Extract	5
	Geranium	Pelargonium Roseum Leaf Oil	2
	Juniper	Juniperus Oxycedrus Fruit Oil	1
	Lemon	Citrus Limon Seed Oil	4
Alternative C2	D-Panthenol	Panthenol	1
	Magnesium ascorbyl phosphate	Magnesium Ascorbyl Phosphate	1
	Jojoba oil	Simmondsia Chinensis Seed Oil	2
	Protalgel	Algae Extract, Hydrolyzed Algin, Carrageenan	4
	Kelp	Laminaria Digitata	3
	Fucus	Fucus Vesiculosus	4
	Lemon	Citrus Limon Seed Oil	4
	Dead sea salt	Sea Salt (Dead Sea Salt)	1

Table 8.17 Formulation of an anticellulite cream (according to alternative C1).

No	Ingredient	Amount (%)	INCI name	EC number
1	Water (cosm.)	43.05	Aqua	231-791-2
2	Mangosteen	5	Garcinia Mangostana Fruit Extract	289-884-9
3	Shea butter	5	Butyrospermum Parkii Butter	293-515-7
4	D-Panthenol	4.3	Panthenol	201-327-3/ 240-540-6
5	Lemon	4	Citrus Limon Seed Oil	296-174-2/284-515-8/285-359-3
6	Sweet almond oil	4	Prunus Amygdalus Sativa Kernel Oil	291-063-5
7	Urea	4	Urea	200-315-5
8	Cetyl stearyl alcohol	3.5	Cetearyl Alcohol	267-008-6
9	Evening primrose oil	3	Oenothera Biennis Oil	289-859-2
10	Hemp oil	3	Cannabis Sativa Seed Oil	289-644-3
11	Glycerol	3	Glycerin	200-289-5
12	Sorbitol	3	Sorbitol	200-061-5
13	Geranium	2	Pelargonium Roseum Leaf Oil	290-144-2
14	Nicotinamide	2	Niacinamide	202-713-4
15	Ethanol	2	Alcohol	200-578-6
16	Cetyl stearyl ether 30EO	1.5	Ceteareth-30	68439-49-6 CAS
17	Silicone oil M350	1.5	Dimethicone	63148-62-9 CAS/9006-65-9 CAS/9016-00-6 CAS
18	Juniper	1	Juniperus Oxycedrus Fruit Oil	289-969-0
19	Vitamin E (natural, D-α; 0.7)/sunflower oil (0.3)	1	Tocopherol/ Helianthus Annuus	59-02-9 (D) CAS/232-273-9
20	Magnesium ascorbyl phosphate	1	Magnesium Ascorbyl Phosphate	114040-31-2/ 113170-55-1 CAS
21	Rosemary	0.9	Rosmarinus Officinalis Leaf Oil	283-291-9
22	Trisodium citrate dihydrate	0.64	Sodium Citrate	200-675-3
23	Glycerol monostearates	0.5	Glyceryl Stearate	250-705-4/ 286-490-9
24	Xanthan	0.3	Xanthan Gum	234-394-2

(Continued)

Table 8.17 (Continued)

No	Ingredient	Amount (%)	INCI name	EC number
25	Potassium sorbate	0.25	Potassium Sorbate	246-376-1
26	Hyaluronic acid (degraded)	0.2	Hyaluronic Acid	232-678-0
27	Citric acid monohydrate	0.16	Citric Acid	201-069-1
28	Allantoin	0.1	Allantoin	202-592-8
29	Hyaluronic acid	0.1	Hyaluronic Acid	232-678-0
	Sum:	56.95		

8.11 Foot Care Cream Against Athlete's Foot

A foot cream can be effective against athlete's foot with gentle agents, without neglecting the care. It combats athlete's foot by creating, on the one hand, an unpleasant environment for the fungus. On the other hand, an effective substance should be used that definitely show positive properties on the skin but can kill the fungus. After studies, the lavender oil represents an appropriate substance. Therefore, the formulation is based on lavender oil, furthermore a significantly acidic pH and in addition (sea) salt and sorbic acid, which also contribute to the destruction of the fungi. The foot cream should have nourishing, horn softening properties. The formulation is based on the 80% module (Tables 8.18 and 8.19).

8.12 Hand Cream

The formulation in Table 8.20 is suitable for all skin types and forms a basis for optimizations. Moisture provides urea and glycerin. A long-chain fatty alcohol and two selected vegetable oils, namely shea butter and safflower oil, ensure the lipid supply. Other fast-absorbing vegetable oils can also be used. The oil phase is kept deliberately low; there can also be used higher amounts. To protect the hand cream from changes during storage and gradual consumption, a good formulation should contain antioxidant acting vitamins. For this purpose, the combination of vitamins C and E creates a strong effect.

8.13 Callus Removal Cream

A suggestion for an anticallus cream is shown in Table 8.21. The cream may only be used on callus and contains the three effective ingredients in medium concentrations, namely urea, AHAs and their salts, as well as allantoin. As the substances also act as a moisturizer, the moisturizing effect of the cream, taking into account glycerol, is unusually high. In addition to the vegetable oils, shea butter, and wheat germ oil, the fatty alcohols cetearyl alcohol and octyldodecanol are available for lipid supply. With panthenol and the vitamins C and E, the cream

Table 8.18 Composition of the "active ingredient module" for an athlete's foot cream.

Module	Ingredient	INCI name	Amount (%)
Ingredients for the care of the feet (part of the modules)			
Lipids	Sweet almond oil	Prunus Amygdalus Dulcis Oil	4
	Evening primrose oil	Oenothera Biennis Oil	3
	Hemp oil	Cannabis Sativa (Hemp) Seed Oil	3
	Shea butter	Butyrospermum Parkii Butter	3
Vitamins	D-Panthenol	Panthenol	2.5
	Nicotinamide	Niacinamide	2
	Vitamin E (natural)/sunflower oil (70 : 30)	Tocopherol/ Helianthus Annuus	1
Moisturizing substances	Glycerol	Glycerin	3
	Sorbitol	Sorbitol	3
	Hyaluronic acid	Hyaluronic Acid	0.1
	Allantoin	Allantoin	0.1
Module "Active ingredients" (athlete's foot)			
Alternative AF	Urea	Urea	7
	Citric acid	Citric acid	0.2
	Allantoin	Allantoin	0.3
	Magnesium ascorbyl phosphate	Magnesium Ascorbyl Phosphate	1
	Sea salt	Sea Salt	3.5
	Lavender oil	Lavandula Angustifolia Oil	6

AF, antifoam.

has important antioxidants with healing properties. To increase the effectiveness of corneal detachment, the pH is reduced to 4.5 by citric acid.

8.14 Body Lotion

The formulation is based on the standard modules. For moisturizing, the body lotion additionally contains urea and aloe as well as geranium for scenting. The lipid supply is taken over by vegetable oils and fatty alcohol. For vitalization, the vitamins E, panthenol, and nicotinamide are responsible. This compilation gives the formulation the required properties of a body lotion. Two ways can be used

Table 8.19 Formulation of an athlete's foot cream.

No	Ingredient	Amount (%)	INCI name	EC number
1	Water (cosm.)	46.45	Aqua	231-791-2
2	Urea	7	Urea	200-315-5
3	Lavender oil	6	Lavandula Angustifolia Oil	289-995-2
4	Sweet almond oil	4	Prunus Amygdalus Sativa Kernel Oil	291-063-5
5	Cetyl stearyl alcohol	3.5	Cetearyl Alcohol	267-008-6
6	Sea salt	3.5	Maris Sal (Sea Salt)	231-598-3
7	Shea butter	3	Butyrospermum Parkii Butter	293-515-7
8	Evening primrose oil	3	Oenothera Biennis Oil	289-859-2
9	Hemp oil	3	Cannabis Sativa Seed Oil	289-644-3
10	Glycerol	3	Glycerin	200-289-5
11	Sorbitol	3	Sorbitol	200-061-5
12	D-Panthenol	2.5	Panthenol	201-327-3/ 240-540-6
13	Nicotinamide	2	Niacinamide	202-713-4
14	Ethanol	2	Alcohol	200-578-6
15	Silicone oil M350	2	Dimethicone	63148-62-9 CAS/9006-65-9 CAS/9016-00-6 CAS
16	Cetyl stearyl ether 30EO	1.5	Ceteareth-30	68439-49-6 CAS
17	Vitamin E (natural, D-α; 0.7)/sunflower oil (0.3)	1	Tocopherol/ Helianthus Annuus	59-02-9 (D) CAS/232-273-9
18	Magnesium ascorbyl phosphate	1	Magnesium Ascorbyl Phosphate	114040-31-2/ 113170-55-1 CAS
19	Trisodium citrate dihydrate	0.64	Sodium Citrate	200-675-3
20	Glycerol monostearates	0.5	Glyceryl Stearate	250-705-4/ 286-490-9
21	Allantoin	0.4	Allantoin	202-592-8
22	Citric acid monohydrate	0.36	Citric Acid	201-069-1
23	Xanthan	0.3	Xanthan Gum	234-394-2
24	Potassium sorbate	0.25	Potassium Sorbate	246-376-1
25	Hyaluronic acid	0.1	Hyaluronic Acid	232-678-0
	Sum (without water):	53.55		

Table 8.20 Proposal for a nourishing hand cream with vegetable oils; for the natural cosmetics replace the silicone oil with octyldodecanol or cetyl palmitate.

No	Ingredient	Amount (%)	INCI name	EC number
1	Water	68.60	Aqua	231-791-2
2	Shea butter	4	Butyrospermum Parkii Butter	293-515-7
3	Urea	4	Urea	200-315-5
4	Glycerol	4	Glycerin	200-289-5
5	Myristyl myristat	3	Myristyl Myristat	221-787-9
6	Safflower oil	3	Carthamus Tinctorius Seed Oil	232-276-5
7	Imwitor®375	3	Glyceryl Citrate, Linoleate, Oleate	—
8	D-Panthenol	2.5	Panthenol	201-327-3/ 240-540-6
9	Octyldodecanol	2	Octyldodecanol	226-242-9
10	Ethanol	2	Alcohol	200-578-6
11	Silicone oil M350	1	Dimethicone	63148-62-9 CAS/9006-65-9 CAS/9016-00-6 CAS
12	Vitamin E (natural, D-α; 0.7)/sunflower oil (0.3)	0.7	Tocopherol/ Helianthus Annuus	59-02-9 (D) CAS/232-273-9
13	Trisodium citrate dihydrate	0.64	Sodium Citrate	200-675-3
14	Magnesium ascorbyl phosphate	0.5	Magnesium Ascorbyl Phosphate	114040-31-2/ 113170-55-1 CAS
15	Geranium	0.35	Pelargonium Roseum Leaf Oil	290-144-2
16	Xanthan	0.3	Xanthan Gum	234-394-2
17	Potassium sorbate	0.25	Potassium Sorbate	246-376-1
18	Citric acid monohydrate	0.16	Citric Acid	201-069-1
	Sum (without water):	31.40		

for the implementation of the recipe in the production. The first possibility is the direct production according to the percentages of the recipe. The second possibility would be to add the three additional components and the required water to the mixture of the various modules for the 100% recipe. Thereafter, the recipe is diluted by adding 15% water to produce the desired body lotion (Table 8.22). The additional substances and the amounts of water are stirred at 30–35 °C. Subsequently, a further homogenization makes sense. Despite the reduced amounts of ingredients, safe preservation is guaranteed.

Table 8.21 Proposal for a corneal dissolving cream.

No	Ingredient	Amount (%)	INCI name	EC number
1	Water	52.61	Aqua	231-791-2
2	Urea	10	Urea	200-315-5
3	Shea butter	6	Butyrospermum Parkii Butter	293-515-7
4	Sodium lactate	5	Sodium Lactate	200-772-0/ 212-762-3
5	Glycerol	4	Glycerin	200-289-5
6	Cetyl stearyl alcohol	3.5	Cetearyl Alcohol	267-008-6
7	Wheat germ oil	3.5	Triticum Vulgare Oil	281-689-7
8	D-Panthenol	2.5	Panthenol	201-327-3/ 240-540-6
9	Lactic acid	2	Lactic acid	200-018-0
10	Octyldodecanol	2	Octyldodecanol	226-242-9
11	Silicone oil M350	2	Dimethicone	63148-62-9 CAS/9006-65-9 CAS/9016-00-6 CAS
12	Ethanol	2	Alcohol	200-578-6
13	Cetyl stearyl ether 30EO	1.5	Ceteareth-30	68439-49-6 CAS
14	Vitamin E (natural, D-α; 0.7)/sunflower oil (0.3)	0.7	Tocopherol/ Helianthus Annuus	59-02-9 (D) CAS/232-273-9
15	Trisodium citrate dihydrate	0.64	Sodium Citrate	200-675-3
16	Magnesium ascorbyl phosphate	0.5	Magnesium Ascorbyl Phosphate	114040-31-2/ 113170-55-1 CAS
17	Chamomile	0.4	Anthemis Nobilis Flower Oil	283-467-5
18	Allantoin	0.3	Allantoin	202-592-8
19	Xanthan	0.3	Xanthan Gum	234-394-2
20	Citric acid monohydrate	0.3	Citric Acid	201-069-1
21	Potassium sorbate	0.25	Potassium Sorbate	246-376-1
	Sum (without water):	47.39		

8.15 Eye Area Formulation with Sun Protection

A cream for the eye area should be low in fat but contain moisturizer, vitalizer, and antiaging ingredients, preferably only oxidation stable and photostable substances. Especially, this applies to dry, wrinkle-prone skin. Here, a tailor

Table 8.22 Suggestion of a body lotion.

No	Ingredient	Amount (%)	Amount (%) after 15% dilution	INCI name	EC number
1	Water	60.4	66.3	Aqua	231-791-2
2	Sweet almond oil	4	3.4	Prunus Amygdalus Sativa Kernel Oil	291-063-5
3	Urea	4	3.4	Urea	200-315-5
4	Cetyl stearyl alcohol	3.5	2.98	Cetearyl Alcohol	267-008-6
5	Evening primrose oil	3	2.55	Oenothera Biennis Oil	289-859-2
6	Hemp oil	3	2.55	Cannabis Sativa Seed Oil	289-644-3
7	Shea butter	3	2.55	Butyrospermum Parkii Butter	293-515-7
8	Glycerol	3	2.55	Glycerin	200-289-5
9	Sorbitol	3	2.55	Sorbitol	200-061-5
10	D-Panthenol	2.5	2.13	Panthenol	201-327-3/ 240-540-6
11	Nicotinamide	2	1.7	Niacinamide	202-713-4
12	Ethanol	2	1.7	Alcohol	200-578-6
13	Cetyl stearyl ether 30EO	1.5	1.28	Ceteareth-30	68439-49-6 CAS
14	Silicone oil M350	1.5	1.28	Dimethicone	63148-62-9 CAS/9006-65-9 CAS/9016-00-6 CAS
15	Vitamin E (natural, D-α; 0.7)/sunflower oil (0.3)	1	0.85	Tocopherol/ Helianthus Annuus	59-02-9 (D) CAS/232-273-9
16	Trisodium citrate dihydrate	0.64	0.54	Sodium Citrate	200-675-3
17	Glycerol monostearates	0.5	0.43	Glyceryl Stearate	250-705-4/ 286-490-9
18	Geranium	0.5	0.43	Pelargonium Roseum Leaf Oil	290-144-2
19	Xanthan	0.3	0.26	Xanthan Gum	234-394-2
20	Potassium sorbate	0.25	0.21	Potassium Sorbate	246-376-1
21	Citric acid monohydrate	0.16	0.14	Citric Acid	201-069-1
22	Aloe vera powder (1 : 200)	0.15	0.13	Aloe Barbadensis Leaf Juice Powder	287-390-8/ 305-181-2
23	Allantoin	0.1	0.085	Allantoin	202-592-8
	Sum (without water):	39.6	33.7		

to the problem, module-independent formulation will be presented with low but sufficient preservation, comparable to the baby cream. This cream is very mild and tailored to the eye area but of course also appropriate for the entire face. In addition, a mixture of organic UV filters provides the sun protection (Mexoplex®); however, the formulation also works without UV protection as a cream for the eye area and face. If necessary, a further viscosity adjustment can be done by adding synthetic waxes (Section 5.3.4). A cream is usually applied at home. Therefore, in contrast to lotions, the content is not exposed to such high temperatures. For lotions, the almond oil should be exchanged for the more stable squalane. In addition, a further temperature-stable emulsifier could provide additional stability at higher temperatures, if necessary. Limit the shelf life to one year and after opening to three months because of the reduced preservative (Table 8.23).

Table 8.23 Cream for the eye area with UV filters.

No	Ingredient	Amount (%)	INCI name	EC number/CAS
1	Water (cosm.)	60.76	Aqua	231-791-2
2	Sweet almond oil	7	Prunus Amygdalus Sativa Kernel Oil	291-063-5
3	Mexoplex® (mixture of UV filters)	5	Drometrizole Trisiloxane, Terephthalylidene Dicamphor Sulfonic Acid, Bis-Ethylhexyl-oxyphenol Methoxyphenyl Triazine	155633-54-8 CAS, 410-960-6, 187393-00-6 CAS
4	D-Panthenol	3.5	Panthenol	201-327-3/ 240-540-6
5	Cetyl stearyl alcohol	3.5	Cetearyl Alcohol	267-008-6
6	Glycerol	3	Glycerin	200-289-5
7	Sorbitol	3	Sorbitol	200-061-5
8	α-Lipoic acid	3	Thioctic Acid	214-071-2
9	Rapeseed sterols	2	Brassica Campestris (Rapeseed) Sterols	292-737-1
10	Hydrolyzed soy protein	2	Hydrolyzed Soy Protein	271-770-5
11	Silicone oil M350	1.5	Dimethicone	63148-62-9 CAS/9006-65-9 CAS/9016-00-6 CAS
12	Nicotinamide	1	Niacinamide	202-713-4
13	Magnesium ascorbyl phosphate	1	Magnesium Ascorbyl Phosphate	114040-31-2/ 113170-55-1 CAS
14	Vitamin E (natural, D-α; 0.7)/sunflower oil (0.3)	1	Tocopherol/ Helianthus Annuus	59-02-9 (D) CAS/232-273-9
15	Cetyl stearyl ether 30EO	1.3	Ceteareth-30	68439-49-6 CAS

(Continued)

Table 8.23 (Continued)

No	Ingredient	Amount (%)	INCI name	EC number/CAS
16	Trisodium citrate dihydrate	0.32	Sodium Citrate	200-675-3
17	Chamomile	0.3	Chamomilla Recutita (Matricaria) Flower Extract	282-006-5
18	Xanthan	0.3	Xanthan Gum	234-394-2
19	Hyaluronic acid (degraded)	0.15	Hyaluronic acid	232-678-0
20	Aloe vera powder (1 : 200)	0.15	Aloe Barbadensis Leaf Juice Powder	287-390-8/ 305-181-2
21	Potassium sorbate	0.14	Potassium Sorbate	246-376-1
22	Citric acid monohydrate	0.08	Citric Acid	201-069-1
	Sum:	39.24		

8.16 Sunscreen Lotion

A sunscreen cream should contain oxidation and photostable ingredients, and in particular, UV filters that comply with the latest regulatory recommendations (sun protection factor [SPF] = 3 persistent pigment darkening [PPD]), labeled with the UVA symbol or PA + to +++ (see Section 7.16). Water-resistant creams are preferably formulated as w/o emulsions. The emulsifiers should be characterized by a high-temperature resistance because the dispensers are often exposed to the sun and easily warm to temperatures of 50–60 °C. This means that high-molecular-weight emulsifiers are preferred. The requirement "water-resistant" can be fulfilled by lipophilic ingredients, in particular by using pharmaceutical-grade silicone oils as well as silicon-coated pigments and an appropriate emulsifier. Because of the buffer system, a pH of 5 is set here. In addition to the UV filter, the cream contains various moisturizing and nourishing ingredients (Table 8.24).

Table 8.24 Proposal for a water-resistant sunscreen lotion without alcohol.

No	Ingredient	Amount (%)	INCI name	EC number/CAS
1	Water (cosm.)	48.55	Aqua	231-791-2
2	Squalane or caprylic/capric triglyceride	8	Squalane or Caprylic/Capric Triglyceride	203-825-6; 277-452-2/ 265-724-3
3	Mexoplex® (mixture of UV filters)	8	Drometrizole Trisiloxane, Terephthalylidene Dicamphor Sulfonic Acid, Bis-Ethylhexyl-oxyphenol Methoxyphenyl Triazine	155633-54-8 CAS, 410-960-6, 187393-00-6 CAS

(Continued)

Table 8.24 (Continued)

No	Ingredient	Amount (%)	INCI name	EC number/CAS
4	SoFiTix broadband HT	8	Titanium Dioxide, Simethicone, Zinc Oxide	13463-67-7
				8050-81-5
				1314-13-2
5	Abil®WE 09	5.5	Polyglyceryl-4 Isostearate, Cetyl PEG/PPG-10/1 Dimethicone, Hexyl Laurate	—
6	D-Panthenol	3.5	Panthenol	201-327-3/ 240-540-6
7	Myristyl myristat	3.5	Myristyl Myristat	221-787-9
8	Glycerol	3	Glycerin	200-289-5
9	Sorbitol	3	Sorbitol	200-061-5
10	Hydrolyzed soy protein	2.5	Hydrolyzed Soy Protein	271-770-5
11	Silfar®1000	2.5	Dimethicone	63148-62-9 CAS/9006-65-9 CAS/9016-00-6 CAS
12	Magnesium ascorbyl phosphate	1	Magnesium Ascorbyl Phosphate	114040-31-2/ 113170-55-1 CAS
13	Vitamin E (natural, D-α; 0.7)/sunflower oil (0.3)	1	Tocopherol/Helianthus Annuus	59-02-9 (D) CAS/232-273-9
14	Trisodium citrate dihydrate	0.64	Sodium Citrate	200-675-3
15	Sodium acrylates/C10-30 alkyl acrylate crosspolymer	0.4	Sodium Acrylates/C10-30 Alkyl Acrylate Crosspolymer	—
16	Chamomile	0.3	Chamomilla Recutita (Matricaria) Flower Extract	282-006-5
17	Potassium sorbate	0.25	Potassium Sorbate	246-376-1
18	Citric acid monohydrate	0.16	Citric Acid	201-069-1
19	Hyaluronic acid	0.1	Hyaluronic Acid	232-678-0
20	Aloe vera powder (1 : 200)	0.1	Aloe Barbadensis Leaf Juice Powder	287-390-8/ 305-181-2
	Sum:	51.45		

PEG, polyethylene glycol; PPG, polypropylene glycol.

9
Perfumes

9.1 Importance of the Perfume for Cosmetic Creams

Perfumes play a big role in cosmetic creams. Often composed of different essential oils or oil components, perfume oils provide a pleasant blend of many fragrances. Fragrances promote well-being and can have a positive impact on the psyche. Depending on the composition, they exert a calming or stimulating effect on the users and the environment. Perfume is important for many people because everyone wants to smell good and give off a sympathetic fragrance. The scent type and strength, reserved or intrusive, is of course a matter of taste. Fragrances can make people attractive or repugnant, depending on the scent type, strength, and rating of the odor by surrounded persons. For the selection, myriad scents are available. Therefore, people use an individual perfume of their choice. A personal perfuming takes place by the applied cosmetic cream and/or by a pure perfume.

When it is allowed in the stop to try a sample or a sample dispenser, many consumers choose a well-scented cosmetic cream, which feels good to the touch, preferably from a famous brand. The strength of the desired effect, however, plays only a minor role at the purchase decision. Also, it cannot be judged within minutes. Therefore, the decision shows particularities because the esthetics (fragrance, haptic, packaging, and brand name) comes here first and not the performance as with other consumer goods (Figure 9.1). Also, of course, the price influences the purchase decision.

The word "smell" comes from the Latin term "olfactus" and represents the basis of the terms "olfactory perception and an olfactory memory." Because of its complex composition and different fragrance orientations, perfume oils allow particularly diverse design. The targeted making requires a precise knowledge of the fragrance notes of each component, including origin, chemical composition, and isolation/synthesis of fragrances.

In cosmetics industry, major manufacturers are trying to establish brands worldwide. Mostly, the name of a famous person stands for the brand, rarely the name of the company. It is advisable to use the name of a particularly positive tested perfume for the brand or brand family. This measure improves the perception and might represent the brand core. A positive acceptance of the perfume causes a high number of further purchases. Against the use of a perfume speaks

Cosmetic Creams: Development, Manufacture and Marketing of Effective Skin Care Products,
First Edition. Wilfried Rähse.

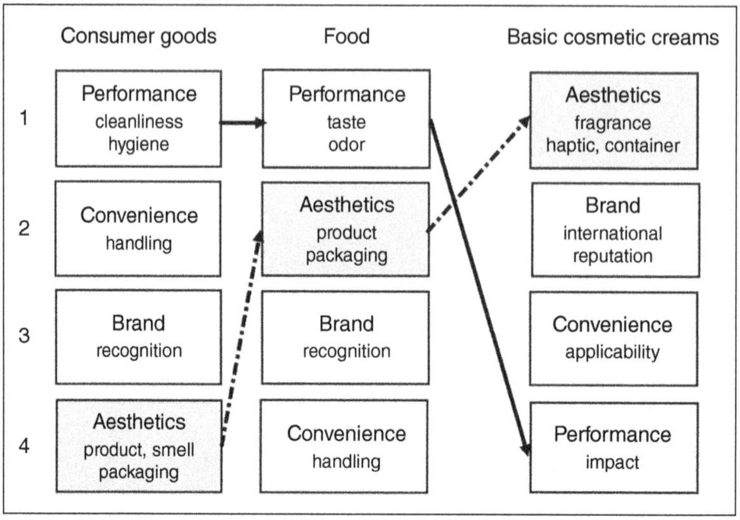

Figure 9.1 Importance of the individual elements of product design for the consumer's purchase decision.

that they can trigger allergies in sensitive individuals. Because of the large number of chemical substances in the perfume, allergic reactions cannot be excluded. The probability might increase with increasing number of components.

Other manufacturers rely on the addition of only a single, but natural fragrance. Here, the composition is generally known and manageable, and allergic effects are easier to avoid. Above all, mutual reinforcements, caused by the interaction of several components, can be avoided. The selected fragrance, an essential oil such as lavender oil, has significant antibacterial and stimulating properties in addition to the pleasant scent. However, the individual fragrance may leave a medicinal or herbal impression, so a lot of customers prefer a well-balanced perfume for basic creams. In contrast, the scent has only minor purchase-critical importance in the cosmeceuticals. The market relevance of the perfumes for creams can be seen from the fact that more than $1 billion dollars flow into this area, valued according to the purchase price of the cosmetic companies.

9.2 History of Perfume Oils

The word perfume derives from the Latin word "per fumum," which means, by smoke, and incense from "incendere," which means, to burn. It was believed that the smoke of incense burnt as part of religious ceremonies transported the prayers to heaven. In history [1–4], the earliest evidence for the making of perfumes dated approximately from 2000 up to 5000 CE. The priests in Mesopotamia used perfumes, scented ointments, and resins for honoring the gods and anointing the dead. The development and manufacturing of perfumes began in ancient Mesopotamia and Egypt. Women known as the Assyrians manufactured perfumes by crushing of various plants. In hot salted water, the fragments stood overnight. Subsequently, the perfume arose on filtering the liquids and mixing them with hot oil. In India, people used perfumed ointments and oils as well as scented plant parts especially for medical purposes and for cleansing the body as well as for rituals.

In the second millennium CE in Mesopotamia, a female chemist (Tapputi) distilled various flowers with oils and other aromatics for the extraction of fragrances. Around this time, the upper class of Egyptians had already discovered the perfume for personal care and therapy. The world's oldest perfume factory, discovered in Pyrgos/Cyprus, dated from this time. They processed flowers, herbs, and spices. Still in use today in the perfume industry, the word "Chypre" indicates the importance of Cyprus in the scent culture. In the ninth century, an Arab chemist (Al-Kindi) described in the book Chemistry of Perfume and Distillations more than 100 recipes and manufacturing equipment, such as the alembic. The Persian Ibn Sina (Avicenna) invented steam distillation about 1000 BCE, a great step in chemistry. He processed petals of roses. Before that, the perfume oils originated from distillation of crushed mixtures of petals and herbs in oil as well as by mixing and squeezing.

The Western culture encountered fragrances through the Crusades and required raw materials and mixtures of the Orient. Until then, only simple lavender water from Charlemagne was known (end of the eighth century). Monks' recipes (1221) demonstrated the knowledge of perfumery in Florence, Italy. In the east, Hungarians (1370) made a perfume from scented oils, blended in alcohol, best known as Hungary Water. In addition, Venice became an important trading center because large amounts of new herbs, spices, and other goods came to Europe through this route. The development of skills and the technical equipment for distillation allowed the manufacturing of essential oils for the market in the fifteenth century. The emergence of perfumeries may date with the arrival of Catherine de Medici (1519–1589) to the court of Henry II. In 1580, the alchemist and apothecary Francesco Tombarelli came to Grasse (France). There, he established a laboratory for the manufacture of perfumes. Thereby, Grasse became the incubator of the European perfume industry.

Lémery, a French chemist and pharmacist, distinguished in 1709 between a royal perfume for scenting and the perfume for the bourgeois to "disinfect" the air. At that time, the perfume was attributed a therapeutic effect to invigorate the mind as well as to cleanse and strengthen the body and as a remedy for plague. The assumption that harmful germs could be transmitted during bathing in shared bathrooms led to a mass use of the perfume waters. They were an indispensable part of the daily toilet (Eau de Toilette) for body cleansing, which was carried out largely without water. The perfume also served to cover up (sweat) odors. Furthermore, the scent pointed to the wealth of the person.

In Germany, the Italian barber Giovanni Paolo Feminis created perfume water, known as Eau de Cologne. His nephew Johann (Giovanni) Maria Farina took over the business in 1732. By the eighteenth century, aromatic flowers and herbs were planted in the Grasse region of France, in Sicily, and in Calabria, Italy, to provide the growing perfume industry with raw materials. After the French Revolution, perfuming was frowned upon for some time. Napoleon, who liked to use Eau de Cologne, made perfume socially acceptable again. Toward the end of the nineteenth century, a perfume industry began to develop. Even today, Italy and France are the center of European trade for scented plants.

9.3 Perfume Composition and Markets

Depending on the origin, different types of fragrance oils exist. In general, the oils are classified into natural and synthetic oils, whereby most fragrances arise from nature-identical syntheses. The market products mostly consist of a mixed form. For the creation of a perfume for personal care, the perfumer selects from over 3000 fragrances about 30–200 fragrances [5, 6]. His art is the combination of individual oils in kind and amount, creating a pleasant composition with respect to the fragrance impression, intensity, and duration of fragrance on the skin or on a scented product such as a cream. Each perfume oil contains three different fragrance types, namely the head, heart, and base notes [7]. They differ in their volatility and aroma. The top note (head, Tête) is characterized by a light, less adherent fragrance and only perceptible in the first 10 minutes. This note gives the initial impression and is thus important for selling of perfumes. Thereafter, the middle note (heart, Bouquet, Coeur) appears perceptibly, which determines the character of the perfume. The base note (fixator) comprises long adherent, low volatile odorants. The scent is still noticeable after one hour and reduces the volatility of heart and top note. Dominated by the middle note, the following terms characterize the typical fragrance of a perfume: aldehydic, animal, aquatic, aromatic, balsamic, floral, earthy, fresh, fruity, green, woody, herbaceous, oriental, powdery, sweet, spicy, or citrus [8].

The flavor and fragrance industry needs aroma chemicals, an important and interesting group of various organic molecules [9]. The synthetic, nature-identical, and artificial ingredients play an important role because of their convenient availability and the relatively lower costs compared to the isolation of natural fragrances from plants whose amount is limited. Aromas and fragrances can often only be differentiated according to the intended use because the same products are applied in both fields. Examples are the essential oils derived from rosemary, sage, or thyme. The nose perceives not only fragrances but also the flavors (taste and odor) used in food and medicines. Incorporated as liquids in food, the flavors should not affect the taste. For increasing the stability and controlling the release of flavors, the addition may occur in dry state [10], for example, with carriers or in capsules. In the production of creams, the perfume oil is stirred into the emulsion at 30 °C, usually undiluted.

There are more than 500 manufacturers in the world market of "Flavors & Fragrances" (F&F). Calculated in sales of manufacturers, the F&F market amounts to about €22.3 billion in 2016 [11], of which €10.1 billion is attributable to the fragrances (Figure 9.2). Major manufacturers are Givaudan (Switzerland) with about 16% of the total market, and International Flavors & Fragrances (USA), Firmenich (Switzerland), and Symrise (Germany) each with around 11%. The big four hold a share of approximately 50% on the world market. The German perfume market, calculated by selling price, is €1.6 billion in 2016, the world

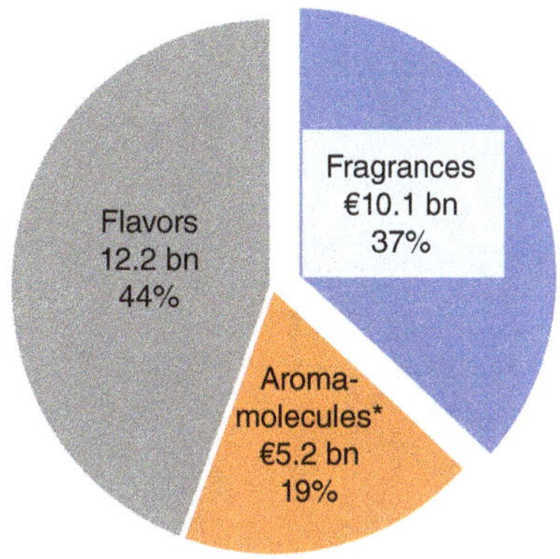

Figure 9.2 World market of fragrances and flavors according to sales figures of the manufacturers. Source: Data from Fa. Symrise, Corporate Report 2016 [11]. https://www.symrise.com/investors/reports/2016.

market is around $40 billion and $42 billion in 2018. This should increase to $50 billion by 2024 [12] with a Compound Annual Growth Rate (CAGR) of 3%. From these figures, it can be seen that there is a factor of 3–7 between the manufacturers' revenues for perfume oils and the market prices of perfume flacons. The largest quantities of perfume (33% each) are consumed in USA and Asia, as well as with 28% in Western Europe. The rest of the world is almost irrelevant (6%) [13].

The perfumers of the big four manufacturers develop in their laboratories most of the new feelgood fragrances for the well-known perfume brands as well as many fragrances for famous manufacturer of **cosmetic creams**. If desired, the perfumers will develop a brand-specific perfume for a cream or a cream family. In addition, small manufacturers can access to existing perfume oils, for example, from suppliers such as Deike Chemie. In total, there are about some hundred perfumes to buy. Also, the mixture of two or three finished oils can create a new characteristic fragrance at an interesting price. This way is only suitable for experienced developers with a sensitive nose and good taste.

There are some perfume houses that produce perfumes exclusively for the scenting of consumer products. One of them is the Henkel Fragrance Center (Germany). An extensive stock of the individual fragrances with sophisticated logistics characterizes a production plant for perfume oils. The manufacture of each batch takes place by precisely working at a dosage station [5, 6], in which the individual components are weighed and mixed in the presence of an inert gas.

Today, consumers perfumed their face or body in different ways, if they like a scenting. The **perfuming of the body** can take place with various concentrated, alcoholic solutions of perfume oils in flacons with a spray device or by direct application (Figure 9.3). Depending on the temperature and the goodness of the sense of smell, fragrances with concentrations above 0.1 ppm can be perceived with the nose. The perception threshold depends on the chemical structure of the fragrance and is generally between 0.2 and 0.8 ppm at room temperature. At higher temperatures, the threshold drops, whereas at low temperatures, it rises significantly. Other options for scenting include the use of perfumed body cleansing and care, protection, and hygiene products. Examples represent soaps, shower lotions, shampoos, bath products, deodorants, aftershave, hair gels, skincare, and sunscreen.

(a) EdC (b) EdT (c) EdP (d) Extract

Figure 9.3 Supply forms of perfume oils in alcoholic solutions: (a) Eau de Cologne (EdC) with 2–5% perfume oil, (b) Eau de Toilette (EdT) with 5–10%, (c) Eau de Parfum (EdP) with 10–15%, and (d) Extrait de Parfum with 15–30% perfume oil. Source: Courtesy of Douglas.

Aromatherapy with essential oils can remedy sensitivity disorders (mood swings) and alleviate diseases. For this purpose, an enrichment of space/breathing air with evaporating oils or the inhalation of fragrance over a hot water bath or oral intake or rubbing/massaging into the skin is necessary [14]. Firstly, an aromatherapy massage, performed by doctors or health practitioners, achieves positive effects on the nervous system. Secondly, there exist fragrance recipes for almost all physical problems [15]. Numerous essential natural oils, alone or in admixture with other, are suitable for the alleviation of disease conditions, particularly of colds. Essential oils can help in cases of psychosomatic disorders, symptoms in the digestive tract, inflammation, pain relief in the mouth and throat. In much diluted form, they stimulate the function of the biliary tract successfully. Some essential oils have antiseptic and antibiotic properties and help in skin injuries. It is also known that some oils accelerate the cure after burns.

The **scenting of rooms** ensues by vaporizing perfumes, distributed with adjustable fan airflow in the air of warehouses, shops, showrooms, and sales and exhibition stands. In private rooms, controlled release allows a constant scenting. Release control is achieved by differences in the consistency of the perfume, which means the oils evaporate from the liquid, a thickened liquid, or out of solidified bars. Special spray devices enable short-term intensive air freshening. For a long-lasting slight scenting, fragrance-containing blocks in open plastic containers exist.

The **perfuming of products**, used at home and in trade, such as soaps, detergents, and cleaning agents, occurs predominantly with synthetic fragrance mixtures. It can be realized by incorporation of the liquids or with scented solids or by spraying. The products are used preferably in the kitchen, bathroom, and toilet areas as well as to wash and care for the laundry. Added perfumes cover up specific odors of the product and increase the attractiveness by sympathetic scenting. Fragrances improve the recognition rate. A chosen fragrance mixture takes into account the application. It should harmonize with the product (product design). Type, quantity, and form of scenting depend on the physical state and interactions with other constituents of product. Molecular interactions [16] may affect the solubility, vapor pressure, and fragrance.

9.4 Extraction of Fragrances from Plants

Natural fragrances can be of vegetable or animal origin (Figure 9.4). The scents from animals are banned and have only historical significance. They are derived from glands of beaver and musk deer or from salve-like secretions of the civet cat. Further processing of agents usually ensues through an extraction with alcohol. However, the odorant Ambergris, sometimes pathologically produced in the digestive system of sperm whales, can be largely created by chemical syntheses. For imitation of this scent mixture, the chemists synthesize Ambroxan natural identical; other necessary fragrances of the mixture arise synthetically without a natural model.

Today, scenting occurs with a mixture of synthetic and vegetable substances. Most natural perfumes include essential (volatile) oils, insoluble in water, and gently obtained by steam distillation (<100 °C) from the plant parts [17]. The distillation produces a concentrate that contains approximately 100-fold amount of active ingredient compared to the plant. Essential oils distilled with this method are described in Table 9.1. Steam distillation generates a number of well-known fragrances from the plants: lavender from the flowering panicles with stems, clove from the leaves and buds, bay leaf and patchouli from the leaves, juniper from the

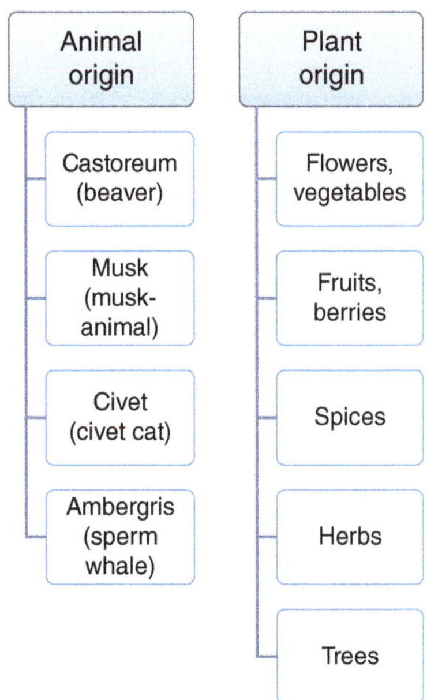

Figure 9.4 Origin of natural scents; today, only from plants.

Table 9.1 Distillation of essential oils (fragrances) from different parts of plants.

Part of the plant	Fragrance
Blossoms	Rose, ylang ylang
Panicle with stem	Lavandin
Leaves, buds	Carnation
Leaves	Laurel, patchouli, peppermint, rhododendron
Berries	Juniper
Fruits, grains	Pepper
Needles, twigs, branches	Pine, stone pine
Branches, (pine) cones	Cypress
Wood, leaves	Camphor
Wood chips	Rose, sandalwood, cedar, amyris
Herb	Chamomile, rosemary, tarragon, basil, sage, yarrow
Grass	Lemongrass, palmarosa
Seed	Coriander, carrots, celery, anise, cardamom, cumin, fennel
Root	Ginger, iris, angelika
Resin	Myrrh, frankincense, galbanum
Bark	Cinnamon

berries, pine needles from the needles and twigs, stone pine from the branches and twigs, rosewood and sandalwood from wood chips, tarragon and rosemary from the herb, lemongrass and palmarosa out of grass, coriander and carrots from the seeds, ginger and iris from the roots, and myrrh from the resin.

The technical implementation of fragrance extraction takes place in more or less simple systems (Figure 9.5). Firstly, the vegetable components are stacked dry in a container on a rack. After contact with water vapor, the volatile oils evaporate from plants and condense in a heat exchanger, wherein the essential oil forms the upper phase of the condensate. The steam distillation is carried out in an alembic (alambic or alambique) that represents a small, simple equipment, mostly made of copper. Large systems of steel often use a distillation still, containing suspended ground plant parts in hot water and equipped with a recirculation system for the water phase. By direct injection of superheated steam into the distillation still and/or by heating of the suspension via a heating/cooling jacket, the evaporation of the mixture takes place, consisting of water and essential oils. After condensation in a heat exchanger, the liquid forms two phases at the end of the cooler, which flow into a separation vessel. The essential oil floats on the water phase, and thence drains off. The recycled water phase contains some essential, soluble substances from the plants (hydrolate) that accumulate after repeated cycles. For instance, these agents help in aromatherapy [15].

The fragrances of some plants (stinging nettle) cannot evaporate with steam alone. In these cases, a second mixed plant supports the distillation of essential oils (codistillation). The pure fragrances, gained through steam distillation, show

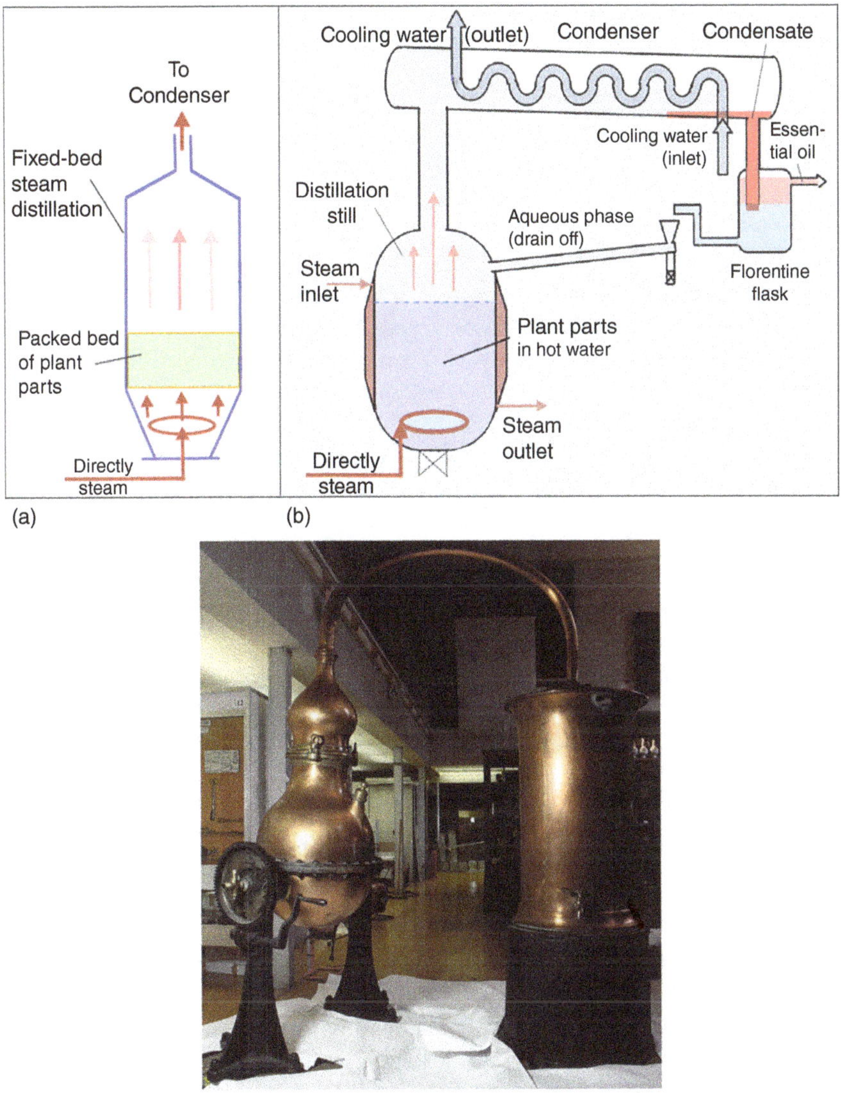

(a) (b) (c)

Figure 9.5 Extraction of essential oils from plant parts with steam distillation: (a) Alembic with a packed bed (g/s); (b) batch distillation of suspended plant parts with steam/evaporating water (g/l/s); and (c) historical alembic. Source: https://commons.wikimedia.org/wiki/File:Alambic_de_pharmacie,_H%C3%B4tel_Dieu_de_Lyon_3.JPG. Licensed Under CC BY-SA 3.0.

an evaporation temperature of 130 °C (eucalyptus) to about 190 °C (sage and thyme), and in exceptional cases even more. The hemp plant contains three different oils: firstly, the vegetable oils (triglycerides) from the seeds, set free by cold pressing in oil mills; secondly, the essential oils (terpenoids), isolated by steam distillation of flowers and leaves, and thirdly, the THC-containing "hash oil" (tetrahydrocannabinol) via a solvent extraction of the "resin" or mostly

Table 9.2 Solvent extractions of plant parts.

Method	Essential oils
Extraction with ethanol (80–100%)	Vanilla (pods), honey (honeycomb), Cocoa (beans), oak moss (herb), Siam benzoin (resin), Tolu (resin), Tonka (resin)
Extraction of flowers with hexane	Champaca, Frangipani, broom, jasmine, mimosa, osmanthus, rose absolue

through pressing of the female flowers and leaves. (Marijuana or grass are derived from the dried flowers of the female hemp plant; THC-content about 2–20%.) As a fragrance for aromatherapy, the essential hemp oil has a certain importance. The seed oil is valuable, used for skin care and cure, and utilized as food oil similar to olive oil.

The rose scent arises from rose petals by steam distillation and preferably by solvent extraction with hexane (Table 9.2). The same extraction method ensues for the production of scents from jasmine, juniper (absolute), and the mimosa. After the extraction of flowers, the solvent is distilled off. A creamy mass remains, the **Concrete**. Dissolved in alcohol, the waxes can be removed by reducing the solubility via low-temperature treatment (−24 °C). After isolation and evaporation of alcohol, the **Essence absolue** (highest purity) emerges.

For the extraction with organic solvents, manufacturers use hexane (boiling point, bp = 69 °C), petroleum (depending on composition: bp = 35–65 °C), or butane (bp = −0.5 °C) as water-insoluble solvents. In some cases, the extraction requires a water-soluble extraction agent. For this, ethanol (bp = 78 °C) is suitable, as in the extraction of fragrances from fermented, roasted, and crushed cocoa beans or from vanilla pods. The supercritical fluid extraction of crushed flowers or plant parts with supercritical carbon dioxide for the production of essential oils represents a relatively new process used in the industry, especially for flavors and coffee.

An old extraction method, applied up to 1980s and for extremely expensive oils (jasmine) until today, mostly in Grasse/France, works with animal fats (pork and/or beef) as an extractant. The fats are spread on glass plates and thereon the petals lay. Exchange of the placed petals happens daily. Over time, the fats solve the essential oils (**Enfleurage**). In some cases, the process accelerates by the use of molten fat at 60 °C (**Mazeration; Enfleurage á chaud**). In a "cold" extraction process, fats are saturated after 10–14 days. The release of essential oils from the pomade ensues with ethanol. The pure fragrance (**Essence absolue d'enfleurage**) remains after distillation of the alcohol [7].

Cocoa butter originates from the seeds by cold pressing. Isolating the fragrances requires cold pressing of shredded fruit peels from lime, lemon, bitter orange, mandarin, grapefruit, and bergamot. After mixing with water, the squeezed liquid forms two phases. The essential oil separates in a centrifuge and flows into a tank. Galbanum is a native, fennel-related plant in Asia Minor. After cutting, a gummy liquid containing the fragrance runs from the root and the lower stem. Figure 9.6 depicts the different processes for production of fragrances from the plants.

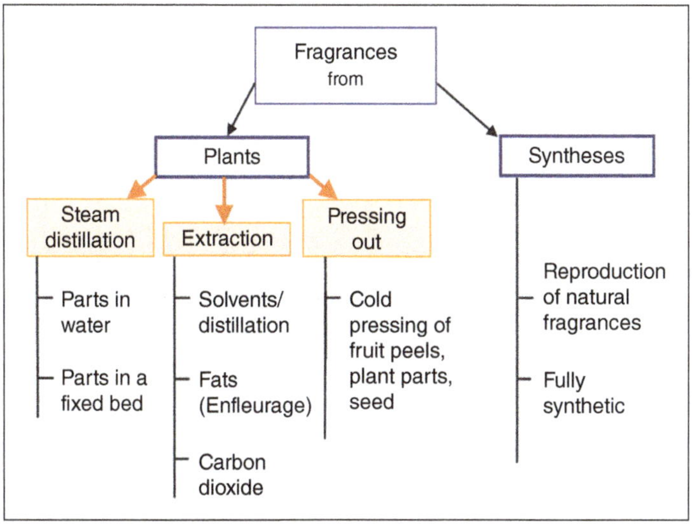

Figure 9.6 Processes for the production of fragrances.

Because the needed quantities of natural fragrances are not available in the world, the large quantities of fragrances required to emerge in syntheses (natural identical or full synthetic). The raw material and synthetic routes depend on the desired scent. About 80% of the fragrances come from syntheses [5]. For example, the industry produces 12 000 t/a of linalool (2000) and 40 000 t/a of citral (BASF 2004). Citral is a mixture of stereoisomers geranial (citral A) and nerval (citral B) and exhibits a fresh, intense lemon fragrance. These acyclic monoterpene aldehydes provide raw materials both for synthesis of other substances and for scenting of household cleaners. Many other synthetic fragrances are not based on the terpene chemistry. These substances also serve as raw materials in a subsequent fragrance synthesis (benzaldehyde, phenylethanol, phenylacetic, and cinnamic aldehyde). The examples in Table 9.3 give a little insight into the numerous, synthetically accessible fragrances with relevance to perfumery. For technical or economic reasons, some scents are available only via chemical synthesis, such as green apple, peach, apple blossom and violets, lilac, lily of the valley, freesia, almond blossom, and lily [14].

9.5 Chemical Composition of Natural Fragrances

Essential oils contain phytochemicals that protect plants from pathogens and predators. Chemically, they are predominantly terpenoids. The chemical compositions, their structures, and syntheses can be found in the literature [9, 18, 19]. The structure of molecules determines the scent impression and in this connection, preferably, the functional groups. On the basis of some substances that were synthesized in a nature-identical way, this is evidenced in Table 9.3. The selected substances occur in nature, but are not terpenoids, and arise in syntheses. Regardless of the origin of the substance, various

Table 9.3 Simple description of the odor impression depending on the functional group by way of examples [9].

No	Name	INCI name	CAS/EC number	Functional group	Odor description
1	Diphenyl ether	DIPHENYL ETHER	101-84-8; 202-981-2	Ether	Aromatic, floral on dilution, rose-like
2	Hexanal	HEXANAL	66-25-1; 200-624-5	Aldehyde	Aldehydic green, slightly fruity; somewhat green apple-like
3	β-Damascone	TRANS-ROSE KETONE-2 (Perfume name)	23726-91-2; 245-842-1	Ketone	Fruity-floral, slightly woody, herbal; somewhat raspberry connotation
4	5-Methyl-2,3-hexanedione	5-METHYL-2,3-HEXANEDIONE	13706-86-0; 237-241-8	Diketone	Buttery, cheesy, “oily,” somewhat fruity
5	*cis*-Geranic-acid	GERANIC ACID	459-80-3; 207-299-9	Unsaturated carboxylic acid	Green, floral, weedy, woody aroma
6	*cis*-3-Hexenyl acetate	CIS-3-HEXENYL ACETATE	3681-71-8; 222-960-1	Carboxylic acid ester	Powerful, strongly green; slightly floral top-note
7	Benzyl acetate	BENZYL ACETATE	140-11-4/101-41-7; 205-399-7/202-940-9	Benzylic ester	Green, dry-powdery, fruity, somewhat milky and estery
8	Methyl benzoate	METHYL BENZOATE	93-58-3; 202-259-7	Benzoate ester	Heavy sweet, slightly floral-fruity, berry-like
9	Isopentylamine	—	—	Amine	Fishy, ammonia-like, in low concentrations somewhat fermented
10	Musk ketone	MUSK KETONE	81-14-1; 201-328-9	Nitroaromatic compound	Dry, powder, nitro-musk; somewhat floral-fruity connotations
11	Diallyl trisulfide	—	—	Thio compound	Sulfurous, characteristic garlic
12	Furfural	FURFURAL	98-01-1; 202-627-7	Furan	Sweet caramel-like, nutty, baked bread, almond
13	Indole	INDOLE	120-72-9; 204-420-7	Indole	Animalic, musk, cheese, slightly fecal on dilution
14	6-Methylquinoline	6-METHYLQUINOLINE	91-62-3; 202-084-6	Quinoline	Narcotic; earthy, green
15	2,3,5-Trimethyl-pyrazine	TRIMETHYLPYRAZINE	14667-55-1; 238-712-0	Pyrazine	Burnt roasted, earthy, tobacco-like
16	4-Methyl-5-vinyl-thiazole	—	—	Thiazole	Nutty, musty, earthy, cocoa powder-like

All substances are synthesized identical with nature, but no components of the essential fragrance oils.

structure/odor relationships exist [20], also outside the essential oils. In many cases, it is possible to synthesize the targeted molecule by incorporating functional groups. In this way, the desired fragrance and required adhesion arise. The targeted modification of molecular structure is known as molecule- or product (substance) engineering and constitutes an essential part of product design. Because odor changes may occur due to chemical reactions, the perfumer has to anticipate possible reactions in the formulation.

Natural fragrances consist mainly of a mixture of different terpenoids, in most cases about 10–15 compounds without substances in traces. For natural ingredients, the results of the complex chemical analysis depend on the plant, origin, vegetation of soil, and weather during plant growth. The most contained fragrance characterizes the oil. At terpenoids, the chemists interpret the monoterpene $C_{10}H_{16}$ as an isoprene dimer (two C_5-units) and the sesquiterpenes $C_{15}H_{24}$ as a trimeric isoprene. Terpenoids are divided into three main groups: aliphatic or acyclic (I), mono-(II), and bicyclic (III) terpenoids. Terpenes are pure hydrocarbons, or as terpenoids bear a functional group (as aldehyde, ketone, and alcohol) and may be etherified or esterified. For example, patchouli oils consist mainly of sesquiterpenols, sesquiterpene, and monoterpene as well as ketones and oxides. In the following Tables 9.4 and 9.5, some well-known fragrances are presented that occur in nature, generated from plants.

Monoterpene hydrocarbons include myrcene (I), limonene (II), and α-pinene (III) $C_{10}H_{16}$. Limonene is the most frequently occurring monoterpene in plants. Typical terpene alcohol scents are geraniol, nerol, citronellol $C_{10}H_{18}O$ (I), menthol $C_{10}H_{20}O$ (II), and borneol (III). Frequently used terpene aldehydes and ketones are citral $C_{10}C_{16}O$ (I), menthone $C_{10}H_{18}O$ (II), and camphor $C_{10}H_{16}O$ (III). Farnesol $C_{15}H_{26}O$ (I), a sesquiterpene found in several essential oils, is appreciated because of the lily of the valley-like pleasant smell. On the other hand, chemists synthesize farnesol from linalool in a mixture of isomers, such as many other fragrances. The ether anethole (anise flavor) and the 1,8-cineol (II), distilled from eucalyptus, represent substances with a monoterpene structure. Other examples are geranyl acetate, linalyl acetate (I), and the phenol thymol (II).

The fragrances contained in perfumes are today almost all produced synthetically (Table 9.6). Most perfumes contain natural substances only in minor

Table 9.4 Main components of natural fragrances (examples).

Compound / Product	Monoterpene C_{10}	Sesquiterpene C_{15}	Terpenole (alcohol) $C_{10,15,20}$	Terpenaldehyde/-ketone	Ester
Lavandin	+		++	+	++
Cistus	+++		+		+
Spruce needles	++		+ (Di-)		++
Sandal-wood		++	+++ (Sesqui-)		
Rosemary	+++	+	+	++	+
Limette	+++		+	+	+
Patchouli	+	+	++ (Sesqui-)	+	

Table 9.5 Composition of selected essential oils from plants [19].

No	Name of the oil	Plant name	INCI name	CAS/EC number	Main components
1	Eucalyptus	Eucalyptus globulus Labill.	Eucalyptus Globulus Leaf Oil	8000-48-4/ 84625-32-1; –/283-406-2	1,8-Cineole: minimum 70% Limonene: ≥12% α-Pinene: ≥9% α-Phellandrene: ≤1.5% β-Pinene: ≤1.5% Sabinene: ≤0.3% Camphor: ≤0.1%
2	Juniper	Juniper communis L. false fruits	Juniperus Communis Fruit Oil	8002-68-4/ 73049-62-4/ 84603-69-0; –/283-268-3	α-Pinene: 20–50% Sabinene: ≤20% β-Myrcene: 1–35% β-Pinene: 1–12% Limonene: 2–12% Terpinen-4-ol: 0.5–10% β-Caryophyllene: ≤7% Bornyl acetate: ≤2% α-Phellandrene: ≤1%
3	Lavender	Lavandula angustifolia P. Mill. (L. officinalis Chaix.)	Lavandula Angustifolia Herb Oil	90063-37-9; 289-995-2	Linalyl acetate: 25–46% Linalool: 20–45% 1,8-Cineole: ≤2.5% Camphor: ≤1.2% 3-Octanone: 0.1–2.5% Terpinen-4-ol: 0.1–6% α-Terpineol: ≤2% Limonene: ≤1% Lavandulyl acetate: ≤0.2% Lavandulol: ≤0.1%
4	Lemon	Citrus limon (L.) Burman fil.	Citrus Limon Leaf Oil	92346-89-9/ 84929-31-7; 296-174-2/ 284-515-8	Limonene: 56–78% β-Pinene: 7–17% γ-Terpinene: 6–12% Sabinene: 1–3% Geranial: 0.5–2.3% Neral: 0.3–1.5% Neryl acetate: 0.2–0.9% Geranyl acetate: 0.1–0.8% α-Terpineol: ≤0.6% β-Caryophyllene: ≤0.5%

Table 9.5 (Continued)

No	Name of the oil	Plant name	INCI name	CAS/EC number	Main components
5	Mandarin	Citrus reticulata Blanco	Citrus Reticulata Leaf Oil	8008-31-9/ 8014-17-3; 284-521-0	Limonene: 65–75% γ-Terpinene: 16–22% α-Pinene: 1.3–3% β-Myrcene: 1.5–2% β-Pinene: 1.2–2% *p*-Cymene: ≤1% Methyl anthranilate: 0.3–0.6% Sabinene: ≤0.3%
6	Mint	Mentha canadensis L. (M. arvensis L.)	Mentha Arvensis Herb Oil	90063-97-1; 290-058-5	Menthol: 30–50% Menthone: 17–35% Isomenthone: 5–13% Limonene: 1.5–7% Menthyl acetate: 1.5–7% Pulegone: ≤2.5% Isopulegol: 1–3% Carvone: ≤2% 1,8-Cineole: ≤1.5%
7	Pine	Pine mugo Turra	PINUS MUGO TWIG OIL	90082-72-7; 290-163-6	α-Pinene: 10–30% δ-3-Carene: 10–20% β-Phellandrene: 10–19% Limonene: 8–14% β-Pinene: 3–14% β-Myrcene: 3–12% Terpinolene: ≤8% Bornyl acetate: 0.5–5% β-Caryophyllene: 0.5–5% *p*-Cymene: ≤2.5% Camphene: ≤2%

(Continued)

Table 9.5 (Continued)

No	Name of the oil	Plant name	INCI name	CAS/EC number	Main components
8	Rosemary	Rosmarinus officinalis L.	Rosmarinus Officinalis Flower Oil	84604-14-8; 283-291-9	Moroccan, Tunisian type 1,8-Cineole: 38–55% α-Pinene: 9–14% Camphor: 5–15% β-Pinene: 4–9% Camphene: 2.5–6% Borneol: 1.5–5% Limonene: 1.5–4% β-Myrcene: 1–2% α-Terpineol: 1–2.6% *p*-Cymene: 0–2.5% Bornyl acetate: 0.1–1.5% Verbenone: ≤0.4%
9	Sage (Spanish)	Salvia lavandulifolia Vahl.	Salvia Lavandulifolia Herb Oil	90106-49-3; 290-272-9	1,8-Cineole: 10–35% Camphor: 11–30.5% α-Pinene: 4–11% Sabinyl acetate: 0.5–9% α-Terpinyl acetate: 0.5–9% Borneol: 1–7% Limonene: 2–6.5% Linalyl acetate: ≤5% Linalool: 0.3–4% Sabinene: 0.1–3.5% Terpinen-4-ol: ≤2% Thujone: ≤0.5%
10	Thyme	Thymus vulgaris L.	Thymus Vulgaris Oil	84929-51-1; 284-535-7	Thymol: 36–55% *p*-Cymene: 15–28% γ-Terpinene: 5–10% Linalool: 4–6.5% Carvacrol: 1–4% β-Myrcene: 1–3% Terpinen-4-ol: 0.2–2.5%

Table 9.6 Natural fragrances contained in many essential oils and their chemical structures ([9] and Wikipedia).

No	Name of the fragrance	INCI name	CAS/EC number	Functional group	Odor impression	Chemical structure
1	Limonene	Perfume name	5989-27-5; 227-813-5	Monoterpene	Citrus	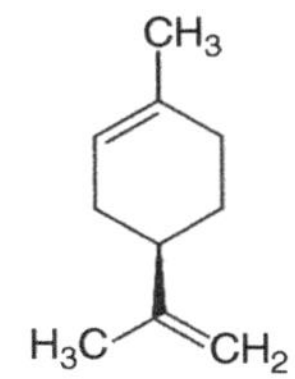
2	α-Pinene	L-ALPHA-PINENE	7785-26-4; 232-077-3	Monoterpene	Terpene-like	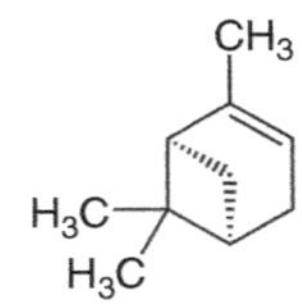
3	Geraniol	GERANIOL	106-24-1; 203-377-1	Terpenic alcohol	Flowery rose-like	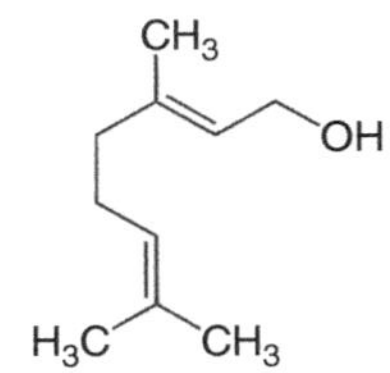
4	Menthol	MENTHOL	1490-04-6/2216-51-5/ 89-78-1/15356-70-4; 216-074-4/218-690-9/ 201-939-0/239-388-3	Terpenic alcohol	Peppermint	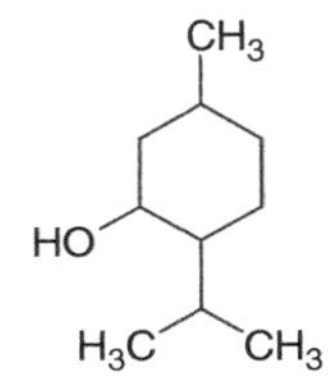

(Continued)

Table 9.6 (Continued)

No	Name of the fragrance	INCI name	CAS/EC number	Functional group	Odor impression	Chemical structure
5	Borneol	BORNEOL	507-70-0; 208-080-0	Bi-cyclic terpene alcohol	Camphor-like, fresh	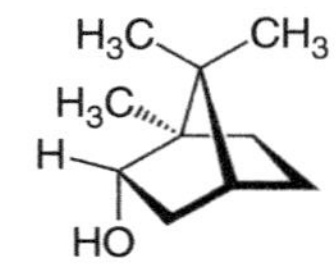
6	Citral	CITRAL	5392-40-5; 226-394-6	Terpenic aldehyde	Fresh, citrus fruits	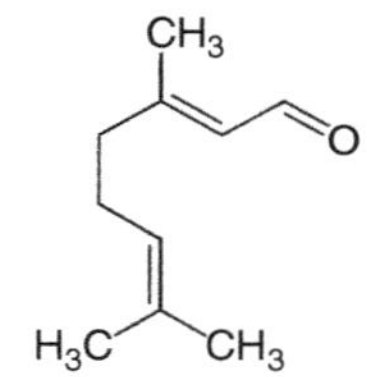
7	Farnesol	FARNESOL	4602-84-0; 225-004-1	Sesquiterpene alcohol	Lily of the valley-like	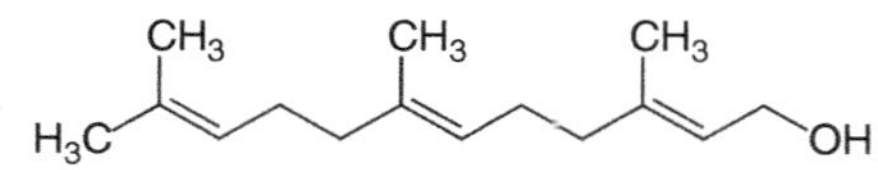
8	1,8-Cineole	CINEOLE	470-67-7; 207-428-9	Monoterpenic oxide	Fresh, strong eucalyptus-like, camphoraceous, minty	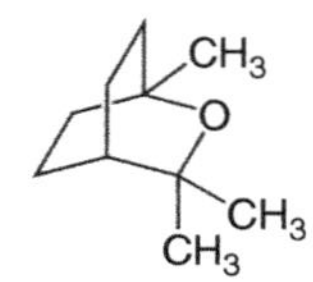
9	α-Cedrene	CEDRENE	11028-42-5; 234-257-7	Sesquiterpene	Woody, cedar-/sandal-wood	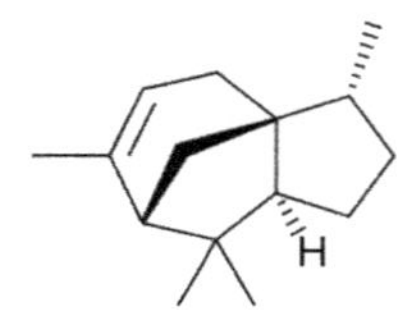
10	Thymol	THYMOL	89-83-8; 201-944-8	Monoterpene phenol	Herbal, spicy, aromatic, medicinal, characteristic thyme	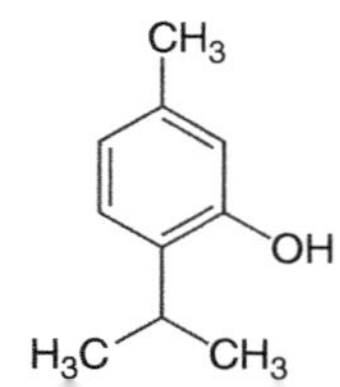

amounts, which is due to the relatively high cost of natural fragrances. As quality differences between natural and synthetic perfumes are not perceived by the consumer, the price differences decide in favor of the synthetic fragrances. Purely natural fragrances can be achieved by adding a single essential oil such as lavender, jasmine, geranium, rose, sandalwood, or chamomile oil. This method is more expensive but recommended. Chemical structures, fragrances, and physicochemical characteristics of customary scents based on nature and synthesis are described in detail [9, 18, 19, 21].

9.6 Possibilities in Product Design of Perfume Oils

The variety of natural and synthetic fragrances in graduated levels of quality and from different places of origin offers the possibility of creating diverse perfume formulations, which differ on the skin in the fragrance delivery (Figure 9.7). Although a synthetic fragrance consists of a single compound, a natural scent of an essential oil contains more than 10 components, so that a pure natural perfume oil can comprise in total more than 200 individual compounds. Depending on the task, an adjustment of the scent experience takes place through changes in the mixture. The composition, head, heart, and base notes, in conjunction with a fixation (such as cinnamyl acetate), determines the fragrance release times and provides the opportunity to prolong these (Figure 9.8). For perfuming of products such as creams, these properties can be important. A uniform release over a longer period without significant changes in the fragrance impression stands in focus at the formulation of liquid perfume oils for creams. This task is simple when adding a single essential oil or synthetic fragrance but difficult in perfume oils.

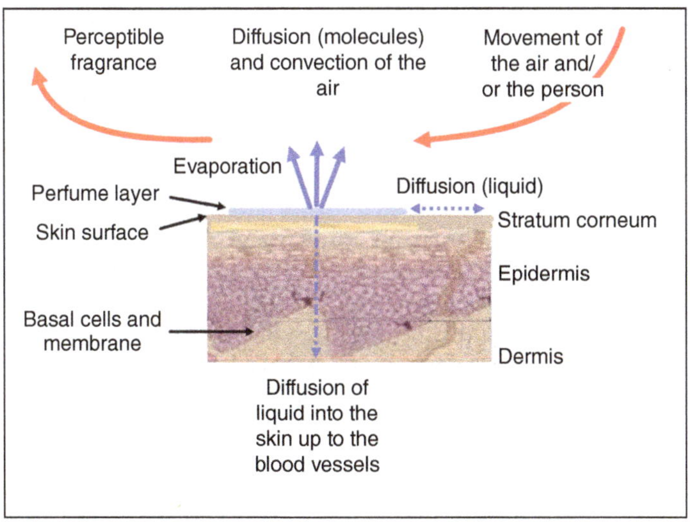

Figure 9.7 Equilibria of the perfume, adsorbed on the skin surface, in dependence of the composition, the temperatures (skin, ambient), and weak airflows.

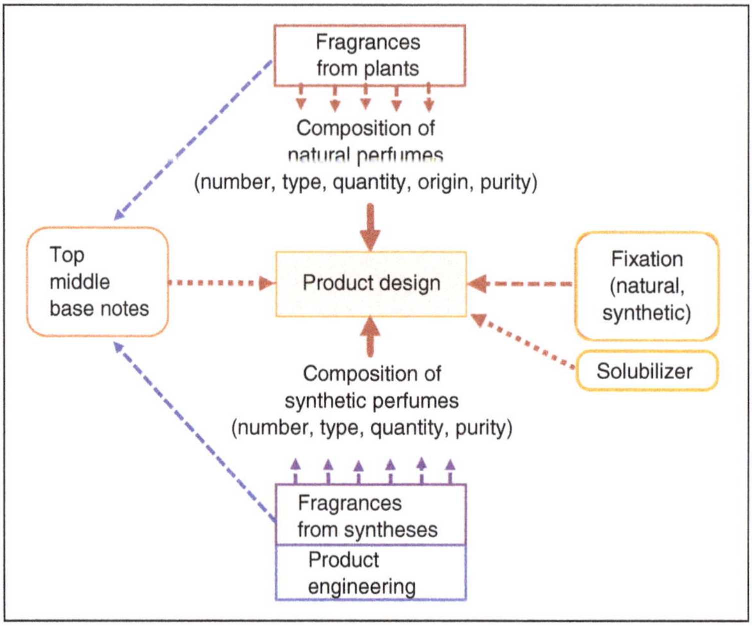

Figure 9.8 Adjustment of the fragrance impression of perfume oils by a choice of synthetic and/or natural scents.

Scenting of **consumer goods** in the detergent and home care industry takes place primarily through the use of synthetic fragrances, which are cheaper than the essential oils and dominate the area of consumer products. Appropriate product fragrances not only cover the intrinsic odors but also deliver interesting scents, corresponding to the product. Here, the synthetic-based fragrance blends have a simpler composition than the perfume oils for the human body and consist of about 5–50 individual fragrances. The product perfumes cost about €8–20 per kg while synthetic/natural-based perfumes for humans are around €35 per kg (€20–70, in some cases more). Good natural essential oils are usually more expensive. For example, the genuine jasmine oil from Egypt costs almost €3000 per kg.

For cleansers and similar products, an advertising core message guarantees hygiene and cleanness, such as "with citrus scent." Hence, in these cases, the presence of one odor note suffices. Perfume oils are processed predominantly as liquids. The addition of perfumes to products ensues product-dependent and application-related. On the one hand, the perfumes can be mixed in surfactant-containing liquids and emulsions or poured into various shaped bodies. On the other hand, many perfumes are sprayed on solid products. Third, one or the other fusible substance allows solidification of oils such as fatty alcohols.

Scenting of products sometimes requires organic solubilizers (dipropylene glycol, diethylene glycol, diethyl phthalate, and isopropyl myristate). For other compositions, the addition of water, mixed with emulsifiers, helps. Because of diffusion and chemical reactions, the fragrance impression may change overtime. Long storage allows perfume molecules to react with other perfume or product components. Examples of reactions between fragrances include acetal formations, aldol condensations, ester cleavages, transesterifications, and (partial) oxidations of alcohols, aldehydes, and ketones to acids [12]. The reactions start at higher temperatures and in the presence of oxygen, and/or by initiation with UV radiation, as well as with moisture. Adaptation of storage temperatures, stable fragrance mixture, and inertizing with nitrogen as well as the use of UV-light impermeable packaging reduce/prevent chemical reactions.

9.7 Personal Care and Other Products

The perfuming of care products for the skin [22] and the hair occurs by setting an adequate fragrance, preferred by the target group in type and strength. For skin care, especially the creams and lotions are used. To a great extent, the product composition and the diameter of the droplets influence the fragrance impression. Components of the selected perfumes are absorbed in the cream and interact with product components, which change and prolong the scent impression. In addition, the consistency exerts a perceptible influence. In manufacturing process, about 0.2–1% (0.35–0.6%) pure perfume oil flow directly into the organic phase of the emulsion, preferably after cooling the emulsion to about 30 °C. Then, the homogenization starts again for some time, incorporating the fragrance.

The selection of "best" fragrance represents a lengthy, difficult-iterative process, both in cosmetics and in the detergent sector. In the end, the determination of a best fragrance may be carried out by an odor panel (sensory assessment by 30–40 people; OEB = Odor Evaluation Board). The decision falls usually after several tests in practice, either by the panel or in case of large products in a wider circle of 100–250 people in a household survey and test (Home Use Test, HUT [23]). The expense can be justified in large companies because the scent as an essential parameter of product design significantly affects the market success.

A check for skin tolerance of fragrances, used in creams and lotions, takes place usually at 10–20 people with sensitive skin in a patch test. Well-known products for sensitive skin, which should have a scent, contain only skin-friendly proven fragrances. Perfume-free products for sensitive skin are still the exception.

In addition to the personal care products, many body cleansers are perfumed. Examples represent soaps, shower gels, alcoholic solutions for the face, and toothpastes. In other areas, leather, plastic, and furniture polish exude an intense fragrance. The moth-repellent sachets contain perfume (lavender or cedar), incorporated into granules. After spraying, windshield washer fluid for cars unfolds a typical (alcohol) odor. Occasionally, people use fragrant stationery in different variants. Some leather products gain more appeal through a leather/wood scent. The consumer meets other examples in everyday life. For increasing the sales, the application of perfumes gets more and more importance.

9.8 Safety

When transporting and storing pure perfume oils or essential oils, hazard labeling is always required because the liquids and vapors may ignite (H 226). Caution, ingestion, and entry into the respiratory tract can be fatal (H 304), and in some cases, the substances cause skin irritation and sensitization (danger of allergies; H 315, 317). They are hazardous to the aquatic environment with chronic aquatic toxicity (H 411; Figure 9.9). Check the required hazard warnings for each oil.

In serial examinations with patch tests, more than 15% of the German population show sensitizations to at least one of the most important contact allergens (Information Network of Departments of Dermatology; in Germany: IVDK [24]). A major problem is that fragrance oils trigger at more than 7% of persons allergies [25, 26]. Unless the substance is not identified by tests of single substances, the allergic person should avoid any skin contact with fragrances because it is not possible to estimate the potential and the danger of allergic substances.

Fragrances divide into four classes, from class 1 (frequently sensitizing) to class 4 (nonsensitizing). Class 4 includes, for example, almond, mandarin, and lime oil, and class 3 (rarely sensitizing) includes rosewood, caraway, and bergamot. The latter stands out with a unique feature: After application to the skin and simultaneous exposure to sunlight, Bergamot (e.g. in Eau de Cologne) can trigger phototoxic reactions. The effect causes skin discoloration and occasionally bubble changes, known as "Berloque Dermatitis." A few other oils can also trigger photoallergic or phototoxic reactions. Attention, avoid such fragrances in sunscreens!

Hazard warnings

H 226 H 304 H 315, 317 H 411

Safety instructions - prevention:

P 210 Keep away from heat, hot surfaces, sparks, open flames, and other sources of ignition. Do not smoke.

P 273 Avoid the release into the environment.

P 280 Wear protective gloves/eye protection.

Safety instructions - reaction:

P301 + P310 IF SWALLOWED: Immediately call a POISON CENTER/doctor.

P303 + P361 + P353 IF ON SKIN (or hair): Remove/Take off immediately all contaminated clothing. Wash skin with water/shower.

P333 + P313 If skin irritation or rash occurs: Get medical advice/attention.

Figure 9.9 Required Hazard Statements for most perfumes and essential oils.

Table 9.7 Required declaration of components in perfume oils according to the Cosmetic Directive (strong allergens emphasized).

INCI name	Chemical description
Amyl cinnamal	2-(Phenylmethylene)heptanal
Amylcinnamyl alcohol	2-(Phenylmethylene)heptanol
Anise alcohol	4-Methoxy-benzyl alcohol
Benzyl alcohol	Benzyl alcohol
Benzyl benzoat	Benzyl benzoat
Benzyl cinnamate	Benzyl cinnamate
Benzyl salicylate	Benzyl salicylate
Cinnamal	Cinnamaldehyde, 3-phenyl-2-propenal
Cinnamyl alcohol	Cinnamyl alcohol, 3-phenyl-2-propen-1-ol
Citral	3,7-Dimethyl-2,6-octadienal
Citronellol	DL-Citronellol, 3,7-dimethyl-6-octen-1-ol
Coumarin	2H-1-Benzopyran-2-one
Eugenol	2-Methoxy-4-(2-propenyl) phenol
Evernia prunastri extract	Oakmoss extract
Evernia furfuracea extract	Treemoss extract
Farnesol	3,7,11-Trimethyl-2,6,10-dodecatrien-1-ol
Geraniol	2,6-Dimethyl-trans-2,6-octadien-8-ol
Hexyl cinnamal	2-(Phenylmethylene)octanal
Hydroxycitronellal	7-Hydroxycitronellal, 7-hydroxy-3,7-dimethyloctanol
Hydroxyisohexyl-3-cyclohexene-carboxaldehyde	4-(4-Hydroxy-4-methylpentyl)-3-cyclohexene-1-carboxaldehyde
Isoeugenol	2-Methoxy-4-(1-propenyl) phenol
Limonene	1-Methyl-4-(1-methylethenyl)-cyclohexene
Linalool	3,7-Dimethyl-1,6-octadien-3-ol
Methyl 2-octynoate	2-Octynoic acid, methyl ester
Alfa-isomethylionone	3-Methyl-4-(2,6,6-trimetyl-2-cyclohexene-1-yl)-3-butene-2-one
Butylphenyl-methylpropional	2-(4-*tert*-Butylbenzyl)-propionaldehyde

In Europe, the SCCNFP (The Scientific Committee on Cosmetic Products and Non-Food Products Intended for Consumers) evaluated fragrances according to their potential for contact allergy [27]. They found 26 major allergens (Table 9.7 [28]), which are annexed to the Cosmetics Directive (§ 5a, 3 Appendix 2 A, 45 and substances from 67 to 92 [29]). From a certain amount (100 ppm for rinse-off and 10 ppm for the leave-on fragrances), the manufacturer has to declare the use of these compounds on the product package in the International Nomenclature of Cosmetic Ingredients (INCI)-list (see Section 1.5).

Applied in larger quantities on the skin, perfume oils may diffuse, depending on composition, in about 10–20 minutes through the epidermis into the dermis and enter the bloodstream. Thereafter, the components are in the whole body. Some scents and degradation products can be detected later in the urine.

For respiratory intake of fragrances, however, little experimental data are available. After inhalation of perfumes and fragrances, in rare cases, both irritating and allergic reactions occur, as the clinical experience demonstrates. The intensity reaches from a mild mucosal irritation with cough and mucus production to severe asthmatic attacks [30].

There are hundreds of ready-made perfume mixtures to buy. Almost all perfume oils contain more or less declarable ingredients. The supplier must name the purchaser of the perfume oils the ingredients to be declared. The responsibility, however, lies with the manufacturer of the perfume and the cream, or the person placing the product on the market. Cream manufacturers, who want to create a scented product without allergic potentials, may use special composed oils that can be bought today. Some allergen-free perfume oils without declarable ingredients are (Deike Chemie [31]; INCI: Parfum) as follows: Aloe Flower, Chocolate, Fresh and Clean, Men's Care, Palais, Pitaya-Coconut, Powder Fruit, Powder Melon, Raspberry-Mint, Red Grape-Mint, Soft and Fruity, Strawberry, and Unisex.

9.9 Learnings

- The most important perfume manufacturers are Givaudan (Switzerland) with about 16% of the total market, and International Flavors & Fragrances (USA), Firmenich (Switzerland), and Symrise (Germany) each with around 11%.
- Perfumers distinguish between natural and synthetic oils as well as mixed forms. The syntheses take place nature-identical or artificial.
- There are about 3000 different scents from which the perfumer can choose.
- Each perfume oil contains three different fragrance types, namely the head, heart, and base notes. They differ in their volatility and aroma.
- Natural fragrances emerge from parts of plants by steam distillation, extraction, or cold pressings. Each essential oil comprises about 10 fragrant compounds (without traces).
- Many conventional perfume oils contain about 80% synthetic fragrances. They are cheaper and easier to process.
- Perfume oils in bottles contain 70–95% alcohol.
- In a cosmetic cream, about 0.5% (±0.2) of pure perfume oil is incorporated.
- In general, synthetic fragrances are produced identically to nature.
- Perfume oils allow a scenting of the human body, of products and spaces. In addition, the aromatherapy utilizes fragrant oils.
- Scented emulsions, foams, granules, and tablets arise from perfume oils by the addition of excipients.
- Perfume oil may cause allergy. Therefore, the fragrance oils divide into four classes, from class 1 (frequently sensitizing) to class 4 (nonsensitizing).
- Annexed to the Cosmetics Directive, there are 26 fragrances with allergenic potential, which the manufacturer of product must declare in the INCI.

References

1 Rimmel, E. (1867). *The Book of Perfumes*, 5e. London: Chapman & Hall (digitized by Google).
2 Morris, E. (2002). *Fragrance: The Story of Perfume from Cleopatra to Chanel*. Dover Publications.
3 Ohloff, G. (1992). *Irdische Düfte—himmlische Lust: eine Kulturgeschichte der Duftstoffe*. Birkhäuser.
4 Ohloff, G. (2004). *Düfte-Signale der Gefühlswelt*. Zürich: Helvetica Chimica Acta.
5 Boeck, A. and Fergen, H.-U. (1994). Compounding, Chapter 15. In: *Perfumes: Art, Science and Technology* (eds. P.M. Müller and D. Lamparsky), 421–440. London: Chapman & Hall.
6 Meine, G. (2004). Herstellung von Parfümölen, Chapter 18. In: *Kosmetik und Hygiene*, 3e (ed. W. Umbach), 493–500. Weinheim: Wiley-VCH.
7 Schwedt, G. (2008). *Betörende Düfte, sinnliche Aromen*, Chapter 4, 178 ff. Weinheim: Wiley-VCH.
8 Schreiber, W.L. (2013). Perfumes, Chapter 4. In: *Chemical Technology of Cosmetics* (ed. Kirk-Othmer), 123–160. Hoboken: Wiley.
9 Zviely, M. (2013). Aroma chemicals, Chapter 7. In: *Chemical Technology of Cosmetics* (ed. Kirk-Othmer), 207–246. Hoboken: Wiley.
10 Uhlemann, J. and Reiß, I. (2009). Produkteigenschaften und Verfahrenstechnik am Beispiel der Aromen. *Chem. Ing. Tech.* 81 (4): 393.
11 Fa. Symrise (2016). Annual report, financial report, Fa. Symrise, p. 92. https://www.symrise.com/investors/reports/2016 (accessed 11 May 2018).
12 Statista, Prognostiziertes weltweites Marktvolumen für Parfum bis 2025, 2019. https://de.statista.com/statistik/daten/studie/255853/umfrage/prognostiziertes-weltweites-marktvolumen-fuer-parfum (accessed 11 May 2018).
13 Teixeira, M.A., Rodriguez, O., Gomes, P. et al. (2013). *Perfume Engineering: Design, Performance and Classification*. Elsevier: Butterworth-Heinemann.
14 Werner, M. and von Braunschweig, R. (2006). *Praxis Aromatherapie: Grundlagen - Steckbriefe - Indikationen*. Stuttgart: Georg Thieme Verlag.
15 Wabner, D. and Beier, C. (eds.) (2012). *Aromatherapie*, 2e. München: Elsevier, Urban & Fischer Verlag.
16 Perring, K.D. (2006). Volatility and substantivity, Chapter 11. In: *The Chemistry of Fragrances* (ed. C. Sell), 199. RSC Publishing.
17 Schwedt, G. (2008). Chapter 3: Zur Chemie der Düfte. In: *Betörende Düfte, sinnliche Aromen*, 99 ff. Weinheim: Wiley-VCH.
18 Sell, C.S. (2013). Terpenoids, Chapter 8. In: *Chemical Technology of Cosmetics* (ed. Kirk-Othmer), 247–373. Hoboken: Wiley.
19 Baser, K.H.C. and Demirci, F. (2013). Essential oils, Chapter 9. In: *Chemical Technology of Cosmetics* (ed. Kirk-Othmer), 375–408. Hoboken: Wiley.
20 Kraft, P., Baigrowicz, J.A., Denis, C., and Fráter, G. (2000). Recent developments in the chemistry of odorants note on trademarks. *Angew. Chem. Int. Ed. Engl.* 39 (17): 2980–3010.
21 Surburg, H. and Patten, J. (2006). *Common Fragrance and Flavor Materials*, 5e. Weinheim: Wiley-VCH.

22 Rähse, W. and Dicoi, O. (2009). Produktdesign disperser Stoffe: Emulsionen für die kosmetische Industrie. *Chem. Ing. Tech.* 81 (9): 1369–1383.
23 Lawless, H.T. and Heymann, H. (2010). *Sensory Evaluation of Food: Principles and Practices*. New York, NY: Springer.
24 Schnuch, A., Uter, W., Lessmann, H. et al. (2008). Klinische Epidemiologie der Kontaktallergien. Das Register und das Überwachungssystem des Informationsverbundes Dermatologischer Kliniken (IVDK). *Allergo J.* 17: 611–624.
25 Schnuch, A., Uter, W., Geier, J. et al. (2005). Überwachung der Kontaktallergie: zur "Wächterfunktion" des IVDK. *Allergo J.* 14: 618–629.
26 Uter, W., Balzer, C., Geier, J. et al. (2005). Ergebnisse der Epikutantestung mit patienteneigenen Parfüms, Deos und Rasierwässern. Ergebnisse des IVDK 1998–2002. *Dermatologie in Beruf und Umwelt* 53: 25–36.
27 Herman, S. (2002). *Fragrance Applications: A Survival Guide*. Carol Stream, IL: Allured Publishing Corporation.
28 Schnuch, A., Uter, W., Geier, J. et al. (2007). Sensitization to 26 fragrances to be labelled according to current European regulation. Results of the IVDK and review of the literature. *Contact Dermatitis* 57: 1–10.
29 Regulation (EC) No 1223/2009 of the European Parliament and of the Council of 30 November 2009 on cosmetic products. https://eur-lex.europa.eu/legal-content/EN/TXT/PDF/?uri=CELEX:32009R1223&from=EN (accessed 13 May 2018).
30 Schnuch, A., Oppel, E., Oppel, T. et al. (2010). Experimental inhalation of fragrance allergens in predisposed subjects: effects on skin and airways. *Br. J. Dermatol.* 162: 598–506.
31 Deike-Chemie, Produktübersicht, 2008. http://www.deike-chemie.de/produktuebersicht (accessed 13 May 2018).

10

Production of Cosmetic Creams

10.1 Method

For the production of creams, many variants exist, depending on the recipe and production equipment. Furthermore, each developer has his own experience-based method of making cosmetic products in the laboratory and in production. The method below describes a proven proposal for the production of emulsions (creams), which contain temperature-sensitive ingredients. In the first step, the recipe is divided into water-soluble, oil-soluble, and temperature-sensitive substances. The temperature-sensitive ingredients include all substances that chemically change (usually cleave) under the working conditions (temperature and vacuum). In addition, to avoid product loss, the volatiles belong to this group because the whole process runs under vacuum. Examples for temperature-sensitive and volatile substances represent the vitamins (depending on the compound), urea, and all essential oils and perfumes. Because of the buffer, the preparation of the water phase always takes place at pH 5. Therefore, the ingredients must be stable in this pH range. Besides the pH and temperature stability, sufficient solubility in the mixture is required at room temperature. The procedure for the allocation of the ingredients is exemplified in Table 10.1.

Before starting production, the entire equipment is disinfected with isopropanol and dried with hot air. To prevent oxidation reactions, the plant should be purged with nitrogen. Subsequently, the gas removal takes place in the emulsifier by generating vacuum because the preparation of emulsions should always be carried out under vacuum (in some cases under nitrogen) to prevent oxidations and foaming. After provision of the weighed raw materials, the water phase arises in a stirred premix vessel, beginning with the generation of the pH buffer by dissolving the citrate and then the citric acid in the water. The other substances are added successively to the buffer solution. Finally, potassium sorbate and the viscosity-forming substances (hyaluronic acid and xanthan gum) follow.

Often, the water phase is prepared directly in the emulsifier step by step. For this purpose, the weighed substances are individually or often together sucked via a small premix tank into the reactor. This is carried out at room temperature or elevated temperatures. By heating the water phase, the working temperature of 50 °C is set in the stirred emulsifying vessel. (The temperature of 45 °C represents the lower limit to avoid later flocculation.) In most cases, a higher temperature is not required. Thus, the thermal load of the substances can be kept low. The

Cosmetic Creams: Development, Manufacture and Marketing of Effective Skin Care Products,
First Edition. Wilfried Rähse.

Table 10.1 Classification of water-soluble, oil-soluble, and temperature-sensitive ingredients (formulation according to Table 8.4).

Water-soluble ingredients	Oil-soluble ingredients	Temperature-sensitive ingredients
Water		
Glycerol	Jojoba oil	D-Panthenol
Sorbitol	Myristyl Myristate	Urea
Magnesium ascorbyl phosphate	Polyglyceryl-3 dicitrate/stearate	Vitamin E
Nicotinamide	Wheat germ oil	Geranium
Glycine	Safflower oil	
Ethanol	Borage seed oil	
Trisodium citrate	Retinaldehyde	
Dihydrate		
Xanthan		
Potassium sorbate		
Hyaluronic acid		
Citric acid monohydrate		
Allantoin		

macromolecules take some time for complete dissolution. Therefore, the temperature should be kept constant in the range of 50 °C until complete dissolution.

Preferably, the preparation of the oil phase takes place in a separate stirred vessel with heater. There, the vegetable oils and the substances solid at room temperature are heated to 60 °C under nitrogen and/or vacuum. For the formulas described in Chapter 8, this temperature is sufficient to melt and/or to dissolve the presented solids. When choosing other ingredients (emulsifiers and thickeners) with high melting points and bad oil solubilities, higher temperatures may be required. However, these are not recommended if the formulation contains sensitive, polyunsaturated oils because high temperatures increase the risk of quality losses. Normally, the dissolution and melting of the solid parts occurs after reaching the predetermined temperature (60 °C). Then, the clear oily solution is sucked into the rotor/stator tool (homogenizer and disperser) of the emulsifying vessel and emulsified quickly with the aqueous phase (two minutes in the laboratory). Subsequently, the cooling of the emulsion takes place with stirring. At a temperature of 30 °C, the temperature-sensitive and slightly vaporizing substances such as perfume can be added by means of the vacuum (Figure 10.1). If only water-soluble ingredients are present, stirring into the o/w emulsion is sufficient. Otherwise, the dispersing machine has to be used again for some time. Cooling is usually completed after reaching about 25 °C. The emulsion is stirred vigorously again and then transferred to a nitrogen-purged buffer tank. After the quality tests are completed and the filling line is ready, the bottling into the nitrogen-filled dispensers begins.

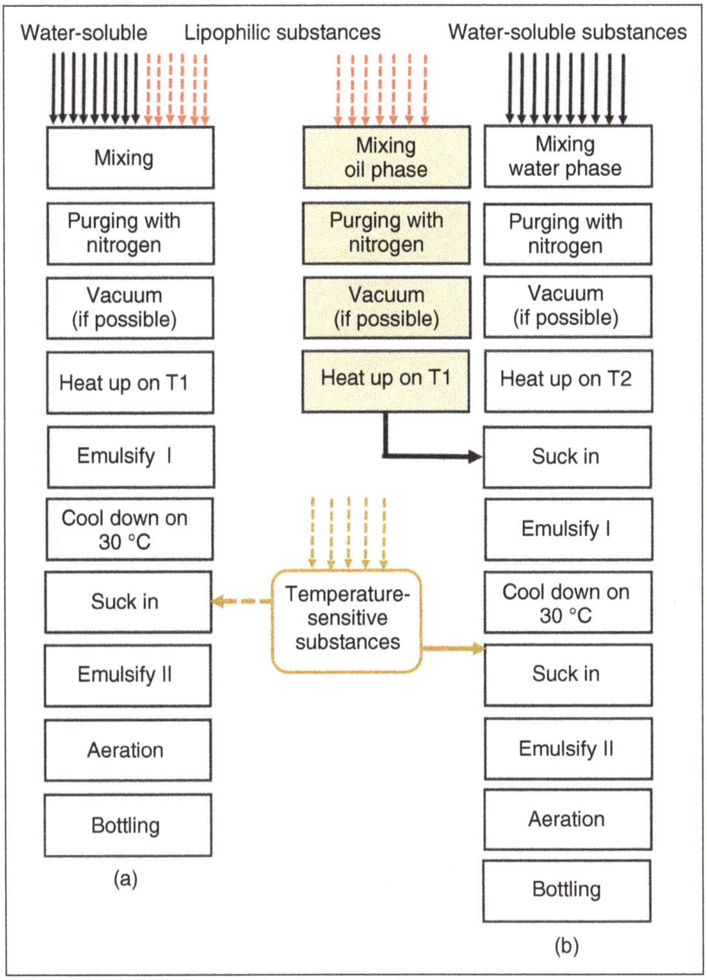

Figure 10.1 Sequence of the process steps in the cream production according to the "one-pot" method (a) and the usual separation of oil and water phase before emulsification (b).

Often, the preparation method is adapted to the existing equipment. The simplest method is the "one-pot" variant, in which all steps are done in the emulsifying vessel. First of all, the water-soluble substances are sucked in as liquid, solution, and suspension. After reaching the necessary oil-phase temperature, the slow addition of the lipophilic ingredients takes place into the emulsifying tool. After the emulsifying of the complete content, the cooling of the emulsion begins.

10.2 Stirring and Homogenizing Tools

The most important element of the cream production line is the emulsifying vessel, which is normally operated in vacuum. Figure 10.2 represents a typical stirred tank with homogenizer. Made of polished stainless steel, the

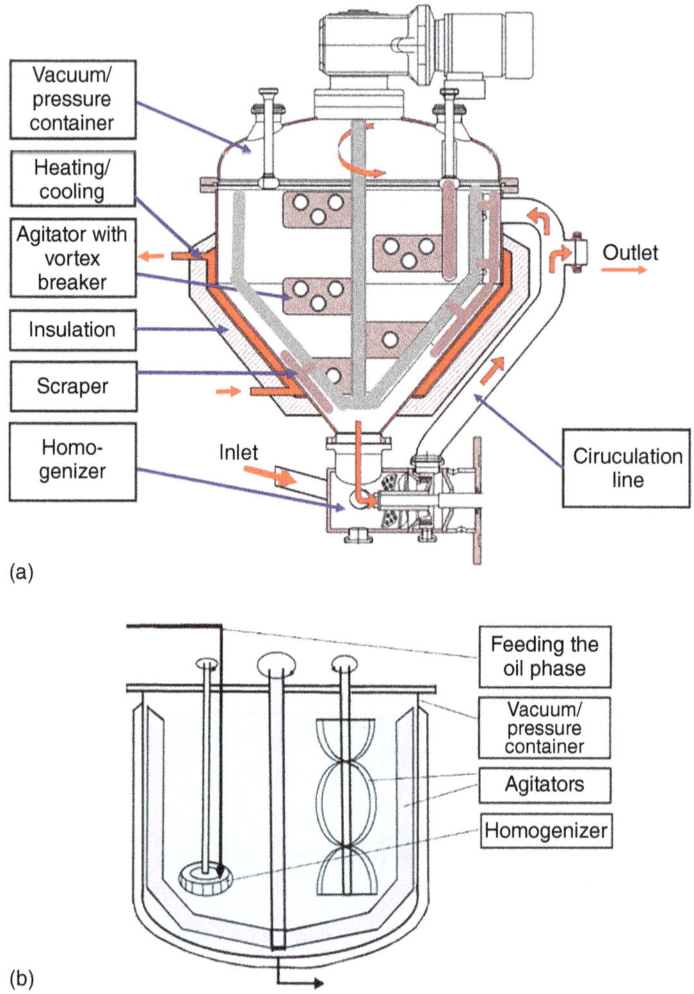

Figure 10.2 Emulsifying container with anchor agitator and homogenizer. (a) Modern equipment, similar emulsifiers use almost all manufacturers. Source: Courtesy of k-process Kaltenbach [14]. (b) An alternative in a simple design for the "one pot method."

temperature-controlled tank is equipped with an agitator close to the wall. Scrapers are needed to completely remove deposits on the container wall that can be formed through viscous emulsions. The constant renewal of the boundary layer improves the heat transfer and shortens the time of heating and cooling. Furthermore, the removal of product residues on the wall prevents local overheating, which leads to quality losses. In order to increase the mixing in low-viscosity phases, baffles are required. In some systems, it is possible to open the lid hydraulically so that the entire stirrer is freely accessible for cleaning the emulsifier container.

Most plant manufacturers offer an anchor stirrer as agitator with additional stirring elements, which ensure mixing even at high viscosities. An example of this embodiment is shown in Figure 10.3a and recommended for tanks

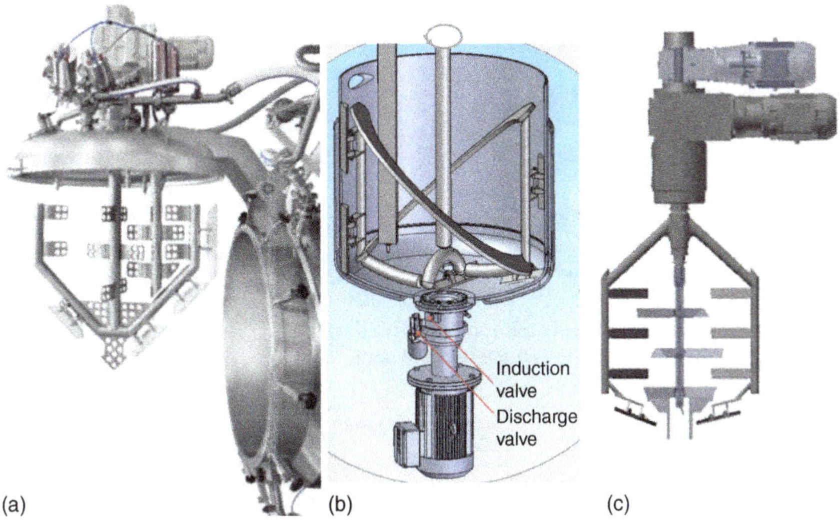

Figure 10.3 Stirrers with scrapers for the production of cosmetic creams. (a) Typical anchor stirrer. Source: Courtesy of IKA.com [13]. (b) Paravisc agitator. Source: Courtesy of EKATO SYSTEMS [10]. (c) Coaxial agitator. Source: Courtesy of Symex [18].

tapering downward. Cylindrical tanks can be operated with a special top-down agitator, as displayed in Figure 10.3b. For difficult stirring tasks, in particular, for high-viscosity emulsions/pastes, the coaxial agitator with two counter-rotating stirrers is suitable (Figure 10.3c). The one with the scrapers cleans the wall and the other conveys the liquid mass horizontally and vertically. Short mixing times and low local temperature differences characterize this complex stirring arrangement. As the others, the counterrotating system operates with infinitely variable ranges of speeds.

In addition to the stirrer, in most cases, the flanged-down disperser supports mixing intensity. Some of these machines can be switched from homogenization to pumping (transport). However, in any case, the homogenizer promotes mixing because these machines also possess, in addition to the dispersing effect, a conveying effect so that, driven by the homogenizer, the liquid can flow back via the circulation line. For a fixed installed power, it is valid that the lower the throughput and the higher the viscosity of the continuous phase, the more energy dissipates by the gear rims of the homogenizer, lowering the droplet sizes and the danger of coalescence. The stirring arrangements for high viscosities in conjunction with commercially available homogenizers normally meet the requirements for the manufacture of skin creams or lotions with viscosities of about 5 ± 3 Pa s, measured at room temperature.

The main task of the disperser, however, is to produce an emulsion from the water and the oil phase, sometimes in the presence of solids. This is done by the common transport of the initially separated liquids through a rotor/stator system, in which the two phases are mixed intensively with each other. Simultaneously, the resulting oil drops are crushed through the coaxially interlocking rings. The introduction of energy and the action of the emulsifiers produce droplets in the size between 0.8 and 10 μm (see Figure 5.6). The process is

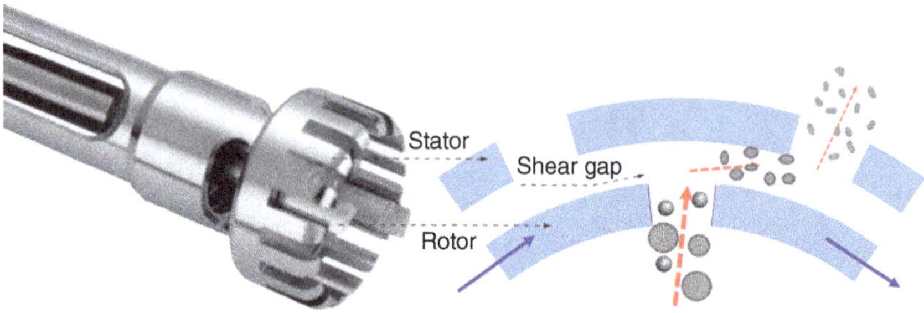

Figure 10.4 Rotor/stator homogenizer tool for use in the laboratory and a schematic description of the process. Source: Courtesy of IKA.com [19].

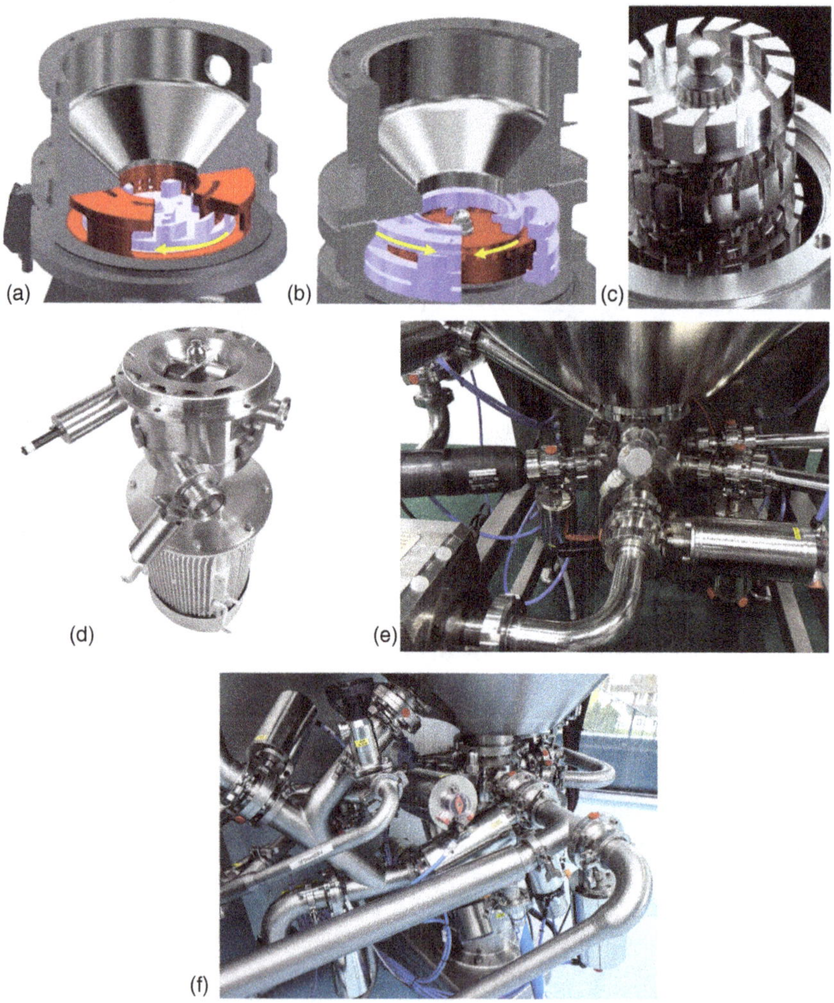

Figure 10.5 Flanged down homogenizer. (a) Rotor/stator principle. Source: Courtesy of Symex. (b) Cotwister, counterrotating, slotted tools. Source: Courtesy of Symex [18]. (c) Multistage rotor/stator machine. Source: Courtesy of IKA.com [13]. (d) Example for the design of a vertical homogenizer. Source: Courtesy of EKATO SYSTEMS [10]. (e) Installed horizontal homogenizer. Source: Courtesy of k-process Kaltenbach [14]. (f) Installed vertical homogenizer with several feed and circulation lines. Source: Courtesy of FrymaKoruma [11].

illustrated schematically for a laboratory homogenizer shown in Figure 10.4. A proven disperser for the laboratory, suitable for typical batch volumes up to 2 l, is also shown in the figure. Check the droplet size distribution from the laboratory, via pilot plant to production. As the specific energy input in the laboratory is by far the highest, the emulsification times in production must be adjusted (see Section 10.6).

The technical design of dispersers (homogenizers) can be seen in Figure 10.5. In most cases, they are rotor/stator machines in different designs, rarely rotor/rotor machines. The peripheral speeds are often in the range of 3 to about 35 m/s, special designs to up 100 m/s. Many machines can be switched so that they only pump. The homogenizers of the rotor/stator machines are characterized by at least two toothed rings, which can be designed differently and have only a small distance of 0.5 mm. Differences between the various manufacturers are in the gaps of the teeth and in the multiple radial or axial arrangement of the rings as well as in the flow direction (Figure 10.6). Rotor/rotor machines represent

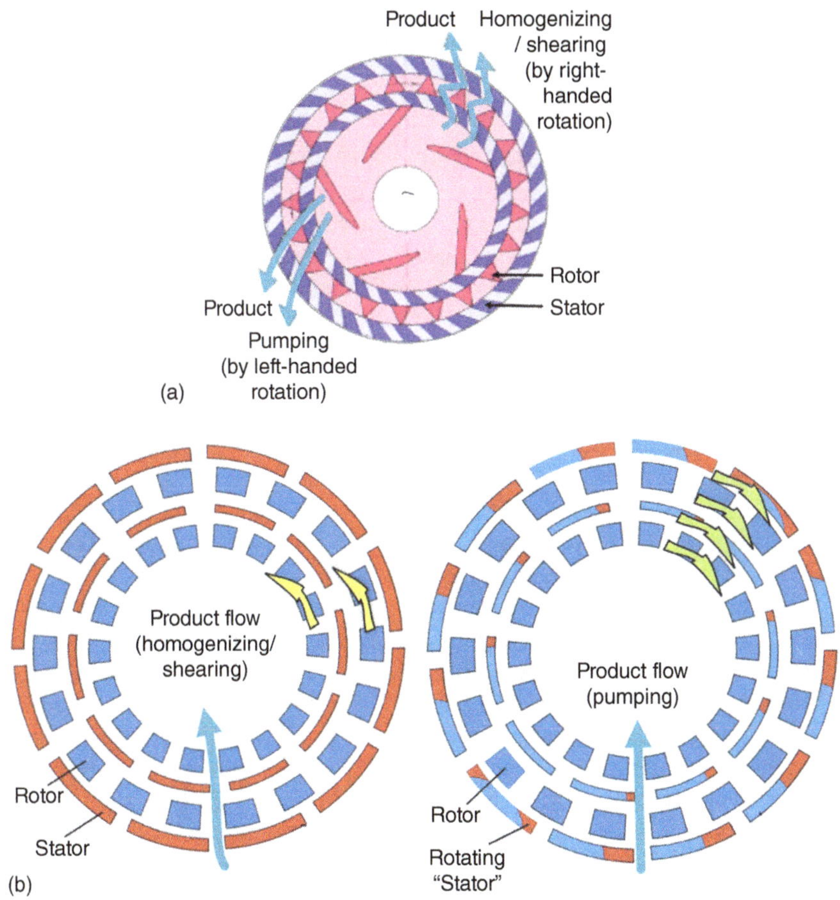

Figure 10.6 Mechanism of homogenizing in rotor/stator machines. (a) By changing the direction of the rotation in the pump mode switchable. Source: Courtesy of Becomix [9]. (b) By rotating the "stator" in the same direction. Source: Courtesy of Hebold Systems [12].

homogenizers in which the “stator” also rotates, switchable in the same or in opposite direction.

Depending on the design, the emulsification and the liquid transport ensue usually coupled in these machines. In many cases, it makes sense to separate the homogenization from the conveying of the cream. For this, a metering pump should follow the homogenizer (rotor/stator or rotor/rotor). Through the pump, the flow rate can be influenced in broad ranges. A reduction in the flow rate compared to the conveying of the homogenizer leads to an increase in the residence time in the homogenizer with positive influences on the droplet sizes. In addition, the pump ensures reproducible conditions with changing formulations.

Because of the extreme energy input of the counter-rotating rotor/rotor homogenizers, the production of nanoemulsions is possible. As the others, these homogenizers can be installed horizontally (motor laterally) or vertically with an underlying electric motor to save space. In the latter case, three circulation lines can be realized. The first leads back to the homogenizer, the second to the lower part of the stirred tank, and the third to the upper part. If the recipe contains solids, the disperser can be converted into a crusher, depending on the particle size of the solids present. This is not necessary for the normal cream production.

10.3 Laboratory Equipment and Pilot Plant

For a first lab test, an apparatus is rapidly constructed, consisting of two vessels (e.g. 500 ml and 1 l beakers) with stirrers, two heating plates with temperature control and display, a nitrogen gas bottle with tubes, and a dispersing tool and a stopwatch to control the emulsification time. Many formulas do not foam strongly, so that the vacuum can be dispensed within the first attempts. The simplest and cheapest method for preparing an emulsion is the “one-pot” method. In a glass vessel, the water-soluble and oil-soluble ingredients are added sequentially in the order described. With continued stirring, the content warms up under nitrogen to the temperature required for the oil phase (here: 60 °C). After reaching the temperature, the emulsion is produced with the disperser tool (rod, Figure 10.7a). However, the batch two-pot approach with a separate oil-phase preparation is safer, more flexible, and therefore preferable.

A better but more expensive alternative is to buy finished laboratory equipment made of stainless steel and glass (Figure 10.7b). In this apparatus, the input of the raw material via vacuum can be optimized. For emulsification, the rotor/stator tool (rod) is inserted vacuum-tight from above into the vessel. In the production plants, today, the addition of the oil takes place directly into the flanged homogenizer, a method that is not adjustable with this laboratory equipment. However, the achievable temperatures of the liquid, the mixing tool, and the vacuum generation correspond to the conditions of a production plant. An even better approach to production conditions is provided by the polished stainless-steel laboratory equipment, characterized by an anchor stirrer with scrapers (polytetrafluoroethylene, PTFE) and especially by a flanged down homogenizer, in a basic or professional design (Figure 10.8). These systems allow to understand

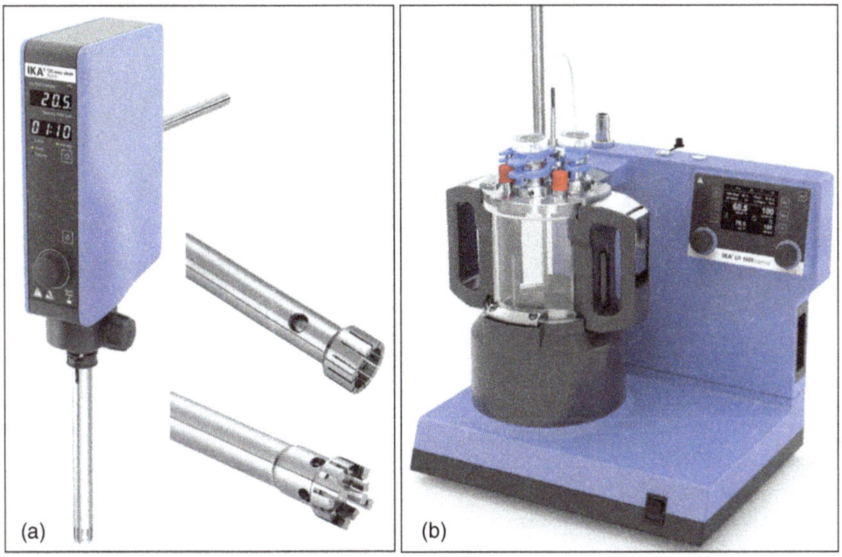

Figure 10.7 Effective laboratory disperser for the preparation of emulsions (a; Ultra-turrax®; power input: 0.5 kW, stepless speed control from 3.000 to 25.000 rpm for working volumes of 1–2000 ml) and complete laboratory equipment (1 l; b) including vacuum, anchor stirrer with PTFE scrapers, and adjustable dispersing tool for first experiences. Source: Courtesy of IKA.com [19].

the homogenization and mixing effects of the disperser by conveying the cream through the various recirculation lines and identify optimal conditions. They are particularly suitable for optimizing formulation and production conditions.

The scale-up as well as sample and small-scale production requires pilot plants with 15, 25, or 50 l of usable working volume (Figure 10.9). Most pilot plants can be automated and represent small production plants in all functions. In the stirred tank made of polished stainless steel, the water phase is usually generated and heated in vacuo. It is an advantage to use a separate, temperature-controlled, stirred vessel for the preparation of the oil phase. Before sucking the finished oil phase into the elements of the dispersing tool, the emulsification is started by switching on the disperser. In this way, it is ensured that the entire quantity of oil passes at least once through the homogenizer. Pilot plants are equipped with described disperser tools working at the bottom of the vessel, with the electric drive underneath. The disperser tools pump the emulsion back into the stirred tank through circulation lines. An estimation of the necessary dispersion time ensues according to the volume of the emulsion and the pumping capacity. The entire content should pass through the homogenizer at least three, preferably five, times. To generate mini-emulsions, the plant must be expanded by a high-pressure homogenizer, which can be integrated in the circulation pipe behind the existing disperser (Section 10.7).

The emulsification and filling pilot plants should be located in a room equipped according to the Good Manufacturing Practice (GMP) regulations for cosmetics. As experiments in the pilot plant can be understood as small-scale productions,

Figure 10.8 Laboratory plants with anchor stirrer and homogenizer flanged down. (a) Modular system. (b) Possible variants for recirculation, pumping, and homogenizing mode. Source: Courtesy of IKA.com [13]. (c) Complete equipment for laboratory experiments with high viscous emulsions (coaxial agitator). Source: Courtesy of Symex [18]. (d) Laboratory plant. Source: Courtesy of Netzsch Vakumix [17].

the regulations also apply to the personnel and the documentation. On the one hand, the recipe as well as the manufacturing and quality specifications can be checked in the pilot plant, but above all, however, a sufficient amount can be made to fill one hundred to five hundred 100 ml bottles. Therefore, the emulsification should be followed by a small bottling plant. Bottles with a cream quality

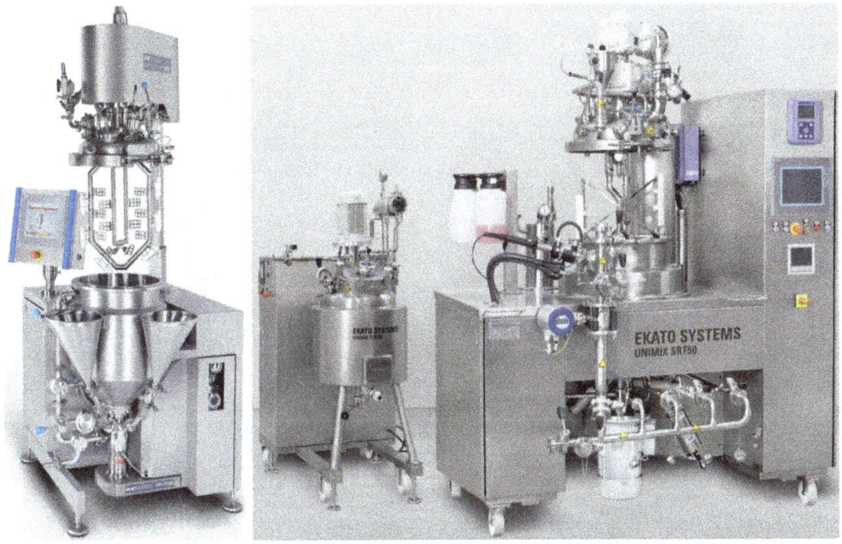

(a) (b)

Figure 10.9 Pilot plants with open lid for determining the production conditions of macroemulsions and for small-scale production. Source: Courtesy of IKA.com [13] and EKATO SYSTEMS [10].

comparable to the production are needed to perform required compatibility and consumer testing.

10.4 Batch Production

In the case of small plants (80–1000 l), the addition of hydrophilic substances, mostly liquids, is made by weight or volume from containers. Balanced solids are sucked in via a hopper or previously dissolved/suspended in a portion of the water and added as a concentrate. The mixture, the heating, and the final dissolution of the solids take place in the stirred emulsifier container. The oil phase is prepared separately in vacuo or in nitrogen atmosphere, brought to temperature, and sucked into the stirred tank (Figure 10.10). Later, upon completion of the emulsification and cooling, the temperature-sensitive ingredients are sucked in via the hopper and emulsified into the cream/lotion.

In contrast, large production plants (1–10 m^3) have premix containers in which the water and oil phases are prepared separately and brought to different temperatures. However, using stirred premix containers, after metering the liquids, the water and oil phases can already be prepared for the next batch, saving time and increase flexibility. First, only the water phase is conveyed into the emulsifier. For this purpose, a pump is available, or the vacuum is used for sucking. Subsequently, the addition of the oil phase takes place directly into the disperser tools, as described in Figure 10.11. Alternatively, the two phases can be metered into the homogenizer at the same time. This effective and time-saving procedure is preferably used.

Figure 10.10 Medium-sized production plants. (a) Schematic flow sheet of a production. Source: Courtesy of IKA.com [13], image modified; small image: Courtesy of Netzsch Vakumix [17]. (b–d) Technical designs. Source: (b) Courtesy of FrymaKoruma [11]. (c) Courtesy of k-process Kaltenbach [14]. (d) Courtesy of Symex [18].

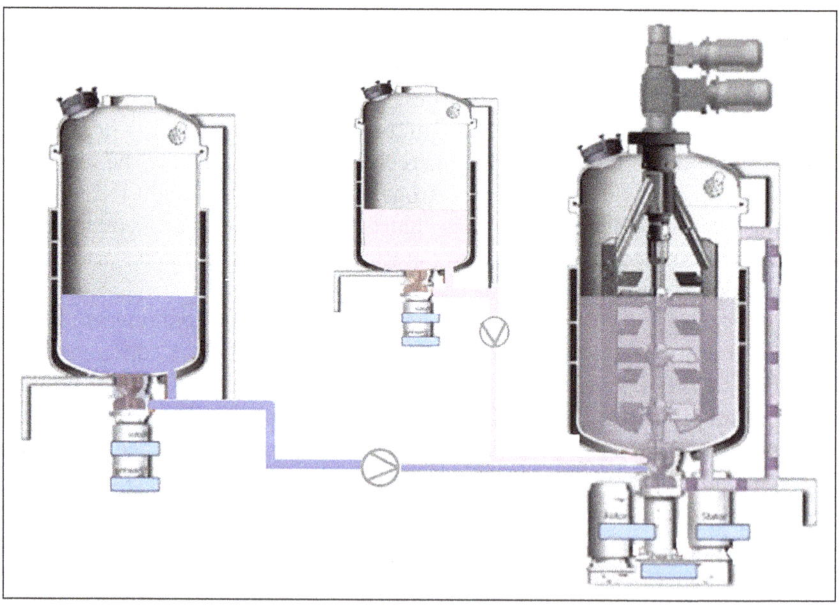

Figure 10.11 Tempered premix containers with circulation pumps for supplying the emulsifier tank. Source: Courtesy of Symex [18], modified image.

The premix tanks contain a device for mixing the content by pumping over via a circulation line or an agitator, are heatable, and can be operated under vacuum. When working with modular formulations (Section 5.2 and Chapter 8), several production lines can be fed economically from two large premix tanks (Figure 10.12). The final formulations are adjusted via further premix containers. After completion of the cream, the batch is transferred to a buffer tank (reservoir) by the homogenizer pump, so that no further pump is needed. The buffer tank serves as a supply for bottling, which takes place after the quality control of the cream. In this way, the stirred emulsifier tank can be used immediately for the next batch.

Emulsifying plants are offered by many manufacturers. The main differences lie in the stirring elements. Usually, a central anchor or a similar stirrer with scrapers is used in the stirred tank. In addition, there may be second stirrer. This also centrally arranged stirrer works in opposite direction (counterclockwise). In another arrangement, a small stirrer with a diameter of less than half the radius is used in the space between shaft and arms of the anchor agitator. The stirred tank variants also differ in the construction of the homogenizer. On the one hand, the disperser similar to the laboratory tool works from above via a long drive shaft in the liquid. This arrangement is rarely sold today. On the other hand, in the prior art, a large rotor–stator disperser flanged to the stirred tank outlet emulsifies the two phases and transports them back via a circulation line. This method is best suited for emulsifying two phases in the preparation of creams because the oil phase or both phases can be fed directly into the intake of the homogenizer.

Figure 10.12 Technical design of a plant for the production of cosmetic creams; on the right, the emulsifier tank, in the middle and on the left the two premix tanks; in addition, the figure shows the measures on the floor, walls, and ceiling for hygienic production. Source: Courtesy of FrymaKoruma [11].

(a) (b) (c)

Figure 10.13 (a) 1250 l, (b) 5000 l, and (c) 10 000 l vacuum process tanks for cream production with homogenizer, flanged down ("UNIMIX"). Source: Courtesy of EKATO SYSTEMS [10].

There are many machine and plant manufacturers who supply, for example, a suitable reactor and/or agitator for emulsifying and mixing of low to very viscous liquids. The customer has the choice to either buy complete systems or he can assemble the plant himself from purchased containers, stirrers, and homogenizers. All parts can be mounted in his own or someone's workshop. The customer must then install the piping, valves, and vacuum pumps as well as the electrical installation and ICA (instrumentation, control, and automation). In doing so, many legal guidelines must be taken into account (Chapter 11). Only large companies have the necessary specialists and are able to do so. This complicated way can be cheaper. The better way is to buy a complete system from an experienced company. Some well-known companies offer turnkey production plants with complete ICA, on request, even with scaffolding, premix, and storage tanks. The customer then only has to combine the energies, raw material supply, and the bottling plants.

Fully automated complete batch plants for the cream production under vacuum are available up to 10 m^3 (Figure 10.13) with equipment for sucking in the ingredients and/or for preparing the premixes as well as for cleaning in place (CIP). Most offered complete installations are characterized by special features:

- Design for pharmaceutical and cosmetic applications free of any dead spaces.
- Stainless-steel equipment with polished or electro polished surfaces on all parts in contact with the product.
- Homogenizer with internal and external circulation (via circulation line) and pump mode.

- Anchor stirrer or other near-wall agitators with scrapers guarantee optimum heating and cooling times and effective horizontal and vertical product mixing, even with high viscosity emulsions.
- Selective addition of raw materials for the optimum process design, short production, and cleaning times.
- CIP equipment.
- GMP (for cosmetics).

The turnkey, compactly built production plants are usually offered with or without raw material supply and storage tank for bottling (Table 10.2). As most creams are solid-free, an effective machine for dispersing is sufficient. Additional special equipment for reducing the particle sizes provides no benefits.

Table 10.2 Manufacturer of turnkey production plants in Germany.

No	References	Company	Laboratory equipment (l)	Pilot plant (l)	Batch production	Mini (nano) emulsion	Continuous production
1	[9]	A. Berents, Becomix, various systems	2.5–30	Up to 125	Up to 10 000 l	Intensive shearing	Yes (no information available)
2	[10]	Ekato Systems, *UNIMIX*	3–6	15–70	100–10 000 l	Yes hph[a)]	—
3	[11]	FrymaKoruma (Swiss/German) *Dinex*	6–12	8–60	Up to 4 000 l	—	—
4	[12]	Hebold Systems, *Hebomix*	5–10	15–150	Up to 10 000 l	—	Yes
5	[13]	IKA, various systems	0.5–2	10–50	Up to 4 000 l	Small hph[a)]	—
6	[14]	Kaltenbach k-process, *K-DisHo*	6–10	10–50	Up to 3 000 l	Intensive shearing	—
7	[15]	Lewa	—	—	—	Possible	Yes (no information available)
8	[16]	Lödige Process Technology	5–20	50	>1 000–10 000 l	—	—
9	[17]	Netzsch Vakumix, *KAPPAVITA*	2–10	15–65	115–10 000 l; premix tank up to 10 000 l	hph[a)] and crr[b)]	Yes
10	[18]	Symex, various systems	4–10	30–125	Up to 20 000 l	crr[b)]	Yes, up to 16 t/h

a) hph = high-pressure homogenizer

b) crr = homogenizer with counterrotating rotors (Netzsch Vakumix: ultrashear-TWIN homogenizer; Symex: cotwister).

10.5 Continuous Production

For frequent production of large batches, in particular for the same or similar composition (for example, using the module method of Chapter 5), continuous operation should be considered. There are many arguments in favor of a continuous production plant. The main advantages are

- investment costs are well below batch systems
- higher throughputs possible
- no vacuum needed, no contact to air within the production plant
- lower personnel costs by fully automated production without the help of personnel
- reduction of the manufacturing costs
- simplified supply of the raw materials
- easy adjustment (up to 30 ingredients) for similar recipes
- low-temperature load of the ingredients because of the short residence time
- quality adjustment and control via online measurements.

The strongest arguments against continuous operations are the high expense of greater recipe changes and the lack of flexibility in small orders for different products. The effort is worthwhile only if the plant can be utilized with large product quantities. To increase the flexibility, a batch system should be available in addition to the continuous plant. Furthermore, in continuous processes, the introduction of solids is more difficult, which can be done, for example, via the dynamic mixer. Alternatives are dissolution or suspension in a liquid.

Starting up the system takes only a few seconds (about 20 seconds). The systems are tailored to the recipes and can be designed flexibly via further dosing points. For simplification, the required temperature of the oil phase is usually used also for the water phase. The production runs at normal pressure without contact with the air. It is advisable to overlay the tanks of oxidation-sensitive substances with nitrogen.

A continuous system consists of storage containers for feeding the system, usually one for each substance, dosing pumps, flow meters, static and dynamic mixers, and heat exchangers. A simplified flow chart is shown in Figure 10.14 and elements of a small production plant in Figure 10.15. Emulsification takes place in a dynamic mixer, optionally with recirculation, so that the liquid passes through several times.

10.6 Scale-up

The droplet size distribution in a macroemulsion for cosmetic creams should be in the range of 0.8–10 μm with a narrow distribution. The width of the distribution is a measure of their stability, the closer the more stable are emulsions. Which droplet size distribution is achievable during emulsification depends on several factors. The recipe plays a role because on the one hand the emulsifier is responsible for the surface tension between oil and water phase, and on the other

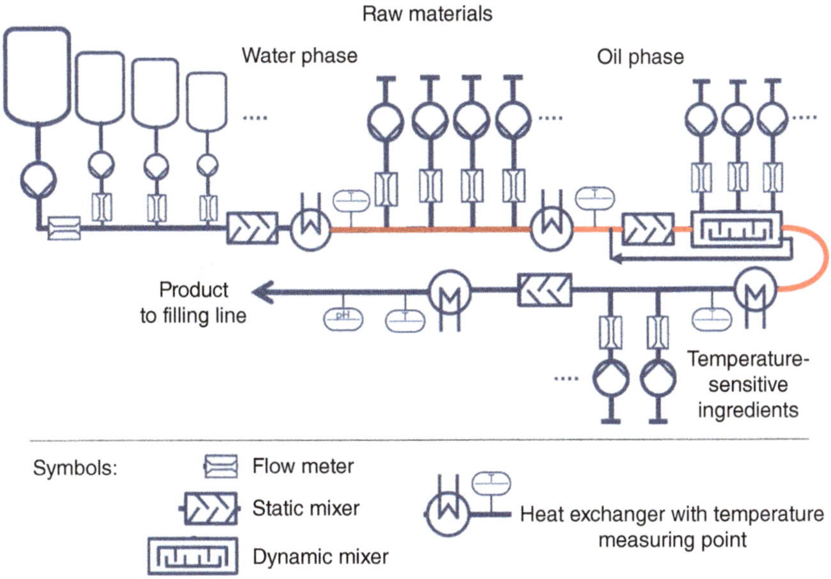

Figure 10.14 Concept of a continuous plant. Source: Courtesy of Symex [18], image modified.

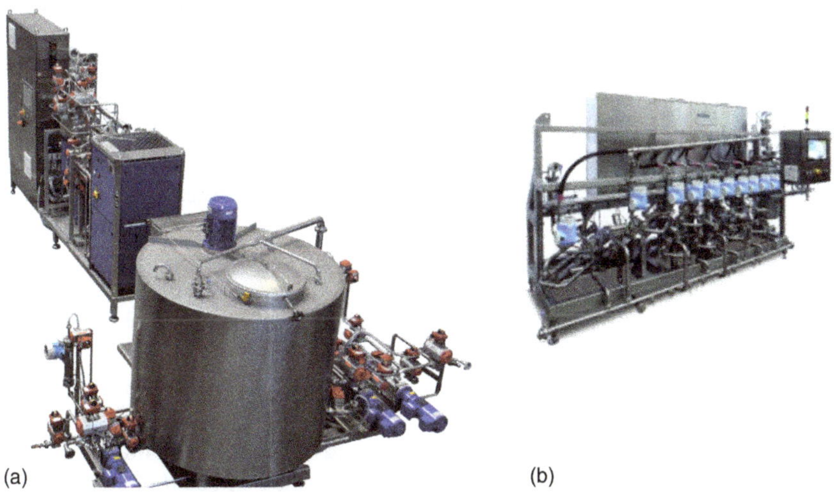

Figure 10.15 (a, b) Elements of a small continuous production plant. Source: (a) Courtesy of Hebold Systems [12]. (b) Courtesy of Netzsch Vakumix [17].

hand, the viscosity-generating substances determine the viscosity in the emulsion and later in the product. The density influences the droplet size in turbulent flow, which is always present in the homogenizer.

In some cases, a pre-emulsion is first generated in the stirred tank via the stirring elements before the disperser starts. The drop diameter d_{32} achieved in the stirred tank depends directly on the volume-specific power input P/V of the stirrer. d_{32} is the Sauter diameter, which results from the drop volume V_p divided by its surface A:

$$d_{32} = 6\, V_p/A \tag{10.1}$$

$$d_{32} = C\,(P/V)^{-b} \tag{10.2}$$

The exponent b depends on the viscosity, which is for low-viscosity liquids at 0.4 and 0.25 for high viscosities of the disperse phase [1]. However, for the achievable droplet diameter, the energy input of the disperser and not of the pre-emulsification is crucial.

In the homogenizer, constantly changing turbulent flows create circulating vortexes that influence the behavior of the drops. In the first case, in the turbulent viscous mechanism [2], for droplets with diameters smaller than the vortices, the shear forces determine their size. The largest, still stable droplet size can be estimated via the surface tension σ, the volume-related power density P/V, and the viscosity η of the continuous phase:

$$d_{\max} \sim \sigma\,(P/V)^{-0.5}\,\eta^{-0.5} \tag{10.3}$$

According to Eq. (10.3), the diameter decreases with increasing energy input and viscosity. In the second, practice-relevant case for drops with diameters in the range or above of the vortex sizes, inertial forces generated by pressure fluctuations act on the drops. These forces that interact with the drops cause deformation and finally the breaking up into small droplets. The dependence of the achievable diameter describes Eq. (10.4) [1, 3].

In almost all processes, the oil phase brought to temperature is dosed directly into the homogenizer and emulsified. This procedure has the great advantage that all portions of the oil must be completely through the homogenizer and represents the common method today. The temperature-controlled water phase in the stirred tank flow into the homogenizer or is metered into the homogenizer at the same time as the oil phase. At emulsifying temperature, the emulsion usually has a low viscosity. As highly turbulent flow conditions prevail in the shear-intensive machine, the following equation applies to drop size under comminution conditions:

$$d \sim \sigma^{3/5}\,(P/V)^{-2/5}\,\rho^{-1/5} \tag{10.4}$$

Low surface tensions σ and high densities ρ favor the desired small droplets. However, these parameters are fixed for the given recipe and cannot be changed. High power input leads to small drops. This process can be supported by corresponding emulsifier mixtures with two or more hydrophilic/lipophilic balance (HLB) values. Decisive is the influenceable specific installed power, which must be considered as a benchmark in scale-up. In the case of stirred tanks with homogenizers and recirculation lines, the necessary installed power of laboratory and pilot emulsifiers can be upscaled to the production plants or from small to large plants and vice versa, as shown in Eq. (10.5). (This relationship is also considered in the prospectus information of FrymaKoruma.)

$$\text{Scale-up rule 1: } P_1 = P_0\,(V_0/V_1)^{-2/5} \tag{10.5}$$

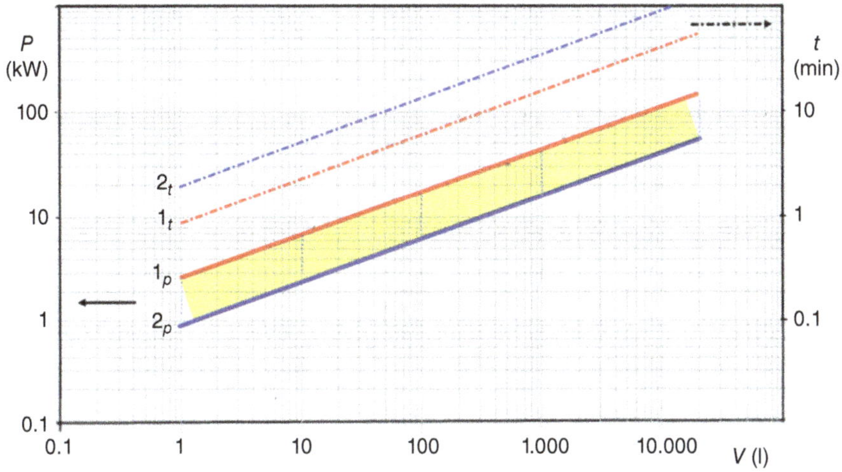

Figure 10.16 Scale-up diagram: Dependence of the installed power of the homogenizer from the usable volume of the stirred tank; right ordinate shows the required emulsification time for small droplet sizes in macroemulsions for the two illustrated power lines (indices: *P* power, *t* emulsifying time); basis is formed by some values from practice.

An emulsifying tank with the volume V_0 and the installed power P_0 serves as a standard. For example, the power at $V_0 = 1\,\text{m}^3$ ranges from 15 (lower limit) to 45 kW (upper limit) according to the manufacturer's specifications. In this area, most designs of the homogenizer for cosmetic creams are likely to lie, as demonstrated in Figure 10.16.

If a narrow residence time distribution and a small drop diameter between 0.8 and 10 μm are required, the emulsion volume must pass several times the homogenizer. This is described by the batch turnover number BTO. The number indicates how many times the usable volume *V* flows through the homogenizer during the emulsifying time *t*. The value of the flow rate has the manufacturer. BTO results by multiplying with the necessary time and divided by the usable volume of the stirred tank.

$$\text{Scale-up rule 2: BTO} = \dot{V}t/V \tag{10.6}$$

From this equation, the emulsification time can be derived, shown in Eq. (10.7). According to experience, the value should be at least 3, if possible at 5. The lower the installed power at constant volume, the greater the BTO number must be. In the example, at least fivefold circulation results for 1 m^3 volume and min. 18 kW power, whereas for 45 kW and more, the triple circulation should be sufficient. The higher energy input causes smaller, faster fluctuating vortices, which have a crushing effect. Therefore, a minimum value may not be fallen below, exceeding a maximum value brings no additional positive effect. BTO numbers less than 3 are not recommended.

$$t(\text{min}) = \text{BTO}\ V/\dot{V} = 5\,\text{V}\,(\text{m}^3)/\dot{V}\,(\text{m}^3/\text{h})\ 60\,\text{min/h} \tag{10.7}$$

The quasi-continuous emulsification with homogenizers does not determine the power density the drop sizes [4], but the energy density. The volume-related energy density E_V represents the product of power density $P_V = P/V$ and residence time (emulsification time t):

$$E_V = E/V = P/\dot{V} = P_V t \tag{10.8}$$

The specific energy input to be calculated via the installed power is usually min. 10 kWh/m^3 (= 3.6×10^7 J/m^3) for cosmetic creams. It is not a fixed value but depends on the effect of the homogenizer, temperature, and recipe. As long as no measured value is present, this number may be used for estimation in systems according to Figure 10.2a. With this information, the minimum emulsification times can be calculated as a function of the stirred tank volume (see Figure 10.16). It should be noted that the BTO number 3 is not undershot; otherwise, use the longer time of Eq. (10.7).

$$\text{Scale-up rule 3: } t = E_V\, V/P \tag{10.9}$$

For explanation, the estimate gives an emulsification time of 24 minutes for a 1 m^3 tank with an installed power of 25 kW and an energy input of 10 kWh/m^3. The verification of the minimum emulsification time can be carried out by measuring the droplet size distribution. As shown in Figure 10.16, large container volumes require long emulsifying times and small volumes used in the laboratory very short times (less than two minutes). In the laboratory, emulsifiers work with installed powers well below the calculated value. However, the times must be increased accordingly (optional up to three minutes).

The scale-up rules 2 and 3 are independent. At a constant power of the homogenizer, the flow through the circulation line can be reduced and/or the number of passes can be adjusted appropriately. The easiest way to determine the optimal conditions is the use of a pilot plant with geometric similarity and with machinery that can be extrapolated to the production. In production, the calculated minimum emulsification time should be increased by 20% as security surcharge on the scale-up calculation. However, further exceeding the emulsification time leads to additional thermal stress, without quality advantages (recipe-dependent). For some ingredients, the shear forces of the homogenizer should be considered.

Depending on the operation mode, feeding and mixing, emulsifying 1 and 2, heating and cooling times as well as the emptying and quality control must be considered for estimating the batch time. This is followed by cleaning, disinfecting, and drying. The calculated emulsification time 1, multiplied by the factor 1.2, gives the duration of homogenization to be used in practice. Furthermore, the emulsification time 2, which is required to incorporate the temperature-sensitive ingredients at lower temperatures, must be taken into account. As there are no literature values, and we did not carry out systematic investigations, the value must be estimated. It should be above 50% of the emulsification time 1. Depending on the recipe, the possible specific power input can be important for emulsification (proved by measurement). For the plant design, it is advisable to install a motor with a power that is the upper range (see Figure 10.16) and on the other hand limit the quantity delivered in the circuit via a metering pump.

10.7 Mini-Emulsions

A mini-emulsion, also called a nanoemulsion, has droplet sizes from 50 to 500 nm, preferably between 50 and 200 nm. The preparation is generally carried out by reducing the droplet sizes of a macroemulsion (pre-emulsion). In order to stabilize the nanodroplets, a hydrophilic and a hydrophobic emulsifier must be present in an adjusted amount (Chapter 5). Nanosized emulsions show a higher stability compared to macroemulsions. The technique knows many ways to reduce droplet sizes. Common to all is the input of much additional specific energy, about an order of magnitude more than in the production of a macroemulsion. In practice, this can also mean that approximately five times more specific energy E_V and twice more time are required. There are many alternatives for the production of nanodroplets [3–5]. In practice, only three variants have prevailed: rotor/rotor and special rotor/stator systems as well as high-pressure machines, used in pharmaceutical and cosmetic production. Additionally, the laboratory still uses ultrasound equipment. However, the use of ultrasound is limited to the laboratory and pilot plants. Because of the low throughputs, no technical application takes place. The development for the adjustment of nanodroplets over membranes is not finished yet.

For the production of nanosized creams, the simplest, most cost-effective, and flexible way to create nanodroplets is to use a flanged rotor/rotor machine that can be installed in place of the homogenizer. Usually, the rotors are driven with separate motors and can be operated in the same or opposite directions (Figure 10.17). In another arrangement, flow and homogenization are separated for rotor/rotor homogenization (Figure 10.18). Instead of the homogenizer,

Figure 10.17 Ultrashear-twin homogenizer with rotors that can run oppositely to each other or synchronously, driven with electrical motors on the right and left side. Source: Courtesy of Netzsch Vakumix [17].

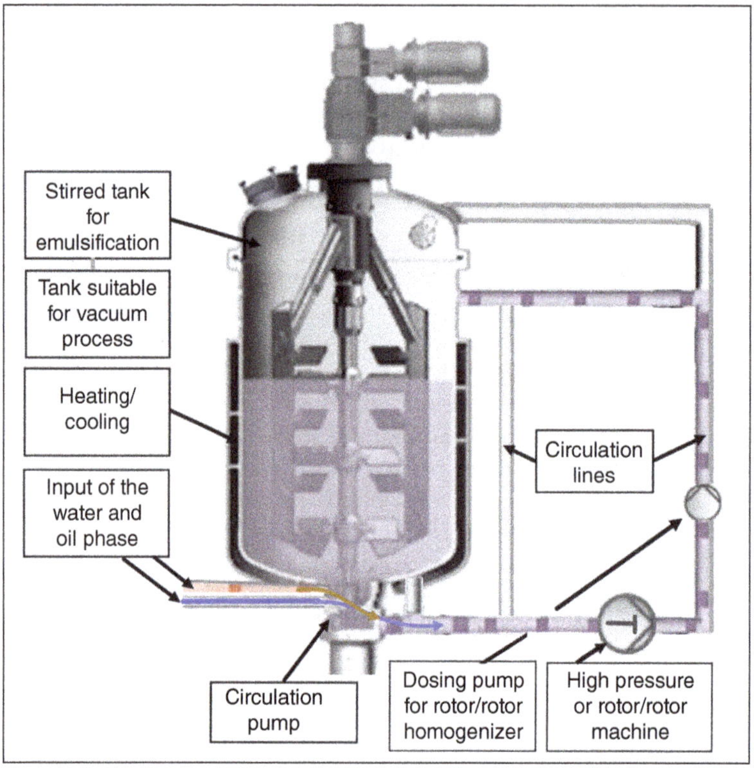

Figure 10.18 Flow sheet for the production of mini-emulsions, either with the high-pressure homogenizer alone or with a rotor/rotor machine followed by a metering pump.

a pump is flanged. In the pump, a pre-emulsion arises from the dosed water and oil input, which is conveyed into the high-pressure homogenizer or in a rotor/rotor machine. There, the nanodroplets originate. In order to obtain a narrow distribution, the emulsion must pass through the homogenizer three to five times. Downstream of the rotor/rotor machine, a metering pump controls the circulation volume and the residence time in the homogenizer. The described methods show several advantages:

- Simple dosing and preparation of the pre-emulsion in the pump or homogenizer or in the stirred tank
- Operate in the stirred tank under vacuum as before
- Subsequent transfer of the macro into the mini-emulsion by higher energy input via the high pressure or rotor/rotor machine
- The amount of specific energy input and the BTO determine the mean droplet size.
- In addition to the introduction of energy, the emulsifier system affects the droplet size distribution
- Finally, incorporation of the temperature-sensitive substances takes place analogously as in the macroemulsions at low temperatures.

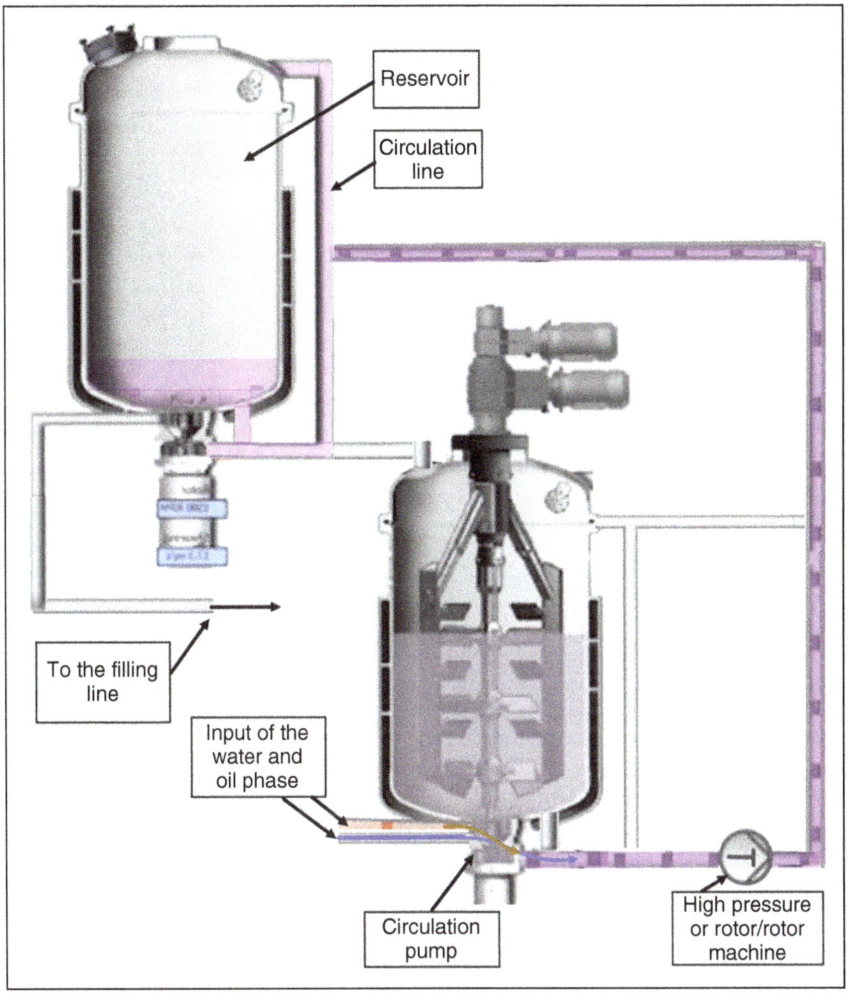

Figure 10.19 Production of mini-emulsions with narrow distribution using a high-pressure homogenizer (or rotor/rotor machine) with three to five complete passes (if necessary reservoir with agitator and scraper).

The residence time distribution causes that with each circulation some volume elements remain in the container. At a BTO number of 3, each volume element passes the high-pressure homogenizer between one to about five times, and only on average three times. If the entire content of the emulsifying container should pass exactly three or five times through the high-pressure homogenizer, two stirred containers must be used, as shown in Figure 10.19. In this proposed arrangement, the pre-emulsion is generated from the simultaneously dosed liquids in the flanged pump. Driven by the pump, this pre-emulsion flows through the high-pressure homogenizer and from there into the upper reservoir. If the reservoir is filled, the discharge takes place into the empty emulsifier tank.

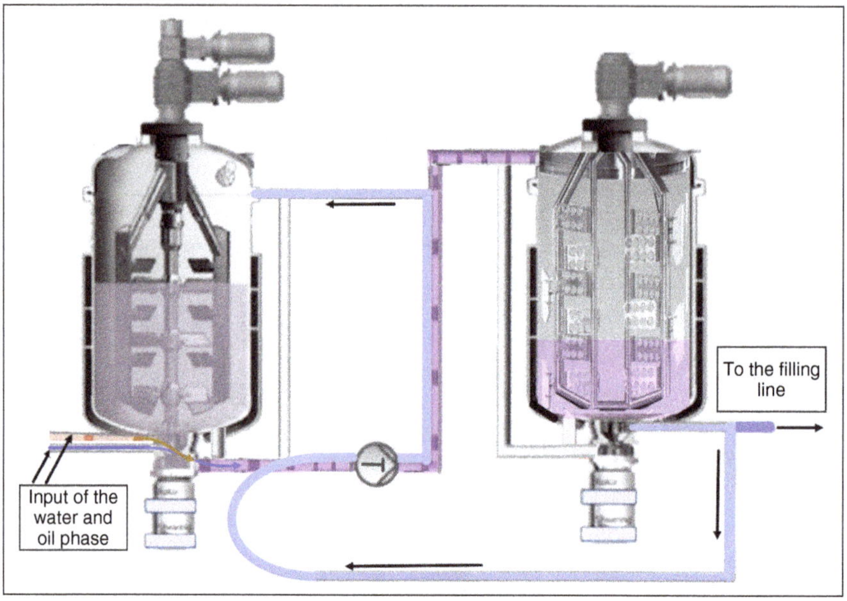

Figure 10.20 Production of mini-emulsions via two alternately working stirred tanks with a high-pressure homogenizer.

From there, the second circulation is carried out via the homogenizer to the reservoir, until it is full and must be drained, etc.

In an alternative arrangement with two stirred containers [6], the emulsion is pumped back and forth via the same pumps under the containers, so that all volume elements pass through the high-pressure machines three or five times (Figure 10.20). The first and the third run through the high-pressure homogenizer start from the emulsifying container, while in the second pass through the homogenizer, the emulsion flows back from the reservoir of the filling line to the emulsifying tank. After the third circulation, the cream is pumped from the stirred reservoir to the filling line, where bottling happens. Before the third pass, the emulsion can be cooled to about 30 °C to vacuum-feed the temperature-sensitive ingredients.

Figure 10.21 displays the drop size distributions after the first and second pass through a 600 bar homogenizer. The third pass can be seen in Figure 5.7 and proves the complete conversion of the macro- into a mini-emulsion. Owing to the high-pressure differences, the microorganisms burst [7]; the product is completely sterile in every case. After the first run, many nanodroplets arise, but numerous drops in the size range of a macro-emulsion are still there. Already in the second run, both the extremely fine droplets below 50 nm and most droplets outside the mini-range disappear. The third pass leads to a close, stable distribution with very few or no micrometer drops. On measuring the volume of the drops, few big drops show a large impact on the distribution.

No significant progress results when the pressure rises from 600 to 800 bar (Figure 10.22). In these experiments, a fresh pre-emulsion was prepared for

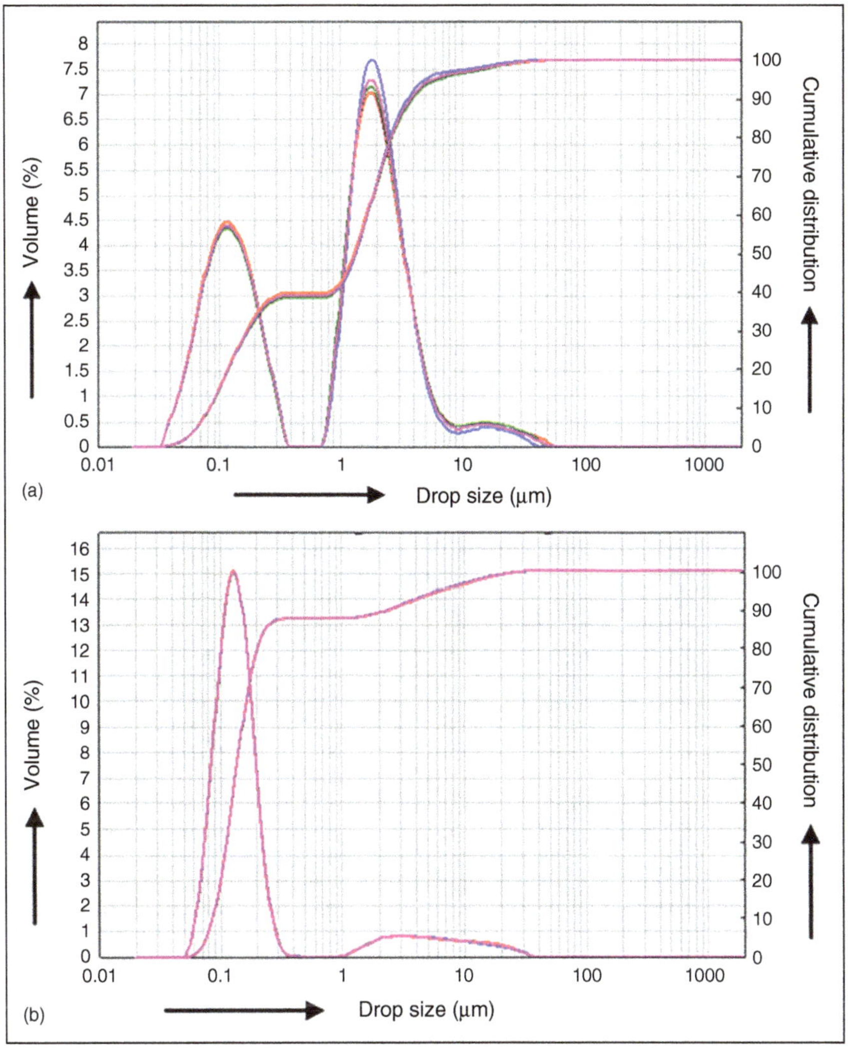

Figure 10.21 Droplet size distributions of a skin care product after (a) one and (b) two runs through a high-pressure homogenizer; triple measurement after dilution by laser diffraction/equipment: Malvern

each pressure setting. The measured values for three passes are within the error limits. Figure 10.23 shows the working principle in a valve of a high-pressure homogenizer. The pressurized (macro-) emulsion is pressed through a narrow gap. There is a sudden pressure letdown, whereby the emulsified drops break up into nanodroplets. The gap (pressure) is adjustable and regulates the throughput and pressure. An industrial high-pressure homogenizer with five pistons, used up to 1500 bar, is shown in Figure 10.24. For the considered working pressure of 600 bar, a throughput of 10 000 l/h is possible for one run, which means about 3300 l/h for the process under consideration of three passes.

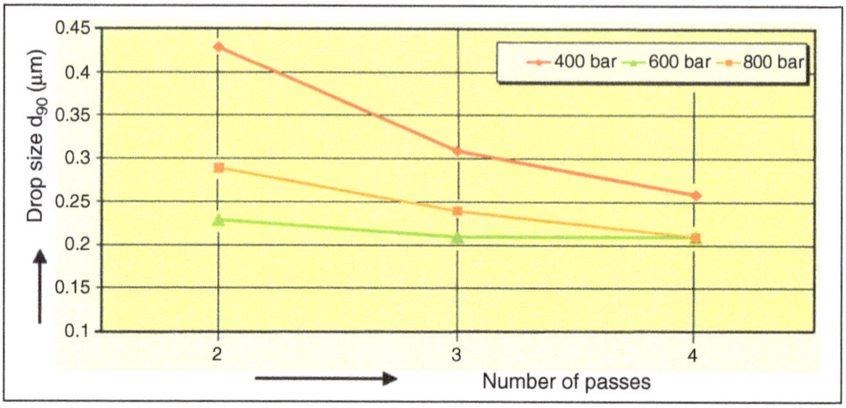

Figure 10.22 Droplet size (diameter d_{90}) as a measure of the width of the droplet size distributions after several runs through a high-pressure homogenizer (series of measurements on different days with freshly prepared macroemulsion).

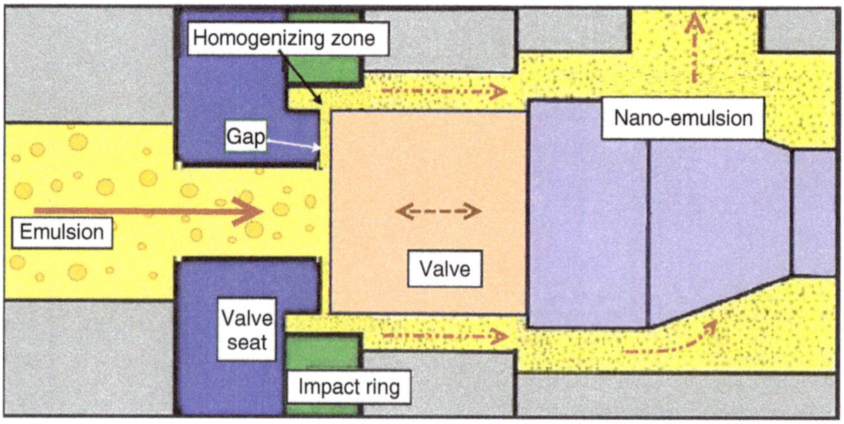

Figure 10.23 Principle of the homogenizing in a valve with adjustable gap in a high-pressure homogenizer.

As stated, in the cosmetics industry, rotor/rotor and high-pressure machines are preferably used for the production of mini-emulsions. As experiments in the laboratory show, ultrasound with the same recipe leads to droplet size distributions (d_{50} about 350 nm) that lie somewhere between the two possibilities Ultra-Turrax (d_{50} about 2 µm) and high-pressure homogenizer (d_{50} about 120 nm). The emulsification techniques, many more or less only applied in pilot plants, can be summarized by the terms rotor–stator machines, colloid mills, pressure valves or aperture or microchannels, membranes, ultrasound, and high-pressure machines/equipment [3]. Owing to the rotor–stator principle, different mills emulsify also well. Recent developments include the microfluidizer and the jet disperser [5]. The specific energy input is about 5–80 kWh/m^3 (1.8×10^7 to 3×10^8 J/m^3). The range of lower values corresponds to macroemulsions and higher values to mini-emulsions.

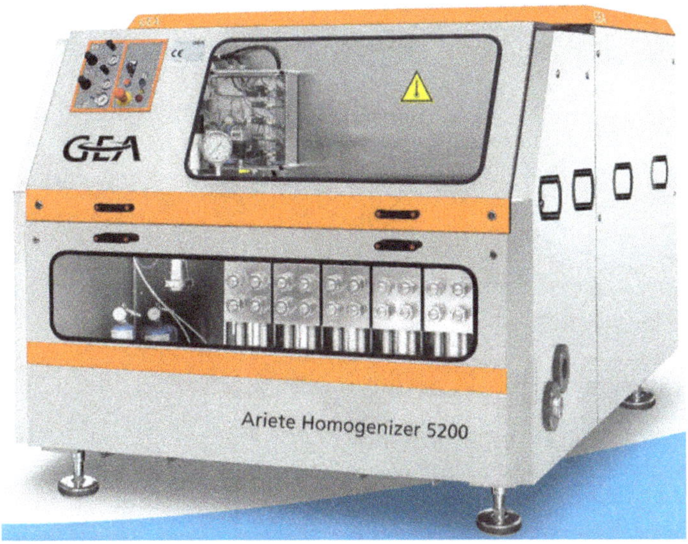

Figure 10.24 Industrial high-pressure homogenizer with five plungers for the production of nanoemulsions. Source: Courtesy of GEA [20].

10.8 Bottles and Filling Lines

In the cosmetic cream market, the following container types (Figure 10.25) are common: for small quantities (10–50 ml), tubes, little plastic bottles, or pots; for medium size (25–200 ml), dispensers, also tubes and pumps or squeeze bottles; and for large sizes (100–500 ml), squeeze bottles with a round opening, also available as an upside standing bottle. The large bottles are sealed with a flip cap or a screw cap. A dosage is possible by shaking and squeezing the bottle.

With the airless dispensers, a new generation of hygienic bottles has been established. The principle is that a piston or a bag conveys the cream to the outlet by reducing the content volume. A vacuum forces the movement of the product. Air contact of the contents with the air is excluded. Precious formulations, containing high levels of polyunsaturated oils and vitamins, can be protected in this way. These substances need effective protection against chemical reactions. UV radiation initiates not only oxidations in the cream, which take place even in the presence of only small amounts of oxygen, but other chemical reactions also occur, such as ester cleavage and condensation. Therefore, the use of UV-light-impermeable bottles is necessary, at least an opaque outer packaging and no sun rays on the bottle. Depending on the type of the container, for cosmeceuticals, transparent or translucent materials such as glass or polyethylene terephthalate (PET) are suitable only in exceptional cases because the sensitive ingredients must survive a storage and consumption time of two years and more.

Against UV rays, coloration with plastic dissolvable paints or with incorporated pigments protects the content of the containers. As pigments, fine solids such as titanium dioxide are suitable. One of the best materials for the bottles is the white, TiO_2-containing, opaque polypropylene (PP) because it is chemically

Figure 10.25 Bottling of cosmetic creams in plastic bottles, jars, and airless dispensers (15–500 ml). Source: Courtesy of Pohli [21].

neutral, physiologically harmless, and shows a smooth, haptically, and visually attractive surface. For the properties of the plastics used, gas permeability plays an important role. Depending on the temperature, through high-density polyethylene (HD-PE), only 20–33% of oxygen is diffused, when compared with an equally strong wall of low-density polyethylene (LD-PE) [8]. PP exhibits comparable values to HD-PE. In both plastics, HD-PE and PP, the water vapor permeability is low. Because of the dimensions and functionality of the piston,

PP airless dispensers require thicker walls compared to the bottles, so the gas permeability drops accordingly. After a careful filling of the almost germ-free cream under nitrogen into the containers, the microorganisms do not multiply until use, even in weakly preserved products.

After opening the container, there are various risks for infection of the cream with germs. High-priced creams, filled in attractively shaped crucibles, meet the ambient air on a large cream surface. In addition, a major disadvantage is the removal of the cream with the finger. Microorganisms not only reach the cream over the air but especially through contact with the fingertips. Therefore, creams in pots must contain higher contents of preservatives. Similar conditions, but with reduced contaminations, occur in the big bottles. In the smaller pump bottles, only the ambient air contacts the cream during ordinary use. Therefore, from a hygienic point of view, pump dispensers are better, compared to crucibles and to big bottles. Using airless dispensers, contact of the content with the ambient air and with finger tips is excluded.

Thus, the airless dispensers offer the best and most expensive choice in terms of hygiene and comfort, as shown in Figure 10.25 below and in Figure 10.26. Available in sizes from 5 to 100 ml, in a few forms up to 200 ml, in round or oval shape, they provide the greatest protection against contamination during both storage and application. The available airless dispensers differ in their construction and function. The motion of the cream ensues mostly by vacuum. When the pump of the airless dispenser is actuated, the cosmetic cream flows to the exit, without air

Figure 10.26 Example of an airless dispenser (a), explanation of the function via the cross section: (1) cap (PP), (2, 3) head (PP), (4) upper valve (EVA, ethylene vinyl acetate), (5) bellows (LD-PE), (6) lower valve (POM, polyoxymethylene), (7) container for the cream (PP), (8) piston (HD-PE), and (9) bottom (PP). Source: Courtesy of Pohli [21]. (b) 50 and 100 ml market products after filling from the top: cosmeceutical skin care creams in airless dispensers. Source: Courtesy of GERCOS [22].

getting into the dispenser. The cream delivery creates a vacuum, which pulls up the piston under the contents. This system ensures almost complete emptying. An example is shown in Figure 10.26.

Filling of airless dispensers takes place free of air in nitrogen at room temperature, in most cases from above. Thereafter, the head with the valves is pressed on the filled container. The alternative bottom filling requires mostly a separate sealing station. During the cream removal, the piston moves slowly upward. For stopping the outlet flow, the pump cap must be unleashed. Then, not only the valves but, in the case of some types, the lips at the outlet also close. These airless dispensers with the designation SC = "self-closing" are equipped with a self-closing opening. A contact of the cream inside the dispenser with the air is virtually impossible. Such dispensers can be optimally used for preservative-reduced or preservative-free formulations. In Germany, typical prices for a purchase of 1000 pieces are between 0.5 and 1.20 €/piece, depending on size, design, color, and pumping equipment. It is notable that the large 100 ml dispensers are often cheaper than the smaller ones. Prices drop significantly if ordered over 10 000 pieces.

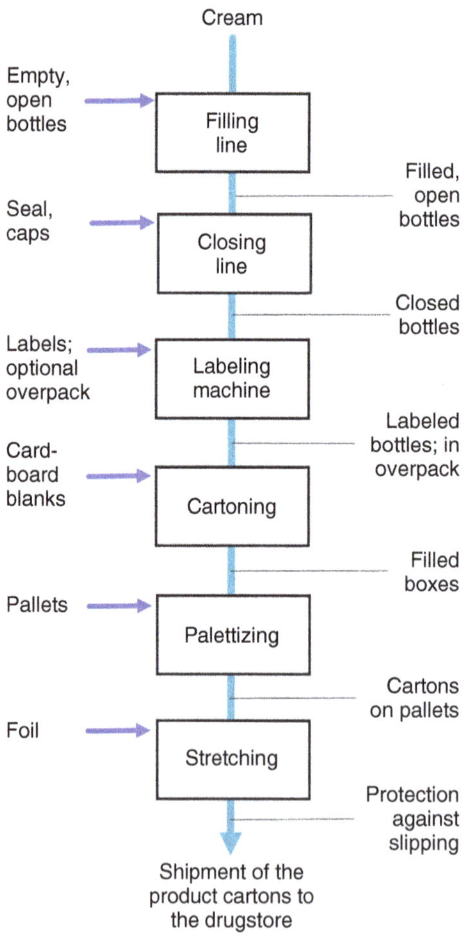

Figure 10.27 Steps of the emulsion after production until shipment.

After production, the cream goes through several stations before reaching the drugstore and finally the customer (Figure 10.27). The first station represents the bottling. Dispensers are divided into two parts. The lower part of the dispenser consists of the empty, open-topped cylinder, which is closed at the bottom by the movable piston. The empty cylinders are fed to a bottling machine and filled there with a cream, for example, with 100 ml. In the following stations, the upper part of the dispenser with the valve system and outlet is pressed into the cylinder as well as the label glued. Figure 10.28 shows a filling and a separate closing machine with a capacity of 70–90 containers/min. For 100 ml bottles, this means can fill about 6 t of cosmetic cream within 15 hours (two shifts, 1 hour left for cleaning and disinfection). A much larger filling and closing line is shown in Figure 10.29, which processes 120–200 containers/min. This machine already fills more than 5 t in 50 ml pots in two shifts or 16 t in 150 ml bottles or 27 t in 250 ml containers. Figure 10.30 (area 4) demonstrates the most common type of product packaging. The automatic sorting of the bottles into a carton, closing the carton, and stacking the cartons on a pallet is shown. Followed by foil-stretching, the entire range is then brought with the forklift to the warehouse or for shipping.

The process is divided into spatially separated areas, as shown in Figure 10.30. The highest attention in terms of hygiene measures applies to the parts in which the aqueous liquid and emulsion come into contact with the room air. This is the case for the premix plant up to the closure of the bottles in the filling line. During the ventilation of the containers under vacuum, room air flows inward. Therefore, this air should be freed from microorganisms by filters, possibly even the entire room air by circulating the air transported by fans through appropriate filters (see Chapter 11). In a few extreme cases, the ventilation of the emulsifier can be done with nitrogen.

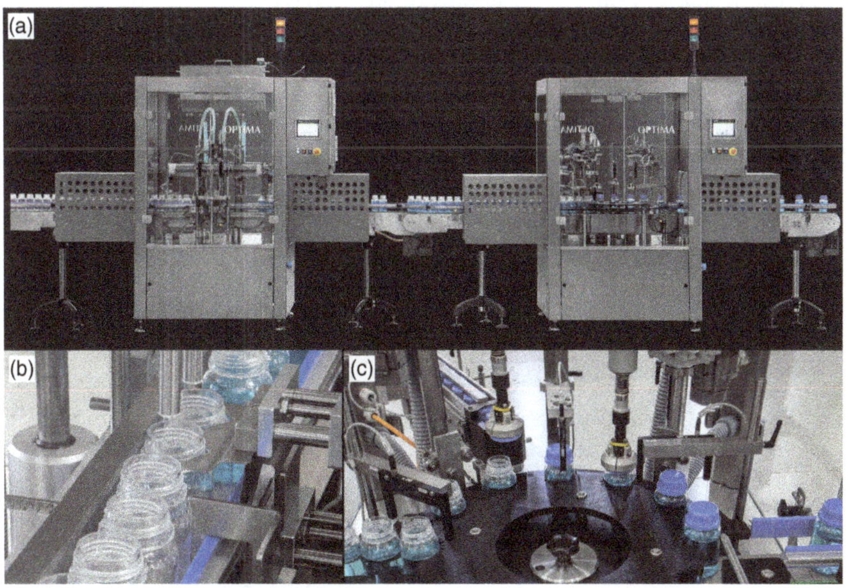

Figure 10.28 Medium-sized filling line with up to 6 working stations for 90 containers/min. and a downstream closure machine, adapted in performance. Source: Courtesy of Optima [23].

Figure 10.29 Filling and closing machine with flexible number of modules from 120 to 200 containers/min. Source: Courtesy of Optima [23].

10.9 Learnings

- For the flexible production of cosmetic creams, predominate batch plants ranging from 100 to 10 000 l.
- Conti systems are only suitable for large product quantities of one quality.
- Center of the system is the emulsifier container, which operates in vacuum and is equipped with an anchor stirrer/scraper and a flanged down rotor/stator homogenizer.
- The rotor/stator homogenizer emulsifies the two phases via energy input and conveys the liquid through a circulation line back to the emulsifying container.
- For the preparation of a cosmetic cream, the formula is divided in water-soluble, oil-soluble, and temperature-sensitive/volatile substances.
- The water phase (50 °C) and the oil phase (60 °C) are prepared separately.
- Solid components of the oil phase are melted and the forming clear oil emulsified in the temperate water phase.
- After cooling, the addition and emulsification of the temperature-sensitive and volatile substances takes place.
- The finished cream is then bottled. The bottles are sealed, labeled, optionally provided with an outer packaging, boxed, palletized, and prepared for shipping.
- The scale-up must consider the specific power of the homogenizer and BTO number. Macroemulsions require an energy input of about 10 kWh/m^3 and mini-emulsions about an order of magnitude more.
- From the specific energy and the minimum BTO number, the emulsification time can be calculated, which increases significantly with increasing batch volume.

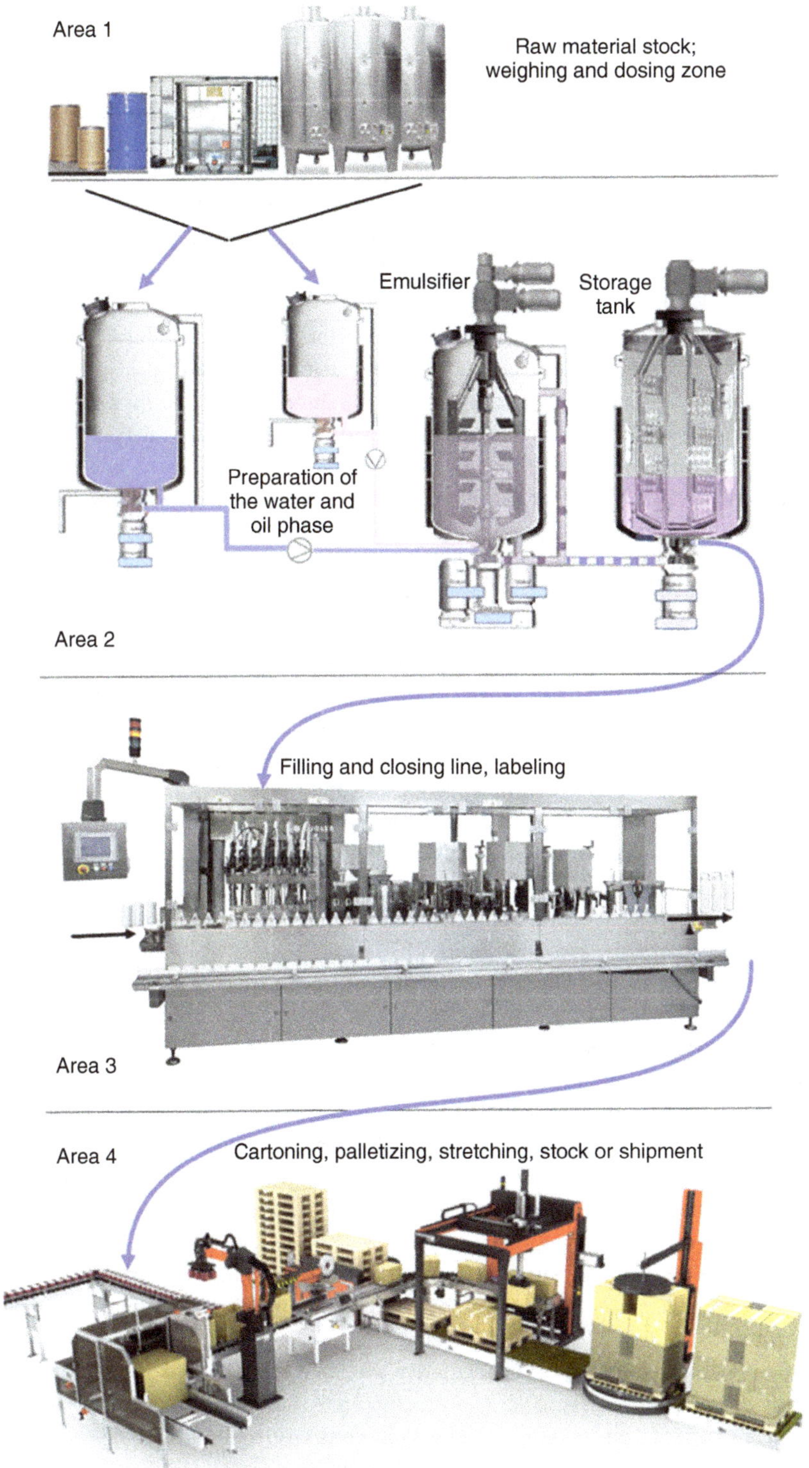

Figure 10.30 Division of the process into spatially separated areas with different hygiene requirements (area 4: Soco System). Source: Courtesy of FAST Technologies [24].

References

1 Ekato (2012). *The Book*, 3e, 59–66. Ekato Holding GmbH.

2 Walstra, P. and Smulders, P.E.A. (1998). Emulsion formation, Chapter 2. In: *Modern Aspects of Emulsion Science* (ed. B.P. Binks), 56–99. Cambridge: Royal Society of Chemistry.

3 Schuchmann, H.P. (2012). Tropfenaufbruch und Energiedichtekonzept beim mechanischen Emulgieren, Chap. 7. In: *Emulgiertechnik: Grundlagen, Verfahren und Anwendungen*, 3e (eds. K. Köhler and H.P. Schuchmann), 133–163. Hamburg: Behr's Verlag.

4 Karbstein, H.P. (1994). Untersuchungen zum Herstellen und Stabilisieren von Öl-in-Wasser-Emulsionen. Dissertation. Universität Karlsruhe.

5 Schuchmann, H.P. (2005). Emulgieren und Schäumen, Chap. 4. In: *Lebensmittelverfahrenstechnik, Schuchmann* (eds. H.P. Schuchmann and H. Schuchmann), 219–226. Weinheim: Wiley-VCH.

6 Rähse, W. (2014). Design of skin care products. In: *Industrial Product Design of Solids and Liquids*, Chapter 13, 319–367. Weinheim: Wiley-VCH.

7 Christi, Y. and Moo-Young, M. (1986). Disruption of microbial cells for intracellular products. *Enzyme Microb. Technol.* 8: 194–204.

8 Nentwig, J. (2006). Folien-eigenschaften, Chap. 4. In: *Kunststoff-Folien: Herstellung-Eigenschaften-Anwendung*, 100. Munich: Carl Hanser.

9 Becomix-Mischtechnik, BECOMIX – Intelligent and efficient. http://www.becomix.online/en/startseite (accessed 7 May 2019).

10 EKATO, Vacuum processing units UNIMIX. https://www.ekato.com/en/products/process-plants/vacuum-processing-units (accessed 27 June 2018).

11 FrymaKoruma, Dinex – the all-rounder for pharmaceuticals and cosmetics. https://www.frymakoruma.com/ww-en/machine-details/dinex/ (accessed 7 May 2019).

12 Hebold Systems, Processing Plants. https://hebold-systems.de/en/products/processing-plants (accessed 27 June 2018).

13 IKA, Systems & Plants. http://www.ikausa.com/product-type/systems-plants (accessed 27 June 2018).

14 K-process, Homogeniser K-DisHo, Vacuum process plants, 2019. https://www.kma-process.de/en/products-vacuum-process.php (accessed 7 May 2019).

15 Lewa, Manufacturing personal care products with LEWA pumps and plants. https://www.lewa.de/en/industries/personal-care/ (accessed 7 May 2019).

16 LÖDIGE as a Partner for the Cosmetics Industry. https://www.loedige.de/en/global-content/industries/life-science-technology/cosmetics (accessed 27 June 2018).

17 Netzsch, Homogenizing Mixer KappaVita. https://www.netzsch-grinding.com/en/products-solutions/dispersing-homogenizing/homogenizing-mixer-kappavita (accessed 27 June 2018).

18 Symex mixing technologies, Immediately available systems. https://symex.de/sofort-verfuegbare-anlagen/?lang=en (accessed 7 May 2019).

19 IKA, Products, Dispersers. https://www.ika.com/en/Products-Lab-Eq/Dispersers-Homogenizer-csp-177 (accessed 28 July 2018).

20 GEA, GEA Niro Soavi Homogenizers. https://www.gea.com/en/binaries/Homogenizers%20-%20One%20Series_Brochure_tcm11-17449.pdf (accessed 5 May 2019).

21 Pohli, Product groups, 2017. https://www.pohli-partner-fuer-packungen.de/Produktgruppen-product-range (accessed 1 August 2018).

22 GERCOS, Entwicklung & Vertrieb von innovativer Wirkkosmetik. http://www.gercos.de (accessed 1 August 2018).

23 OPTIMA, Filling and packaging machines for cosmetic products. https://www.optima-packaging.com/en/consumer/fields-of-application/cosmetics (accessed 5 May 2019).

24 FAST technologies, Packing Automation. https://fasttechnologies.com/sectors/packing-automation/ (accessed 5 May 2019).

11

Regulations and Guidelines for the Execution of Hygienic Productions

11.1 Good Manufacturing Practice Rules for the Manufacture of Cosmetics

GMP guidelines ensure that cosmetic products are produced in consistently good quality and that the defined quality standard is checked. The term "good manufacturing practice" (GMP) was introduced in 1962 by the Food and Drug Administration (United States) through the current good manufacturing practice initiative [1]. In the Unites States, the cosmetic good manufacturing practice (cGMP) is updated annually. GMP summarizes the guidelines for quality assurance of hygienic production processes. These guidelines provide minimum requirements that a manufacturer must meet to assure that their products are consistently high in quality, from batch to batch, for their intended use [2]. The guidelines apply not only to the production of active ingredients for therapeutics but also in cosmetics, food, and feed as well as in dietary supplements and medical devices. However, each industry has its own GMP rules whose main purpose is to protect the consumer. Quality deviations can have a direct impact on the health of consumers. Additional principles include that the end product is free of impurities, that the process and the manufacture are well documented, that personnel are well trained, and the product has been checked for quality more than just in the final stage.

In 2005, the Good Manufacturing Practice Working Group elaborated the first GMP guidelines specifically adapted to cosmetic ingredients, complying with by ISO 22716 [3]. The cosmetic GMP (abbreviated cGMP) guidelines are based on the guidelines used in the pharmaceutical industry and today recognized worldwide. They guarantee quality and safety in the production of cosmetic products. Furthermore, the GMP-quality management system (QMS) ensures compliance with the requirements of health authorities worldwide. The Cosmetics Regulation (CR) in Europe [4] demands in Article 8 §1 and 2 that for the production of cosmetic products, the GMP rules must be observed. Similar regulations exist worldwide, e.g. for therapeutics by the "International Council for Harmonization of Technical Requirements for Pharmaceuticals for Human Use (*ICH*)." Verbatim in the CR:

Cosmetic Creams: Development, Manufacture and Marketing of Effective Skin Care Products,
First Edition. Wilfried Rähse.

> Article 8: §1 The manufacture of cosmetic products shall comply with good manufacturing practice with a view to ensuring the objectives of Article 1. §2 Compliance with good manufacturing practice shall be presumed where the manufacture is in accordance with the relevant harmonized standards, the references of which have been published in the Official Journal of the European Union.

The detailed cGMP rules take many pages and can vary depending on the continent or country. It is recommended to buy the DIN standard for manufacture of cosmetics and follow the detailed rules. Here is just a simplified, short version compiled to get an impression of the diversity of the essential rules. The overview contains 17 points of the European GMP Guide Cosmetics (Guide to Good Manufacturing Practice, written down in DIN EN ISO 22716:2008-12; last version [5]). The cGMP certification takes place by an internal or external auditor. In addition, GMP guidelines also exist for the ingredients, defined in the ciGMP (ci, cosmetic ingredients; released by the EFfCI, European Federation for Cosmetic Ingredients). The ciGMP is specially formulated for the raw material manufactures, in some aspects also for the suppliers [6]. The guidelines ensure quality. Therefore, raw materials should be purchased from certified manufacturers listed by the responsible organization EFfCI [7].

1. Scope of the cGMP
   Every manufacturer and anyone who markets cosmetic products (importer, distributor) are responsible for the quality of their products. The responsible person must ensure that the production has been carried out in accordance with the GMP rules, comprehensibly documented and that the quality was checked analytically (see also CR §4 – Responsible person, §5 – Obligations of responsible persons, §6 – Obligations of distributors). The proof of production according to cosmetic GMP rules need not to be product related. The products may be divided into classes, so cosmetic creams can form a class. The production manager must confirm that the GMP production guidelines are followed. These include personnel, premises, technical equipment, factory hygiene, raw and packaging materials and products, manufacturing and bottling, quality control, and documentation.
   In the legal sense, the management is responsible for the quality of the manufactured products. This responsibility is usually delegated to the production manager via a job description. The production manager of a cosmetic company must demonstrate a technical or scientific education and a three-year practice in cosmetic production.
2. Terms
   Terms such as starting materials, lot number, process control, quality assurance, and a few others are briefly described. Some of the terms have already been explained in the previous chapters or are known to the person skilled in the art.

3. Staff
   Each business should employ a reasonable number of experienced scientists and engineers to manage the production and carry out the quality control. There is an organizational plan for the plants. Each employee should have an area of responsibility, confirmed in writing. All employees of the company as well as locksmiths and cleaning staff must be trained regularly, especially in GMP principles and hygienic production. No one should be employed who suffers from communicable diseases or has open injuries to uncovered parts of the body. All employees must wash and disinfect their hands and wear suitable protective clothing in the production area, which may be used only there. Similar regulations apply to visitors in terms of protective clothing.
4. Premises
   - The production areas (emulsification and filling) are clearly separated from other areas. The employees there may leave the production area only in exceptional cases, so changing rooms, washrooms, and toilets as well as premises for the services (cleaning rooms for machines and apparatus, workshop) must lie on the edge of the production area.
   - The interior surfaces of the rooms must be smooth and crack-free, easy to clean, and disinfect (usually tiles or plastics, epoxy resins). The penetration of animals (insects, small rodents) should be prevented by appropriate measures.
   - Heating, ventilation, and lighting must not affect the quality of the products.
   - The rooms should ensure a proper production process, without the possibility of confusion of materials or mutual contamination.
   - The storage area should provide enough space for proper, separate storage of raw materials and products, as well as quarantined materials.
   - Separated areas should be available for both combustible as well as for water pollutants (water hazard class, WHC, >1).
   - In the tiled sanitary rooms, which consist of dress changing, washing, and toilet rooms, the separation of street and work clothes should be made possible. Employees may enter the production area only in workwear and protective clothing after washing and disinfecting the hands.
5. Equipment
   The facilities should be designed such that
   - Product-wetted parts must be resistant to the ingredients as well as to cleaning and disinfecting agents.
   - Contamination of the substances and containers are avoided.
   - Confusions or omission of a production stage cannot take place.
   - Easy thorough cleaning and disinfection is possible (smooth surfaces, no dead spaces, and completely drainable).
   - The systems are easy to operate and maintain.

   After each batch, the equipment used should be cleaned and, if necessary, disinfected. For this, the installed automatic cleaning and disinfection systems are used (cleaning in place – CIP, or sterilization in place – SIP). Weighing and measuring equipment should be calibrated and regularly checked by recognized methods. The checkweighers for the packaged product must be calibrated officially.

6. Starting and packaging materials
   The handling of the materials follows written instructions.
   - All materials must meet the specifications and be of consistent quality. The clean containers are properly labeled with regard to content (substance and quantity) and batch number.
   - All starting materials should be checked by sampling for compliance with the specification and released for production.
   - The materials should be properly labeled and stored, i.e. without the possibility of confusion.
   - The water to be used must be regularly microbiologically tested for germs. Pathogenic germs must not be present. The maximum number of available germs determines the company depending on the products themselves. The water should be treated with an ion exchange or reverse osmosis system (Caution: the plants are prone to microbial contamination).
7. Production
   In the production area, all employees should pay attention to good operational hygiene, i.e. it has to be dry and clean everywhere.
   - Production takes place under the direction of a competent person with trained staff.
   - Production plant and packing line are assigned. Packaging material and labels are checked.
   - Before starting, all required documents for the production and packaging must be present, especially the production instructions and recipe as well as the assignment of a batch number.
   - The cleaning and disinfection of the staff, technical equipment, and premises were carried out and controlled.
   - Equipment and devices should be provided with name and batch designation.
   - The planned in-process controls will be carried out.
   - The re-storage of unneeded materials ensues after labeling in the warehouse without danger of confusion.
   - Dust formation should be avoided or eliminated with a strong suction system.
   - Important process parameters and deviations from the production instructions must be documented.
8. Finished Products
   The batch is released from quality laboratory before bottling. A storage of the cartonized containers should be provided separately from the raw materials on labeled pallets ready for shipment.
9. Quality control laboratory
   Each manufacturer should establish a competent quality control department with the following tasks:

- o Elaboration or approval of specifications, test methods and in-process controls, as well as sampling methods and instructions for production hygiene.
- o Testing, approval, or blocking of raw materials and finished products based on chemical/physical, microbiological, and sensory tests. Indication of the test results at the plant.
- o Elaboration of test reports.
- o Contribution to the processing and clarification of quality complaints.
- o Performing regular hygiene checks of the production and recommending actions.

10. Treatment of unspecified products
Materials or bottles rejected by the quality lab will be discussed with the manufacturer and returned. After suggestions of the quality lab, end products with small deviations in the specifications (e.g. viscosity) can be reworked and then released considering the ingredient list on the packaging. For non-correctable errors, the removal is done as waste.
11. Wastes
The waste is disposed of in accordance with local laws.
12. Subcontracting
For contract production, either according to the own cream formulation or a proposal of the contract producer, a contract should be concluded. The certified contract manufacturer is obliged to produce according to the cGMP and to prove this as well as to adhere to the quality specified by the client.
13. Deviations
In the case of quality deviations, the quality laboratory must develop and check a feasible solution (see point 10).
14. Complaint and recall
In the case of complaints by customers, the stored reference sample, which can be found using the batch number, must be analyzed for verification. To clarify possible causes in the production, it is necessary to resort to the production documents. For recalls, the documented batch number makes it possible to find the distribution in the market and the appropriate bottles.
15. Change control
Each change should be approved and documented by the quality laboratory.
16. Internal audit
The company should have a trained auditor. The audit is carried out at a fixed time according to a method, based on the DIN EN ISO 22716: 2008-12. During the auditing, the responsible persons from the production and the quality laboratory are present. The auditor suggests improvements and documents the results. If all aspects of the inspection are completed to the satisfaction, the audit is passed and the company is certified. The audit should be repeated at regular intervals, for example, six months or one year.
17. Documentation
The documentation system should document the individual activities of the manufacturing, testing, storage, and distribution in electronic form and ensure complete traceability. All specifications must be kept up to date with science and technology. Each product should have a recipe, a production

instruction (PI) sheet, the production and test report, as well as the batch and storage documentation with the same batch number.

- Production instructions (PI)
  The PI contains the name and batch number of the cosmetic product, the list and name of the equipment required for production and filling, the formulation with the individual quantities, the manufacturing route, and the in-process controls and specifications.
- Production report (PR)
  Each product contains a production report (PR) sheet, in which the production according to the PI sheets is confirmed. The PR sheet contains the following:
  - Name, quantity, and batch (lot) number of the cosmetic product; the test protocol numbers of the starting materials; and the bulk goods as well as the measurement data of the different production phases.
    Notes on used machines and equipment as well as deviations from the PI.
    Results of the in-process controls carried out with the date and signature of the processor (or name in the PC document).
    Names of the employees and confirmation of the production manager with date.
    Test report

  For each tested starting materials and finished products sample, a test report must be prepared containing the following information:
  - Designation, date of manufacture, and batch number as well as produced quantity.
  - Origin of the batch (production plant) and of the sample.
  - Result of the individual test.
  - Made decision. Date, person who carried out the investigation (electronic log).
- Batch documentation
  The batch documentation consists of the PI sheet as well as the manufacturing and test report.
- Storage of documentation and reference samples
  The batch documentation and reference samples shall be kept for at least one year beyond the expected period of use. The reference samples must be large enough to allow for the necessary analyses in the case of complaints and to ensure traceability. The samples should be stored under the conditions recommended by the manufacturer for the product (usually room temperature, rarely cooled).

Annex 9 of the GMP guidelines comprises additional rules. They can be taken from the following text (copy from [8]):

> MANUFACTURE OF LIQUIDS, CREAMS AND OINTMENTS
>
> **Principle**
> Liquids, creams and ointments may be particularly susceptible to microbial and other contamination during manufacture. Therefore, special measures must be taken to prevent any contamination.

Premises and equipment

1. The use of closed systems for processing and transfer is recommended in order to protect the product from contamination. Production areas where the products or open clean containers are exposed should normally be effectively ventilated with filtered air.
2. Tanks, containers, pipework and pumps should be designed and installed so that they may be readily cleaned and if necessary sanitized. In particular, equipment design should include a minimum of dead-legs or sites where residues can accumulate and promote microbial proliferation.
3. The use of glass apparatus should be avoided wherever possible. High quality stainless steel is often the material of choice for parts coming into contact with product.

Production

4. The chemical and microbiological quality of water used in production should be specified and monitored. Care should be taken in the maintenance of water systems in order to avoid the risk of microbial proliferation. After any chemical sanitization of the water systems, a validated flushing procedure should be followed to ensure that the sanitizing agent has been effectively removed.
5. The quality of materials received in bulk tankers should be checked before they are transferred to bulk storage tanks.
6. Care should be taken when transferring materials via pipelines to ensure that they are delivered to their correct destination.
7. Materials likely to shed fibers or other contaminants, like cardboard or wooden pallets, should not enter the areas where products or clean containers are exposed.
8. Care should be taken to maintain the homogeneity of mixtures, suspensions, etc. during filling. Mixing and filling processes should be validated. Special care should be taken at the beginning of a filling process, after stoppages and at the end of the process to ensure that homogeneity is maintained.
9. When the finished product is not immediately packaged, the maximum period of storage and the storage conditions should be specified and adhered to.

The German Industry Association for Personal Care and Detergents e.V. (IKW) has published a self-assessment checklist (36 pages, only in German), which helps the production manager to prepare for the cGMP audit [9]. After passing audit, the manufacturer of cosmetics should provide a summary of the last internal audit for the customer. This could look like this:

Certifacte

Proof of the use of cosmetic GMP for the manufacture of cosmetic products

In the factory ..X.. (address) all products are manufactured in accordance with the Guidelines of the Good Manufacturing Practice of Cosmetic Products (cosmetic GMP) **DIN EN ISO 22716.** The last internal audit was carried out on ..date .., including

- Staff
- Premises
- Technical equipment
- Company hygiene
- Starting materials and final products
- Intermediates and bulk goods
- Manufacture (from the raw materials to the packaged end products)
- Quality inspection
- Documentation

Compliance is ensured by the production and quality management.
Next audit: ..date..

Date: Signature of the auditor:

---

or, for small companies, by an external auditor

Certificate

The company ... (address) manufactures its cosmetics products according to the standards of

Cosmetic – GMP
DIN EN ISO 22716

after having been audited in *..date..*
The certification is based on the guidelines of
DIN EN ISO 22716:2008-12 (ISO 22716:2007)
in accordance with Cosmetic – GMP.
Valid till: *..date..*
Report No: ...
Next audit: *..date..*

Date: Signature of the auditor (Name, address):

Raw materials from a certified manufacturer or dealer do not need to be analyzed but only verified as an analytic certificate exists. The number of germs should be measured in the water used and in some aqueous solutions, except at high concentrations, as well as in the emulsion. In particular, solids of natural origin show high bacterial counts. Here, a thermal treatment may be required before use.

11.2 EHEDG Guidelines for the Construction of the Facility

In hygienic production plants, such as those used for food, cosmetics, pharmaceutical, and biotechnological products, the quality of the products depends on the design of the system. Individual elements and the entire facility must be designed and constructed so that microorganisms cannot find places to live and multiply. This assumes an appropriate grade of stainless-steel with polished surfaces and a dead space-free system. Furthermore, after each batch, the facility needs to be effectively cleaned and disinfected to minimize the number of microorganisms (Section 11.4). Besides the material and its surface, the quality of the workmanship and the integration of the component into the system must be professional.

The know-how in the optimal construction of components from university, research institutes, and industry were bundled in the European Hygienic Engineering and Design Group (EHEDG), which deals with the correct design of components for the safe, hygienic production of food. EHEDG guidelines and US 3-A standards are harmonized. EHEDG [10], founded in 1989 as a nonprofit consortium, has developed and published a variety of practical guides on adequate hygienic design. These represent the basis of the component manufacturers for the design and construction (valves, pipes, pumps, container, etc.). The proposed solutions of general relevance can be transferred without restrictions to the production of cosmetics, as the food and cosmetic industries produce with low germ counts but not sterile like biotechnology. The instructions to be followed when manufacturing a pump, valve, or container must be purchased from the EHEDG [10, 11].

The manufacturer of cosmetics has other tasks. He should consider some points when purchasing the equipment or the complete production plant. In overview, these are

- Realization of the functional hygiene requirement (against microorganisms).
- Materials of construction (quality of the stainless steel).
- Hygienic design and construction (polished surfaces and proper welds).
- EHEDG certificate from the manufacturer (or the written assurance that the equipment has been manufactured in accordance with the EHEDG – 3 A standard).

In hygienic productions, cGMP and especially EHEDG instructions help to prevent microbiological infections. This target requires a number of costly measures concerning the premises, machinery, materials, and design, which the production manager and operator should be aware of:

- All installations must be free of dead spaces and show a product flow from the top downward (Figure 11.1). Use only dead-space-free valves and butterfly valves (Figure 11.2) and pumps (Figure 11.3).
- Preferably, they are completely drainable at the deepest point (for example, under the emulsifying container). Install all pipes with slight negative gradients toward the lowest point.
  This point is important for plant manufacturers. Traditionally and for aesthetic reasons, this technically meaningful advice is not obeyed.
- Only high-quality stainless steel with smoothed surfaces in all parts should be used (Sections 11.3.3 and 11.3.4).
- There should be precise (automatic orbital) welding of stainless steel to meet the hygienic requirements.
- The premises should be tiled or sealed with (epoxy) resin (positive example in Figure 11.4), rounded off at all corners with resin.
- Simple cleaning and disinfection of the plant in all parts is a must, preferably with an automatized facility (Section 11.4).
- Thorough cleaning and disinfection also apply to the outer surfaces and premises.
- The design of sensors should be hygienic.

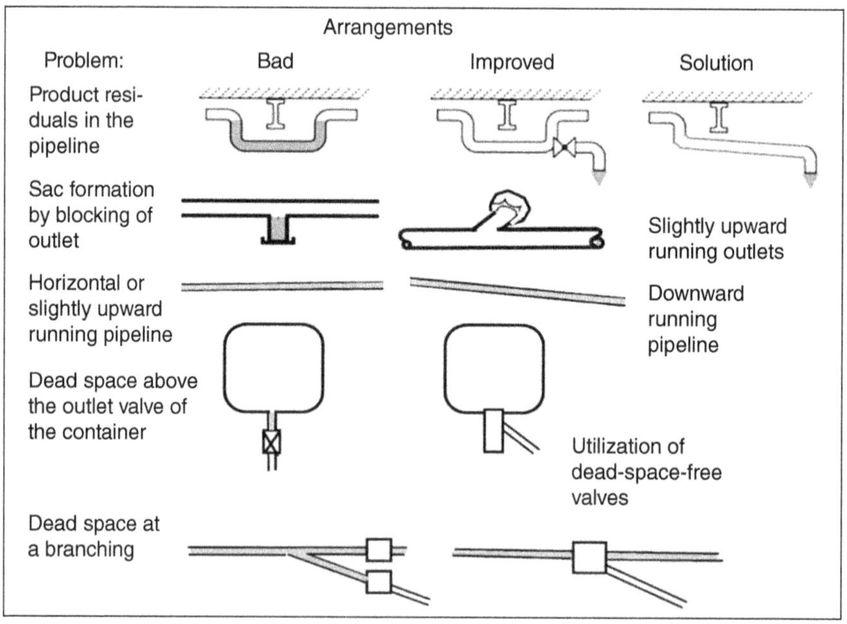

Figure 11.1 Hygienic requirements for the installations; fully drainable piping and dead-space-free valves.

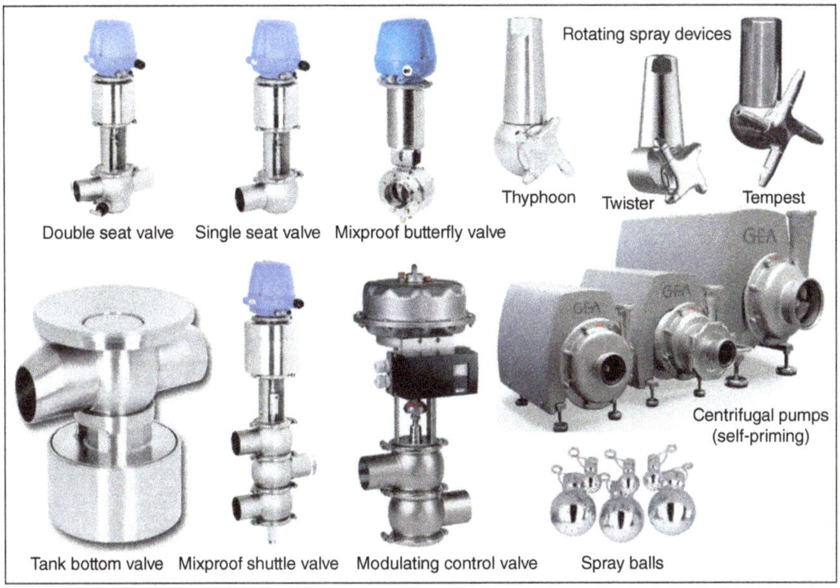

Figure 11.2 Fittings, cleaning heads, and pumps for hygienic production. Source: Courtesy of GEA Tuchenhagen.

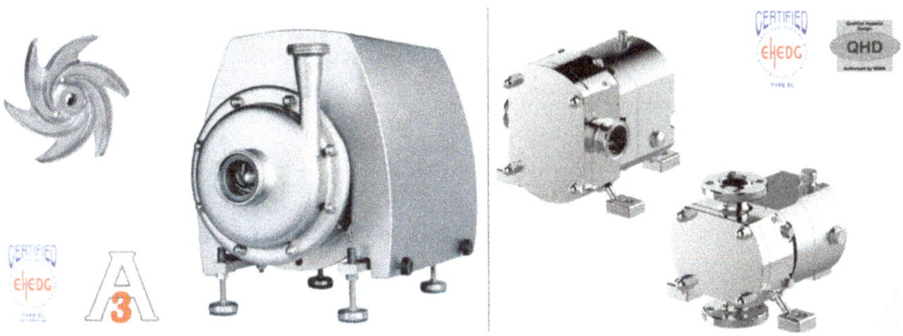

Figure 11.3 EHEDG-certified pumps for low, medium, and high-viscous liquids. Source: Courtesy of GEA Tuchenhagen.

- o Hygienic water should be provided by membrane filtration and UV irradiation.
- o The air inside should be cleaned by sterile filter (recommendation).

Microorganisms can enter the plant and the product via various routes:

- *Air in the production premises*: the air should be cleaned in circuit via HEPA (High-Efficiency Particulate Air) filters.
- *Air from outside*: supply air may only enter the building via HEPA filters; a slight overpressure in the building is recommended.
- *Aqueous solutions*: examples are dilute polysaccharides (sugar) and protein solutions (degraded soy proteins).
- *Water*: the required water for the formula should be sterilized (membranes, UV lamps).

Figure 11.4 Hygienic cosmetic plants with smoothed outer surfaces of the equipment in suitable premises. Source: Courtesy of FrymaKoruma.

- *Solids*: solids of natural origin, plant parts, and herbs contain germs, often in high concentrations, and should be sanitized before use.

The last two points are the most important because dilute solutions are rarely used and clean air contamination is unlikely. In case of infections of the product, analyze the germ counts in the natural solids. If no natural solids are present, the water should be checked because according to an estimate in about 90% of these cases, the infection is caused by the water. Frequently used ion exchangers and systems with organic membranes tend to germinate and should be checked regularly. Another cause of infection can be fixed microorganisms in the system, which live in cracks, gaps, and crevices or abrasive marks and insufficiently smoothed areas of the material as well as dead spaces and biofilms of the components. Therefore, the choice of material and its professional processing is of great importance. The production should be done in carefully made polished stainless-steel equipment. Except for the addition of some raw materials and bottling, these are closed systems.

In addition, EHEDG actively supports European legislation, which demands hygienic processing and packaging of food in hygienic machinery and in hygienic premises (EC Directive 2006/42/EC for Machinery, EN 1672-2, and EN ISO 14159 Hygiene requirement). Furthermore, EHEDG provides instructions (must be purchased) to optimize food processing, also suitable for cosmetics, for example:

- ➢ Integration of hygienic and antiseptic systems.
- ➢ Test methods for determining the hygienic condition of the system.
- ➢ Safe storage and distribution of water.
- ➢ Air quality control for building ventilation.
- ➢ Hygienic packaging of products.
- ➢ Aseptic and hygienic filling machines – installation, qualification, and operation.

- General principles of cleaning validation.
- CIP (in work, 2018).

11.3 Materials for the Equipment of Cosmetic Plants

11.3.1 Problem

During the manufacture of food, pharmaceutical, biotechnology, and cosmetic preparations, and also for some other products (e.g. dietary supplements and liquid detergents), microorganisms must be completely absent or their numbers below specified limits. Certain microorganisms may not be present. Product hygiene requires that the production is carried out in an appropriate facility. The plant should be designed dead space free to avoid the formation of biofilms. On the other hand, a firm adhesion of microorganisms on the walls of equipment must be prevented. Therefore, the materials and their processing are of great importance for hygiene. Furthermore, a suitable plant design with appropriate materials simplifies the system's cleaning and reduces the wastewater pollution.

Although some very large companies have specialists in the field of materials who develop optimal solutions for each process, most of the others are dependent on the knowledge and experience of their employees from process development, project management, and production. During the purchase of new machinery and equipment, the material and its processing is usually defined in a joint discussion with the manufacturer. This chapter is intended to give the developers and operators of plants some evidence on the properties of materials. This knowledge is important when ordering new machinery and for trouble-free production of high-quality products. The remarks come from own experience and background in the biotechnology, cosmetics, and consumer goods industry, supported by reading the literature.

Chemical engineers who work in the field of product and process development for the food, chemical, or pharmaceutical industry, must not only specify the materials when ordering equipment but, in addition, often the finish of wetted surfaces and welds. In many cases, specifications exist because the company prescribes for some or all plants a standard stainless steel (such as AISI 304, Din 1.4301); optionally, it also specifies a surface finish for the inside and outside of the equipment. Therefore, the storage in the company's workshops is reduced; furthermore, the processing in the plant construction is standardized. If a device requires a better material grade against chemical corrosion and abrasive wear, the decision must be made between alternatives. Howsoever, the economy (cost, repair, resistance, and cleaning cycles) of the material stands against the product quality (metal contents and hygiene).

Quality includes product safety and product hygiene, which require extensive plant hygiene. Especially in the pharmaceutical and biotechnology industries, as well as for food and cosmetics, stainless steels are used with ground and polished surfaces. On the one hand, the polishing of the product's wetted surfaces significantly increases corrosion resistance. On the other hand, a residue-free cleaning can be performed, avoiding cross-contamination and infection by microorganisms in the entire production plant; when, as a prerequisite, a dead-space-free, fully drainable plant is available. Examples of avoiding dead spaces can be found in DIN EN 1672-2:2005 and the EHEDG (3-A and NSF International) brochures [10]. The adhesion of microorganisms, the formation of biofilms, the adhesion of product residues and deposits, and the development of crusts depend on the material and surface properties [12]. The same applies to the cleaning and disinfection of the surfaces and the removal of residues with appropriate solutions. These parameters worsen over the years on account of mechanical wear and chemical attack. Typical aging processes affect mainly plastics (bellows and seals). Because rough surfaces, incipient corrosion, and the gradual formation of microcracks offer numerous opportunities for the microorganisms to adhere, these are hygienically questionable [13] and not allowed.

The DIN standard 1672-2 specifies that in the food industry, product contact surfaces must have an average surface roughness $Ra \leq 0.8\,\mu m$. The required processing of stainless steel by grinding/polishing or pickling/electropolishing clearly reflects in the purchase price. The cleanability of the plant is not only an essential quality but also a cost factor because the system's availability depends on the duration of cleaning. The cleaning method is a function of the material, its surface smoothness, the distance between roughness mountains, and the composition of cleaning agents (surfactants, oxidants, and pH value), as well as the cleaning conditions (flow and temperature). The pharmaceutical industry often produces in electropolished equipment. The electropolishing of the surfaces ensures the necessary hygiene and a rather easy cleaning. This allows reducing the duration and consumption of cleaning solutions. In addition, it is possible to disinfect the production plant including the filling lines with 2-propanol, which regularly takes place in the cosmetic industry after cleaning. Even higher demands on sterility meet with chemical and steam sterilization. The choice of material depends on the chemistry and operating conditions as well as the raw materials and product phases (solid, liquid, gas, and multiphase systems). Theoretically, for each part of the production line (tank farms, silos, main plant, filling line, and product storage), another material might come into question. This means that the term plant design includes not just the machinery and equipment of the core facility. For the production of cosmetic creams, the recommended standard grade AISI 316L (Din 1.4404) should be installed for each part of the complete system.

11.3.2 Choice of Material

The materials are chosen on the basis of chemical, thermal, and mechanical stresses. Materials for the production plant can be divided into metals (stainless steel, nickel, titanium, etc.) and inorganic nonmetals (such as graphite, glass, and ceramics) as well as organic polymers (plastics). For equipment, the chemical, pharmaceutical, biotechnology, cosmetic, and food industries utilize mainly alloyed stainless steel of high quality. Typically, they work with standard grades, mostly AISI 304 or 316 qualities. If the following critical substances are involved in the process, another grade or material must be chosen [14]:

- Strong organic and inorganic acids;
- Acidic gases (including chlorine and bromine);
- Chloride, bromide, and fluoride ions;
- Alkalis (especially at high temperatures);
- Molten salts;
- Hydrogen.

Several chemical processes on the material surface can trigger corrosion. This ensues in various forms. The best known are probably surface corrosion, crack, contact, pitting, and stress corrosion. Further, fatigue cracking and intergranular corrosion occur. The corrosion resistance of the material depends on its chemical composition (usually Fe alloy) and the processing conditions. From the outside, chemical substances attack and abrade the surfaces by raw and auxiliary materials, intermediate, secondary, and end products. These attacks determine the lifetime of the machinery in detail via the following:

- Application time and temperature;
- Concentration of problematic substances, gases, liquids, suspensions, melting, and solids;
- pH;
- Presence of oxidizing or reducing substances;
- Presence of abrasives;
- Presence of salts, particularly of chlorides and fluorides (Na, K, and heavy metals).

In the production of cosmetic creams, the influence of temperature (about 60 °C, max. 120 °C) and the pH value (4.5–8) is low, and problematic substances are not present, rarely oxidative or abrasive substances. However, the presence of chloride ions should be noted as formulations may contain a few percent of the salt. The physiological concentration is 0.9% NaCl, 3.5% the concentration of the sea.

11.3.3 Stainless Steel

In the chemical apparatus, stainless steel is by far the most common material (list of stainless steels: DIN EN 10088-1:2014-12). In general, stainless steels have a chromium content (Cr) of >10.5%, contents of sulfur (S) and phosphorus (P) of <0.025%, and content of carbon (C) <1.2%. They are characterized by forming a thin protective layer on the surface in contact with air. After minor damage, the layer regenerates by itself, when the steel is exposed to air and moisture [12]. The weldability increases with decreasing carbon content. Stainless steel can be improved in some properties by further alloy constituents. For this purpose, nickel (Ni) and molybdenum (Mo) are particularly suitable.

Hygienic production plants require high-alloy steels containing at least 17% Cr (up to 30%) and 12% Ni (up to 26%), besides C <0.1%, Si <1%, and Mn <2%. These form in contact with oxygen from the air a thin, invisible passivation layer of chromium (III) oxide (Cr_2O_3), which protects the underlying metal. The addition of Ni improves the corrosion resistance of stainless steel because it oxidizes very slowly (in contrast to iron). However, the standard stainless steels cannot meet all the requirements of the chemical apparatus. For example, the steels are additionally alloyed with 2–5% molybdenum (Mo) to avoid pitting and stress corrosion cracking. Other requirements lead to the addition of titanium (Ti), copper (Cu), aluminum (Al), niobium (Nb), and others.

The compositions of frequently used stainless steels in the manufacturing industries are listed in Table 11.1. The prices depend on the manufactured quality, the design of the produced stainless steel, the alloy surcharges, and the purchase quantity. For hygienic systems, bright steel is preferably used, the surface of which is smooth. To assess the price differences, the percentage column should be used as a basis. It should also be noted that only about 30–66% of the difference between two varieties is included in the final price of the machinery/equipment, depending on the effort of manufacture. Seamless pipes cost about 17% more than cold-drawn steel.

The standard grade AISI 304 (1.4301) is probably the first commercial stainless steel and currently often processed in industry. This austenitic, nonmagnetic

Table 11.1 High-quality stainless-steel compositions and estimated prices for cold drawn steel (August 2018; prices are quantity-dependent; alloy composition according to DIN EN 10088-1: X5CrNi18-10, number after X = gives the carbon content after division by 100, 5 = 0.05% C; CrNi18-10 = 18% Cr, 10% Ni).

Material number (Din)	AISI standard, (and ASTM International)	Steel composition DIN EN 10088-1	PREN[a)]	Estimated prices[b)] for stainless steels (€/t) (cold-drawn steel)	Relative price (standard steel = 1.4404/316L = 100%)
1.4301 (V2A)	304	X5CrNi18-10	19	3300	72%
1.4307	**304L**[c)]	**X2CrNi18-9**	**19**	**3300**	**72%**
1.4401	316	X5CrNiMo17-12-2	25	4600	100%
1.4404	**316L**[c),d)]	**X2CrNiMo17-12-2**	**25**	**4600**	**100%**
1.4435	**316L**[c),d)]	**X2CrNiMo18-14-3**	**27**	**5300**	**115%**
1.4439	317LMN	X2CrNiMoN17-13-5	38	–	
1.4462	A182 F51 (329LN)	X2CrNiMoN22-5-3	35	3800	83%
1.4539	904L[c)]	X1NiCrMoCu25-20-5	35	8400	183%
1.4541	321	X6CrNiTi18-10	18	3500	76%
1.4571 (V4A)	316Ti	X6CrNiMoTi17-12-2	25	4700	102%

a) PREN, Pitting Resistance Equivalent Number [15], from 32 on, the (unpolished) steels are seawater resistant.
b) Alloy surcharges according to published table for August 2018; prices estimated by factors with reference to [16, 17].
c) L stands for low carbon.
d) AISI 316L = 1.4404/1.4432/1.4435/1.4436.

material and the quality with less carbon (1.4307), which is easier to weld and polish, show good resistance to water, steam, moisture, fruit acids, and weak organic and inorganic acids. The food, beverage, pharmaceutical, and cosmetic industries use these steel grades commonly in the processing of liquids. For improved cleaning and corrosion resistance, the product contact surfaces are usually ground and/or polished. This steel quality is not suitable for permanent contact with chloride-containing solutions with Cl^- >50 ppm.

The stainless-steel AISI 316Ti (1.4571) includes titanium to improve the stability and weldability. Together with the added molybdenum, this steel shows increased chemical resistance, especially against intergranular corrosion, pitting, and stress corrosion cracking. As this alloy covers most requirements, it is widely used in chemical apparatuses, providing the standard material in many companies (chemicals and consumer goods). Owing to the stabilization of the stainless steel with titanium, mechanical smoothing and electropolishing are difficult. Mechanical grinding of the surfaces causes many scratches by the very hard particles from titanium carbide, torn from the surface, which also do not dissolve in acids. This concerns the Ti, Nb-stabilized steels (AISI 316Ti, 316Nb, and 321) that are less suitable for use in the pharmaceutical, food, biotechnology, and cosmetic industries.

For hygienic production, the equipment surface must be polished inside up to mirror finish. The engineers use for these applications some low-carbon grades, such as AISI 316L (1.4404 and 1.4435) or for lower requirements AISI 304L (1.4307). In the United States, no distinction is made between Din 1.4404 and 1.4435, both are AISI 316L. New plants for the cosmetic industry are today practically only manufactured in Din 1.4404 (AISI 316L). This well-processable standard quality shows a medium, mechanical stress tolerance (e.g. as material in mills), and a mean resistance to chemical agents, which increases by polishing. It meets all requirements for the production of cosmetic creams, from the premix containers to bottling lines. Even the usual cleaning solutions do not attack the material. In some cases, only the wetted parts are made of this steel quality, all others in 1.4301 (AISI 304).

For high-salt formulations, the salt is added after cooling together with the temperature-sensitive ingredients directly before the second emulsification (previously tested in the laboratory). Then, low temperatures prevail (<30 °C) and an oil film protects the steel walls. Alternatively, after measuring the corrosion resistance of the saline emulsion, addition can be made after the start of the emulsification 1. This path should only be followed if low-temperature procedure does not bring the desired result. Chloride ion-containing liquids must be in constant motion. If higher chloride ion stability is required, it is possible to switch to other grades (1.4435, better 1.4439, or after testing to the economical 1.4462).

11.3.4 Smoothing the Metal Surfaces

The manufacture of machinery and equipment requires the processing of stainless steel in the form of sheets, seamless pipes, and partly steel bars (for shafts). These materials are available in different qualities with surfaces that are smooth, polished, brushed, or shiny polished [18]. The average Ra-value of cold-rolled sheets is in the range of 0.2–0.5 μm [19], depending on the pretreatment. This starting material facilitates further operation for the production of machinery with surfaces in mirror finish. For cleaning and smoothing the steel surfaces, several methods (Table 11.2) exist. The procedures are usually applied individually or in combination. Absolutely necessary steps represent the precleaning of the machinery and the pickling of all welds. A smoothing of the surfaces may be carried out chemically or mechanically. If subsequently an electropolishing follows, the equipment always must be completely pickled. For a perfect result of the pickling, a previous cleaning of the surfaces is required. In the procedure, the dirt and residue from the stickers are removed and a pretreatment for degreasing executed. Alkaline solutions at elevated temperatures provide good results. Clinging dirt, slag residues, or scale layers can be removed with stainless-steel brushes or

Table 11.2 Improving stainless-steel surfaces by cleaning and smoothing in an overview.

Procedure	Action	Execution	Achievable roughness, mean value, Ra (μm)
Cleaning as an upstream process step	Removing coarse dirt	High-pressure cleaning, steam cleaning (140 °C)	
	Removing dirt and grease, possibly supported by ultrasound	Dipping into a surfactant–alkaline cleaning solution or spraying with a mixture of phosphoric acid/surfactant	
	Removing tough dirt	Stainless-steel brush, electrolytic processes	
	Removal of caking	Ultrasound	
Mechanical smoothing	Making of a clean, matte layer	Blasting of surfaces with quartz sand or beads of glass or of stainless steel or carbon dioxide	<1.2
	Satinized matte structure	Brushes (coarse to very fine)	
	Shiny, planar surface	Buffing (felt disc with decreasing grain sizes of the pastes)	
	Smooth surface (up to mirror finish)	Grinding or polishing (with decreasing grain size)	Up to 0.1
Chemical smoothing	Passivation layer	Accelerated formation with dilute nitric acid	0.5–1.0
	Removal of scale layers after heat treatment and of tarnishing after welding, further of external rust	Pickling with acids (mixture of nitric and hydrofluoric) in the bath or by spraying	
	Smooth surface (up to mirror finish)	Electropolishing, plasma polishing	0.2–0.5 <0.01

by electrolytic processes or ultrasound. Only completely purified surfaces can form the required continuous passivation layer.

Through blasting, brushing, grinding, buffing, pickling, and **polishing**, esthetic structures on stainless-steel surfaces arise for goods in the private and public sector (gates, railings, wall coverings, household items, and appliances). In the hygienic sector, however, the surfaces must be corrosion resistant and easily cleanable. Size, accessibility, and application of the metal surface determine the method for smoothing. On large areas, such as the interior walls of stirred tanks, the smoothing is performed mechanically by hand with the aid of grinding and polishing machines. Fissured surfaces are preground in a first step with coarse sandpaper (grade 60 or 80). Each subsequent step requires a finer grain size in the gradation up to the maximum factor of 2, that is, with pastes grit 120, 240–320 and with sanding belts grit P120, P240, and P320–600 (Table 11.3) until the desired target (approximately Ra = 0.2–0.4 μm) is reached. The coarse abrasives eliminate the imperfections in the metal surface, such as sanding marks, nicks, and hairlines. With progressively finer grits, all scratches disappear, even the invisible ones. The high-gloss finish requires appropriate polishing pastes and sanding belts, discs, and high-speed machines. The grinding result is highly dependent on the skill of the polishers, their expertise, and equipment (grinding wheel machine and sanding pastes). During the "dry" applications in steel polishing, waxes and kerosene are used as lubricants and coolants. For the smoothing of stainless steel, the polisher needs in graduated grain sizes not only sanding particles, made of alumina, silica, or zirconia, but also aluminum carbide and silicon carbide and sometimes diamond dust. These mechanical methods remove the scratches, but the base material is not eroded appreciably.

Table 11.3 Classification of the grade for abrasive pastes and abrasive papers/sanding belts.

Abrasive grit for pastes according to FEPA[a]/ANSI[b]	Average grain size of pastes d_P (μm) FEPA/ANSI	Abrasive grit for sandpaper and sanding belts FEPA	Average grain sizes of sandpaper and sanding belts d_P (μm)	Achievable roughness, mean value[c], Ra (μm)
60/–	260	P20	1000	
80/–	185	P60	269	3.5
100/100	129/125	P120	125	1.1
180/180	69/70	P240	58.5	0.2–0.5
220/220	58/58	P320	46.2	0.15–0.4
320/360	29/28	P600	25.8	0.1–0.25
360/400	23/23	P1000	18.3	
400/500	17/18	P2000	10.3	

a) FEPA, Fédération Européenne des Fabricants de Produits Abrasifs.
b) ANSI, American National Standards Institute.
c) [19].

The result of polishing the outer surface is shown in Figure 11.4 for two cosmetic plants. The product-wetted surfaces inside are usually polished finer.

Pickling [20] is the process for removing scale layers after heat treatments, residues from welding, annealing colors, sanding dust, abraded metal, external rust, and chromium carbide (which can arise during drilling without a coolant). By the action of the acid (usually nitric acid mixed with hydrofluoric, HF acid), a metal erosion of 1–5 μm occurs at ~40(+20) °C. For some parts of the plant or for the entire apparatus, the acid bath is applied. Pickling ensures the subsequent formation of a complete passive layer to protect the metal against corrosion. After rinsing with soft water, the formation of the layer begins in contact with air on the clean surface. Two to eight hours later, the passivation layer is completely present in a thickness of about 20 nm. By the addition of oxidizing agents, such as dilute nitric acid or hydrogen peroxide, the layer is formed immediately on all surfaces. In the food, cosmetic and pharmaceutical industries, the **passivation** is preferably carried out with citric acid at temperatures above 40 °C over a period of a few hours. Pickling is advisable before a mechanical polishing but constitutes an indispensable part of electropolishing. It takes place with acid mixtures in a dipping and spraying process, or by pumping through tubes (Table 11.4). All welds must be pickled. If it is not done with the whole apparatus in the bath, the acid is often applied in the form of a paste with a brush. Careful pickling allows the formation of a passive layer on the welds to prevent corrosion. In addition, pickling improves the appearance (as shown in Figure 11.4).

Depending on the diameter and length of tubes, a mechanical smoothing of the inner surfaces is difficult to accomplish. In addition, especially for small parts, such as pipe bends, fittings, valves, sensors, and pumps, pickling/**electropolishing** is the appropriate method. However, large apparatuses, preferably for the pharmaceutical and biotechnology industry, are also electropolished. The method [20] begins with a cleaning of the metal surfaces with alkali followed by pickling. Then, electrolysis expires in a mixture of concentrated sulfuric and phosphoric acid at elevated temperatures (see Table 11.4). The steel parts are placed in a basket, which is connected as the anode. By applying a DC voltage (Figure 11.5), the "hills, mountains, and elevations" on the surfaces dissolve, by forming ions. Either they remain dissolved or fall as mud at the cathode on the ground. During the electropolishing, about 10–40 μm of steel

Table 11.4 Acid treatment of stainless steel [20, 21].

Action	Acid (mixture) (nitric, sulfuric, phosphoric acid)	Hydrofluoric (HF)	Additives	Temperature (°C)	Duration (min)
Pickling "classic"	10–28% HNO_3 (50%)	3–8%	0.1% Surfactants	15–60	20–300
Pickling "nitrate-free"	10–25% H_2SO_4/H_3PO_4	3–5%	1–5% H_2O_2	15–60	20–300
Elektropolishing	1 : 1; H_2SO_4 (96%)/H_3PO_4 (86%)	—	—	40–75	2–20

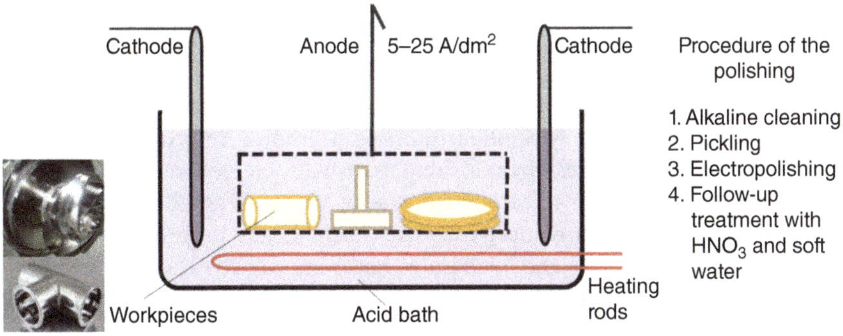

Figure 11.5 Smoothing of stainless-steel surfaces by electropolishing (2–20 minutes at 40–75 °C).

are dissolved from the entire surface, depending on the time, temperature, composition of the acids, and the applied voltage. Thus, not only the craters but also the inhomogeneous superficial layers and ridges disappear. The average roughness is halved by the process. All these advantages especially appreciate the pharmaceutical industry and consequently manufacture their products in electropolished plants.

Microscopy images, as shown in Figure 11.6, demonstrate that edges are converted into roundness. For the achievement of the same smoothness (Ra values of ~0.2 μm), electropolishing is a superior method. It neither leaves residues nor influences the surface mechanically. The method can be completely reproduced worldwide. In contrast to mechanical polishing, the electropolishing leads, besides the smoothing, to a (desired) significant erosion of inhomogeneous layers in the superficial areas. During mechanical polishing, the surfaces are

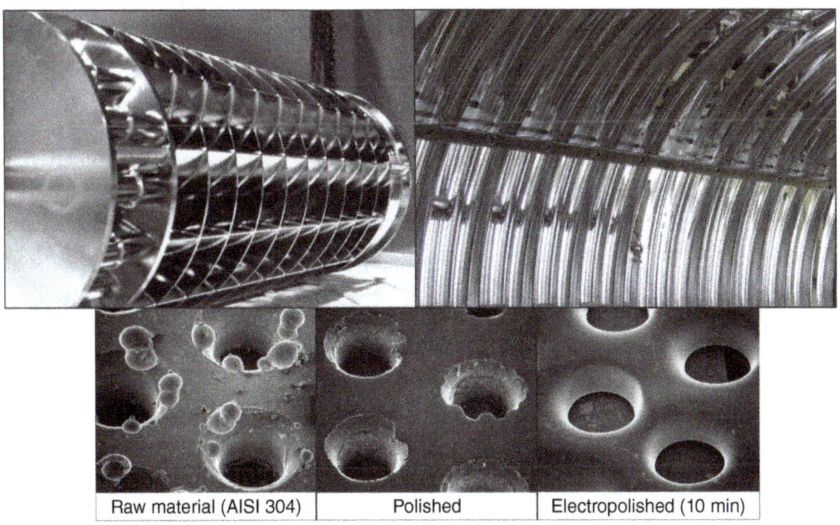

Figure 11.6 Mixers and heat exchanger, electropolished to prevent the formation of deposits, and a filter plate in 50-fold magnification. Source: Courtesy of Poligrat.

(minimal) deformed by heat and pressure. Often, there remain microscopic scratches and traces of metal dust and polishing paste. Advantageously, the mechanical smoothing allows to start with raw material surfaces of much higher average roughness (>3 μm). By using pastes and sanding belts of various grits, especially when successively applied, there is a gradual improvement of the surface. In contrast, the electropolishing is usually carried out with a steel material having an average surface roughness of about 0.5 μm. Therefore, some parts like the welding seams must often be mechanically preprocessed. In the electropolishing process, Fe and Ni dissolve relatively quickly. Thus, the resulting surface is rich in Cr. Cr determines the appearance and corrosion resistance by the passivation layer formed. Electropolishing of titanium-stabilized stainless steel is difficult because TiC does not dissolve, and the carbon content of the steel is relatively high. Therefore, similar alloys without stabilization and with reduced carbon content (AISI 316L, 1.4404, and 1.4435) are preferred. After the electrolysis treatment of work pieces, a rinsing with nitric acid takes place. This converts the hardly soluble heavy metal salts (sulfates and phosphates) in slightly soluble nitrates. The metal nitrates can be rinsed off with soft water, leaving a completely clean metal surface [22].

11.4 Cleaning-in-Place

CIP and SIP systems are designed to optimize and automate essential cleaning and disinfection processes as well as eliminate time-consuming disassembly and assembly work. CIP works with mechanics, chemistry, temperature, and time to achieve synergistic effects in the removal of product residues, deposits, and microorganisms. The optimization of the total production process includes effective cleaning of the plant with the filling lines. The cleaning and disinfection of the entire system takes about 1.5–2.0 hours, in exceptional cases 2.5 hours. Time and effect depend partly on the steel and polishing quality as well as the arrangement of the connections. Second, the recipe of the cosmetic cream plays a role, in particular the viscosity. Third, the cleaning solutions, their composition and concentration, temperature, exposure time, sequence, and velocity generate the positive cleaning result. In addition, the static and rotating spray heads in tanks act mechanically. After optimization of the individual cleaning steps, the process can be automated. This ensures repeatability, allows validation, and minimizes downtime.

For effective cleaning, the production plant is broken down into cleaning cycles. Each circle can be separately cleaned by multiple pumping of the solution through a part of the system, safely locked in the other directions by the special valves (see Figure 11.2). The solutions circulate between the production and CIP system with turbulent flows in the pipelines (velocities about 2 m/s and more). Order of solutions, times, and temperatures are controlled by a program. A typical CIP cycle consists of many steps:

- Prerinse with purified water (PW) to remove residues (if possible, use water from the intermediate rinse).

- Circulation of the heated caustic solution through the facility to remove carbon hydrates, proteins, and fats/oils.
- Intermediate rinse with PW to remove alkali residues.
- Circulation of the heated acid solution used to remove mineral precipitates and protein residues.
- Final rinse with PW to flush out residual cleaning agents.
- Final air blow to remove water drops and moisture remaining after CIP cycle.

First, after cream production, the use of some liters of fresh water (PW) in the product-contaminated tanks and lines allows product recovery. This step is often economical. For the cleaning of the facility, rinsing with hot water shows great effects. The hot water from the intermediate rinse is advantageously used for the prerinsing to save water and energy [23]. Traces of caustic soda do not interfere. The prerinsing water removes product residues and prevents soiling of the detergent tank. This step is followed by the alkaline and acid treatment of the facility, shown in Table 11.5 as scheme of the CIP process for cleaning. The alkali and acid solutions are recycled several times until the activity is reduced. The cleaning solutions can be worked up; for example, there are ready-to-buy solutions with membrane systems. The corresponding CIP facility comprises containers for water, alkaline, and acid solutions, pumps, and heat exchanger, as shown in Figure 11.7. Large CIP plants supply multiple production facilities, creating complex distribution systems, as shown in Figure 11.8.

As an alkaline cleaning agent, 50% sodium hydroxide is suitable (observe the quality level). Or a surfactant-containing alkali solution can be used, available

Table 11.5 Steps of a CIP process; for pharmaceutical products including steam or chemical SIP.

No.	Step	Liquid	Concentration	Temperature[a] (°C)	Contact time (min)
(A) Cleaning					
1	Prerinsing	Some liters PW		20	5
2	Prerinsing	Water from intermediate rinsing		20/75	5
3	Alkali	Sodium hydroxide	0.5–1.5%	70–80	20–30
4	Intermediate rinsing	PW		20/75	5
5	Acid	e.g. Phosphoric or nitric acid	0.5–1.0%	70	20–30
5a	SIP	Steam 1 bar		121	30–40
6	Final rinsing	PW		15	5
7	Final air blow	Sterile air		15–20	2
(B) Disinfection					
8	Rinsing	Propanol-2	Pure liquid	15–20	10–15
9	Drying	Hot air		50–80	5

a) High temperature preferred.

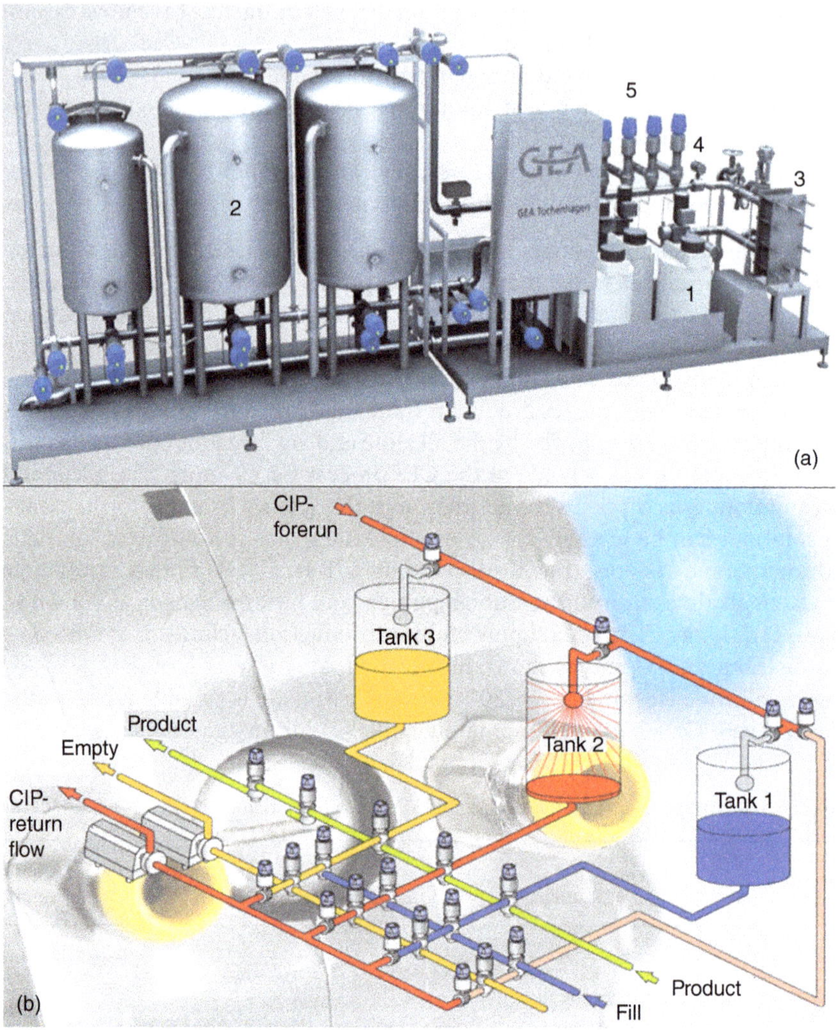

Figure 11.7 Cleaning in place: (a) preparation of cleaning solutions in a small CIP plant; (b) principle of the cleaning during production (tank 2); (1) concentrated solutions; (2) solutions in the concentration of application; (3) heat exchanger; (4) valves; and (5) heated solutions to the emulsification plant. Source: Courtesy of GEA/Tuchenhagen.

as a concentrate in the finished composition (e.g. Ecolab-P3 Cosa® CIP 92) and diluted accordingly in the CIP system. Phosphoric and nitric acid do not attack the stainless steel, have an oxidizing effect, renew the passivation layer on the surface, and remove protein residues or deposits. If these cleaning solutions are not strong enough, the dirt-dissolving effect can be enhanced by oxidizing agents. Suitable for this purpose and for the SIP-Step are hydrogen peroxide, in particular sodium hypochlorite, and in extreme cases peracetic acid, a mixture of acetic acid and hydrogen peroxide, because the oxidizing agents have additionally a strong germicidal effect. A wide range of cleaning concentrates is available on the

Figure 11.8 Complex distribution systems for liquids with dead-space-free valves. Source: Courtesy of GEA/Tuchenhagen.

market. Depending on the intended use, the appropriate agent can be selected. For cosmetic creams, the cleaning effort described in Table 11.5 is usually sufficient. For the cleaning and disinfection of the outer surfaces, there are cleaning solutions that are applied with a cloth or sprayed on. As the solutions evaporate without residue [24], it is a "dry" cleaning/disinfecting.

When buying a cosmetic production line or equipment, attention must be paid on the resistance of the seals against the creams, but especially against the cleaning solutions, their concentrations, and temperatures. Table 11.6 contains

Table 11.6 Resistances of selected sealing materials [25].

Sealing material			EPDM[a)]	FKM[b)]	FFKM[c)]	HNBR[d)]
Maximum application temperature (°C)			135	200	230	140
Medium	Concentration	At permissible operating temperature	*Symbols:* *+ stable* *o reduced service life* *– unstable*			
Alkali	<3%	Up to 80 °C	+	o	+	+
Inorganic acid	<3%	Up to 80 °C	+	+	+	+
Steam		Up to 135 °C	+	o	+	o
Products containing grease	<35%		+	+	+	+
	>35%		–	+	+	+

a) Ethylene propylene diene monomer rubber.
b) About 80% of fluoroelastomers and vinylidene fluoride as a monomer (Viton).
c) Perfluoro rubber (similar to PTFE), perfluoroelastomeric compounds containing an even higher amount of fluoride than FKM fluoroelastomers.
d) Hydrogenated nitrile butadiene rubber.
Source: Data from Ref. [25].

examples of different sealing materials. EPDM cannot be recommended because of the lack of resistance to oils. The other three materials, FKM – FFKM – HNBR, may be installed without any problem because caustic soda is used with a concentration of 1.0% (max. 1.5%). For steam treatment, if necessary, a temperature resistance up to 121 °C is usually sufficient.

Treatment of the water depends on the quality of the available drinking or spring water. The water must be hygienic in any case. Calcium, magnesium, and heavy metal ions can affect product quality and make it difficult to clean the system. Therefore, minimum values should not be exceeded. Little effort is required to treat the water when soft water with low heavy metal contents can be used. Under these conditions, a UV treatment for microbiological protection may be sufficient. Otherwise, an upstream desalination via reverse osmosis (RO) is recommended for the production of the purified water for cosmetics. There are many other technical ways to make purified water. The coupling of RO with downstream UV treatment represent today's standard.

11.5 Learnings

✓ The Cosmetics Regulation prescribes a production in accordance with the GMP guidelines.

✓ The guidelines refer with main focus to the
- Hygienic design of the premises and facilities.
- Training and cleanliness of the staff with exact defined areas of responsibilities.
- Production procedure and quality lab, operating according to written instructions.
- Traceable documentation of the production process and the quality checks.

✓ The design of the system should conform to the EHEDG and 3A standard with respect to the
- Dead-space-free, hygienic design of the plant and the equipment.
- Material – stainless steel (AISI 316L = Din 1.4404) with polished surfaces (mirror finish) should be processed.
- Drainability and cleanness – the system should be easy to empty and clean/disinfect with an automated process (CIP) to reduce microorganisms.

✓ The CIP cleaning consists of the rinsing with water, alkali, and acid solution at elevated temperatures.

✓ Disinfection takes place by rinsing with propanol-2.

References

1 Gute Herstellungspraxis, Wikipedia, 2019. https://de.wikipedia.org/wiki/Gute_Herstellungspraxis (accessed 08 May 2019).

2 Good manufacturing practice, From Wikipedia, the free encyclopedia, 2019. https://en.wikipedia.org/wiki/Good_manufacturing_practice (accessed 08 May 2019).

3 EF*f*CI, Discover the European Federation for Cosmetic Ingredients, 2019. http://effci.com/ (accessed 08 May 2019).

4 Regulation (EC) No 1223/2009 of the European Parliament and of the Council of 30 November 2009 on cosmetic products. https://eur-lex.europa.eu/legal-content/EN/TXT/PDF/?uri=CELEX:32009R1223&from=DE (accessed 26 July 2018).

5 Beuth Publishing DIN, DIN EN ISO 22716:2008-12, 2019. https://www.beuth.de/de/norm/din-en-iso-22716/112877654 (accessed 08 May 2019).

6 EFfCI (2017). EFfCI GMP FOR COSMETIC INGREDIENTS Including the Certification Scheme for GMP for Cosmetic Ingredients, Revision 2017, Prepared by the European Federation for Cosmetic Ingredients; http://effci.com/docs/EFfCI%20GMP%20Guide%20Final%202017.2.pdf (accessed 3 August 2018).

7 EFfCI, List of Certified Companies to EFfCI GMP Standard, 2018. http://effci.com/docs/List%20of%20Certified%20Companies_2018-07-24.pdf (accessed 10 August 2018).

8 GMP guidelines, Annex 9, Manufacture of Liquids, Creams and Ointments. https://ec.europa.eu/health/sites/health/files/files/eudralex/vol-4/pdfs-en/anx09_en.pdf (accessed 3 August 2018).

9 IKW (2017). Kosmetik-GMP, Checkliste zur Selbstbewertung, Industrieverband Körperpflege und Waschmittel e.V., 3. Auflage, Sept. 2017; https://www.ikw.org/fileadmin/ikw/Schoenheitspflege/2017-09_GMP-Checkliste_zur_Selbstbewertung.pdf (accessed 3 August 2018).

10 EHEDG guidelines. https://www.ehedg.org/guidelines/ (accessed 14 August 2018).

11 EHEDG (2018). Overview of EHEDG Guidelines by Topics. https://www.ehedg.org/fileadmin/user_upload/General/Overview_EHEDG_Guidelines_by_Topics_2018_04.pdf (accessed 13 August 2018).

12 Hauser, G. (2008). *Hygienegerechte Apparate und Anlagen*. Weinheim: Wiley-VCH Verlag GmbH.

13 Bode, U. and Wildbrett, G. (2006). Anforderungen an Werkstoffe und Werkstoffoberflächen bezüglich Reinigbarkeit und Beständigkeit. *Chem. Ing. Tech.* 78 (11): 1615–1622.

14 Behrens, D., Kreysa, G., and Eckermann, R. (1988–1992). *DECHEMA Corrosion Handbook – Corrosive Agents and Their Interaction*, vol. 1–12. Wiley-VCH Verlag GmbH.

15 https://www.edelstahl-rostfrei.de/downloads/iser/mb_830.pdf (accessed 24 August 2018).

16 Damstahl, Preisentwicklung, Legierungszuschläge, Rohstoffpreise, Preisentwicklung, 2019. http://www.damstahl.de/Default.aspx?ID=3724&startdate=18%2F08%2F2017&enddate=18%2F08%2F2018&category=4404 (accessed 08 May 2019).

17 Statista, Stahlpreisentwicklung anhand des Stahlpreisindex in Deutschland in den Jahren 2005 bis 2018, 2019. https://www.ikb.de/MediaLibrary/b6a4a61e-a82c-4885-91a3-6e34e223ad15/161206_Q4%202016_Rohstoffpreisinfo.pdf (accessed 08 May 2019).

18 Hess, D., Krämer, P., Pießlinger-Schweiger, S. et al. (2006). *Beizen von Edelstahl*, 3e. Düsseldorf: Informationsstelle Edelstahl rostfrei https://www.edelstahl-rostfrei.de/page.asp?pageID=1582 (accessed 05 May 2019).
19 Hauser, G., Curiel, G.J., Bellin, H.-W. et al. (2004). *Hygienic Equipment Design Criteria*, 2e. Frankfurt/Main: European Hygienic Engineering and Design Group. https://www.goudsmitmagnets.com/data/uploads/Standards%20and%20Guidelines/EHEDG_DOC_08_English.pdf (accessed 19 August 2018).
20 Kosmac, A. (2010). *Elektropolieren nichtrostender Stähle*, Reihe Werkstoff und Anwendungen, 1e, vol. 11. Bruxelles: Euro Inox.
21 Bettenworth, E., Bovensiegen, E.W.A., Buchwald, W. et al. (2012). *Die Verarbeitung von Edelstahl rostfrei*, 4e. Düsseldorf: Informationsstelle Edelstahl rostfrei https://www.edelstahl-rostfrei.de/downloads/iser/MB_822.pdf (accessed 05 May 2019).
22 Rähse, W. (2014). *Industrial product design of solids and liquids*, Chapter 16, 423–448. Weinheim: Wiley-VCH.
23 Tamime, A.Y. (ed.) (2009). *Cleaning-in-Place: Dairy, Food and Beverage Operations*, 3e. Wiley Blackwell.
24 Ecolab, Drysan™ Oxy Reiniger und Desinfektionsmittel, 2019. https://de-de.ecolab.com/offerings/drysan-oxy-cleaner-and-disinfectant (accessed 08 May 2019).
25 GEA, Aseptic valve technology, Brochure. https://www.gea.com/de/binaries/Aseptic-Valve-Technology-2012-10-EN_tcm24-23513.pdf and: Manual for the Design of Pipe Systems and Pumps, https://www.gea.com/de/binaries/manual-design-pipe-systems-pumps-608e-1408_tcm24-13269.pdf (accessed 08 May 2019).

12

Assessment of the Quality of Cosmetic Creams

12.1 Options for Quality Evaluation

The quality of cosmetic creams is checked by a safety analysis of the formulation and by various measurements. First of all, a safety assessment, performed by a specialist (see Chapter13), demonstrates the safe use of the formulation. Second, immediately after production (laboratory, pilot, and production), the pH value, viscosity, and stability of the emulsion; the number of microbes; and the microbiological stability as well as the compatibility are measured by default. Furthermore, physical measurements and visual examination at the skin before and after application characterize the cream effect (Section 12.3). Measurements provide scientifically comprehensible and reproducible results, which can be compared with other creams and should be considered in detail here. Another option is to interview selected customers about their assessment before the market launch (Figure 12.1). Depending on the recipe, this happens once or several times during the application after one week up to three months.

The emulsion stability can be checked by a centrifugal test. An example, frequently used (from us) and proven, is the 40 °C/30 000 g check. For this purpose, the emulsion is warmed carefully to 40 °C with stirring, then poured into the centrifuge tubes of a laboratory centrifuge, and centrifuged for five minutes at 30 000 g. All hygienic samples that pass this test without phase separation have been stable for a long period, usually several years under the temperature conditions of typical storage. As contamination and temperature/humidity influences during use of the cream are unpredictable, the shelf life should be limited to 3, 6, or 12 months after opening, depending on the vessel in which the cream is located (pot 3 m, airless dispenser 12 m).

After production, the first test on humans is the so-called patch tests (epicutaneous test), carried out to ensure the compatibility of the cream. Several cream patches are attached to the back or upper arm of 12–24 voluntary subjects and removed after approximately 48and 72 hours. During the test phase, their backs/upper arms should not come into contact with water. Several subjects should have a rather sensitive skin to cover all skin types and to be on the safe side in the test. The exposed areas are assessed by a physician or another expert. Normally, the skin remains without abnormalities. If changes occur, they are usually due to an allergic reaction (contact dermatitis; [1]). These reactions

Cosmetic Creams: Development, Manufacture and Marketing of Effective Skin Care Products,
First Edition. Wilfried Rähse.

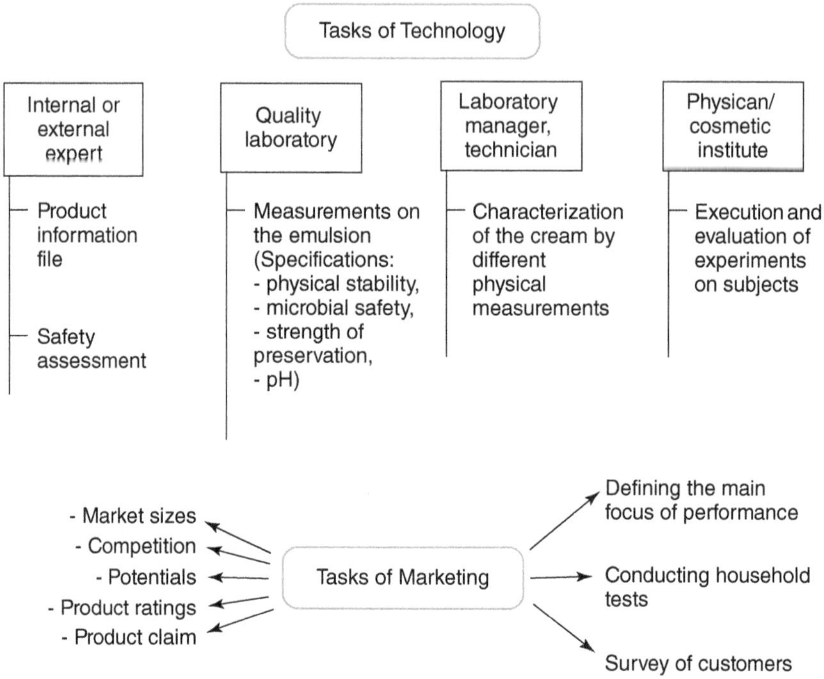

Figure 12.1 Tasks of technology and marketing in the evaluation of a product.

require changes in the formulation. Arising comedones must be evaluated. The appearances are only problematic if the comedones occur in several people and not represent isolated cases. The relevant substances that cause comedones are mostly known and should be avoided.

12.2 Microbial Checks

The production of cosmetic creams is not sterile but takes place with very low germ counts. Therefore, some germs can be detected in the products. The SCCS (Scientific Committee on Consumer Safety) has defined the accepted number of microorganisms and differentiates two cases (Table 12.1). In fact, the application of the cream is predetermined, but the consumer can use the product differently. For this reason, compliance to the stricter values is recommended. As a rule, the microbial load after good manufacturing practice (GMP) production in an European Hygienic Engineering and Design Group (EHEDG) compliant system is below 100 cfu/g. Pathogenic germs must be absent. The SCCS proposals are not legally defined but are considered by the producers as such. The manufacturer is responsible for compliance with the recommendations of the SCCS. If the values are exceeded, the product is out of specification and cannot be used for tests or sold. However, this case is rare.

The stress test according to DIN EN ISO 11930:2013-10 [2] allows the evaluation of the effectiveness of the preservative. In the test procedure, a sample of

Table 12.1 Microbiological limits for cosmetics; European Standard EN ISO 17516:2014 Cosmetics – Microbiology – Microbiological limits (according to SCCS).

Microorganisms	Limits for cosmetics	Limits for Cosmetics at increased risk[a)]
Total aerobic mesophilic Microorganisms (bacteria plus yeast and mold)	1000 cfu/g	100 cfu/g
Escherichia coli	Absence in 1 g or 1 ml	Absence in 1 g or 1 ml
Pseudomonas aeruginosa	Absence in 1 g or 1 ml	Absence in 1 g or 1 ml
Staphylococcus aureus	Absence in 1 g or 1 ml	Absence in 1 g or 1 ml
Candida albicans	Absence in 1 g or 1 ml	Absence in 1 g or 1 ml
Out of limits	>2000 cfu/g	>200 cfu/g

a) Products for use in the eye area, mucous membranes, and injuries as well as in the elderly and children under age 3; cfu: colony forming units.

Table 12.2 Acceptance criteria according to the norm DIN ISO 11930.

	Required logarithmic decrease[a)]			
	0 h	7 d	14 d	28 d
Criterion A				
Bacteria	—	≥3	≥3[b)]	≥3[b)]
C. albicans	—	≥1	≥1[b)]	≥1[b)]
A. brasiliensis	—	—	≥0[b)]	≥1[b)]
Criterion B				
Bacteria	—	—	≥3	≥3[b)]
C. albicans	—	—	≥1	≥1[b)]
A. brasiliensis	—	—	≥0	≥0[b)]

a) A deviation of 0.5 log steps is permitted.
b) No increase in comparison to the last count.

the cream is initially inoculated with specified microorganisms (each germ individually). Then, the storage of the inoculated product takes place under defined conditions. After the specified times, the germ counts are determined, also the germ development. If this expires according to the specifications, the microbial stability is sufficient (Table 12.2). In most cases, the reduction of the microbial count by more than 3 orders of magnitude is achieved after 14 days with the stated amount of sorbic acid at pH 5.

Test strains are as follows:

- *Pseudomonas aeruginosa* ATCC 9027
- *Staphylococcus aureus* ATCC 6538

- *Escherichia coli* ATCC 8739
- *Candida albicans* ATCC 10231
- *Aspergillus brasiliensis* ATCC 16404

Additional strains may be added, in particular germs that have already caused difficulties for the company. The use of mixed cultures cannot be recommended. According to norm, the germ counts are made after 7 (exception: *A. brasiliensis*), 14, and 28 days. If the result is ambiguous because the reduction is slow, the stress test will be continued to provide further information on the long-term development of the germ counts in the product.

Meeting criterion A, the product is sufficiently protected against microbial proliferation. If the counts only fulfill criterion B, further safety measures must be taken to reduce the microbial risk during use and to meet the requirements of the standard (nitrogen, airless-dispensers, etc.).

Microorganisms need an aqueous medium. Therefore, they multiply well in oil in water (o/w), poor in water in oil (w/o) emulsions. Especially, water-rich emulsions at a pH around 6.5 and a temperature between 30 and 40 °C in contact with oxygen form ideal conditions for proliferation. During growth, microorganisms feed on the ingredients of the cream. They excrete enzymes that cause a degradation of (natural) polymers to oligomers and monomers. As a result, the emulsion loses significant viscosity, becomes more fluid, and finally breaks down into the two phases. Because of the consumption of oxygen, the intracellular utilization of nutrients may lead to fermentation processes that form gases or organic acids [3]. This may be recognized by the formation of bubbles and the lowered pH that causes an odor change of the cream, perceptible by the customer. Rarely, pathogens are also found after infections. The risk for the user cannot be estimated, as type, amount, and pathogenicity as well as the application area play a role. The customer complains about infected products during the warranty period and throws mostly changed older products into the dustbin. Both mean a loss of image for the manufacturer. The customer will no longer buy his products.

Therefore, good, compatible preservation is important. The market is increasingly offering "preservative-free" creams because some less recommended preservatives have negatively impacted the image of preservation. The so-called preservation-free products use the antibacterial effect of some ingredients for preservation. Such ingredients show dual effects and are therefore considered as active ingredients, less as a preservative and not as a preservative within the meaning of the Cosmetics Regulation. Of these, there are over a hundred substances, especially extracts and essential oils (thyme, sage, and tea tree). Preservative-free products of the market preferably inhibit microbial proliferation with mono-, di-, and trivalent alcohols (ethanol, 1,5-pentanediol, and glycerin) mixed with the other natural ingredients. The number and amount of required ingredients as well as their smell limit the formulation options. For conventionally preserved products, the choice of ingredients is not limited and the microbial inhibition is better, so that such products are preferable.

12.3 Specifying the Quality of Cosmetic Creams by Physical Measurements

12.3.1 Probes and Devices

The marketing statements should consist of a summary of the physical measurements and then they are verifiable. Scientifically guided measurements enable the determination of the quality of a skin cream. This is shown in an example. First, the untreated facial skin of 24 subjects is characterized using different methods according to Table 12.3. The persons are divided into two equivalent groups. Afterward, the first 12 volunteers use the test cream with firming and revitalizing properties once a day for eight weeks. Under the same conditions, the second group applies a good, well-known competing product. After defined time intervals, the physical measurements are repeated and compared with the baseline values. At the same time, the subjects can note their assessments via questionnaires. The evaluation of the survey permits the comparison of the subjective impressions, which are influenced by the scent and the feel of the cream, with the findings. After completion of the test phase, many data from the measurements of various physical parameters are available for the evaluation, in fact the base values, values of the competition cream and of the own cosmetic cream. These allow a classification of the own cream and a statement about the strength of the effects, in particular the moisture, elasticity, and skin tightening.

The world market leader [7], Courage+Khazaka electronic, offers most of the basic measuring instruments (Figure 12.2) for cosmetic companies and institutes, which are briefly described below. In the scientific field, further analytical methods are used. These often-expensive devices are usually not suitable for routine measurements. Thus, in the scientific research, the skin moisture is measured by various methods: infrared spectroscopy, resonance frequency and nuclear magnetic resonance measurement, desquamation, and impedance measurement.

12.3.2 Moisture and Sebum

In the laboratory of the cosmetics, however, the capacitive measurement with a corneometer most is commonly used because it is simple and precise. The **Corneometer®** provides reproducible values of skin moisture in the stratum corneum. This device detects changes in the dielectric constant (permittivity and dielectric conductivity) by capacitance changes of a precision capacitor due to changed water content. The higher the water content, the higher the values of the corneometer. An influence of substances on the skin surface during the capacity measurement is small. The modern electronics allow fast measurements (one second), so that occlusion effects are avoided. At a penetration depth of 10–20 μm, the measurement takes place in the stratum corneum. Influences due to water deeper skin layers are excluded. Normally, the reading of untreated skin is set to zero, and after cream application, the measured values minus the zero value are

Table 12.3 Standardized in vivo measurements of skin-characterizing properties using various physical methods [4–6].

No.	Measuring device	Physical measuring principle	Skin property
1	Corneometer®	Permittivity, dielectric conductivity	Skin moisture in the stratum corneum
2	Tewameter®	Water evaporation rate via the relative humidity	TEWL, barrier function
3	Sebumeter®	Quantitative grease spot photometry	Skin surface lipids
4	Sebufix®	Qualitative optical method	Sebum production points
5	Corneofix®	Tear-off adhesive tape	Condition of the skin
6	Cutometer®	Suction method, depth, and reset by optical measurement	Skin firmness and elasticity, age of skin
7	CutiScan	Suction method, video	Skin firmness and elasticity, age of skin
8	Frictiometer	Rotating Teflon (PTFE) disc, the torque is measured	Friction of the skin
9	Indentometer	Deformation of the skin, a probe indenter moves the skin, the displacement is measured in millimeters	Skin strength/stiffness
10	FOITS (fast optical in vivo topometry of human skin)	Optical surface analysis of the skin	Skin surface, structure, and wrinkles
11	Visioline®	Image analysis of silicone replica, reflection	Microrelief of the skin
12	Skin visiometer®	3D image analysis of silicone replicas, transmission	Microrelief (fold width and depth)
13	Visioscan®	UVA camera	Skin structure and scaliness
14	MoistureMap	Capacitance-based sensor	Near-surface hydration distribution and the microtopography
15	Skin glossymeter	Reflection of LED light	Shine on skin and hair
16	Colorimeter	Reflected light	Skin and hair color
17	Mexameter®	Absorption/reflection of light	Melanin and erythema
18	Skin pH meter	Combi glass electrode	pH on the skin surface
19	Skin thermometer	Infrared temperature measurement	Skin temperature

TEWL, transepidermal water loss

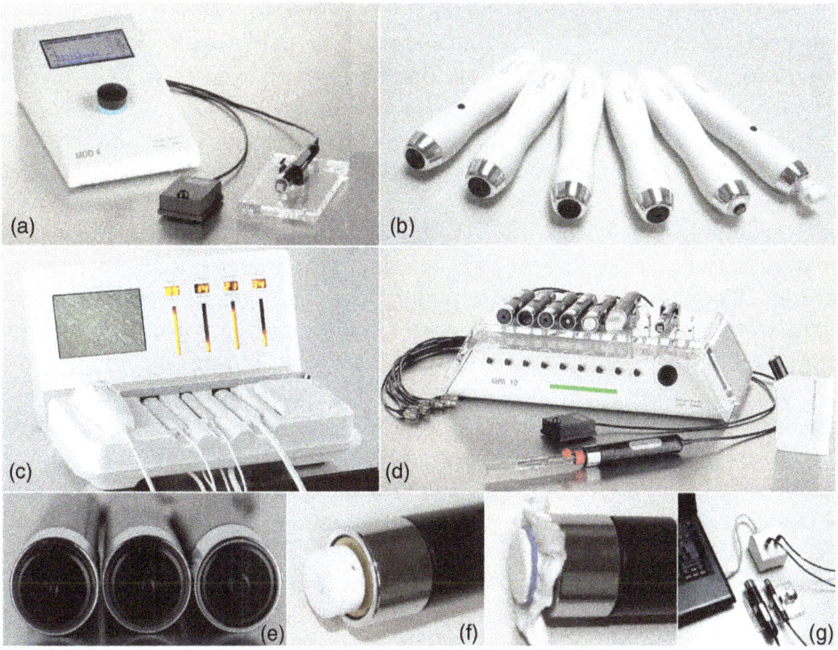

Figure 12.2 Probes and devices for the characterization of the skin for routine measurements: (a) standalone device (Corneometer), (b) wireless probes transmit the data to the receiver, which is connected with a computer, (c) various probes with an evaluation unit, (d) modular multiprobe adapter system, probes, and basic device with connection to a PC, (e) probes of the Indentometer, (f) probes of the Frictiometer, and (g) connection to the computer via USB. Source: Courtesy of Courage+Khazaka electronics.

given in % (see Figure 4.14). The expiry is traced over a period of 24–48 hours. For dry and normal skin, the reading should be above 10% about 24 hours after cream application, with pure moisturizer creams even higher.

Water from the body diffuses through the epidermis and evaporates continuously over the skin. This natural process is called transepidermal water loss (**TEWL**). The triggering takes place via the moisture gradients in the skin. Depending on the location and condition of the skin, the amount of evaporated moisture varies. The TEWL is an indispensable parameter for the evaluation of the barrier function of the skin. Even the slightest damage in the stratum corneum, invisible to the human eye, can be determined at an early stage. In particular, rough skin areas have a high TEWL. The TEWL is measured with a simple, noninvasive method. Here, the flow of the water vapor is determined in g/(m h). Inside a hollow open cylinder, the **Tewameter®** probes measure the density gradient of the water evaporation from the skin indirectly through two superimposed sensor pairs (temperature and relative humidity). This means, each of the two probes measures the partial pressure of the water vapor, with the gradient between the two probes being directly proportional to the degree of evaporation. Taking the length and diameter of the open measuring cylinder into account, the measured values can be converted into the TEWL. This arrangement allows to measure the TEWL continuously without an influence of the microclimate. The sensor will only deliver stable readings after about 15 seconds. Typical measured values on the forearm will be 6–11 g/(m^2 h), respectively, 9–17 g/(m^2 h) for the more heavily loaded backs of the hands. High values (>25) indicate that the stratum corneum is not intact. An effective skin cream should be able to repair the skin barrier within a few days after repeated use.

The **Sebumeter®** is used worldwide for the reproducible quantitative determination of the fat content of the skin surface. Water does not affect the measurement. Upon touching the skin with the probe, a semitransparent plastic film as part of the sebumeter cassette becomes transparent, in dependence on the amount of the sebum. The frosted tape is then inserted into the device. During evaluation by the base station, light is passed through the plastic film and reflected by a mirror. By means of a photocell, the measurement of the light transmission takes place. The intensity is a measure of the sebum content of the skin surface at the measuring point. The measuring cassette releases 64 mm^2 of the matted special tape (0.1 mm) per measurement. With a slider, the band can be easily transported. The cassette contains plastic film for about 400 measurements. The result allows the classification of the skin in different condition: low oil, normal, and oily. This is important for the choice of the cream. Skin with little oil needs night creams (w/o); with a rather oily skin, a night cream is not recommended but an oil-poor day cream (o/w). **Sebufix®**, where the oil on the skin leaves spots on the tape, qualitatively measures sebum production of the various sebaceous glands. The evaluation ensues optically and supplements the statements of the sebumeter.

A special foil (**Corneofix®**) is glued to the skin and torn off. The number, size, and thickness of the corneocytes attached to the film describe the condition of the stratum corneum. From dehydrated or even damaged skin, thick, large corneocytes release and stick to the tape. In contrast, from healthy, moist skin, small regular flakes are peeled off. The desquamation can be evaluated with the Visioscan camera and its software.

12.3.3 Firmness and Elasticity

Cutometer® is the most widely used device in the world for skin elasticity studies and represents a standard measurement in cosmetics. The measuring principle is based on the suction method (Fig. 12.3). Negative pressure in the device, produced with a pump, pulls the skin into the opening of the measuring probe. Depending on the intensity of the incident light, the penetration depth of the skin can be determined without contact. The ability of the skin to resist the suction is a measure of firmness. After aeration of the probe interior, the time dependence of the relapse in the initial state describes the elasticity of the skin and is displayed as a curve (penetration depth in mm/time) in real time on the screen. Evaluating the curves provides information about the elastic and viscoelastic properties of the skin as well as notes on the biological age.

CutiScan measures the temporal skin displacements during circular suction/relaxation with a video camera. The probe contains a suction ring (14 mm

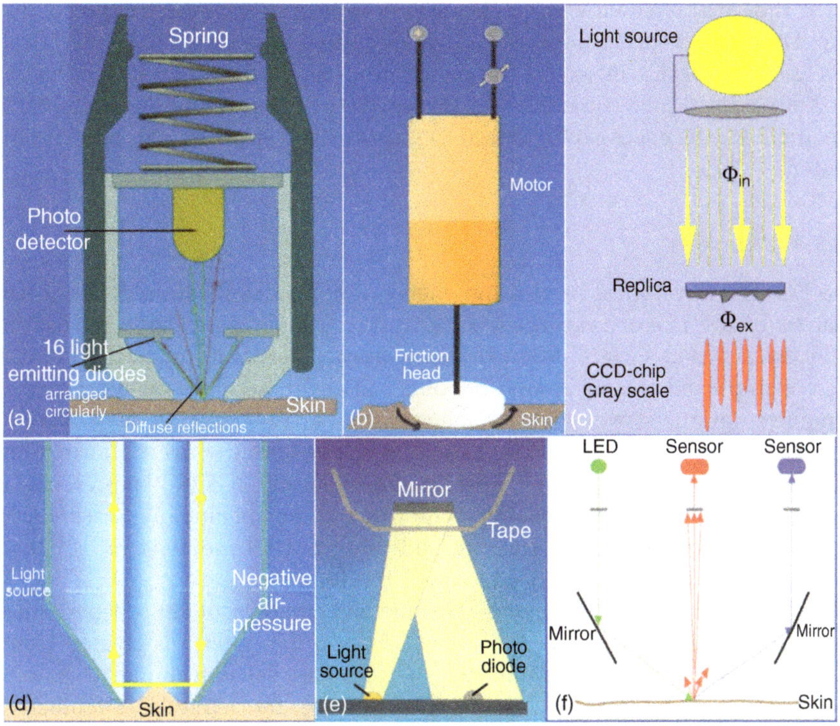

Figure 12.3 Measuring principles: (a) Mexameter®, (b) Frictiometer, (c) Visiometer®, (d) Cutometer®, (e) Sebumeter®, and (f) Colorimeter. Source: Courtesy of Courage+Khazaka electronic.

diameter). An applied constant negative pressure draws the skin uniformly in all directions for a few seconds and relaxes completely again, after releasing the pressure for some seconds. The high-resolution CCD camera in the middle of the suction ring records the displacement of each pixel. Analysis of the movements and creation of graphics with interesting parameters, which characterize the skin, takes place by the connected computer. The higher the skin's ability to resist the displacement, the firmer the skin. After aeration, the skin is not immediately in its original condition because depending on the elasticity, the relaxation time varies.

The probe of the **Frictiometer** consists of a motor, a steering unit, and the friction head with a plain, smooth Teflon disc or other materials. By the friction head, a constant rotational speed is applied onto the skin. The torque is measured and displayed in frictiometer units. Dryness, smoothness, roughness, and scaliness of the skin influence the torque, but the measured values are better suited for comparison than for absolute measurements. The method enables efficacy testing of skin care products. Increasing smoothness and hydration during the cream application can be detected. The process is also sensitive to different kinds of formulations because w/o emulsions reduce the frictionary resistance more than o/w emulsions and the effect lasts longer.

The **Indentometer** allows a quick and easy look at the softness/stiffness of the skin. Driven by a spring force, the skin is deformed through a small cylindrical indenter. The device measures the penetration depth of the pin in millimeter (0–3 mm). The firmer/stiffer the skin, the less deep is the penetration depth. With this method, the success of a cream to improve skin firmness can be demonstrated.

12.3.4 Wrinkles

The **FOITS** (fast optical in vivo topometry of human skin) three-dimensional analysis of the microstructures is a touch-free technique of fringe projection to investigate skin surface. Many technical advancements in the past 20 years, such as improved camera resolution, blue LED light, and laser-supported and computer-optimized overlaying procedures, allow an easy and precise evaluation on the computer. As the scientific and public interest on the wrinkle formation increases worldwide, the new developed generation FOITS-2 represents a one-shot technique with excellent resolutions and time regime [8]. Researchers like to work with the FOITS device and other optical methods [9, 10] that manage without a replica and lead to quick results.

The **Visioline®** device provides an objective analysis of the deeper lines and macro wrinkles such as "crow's feet." First, a skin replica is created from Silflo®. Silflo may be a fast-setting, high-performance calcium aluminate-based, polymer-modified cement. In the Visioline device, the replica is illuminated by oblique lighting at an angle of 35°. In this way, shadows emerge from the mountains of the replica. They represent the wrinkles of the skin that can be digitalized by a high-resolution camera. The camera, which is mounted perpendicular to the replica, allows the arithmetic calculation of the wrinkles (length, depth, and area in micrometer).

Another common tool for determining the depth of small wrinkles represents the Skin-**Visiometer®**, which analyze the topography of the skin surface. For this purpose, a very thin skin replica is made of a special blue dyed silicone. The replica, placed between a parallel light source and a special b/w camera, is irradiated with light. According to the thickness of the silicone material, the replica absorbs the light at known absorption rate of the blue pigments. This copy reproduces the heights and depths of the skin as a negative. Wrinkles appear as mountains that barely let the light pass. The light passed through is proportional to the incoming light, the thickness of the material, and the material constant. As the coordinates of each pixel are known, an accurate digital image can be generated out of the light points using the 256 gray levels. The evaluation happens within one second and provides the standard roughness parameters including further parameters that describe the skin topography. A colored 3D image can be displayed immediately. The method is precise and suitable for all skin irregularities.

For direct examination of the skin surface, the **Visioscan®** device is useful. Visioscan consists of an UVA light video camera with high resolution. The images characterize the structure of the skin and the level of dryness in a short time. The camera has an annular UVA light source for even illumination of the skin, without danger to normal human skin. In the presentation, wrinkles appear dark and the scaliness bright. With this economic and competitive method, pigmentation under the skin surface can be shown very well. On the computer connected via USB, the evaluation method SELS (surface evaluation of the living skin) analyzes the gray level distribution (256 levels) and permits the calculation of four clinical parameters that describe quantitatively and qualitatively the skin surface: skin smoothness, skin roughness, scaliness, and wrinkles. A colored 3D image of the skin can be displayed.

The capacitance-based sensor of the **MoistureMap** device gives graphical information on the near-surface hydration distribution and the microtopography of skin. The sensor measures the penetration of an electromagnetic field. Water in the skin is conductive and reflects the signal making the resulting pixel darker. The nonconductive material shows lighter pixels on a scale of 256 gray levels. MoistureMap displays the distribution of hydration on the skin surface, but not the absolute values. The evaluation of the image is done by a special analysis software and complements the results of corneometer and tewameter.

12.3.5 Gloss and Color

For the acceptance of a cosmetic product, the skin must not look greasy after applying a cream. The shine is measured by the **Skin Glossymeter**. Parallel white light generated by LEDs in the probe head hits the skin via a mirror at a flat angle. Reflected from another mirror, the measurement considers the portion of directly reflected light (gloss), but also the diffused scattered portion from the skin surface. The formulation and claim support for skin care, hair care, and decorative cosmetics (lipsticks, makeup, etc.) require such measurement results. Moreover, it is possible to determine the skin shine reducing ingredients in facial care.

When measuring the skin color with the **Colorimeter**, the probe sends circularly arranged LED light on the skin. The skin reflects a part of the light, which is measured in the probe. The raw data are corrected in the computer to adapt them to standard values. Smallest color changes in the skin can be detected. Therefore, the device is ideally suited for comparisons.

Mexameter® measures the coloring of the skin by reflectance, caused by melanin and hemoglobin (erythema). The probe emits three specific light wavelengths. A receiver measures the light reflected by the skin. The quantity of light absorbed through the skin can be calculated by a computer, as the quantity of emitted light is known. For melanin, the specific wavelengths are chosen to match the different absorption rates by the pigments. The same applies to the erythema, where the wavelengths must correspond to the spectral absorption peak of hemoglobin. The measurement yields in one second the two results: the melanin and erythema index.

12.3.6 Support of Advertising Claims

There are three ways to support advertising claims. First, an active ingredient of the formulation can be highlighted. This only makes sense if the effects are known and the substance is present in comparatively large quantities (for instance, cream with Q10). Second, the effect can be demonstrated by a physical measurement,

Table 12.4 Proposals to back up a meaningful claim by physical measurements.

No	Cream type	Main property	Measuring device (selected from Table 12.3)
1	Moisturizing	Moisturizing the skin	Corneometer®, Tewameter®
2	Vitalizing	Vitalizing the skin	Cutometer®, Indentometer, FOITS, Corneometer®
3	Antiaging	Tighten the skin, reduce wrinkles	FOITS, Cutometer®, Corneometer®
4	Smoothing cream	Smoothes rough skin, also neurodermitic skin	Frictiometer, Tewameter®, Cutometer®, Sebumeter®
5	Sunscreen (also spots)	Protection against UV radiation	Mexameter®, Colorimeter, Corneometer®
6	Firming	Tightening of the skin	Cutometer®, Indentometer, Corneometer®
7	Rejuvenation	Vitalizing and tightening of the skin	FOITS, Cutometer®, Indentometer, Corneometer®

which should be done comparatively. Examples display Table 12.4. In principle, all measuring devices can contribute. However, the quality of the cream should be determined on one or two devices compared to market products. If the claim is not confirmed by the measurements, the recipe should be changed accordingly. The third possibility is to interview customers. For this purpose, 20–200 people of the target group, differentiated by country, age, gender, and habits, get a sample for an application of at least two weeks, up to eight weeks for antiaging creams. Their experiences with the cream, also in comparison to a previously used cream, are registered in a questionnaire and analyzed by the marketing. If the customer group confirms the effect positively, the claim can be used.

12.4 Example of a Cream Test by Customers

In Germany, numerous cosmetic institutes exist to conduct studies on testing newly developed creams. They have a long list of probands from which they can pick out appropriate persons, in most cases 12–30, and modern physical measuring methods. In addition to the statements of the test persons, their experts also assess the appearance, usually by comparing the treated with the untreated skin. Moreover, the study of creams that support therapeutic agents is offered by some dermatologists and university institutes.

An example should demonstrate how a complicated test on customers of the target group can be performed. To be tested is a new cream from the transition zone between therapeutics and cosmetics. Based on considerations of the known effects of atopic dermatitis on the skin, a cream has been developed that can smooth a rough, neurodermitic skin within a few days. As many people suffering from atopic dermatitis stop the use of a cortisone-containing therapeutic ointment after cessation of the exacerbation, this cosmetic cream should then be used for care, smoothing, and calming (not a substitute for the cortisone cream, but a supplement). In a test on subjects with atopic dermatitis, the effect should be insured. A dermatologist with large practice carried out the examination and monitored the patients. As the doctor did not have enough neurodermatitis patients as subjects, others have been added. Table 12.5 shows the results.

Table 12.5 Test of a cream for smoothing neurodermitic skin by a dermatologist.

				Assessment of the subject		
No	**Disease**	**Area**	**Optical rating**	**Itching**	**Smoothness**	**Remark**
1	Atopic dermatitis, nummular form	Neck	+	+	+	Eczema completely healed only by the provided cream!!! Patient very satisfied. Will buy the cream
2	Atopic dermatitis	Crook of the arm	–	+	o	Atopic dermatitis is currently in exacerbation. On acute eczema, the cream burns
3	Atopic dermatitis, sebostasis	Elbow	+	o	+	Got well. Patient very satisfied
4	Neurodermatitis	Elbow	–	o	+	In the test area, redness and pruritus. Deterioration after applying the cream
5	Neurodermatitis	Neck	+	+	+	Patient is very satisfied, helped well. Eczema much better. No more reddened
6	Neurodermatitis	Arm	+	o	o	Keratotic areas are already flattened
7	Neurodermatitis	Crook of the arm	+	+	+	With the cream, the slight eczema on both areas healed well
8	Sebostatic dermatitis	Hand	o	o	o	The subject finds the cream is not good, thinks she has brought nothing
9	Sebostatic dermatitis	Hand	+	+	o	When is the cream to buy? – Quote: patient is very satisfied with the cream
10	Sebostatic dermatitis	Knee	+	o	o	Slight improvement in the test field. Patient finds the lotion too fat
11	Sebostatic dermatitis	Foot	(+)	o	+	—
12	Exsikationsekzem	Body side	(+)	+	o	Flanks very well improved, skin smoother
13	Dyshidrotic hand eczema, atopy	Hand	o	+	o	Skin has become noticeably smoother. Patient very satisfied

(Continued)

Table 12.5 (Continued)

No	Disease	Area	Optical rating	Assessment of the subject		
				Itching	Smoothness	Remark
14	Hyperkeratosis circumscripta	Elbow	+	o	o	Good response of the treated area. The woman finds the cream well
15	Hyperkeratosis circumscripta	Elbow	+	o	o	Significant smoothing and reduction of the inflammatory site
16	Keratosis follicularis	Arm	+	o	–	Skin softer to the treated areas
17	Keratosis follicularis	Upper arm	+	o	+	The keratoses have become soft and the perifollicular inflammation has regressed. Patient very satisfied
18	Keratosis follicularis	Thigh	+	o		—
19	Nummular eczema	Forearm	+	+	o	Flanks very well improved, skin smoother
20	Nummular eczema	Foot	+	o	+	Helped very well. The place has healed. Patient very satisfied with the cream
21	Nummular eczema	Neck	(+)	+	o	Almost healed only with the cream! No cortisone needed
22	Nummular eczema	Arm	o	o	o	No improvement of the findings
23	Nummular eczema	Hollow of the knee	+	+	+	Patient finds the cream very greasy. Eczema healed on the treated side!
24	Psoriasis vulgaris	Hand	+	o	+	Skin smoother, but underlying disease continues to develop, under Enbrel slow regression of efflorescences
25	Psoriasis vulgaris	Knee	+	o	+	Patient finds the product good, skin has become soft
26	Psoriasis vulgaris	Elbow	+	+	+	Skin completely descaled/soft in the test area. Excellent. Patient enthusiastic: "When is the cream to buy?"
27	Dermopan radiation of a T-cell lymphoma of the skin	Forehead	o	o	o	No influence on a radioderm. The patient finds the cream on her face too fat
28	Contact dermatitis	Wrist	+	+	+	Test field much softer than control area
29	Knuckle Pads (pressure points)	Foot joint	+	o	+	The treated area has become much better. The horny layer (thickened) has come off
Sum			23 + 4 o 2 –	12 + 17 o 0 –	14 + 13 o 1 – 1 without indication	

Patient no. 2 should not be evaluated because the cream was expressly designated as unsuitable for the acute phase (on the label under "Application"). In addition, as the cream contains salt, burning on open skin is expected (on the label under "Warnings"). Overall, the result for the other patient is good, especially considering that it is a cosmetic cream with known ingredients, mostly from nature, and not a synthetical, therapeutic agent as a specialist for the disease. The positive result preferably relates to the visual judgment of the dermatologist, who compares the treated with untreated areas. Characterized by a strong smoothing effect on rough skin, the cream is not only suitable for people with skin problems but also improves the skin condition of each person. A further development, similar to the recipe shown in Table 8.6, is considerably better. The fat impression is often based on too much applied amount of cream; otherwise, the oil content in the cream can be easily reduced or the oil composition changed. This is valid for all proposed formulations.

Marketing particularly likes the survey of a large number of people from the target group (Table 12.6). An antiaging cream may be an example. For the test, 150–250 people of the target group are provided with the cream, in some cases with a second dispenser of the competition or another own product (e.g. for the

Table 12.6 Methods for assessing cream quality by comparing treated and untreated skin.

Test target	Compatibility	Subjective assessment by probands and an expert	Objective quality assessment by measurements	Subjective assessment by customer
Execution	Expert of the company; cosmetic institute	Cosmetic institute; dermatologists and institutes of the universities	Cosmetic institute; institutes of the universities	Marketing with agencies
Number of participants	12–24	12–30	12–30	150–250
Action	Patches	Providing the participants with cosmetic cream	Measuring on the skin with various physical probes	Providing the participants with cosmetic cream
Analysis	Evaluation by visual examination	Evaluation by visual examination of the expert in combination with subject opinion	Objective quality evaluation based on measured values	Subjective quality rating after prolonged use including fragrance and feel
Significance	Low; limited to compatibility	High; the statements describe the effect of the cream	High; essential measured values characterize the skin before and after cream use	High; positive customer assessment allows hints on market success
Validity	Country dependence	Similar statements globally	International validity of the measured values	Country dependence

left and right sides of the face). The group consists mainly of women between the ages of 45 and 75, who at the end of the six to eight weeks of application submit their opinion in a questionnaire. Subsequently, the evaluation is carried out by marketing according to various criteria:

- Age group (45–55, 56–65, and 66–75),
- Monthly income (€ <1500, <2500, <4000, and >4000),
- Previous frequency of use of an antiaging cream (never, once a week, every 2–3 days, and daily),
- Self-assessment of skin condition (smooth, few small wrinkles, and noticeable wrinkles).

In addition to the survey of the cream properties, the following questions are particularly valuable:

- Do you find the new cream interesting?
- Do you benefit from the cream?
- Can you confirm the advertising slogan "skin tightening cream"?
- Would you buy the cream?
- For which maximum price in the interval of €12–19 would you buy the cream?

The result of the survey is discussed intensively and finally leads to the decision whether the cream is introduced in this or a slightly modified form in the market or not. In addition, the survey provides information on a market-driven pricing.

12.5 Learnings

- ✓ Immediately after production, the pH value, viscosity, and stability of the emulsion, the number of microbes, as well as the compatibility are measured by default.
- ✓ The number of germs must be less than 1000 cfu/g for most creams and less than 100 for face creams.
- ✓ To ensure the compatibility of a cream, patch tests are carried out. Several patches are attached to the upper arm of 12–24 volunteer subjects and removed after approximately 48 hours. No skin change is positive.
- ✓ To determine the quality of preservation, the microbial stability test is suitable. Inoculated samples are stored under defined conditions. After 14 days, the number of germs should have fallen by 3 powers of 10.
- ✓ Numerous physical measuring methods allow an objective assessment of a cream. Here, measured values of the untreated and treated areas are compared and evaluated.
- ✓ These devices, for example, Corneometer for skin moisture, Tewameter for skin barrier, Cutometer for firmness and elasticity, and FOITS for wrinkles, provide objective values and are used for studies.
- ✓ For marketing test, 150–250 people of the target group are supplied with the cream. After applying, the assessment of the cream takes place via questionnaire.

References

1 Lachapelle, J.-M. and Maibach, H.I. (2012). *Patch Testing and Prick Testing, A Practical Guide Official Publication of the ICDRG*. Berlin Heidelberg: Springer-Verlag.

2 Beuth Publishing DIN, DIN EN ISO 11930:2013-10, 2019. https://www.beuth.de/de/norm/din-en-iso-11930/191730872 (assessed 08 May 2019).

3 Heinzel, M. (2004). Mikrobiologie kosmetischer Mittel, Chapter 13. In: *Kosmetik und Hygiene*, 3e (ed. W. Umbach), 391–407. Weinheim: Wiley-VCH.

4 Fluhr, J.W. (ed.) (2011). *Practical Aspects of Cosmetic Testing*, 1e. Berlin: Springer.

5 C+K, Courage+Khazaka electronic, Scientific Products. https://www.courage-khazaka.de/en/scientific-products/all-products (assessed 08 May 2019).

6 C+K, Courage+Khazaka electronic, General questions on skin measurements at the point of sale. https://www.courage-khazaka.de/en/12-faq/229-cosmetic-consulting-faq-e (assessed 08 May 2019).

7 C+K, Courage+Khazaka electronic, Measurements. https://www.courage-khazaka.de/en/products-for-cosmetic-consulting/measurements (assessed 08 May 2019).

8 Rohr, M. and Schrader, A. (2013). FOITS (fast optical in vivo topometry of human skin). In: *NonInvasive Diagnostic Techniques in Clinical Dermatology* (eds. E. Berardesca, H.I. Maibach and K.-P. Wilhelm), 55–64. Berlin: Springer.

9 Wilhelm, K.P., Elsner, P., Berardesca, E., and Maibach, H.I. (eds.) (2007). *Bioengineering of the Skin: Skin Imaging & Analysis*. Group: Taylor & Francis.

10 Neerken, S., Lucassen, G.W., Bisschop, M.A. et al. (2004). Characterization of age-related effects in human skin: a comparative study that applies confocal laser scanning microscopy and optical coherence tomography. *J. Biomed. Opt.* 9: 274–281.

13
Product Information File (P.I.F.)

13.1 Provisions of the Cosmetics Regulation

The procedure described here for creating a P.I.F. [1] serves as a support for small and medium-sized businesses or is intended for newcomers. Not all elements are required by law. The safety assessment is individually adapted to the cosmetic product.

The Cosmetics Regulation (EU) stipulates in Article 11 that a product information file (P.I.F.) must be available for every product put on the market. The P.I.F. is to be kept by the product owner 10 years after the production of the last batch. The P.I.F. shall contain the following information and data, which shall be updated as necessary [2]:

- A description of the cosmetic product that can be clearly attributed to the product (e.g. by the recipe number);
- The cosmetic product safety report (CPSR) referred to in Article 10(1);
- The production instructions (PIs) and a statement on compliance with good manufacturing practice (GMP) referred to in Article 8 (chapter 11);
- Proof of the effect claimed for the cosmetic product, where justified by the nature or the effect of the cosmetic product;
- Data on animal experiments (e.g. if required in foreign countries, such as partially in China; in Germany, manufacturers of cosmetic products do not carry out any animal tests).

The responsible person, ascertainable by the label of the product, shall make the P.I.F. readily accessible in electronic form to the competent authority in the national language. For the overview, Chapters 1 and 11 complete the P.I.F., on the one side for the correct labeling of the cosmetic cream on the packaging in Section 1.5 and on the other side in compliance with the GMP guidelines (Section 11.1).

13.2 Requirements for the Product Safety Report According to the Cosmetics Regulation

According to ANNEX 1 of the Cosmetics Regulation, the CPSR must contain at least the following points (cited text from [2]):

Cosmetic Creams: Development, Manufacture and Marketing of Effective Skin Care Products,
First Edition. Wilfried Rähse.

PART A – Cosmetic product safety information

1. Quantitative and qualitative composition of the cosmetic product
   The qualitative and quantitative composition of the cosmetic product, including chemical identity of the substances (incl. chemical name, INCI, CAS, EINECS/ELINCS, where possible) and their intended function. In the case of perfume and aromatic compositions, description of the name and code number of the composition and the identity of the supplier.
2. Physical/chemical characteristics and stability of the cosmetic product
   The physical and chemical characteristics of the substances or mixtures, as well as the cosmetic product. The stability of the cosmetics product under reasonably foreseeable storage conditions.
3. Microbiological quality
   The microbiological specifications of the substance or mixture and the cosmetic product. Particular attention shall be paid to cosmetics used around the eyes, on mucous membranes in general, on damaged skin, on children under three years of age, on elderly people, and persons showing compromised immune responses. Results of preservation challenge test.
4. Impurities, traces, information about the packaging material
   The purity of the substances and mixtures. In the case of traces of prohibited substances, evidence for their technical unavoidability. The relevant characteristics of packaging material, in particular purity and stability.
5. Normal and reasonably foreseeable use
   The normal and reasonably foreseeable use of the product. The reasoning shall be justified in particular in the light of warnings and other explanations in the product labelling.
6. Exposure to the cosmetic product
   Data on the exposure to cosmetic product taking into consideration the findings under Section 5 in relation to
   - The site(s) of application;
   - The surface area(s) of application;
   - The amount of product applied;
   - The duration and frequency of use;
   - The normal and reasonably foreseeable exposure route(s);
   - The targeted (or exposed) population(s). Potential exposure of a specific population shall also be taken into account.

   The calculation of the exposure shall also take into consideration the toxicological effects to be considered (e.g. exposure might need to be calculated per unit area of skin or per unit of body weight). The possibility of secondary exposure by routes other than those resulting from direct application should also be considered (e.g. non-intended inhalation of sprays, non-intended ingestion of lip products, etc.). Particular consideration shall be given to any possible impacts on exposure due to particle sizes.
7. Exposure to the substances
   Data on the exposure to the substances contained in the cosmetic product for the relevant toxicological endpoints taking into account the information under Section 6.

8. Toxicological profile of the substances
   Without prejudice to Article 18, the toxicological profile of substance contained in the cosmetic product for all relevant toxicological endpoints. A particular focus on local toxicity evaluation (skin and eye irritation), skin sensitization, and in the case of UV absorption photo-induced toxicity shall be made. All significant toxicological routes of absorption shall be considered as well as the systemic effects and margin of safety (MoS) based on a no observed adverse effects level (NOAEL) shall be calculated. The absence of these considerations shall be duly justified.
   Particular consideration shall be given to any possible impacts on the toxicological profile due to
   - particle sizes, including nanomaterials,
   - impurities of the substances and raw material used, and
   - interaction of substances.

   Any read-across shall be duly substantiated and justified. The source of information shall be clearly identified.
9. Undesirable effects and serious undesirable effects
   All available data on the undesirable effects and serious undesirable effects to the cosmetic product or, where relevant, other cosmetic products. This includes statistical data.
10. Information on the cosmetic product
   Other relevant information, e.g. existing studies from human volunteers or the duly confirmed and substantiated findings of risk assessments carried out in other relevant areas.

PART B – Cosmetic product safety assessment

1. Assessment conclusion
   Statement on the safety of the cosmetic product in relation to Article 3.
2. Labelled warnings and instructions of use
   Statement on the need to label any particular warnings and instructions of use in accordance with Article 19(1)(d).
3. Reasoning
   Explanation of the scientific reasoning leading to the assessment conclusion set out under Section 1 and the statement set out under Section 2. This explanation shall be based on the descriptions set out under Part A. Where relevant, margins of safety shall be assessed and discussed. There shall be inter alia a specific assessment for cosmetic products intended for use on children under the age of three and for cosmetic products intended exclusively for use in external intimate hygiene. Possible interactions of the substances contained in the cosmetic product shall be assessed. The consideration and non-consideration of the different toxicological profiles shall be duly justified. Impacts of the stability on the safety of the cosmetic product shall be duly considered.
4. Assessor's credentials and approval of part B
   Name and address of the safety assessor.

Proof of qualification of safety assessor.
Date and signature of safety assessor"

13.3 Safety Data Sheet

A **safety data sheet** (SDS), material safety data sheet (MSDS), or product safety data sheet (PSDS) is a document that lists information relating to the physical data, transport, application, ecology, toxicology, occupational safety, and health at the use of the substance (see the complete list in Figure 13.1). SDSs are widely used for cataloging information on chemicals, chemical compounds, and chemical mixtures [6] and contribute to the safety assessment. They include instructions for the safe use and potential hazards associated with the ingredient or product, in most cases in pure, undiluted form, but less frequently in solution, as indicated. Some substances are offered in several purity levels (for example, propylene glycol). Logically, the cosmetic or pharmaceutical quality is used. As a rule, the supplier provides not only the MSDS but also an **analysis certificate** for the ordered substance, which can be viewed and printed via download. In the cosmetic cream, the raw materials are present in diluted form as part of the water or oil phase. Natural oils have concentrations between 2% and 15% of the emulsion, vitamins, and other active ingredients between 0.5% and 6%, while the excipients are often between 0.05% and 1%. The safety assessment consists of the individual evaluations of each raw material.

The abbreviated example of potassium sorbate in Figure 13.1 shows the typical structure of the SDS. In Section 1, the material is characterized by the EC and CAS number. Also, the sheet contains all information about the supplier. For the P.I.F., the toxicological profile of each individual component of the cream formulation can be found in Section 11 in the corresponding SDS, in particular the lethal dose for rats and the irritant effects. In addition, the sheet provides information to by-products and solubilities that may be useful for the P.I.F. creation. Note that Section 11 of Figure 13.1 deals with the pure potassium sorbate and the described effects must be transferred in the typical amounts of 0.15–0.35% and also the pH from 8 to 11 to the emulsion pH of 5. The structure of the SDS format is fixed, but the contents can vary from source to source. For example, the important value "lethal dose of 50% rats (LD_{50})," taken orally, varies for potassium sorbate between 3800 and 8000 mg/kg rat in the SDSs of different suppliers, depending on their source. The use of potassium sorbate is restricted to 0.6% (acid) according to the Cosmetic Regulation Annex V/4. A complete, exemplary MSDS is shown in Appendix 2 using the example of nicotinamide (INCI: Niacinamide).

It is a difficult task to extrapolate from rat oral LD_{50} to dermal value in humans. This estimate is based on the margin of safety (MoS), which should be over 100. As the systemic exposure dose (SED) is mostly unknown, a reliable estimate must be sufficient, if the MoS reaches well over 100 (see Section 13.4.5). With unproblematic substances, the values are several hundreds (body lotion) or even several thousands (face cream).

Page 1/9

Safety data sheet
according to 1907/2006/EC, Article 31

Printing date 16.12.2014 Version number 1 Revision: 16.12.2014

SECTION 1: Identification of the substance/mixture and of the company/undertaking

1.1 Product identifier

Trade name: Potassium sorbate

Article number: NC09

CAS Number:
24634-61-5
EC number:
246-376-1

Registration number
A registration number is not available for this substance as the substance or its use are exempted from registration according to Article 2 REACH Regulation (EC) No 1907/2006, the annual tonnage does not require a registration or the registration is envisaged for a later registration deadline.

1.2 Relevant identified uses of the substance or mixture and uses advised against

Application of the substance / the mixture Laboratory chemical , ***also suitable for pharmaceutic use***

1.3 Details of the supplier of the safety data sheet

Manufacturer/Supplier:
Carl Roth GmbH + Co.KG
Schoemperlenstraße 3-5
76185 Karlsruhe
Germany
Telefon: +49/(0)721 5606-0 Telefax: +49/(0)721 5606-149 e-mail: sicherheit@carlroth.de

Further information obtainable from: Department Health, Safety and Environment

1.4 Emergency telephone number:
Poison Centre Munich Telefon +49/(0)89 19240

SECTION 2: Hazards identification
SECTION 3: Composition/information on ingredients
SECTION 4: First aid measures
SECTION 5: Firefighting measures
SECTION 6: Accidental release measures
SECTION 7: Handling and storage
SECTION 8: Exposure controls/personal protection
SECTION 9: Physical and chemical properties
SECTION 10: Stability and reactivity

SECTION 11: Toxicological information

11.1 Information on toxicological effects

Acute toxicity:

LD/LC50 values relevant for classification:		
Oral	LD_{50}	3800 mg/kg (rat) (IUCLID)

Primary irritant effect:

on the skin:
Irritant to skin and mucous membranes.

on the eye:
Irritating effect.

after inhalation:
Slight irritations.
Sensitisation:
No sensitising effects known.

CMR effects:
Germ cell mutagenicity:
No known significant effects or critical hazards.
Carcinogenicity:
No known significant effects or critical hazards.
Reproductive toxicity:
No known significant effects or critical hazards.

Aspiration hazard:
No aspiration toxicity classification.

Specific target organ toxicity - single exposure
The substance or mixture is not classified as specific target organ toxicant, single exposure.
Specific target organ toxicity - repeated exposure
The substance or mixture is not classified as specific target organ toxicant, repeated exposure.

Additional toxicological information:
We have no description of any toxic symptoms.

Further information:
Further hazardous properties cannot be excluded.
The product should be handled with the care usual when dealing with chemicals.

SECTION 12: Ecological information
SECTION 13: Disposal considerations
SECTION 14: Transport information
SECTION 15: Regulatory information
SECTION 16: Other information

Figure 13.1 Structure of a safety data sheet using the example of potassium sorbate (preservative) [3–5].

13.4 Structure of the P.I.F.

The P.I.F. is based on the requirements of Annex 1 of the EU Cosmetics Regulation and consists essentially of the product identification, product description, and safety information as well as the safety assessment.

13.4.1 Product Identification and Description

The product identification could look as shown in Table 13.1. The manufacturer and product owner may be the same company. An importer must be indicated for goods imported from outside the EC. A short product description suffices to characterize a cream. For a moisturizer, a possible description is given in Table 13.2.

13.4.2 Composition of the Cosmetic Cream

The composition is best represented in the form of a table (examples in Table 13.3). The table contains both the common and the INCI name with the EC or CAS number together with the composition and functions of each component of the formulation. It is also appropriate to include the LD50 level of

Table 13.1 Product assignment for secure identification.

Product or Brand name/application area/recipe No./bottle types and sizes	
Manufacturer or contract manufacturer – Company, address	Importer – Company, address
Product owner – Company, address – website	Contact person – Name, address – phone number – E-Mail

Table 13.2 Product description of a moisturizer as an example.

Moisturizer

The cosmetic cream is a moisturizing face cream for daily use. The application ensues in the morning and in the evening. It is a pumpable, easily distributable o/w emulsion based on natural oils such as evening primrose oil, shea butter, hemp, and almond oil. As known moisturizing ingredients, glycerin and sorbitol as well as urea and hyaluronic acid are formulated. Vitamin E operates for antioxidation, panthenol, and nicotinamide for soothing the skin. Thanks to a buffer system, the pH in the finished product is at a constant value of 5.1. The emulsion is offered in an airless dispenser (100 ml) that delivers the cream without air access. All ingredients are described positively in the literature. Label is made in accordance with the Cosmetics Regulation.

Recommended addition 1 (exclusion of problematic substances):

The cream contains no parabens, mineral oils, waxes, dyes, and perfume.

Recommended addition 2:

Copy of the label (Figure 1.2) or write down the list according to Table 1.3.

Table 13.3 Part of the composition of a moisturizer with the Recipe No: type R-number/date.

	I	II	III	IV	V	VI	VII	VIII	IX
No	Ingredient	(%)	INCI name	EC number	Function	Oral toxicity LD_{50} (mg/kg rat)	Irritant	Limit conc.	Supplier/lot No.
1	Water (cosm.)	61.5	Aqua	231-791-2	**S**	—	—	—	Manufacturer/...
2	Sweet almond oil	4	Prunus Amygdalus Sativa Kernel Oil	291-063-5	Emo, **Li**	—	—	—	Lamotte/...
3	Cetyl stearyl alcohol	3.5	Cetearyl Alcohol	267-008-6	**CoE**, Emo, **CA**, RF	2200	Skin, eye	—	BTC, spectrum chemical/...
4	Evening primrose oil	3	Oenothera Biennis Oil	289-859-2	Emo, **Li**, **Nd**	—	—	—	Lamotte/...
5	Hemp oil	3	Cannabis Sativa Seed Oil	289-644-3	Emo, **Li**	—	—	—	Lamotte/...
6	Shea butter	3	Butyrospermum Parkii Butter	293-515-7	**Li**, Emo, CA	—	—	—	Lamotte/...
7	Urea	3	Urea	200-315-5	ASt, Ker, NMF/**Hu**	8200	—	—	Caesar & Loretz/...
8	Glycerol	3	Glycerin	200-289-5	**Hu**, S	>11 500	Eye	—	Caesar & Loretz/...
9	Sorbitol	3	Sorbitol	200-061-5	**Hu**	24 300 NOAEL: 4500	—	—	Caesar & Loretz/...
10	D-Panthenol	2.5	Panthenol	201-327-3/ 240-540-6	**SC**, **PV**	21 000	—	—	Caesar & Loretz/...
	...	...	...	...		...	...		...

ASt, antistatic agents; CA, consistency agents; CoE, coemulsifier; Emo, emollients, skin softener; Hu, humectant; Ker, keraloytic agent; Li, lipids; NMF, natural moisturizing factor; Nd, effective against atopic dermatitis; PV, provitamin; RF, refatting; S, solvent; SC, skin care agents.

oral intake in rats and the potential for irritation, determined on animals (rats, mice, and rabbits). Furthermore, the restricted use according to the Cosmetics Regulation should be mentioned. The last column contains the lot number and the supplier, which provides the ingredient with the MSDS and analysis certificate. The lot number, assigned by the quality lab, links the supplier to the invoice and offers information about the location of the chemical and the quantity available as well as the results of the quality tests. For natural cosmetics, a column should be added that gives information about the certificate of the raw material. Small changes to the formula can be indicated by adding a letter (A–Z). A complete table can be seen in the example (Section 13.5).

By adhering to the correct procedure in the formulation and preparation, the formation of cleavage and other reaction products can be avoided. When formulating, attention must be paid to the stability of the ingredients in the pH range and also to temperatures during the preparation. It is therefore beneficial to provide a buffer system for preparing the water phase directly in the first step, so that all other substances only pass through a pH of around 5. In absolute exceptional cases, ingredients may not be stable at this pH and therefore cannot be formulated. Furthermore, there are advantages to limiting the maximum temperature to 50 °C for water-soluble ingredients and to 60 °C for lipophilic substances in vacuo. Also, reactive substances, especially oxidants, should be avoided.

For storage of the cream, a temperature of 15–20 °C is recommended. To test the compatibility, no animal experiments are carried out.

13.4.3 Toxicological Profile

For cosmetic creams, the emphasis is primarily on the assessment of local toxicity as skin and eye irritation, skin sensitization, and for UV absorption photoinduced toxicity. In case of biologically relevant dermal/percutaneous absorption, systemic effects will also be examined in detail. When certain test results are not available, a scientific justification should be included (proposal of the Scientific Committee of Consumer Safety, SCCS). Toxicological data of the raw materials correspond to the descriptions of the MSDS in Table 13.4 and Figure 13.1 (Section 11, also in Appendix 2). It is advisable to evaluate on the one hand the SDS of several suppliers because they can complement each other, and on the other hand to consider the literature. Always start from a credible result. Most of the raw materials show no primary irritant and CMR effect, aspiration hazard, and specific target organ toxicity (CMR: carcinogenicity, germ cell mutagenicity, and reproductive toxicity). Do not use ingredients with CMR and toxic or highly irritant effects.

In most cases, the skin and eye irritations mentioned refer to the pure substance. These irritations can often be attributed to the concentration and pH value, as in the case of citric acid and potassium sorbate. In the presence of water, the acid produces a very low pH of about 2 or, via the pure salt, arises a high pH of 8–9, which causes irritation. First, the pH value of 5 set by a buffer system characterizes the creams, so that instead of pure citric acid, very little citric is present in a mixture with sodium citrate. On the other hand, the concentrations are only a few per thousand or percent, as with citric or glycerol. Most of the eye-irritating

Table 13.4 Information on the toxicological profile of any raw material from the safety data sheet (example, the structure of the profile is comparable worldwide, while the wording differs from supplier to supplier as well the accuracy and care).

Section 11: Toxicological information

- **Acute toxicity**

LD/LC50 values relevant for classification

Oral LD_{50} value (mg/kg rat), e.g. 2500

Subchronic toxicity (28 d, rat), NOAEL (mg/kg/d)[a]; *(NOAEL may be 1500)*

Percutaneous permeation (%/24 h)[a]; *(P may be 8%)*

– Skin corrosion/irritation
Shall not be classified as corrosive/irritant to skin.
– Serious eye damage/eye irritation
Shall not be classified as seriously damaging to the eye or eye irritant.
– Respiratory or skin sensitization
Shall not be classified as a respiratory or skin sensitizer.
– Summary of evaluation of the CMR properties
Shall not be classified as germ cell mutagenic, carcinogenic, nor as a reproductive toxicant.
– Specific target organ toxicity – single exposure
Shall not be classified as a specific target organ toxicant (single exposure).
– Specific target organ toxicity – repeated exposure
Shall not be classified as a specific target organ toxicant (repeated exposure).

---

a) These lines are often omitted

substances can be completely unproblematic in the dilution with a factor of 20 to thousand, e.g. this applies to glycerol. The third argument relates to the environment of these substances in the emulsion, which has a diluting and dampening effect. Low concentrations and the substance environment suppress a possible impact. When cream gets into the eyes, sensitive individuals can experience eye irritation, even if care creams cause no skin irritation. According to column VII in Table 13.3, a warning should be printed on the label: *Avoid eye contact*. In the example, the care cream is free from solid particles; therefore, considerations of nanoparticles are omitted.

The INCI list can be easily created from the recipe. In the case of essential oils, the INCI information depends on the composition and the application concentration. Some minor components of the essential oil, exceeding a concentration of 0.001% (10 ppm) in the cream, require a declaration in the INCI list (see Table 9.7). Therefore, the length of the list depends on the use concentration and type of the essential oil. As customer service, the supplier usually names the declarable ingredients. Examples for the declaration of higher concentrated essential oils in the INCI list are shown in Table 13.5. Sandalwood (Amyris Balsamifera Bark Oil) and patchouli (Pogostemon Cablin Leaf Oil) might be free of notifiable substances.

In summary, it can be stated that all components of the recipe meet the requirements for a cosmetic use and correspond in type and amount to the Cosmetics Regulation including the SCCS recommendations. Many are also approved for food and some for pharmaceutical products. The evaluation of the certificates of analysis gives no indication of questionable by-products/traces.

Table 13.5 INCI-declarable components of essential oils.

(a) Frequently used oils

Essential oil	Lavender	Rosemary	Eucalyptus	Ylang Ylang
INCI name	Lavandula Angustifolia	Rosmarinus Officinalis Leaf Oil	Eucalyptus Globulus Leaf Oil	Cananga Odorata Flower Oil
Declarable components	Geraniol	Citronellol	Limonene	Benzyl Salicylate
	Limonene	Limonene		Eugenol
	Linalool	Linalool		Isoeugenol
	Cinnamal	Eugenol		Geraniol
	Citral	Geraniol		Farnesol
	Citronellol	Citral		Linalool
	Coumarin			Benzyl Benzoate
				Benzyl Alcohol
				Limonene

(b) Oils from fruit peels

Essential oil	Blood orange	Grapefruit	Lemon	Mandarin
INCI name	Citrus Aurantium Dulcis Peel Oil	Citrus Grandis Peel Oil	Citrus Limon Peel Oil	Citrus Nobilis Peel Oil
Declarable components	Citral	Limonene	Limonene	Citral
	Limonene	Citral	Citral	Linalool
	Linalool	Linalool	Linalool	Limonene
			Geraniol	Geraniol
			Citronellol	Citronellol

13.4.4 Production Instruction

The recipe should follow a short, more generally written PI appropriate to this formulation (Table 13.6). The Cosmetics Regulation requests the production of a PI, which provides information about the production according to the prior art and the GMP standard, for example, that the system has been previously cleaned and disinfected. For the recipes presented here should be mentioned that a constant pH of about 5 already prevails during the preparation of the water phase, which excludes possible side reactions or instabilities of some raw materials. To assess the stability of a raw material and the product, it suffices to consider the pH range of 4.8–5.5 and the temperatures between 4 and 50 °C, for lipophilic substances up to 60 °C.

13.4.5 Analysis Report of the Produced Cream

For each batch, a table should be prepared containing the visual, physical, and microbiological properties of the cream and packaging. To enable traceability

Table 13.6 Plant-independent production instruction for a cream formulation or for a formulation group.

General production instruction for a face cream (recipe group no.: type R-number/date)

The cleanliness of the disinfected production plant is verified.

- First of all, the sensitive substances are separated.
  - List of the sensitive substances.
- Then, the formulation is divided into a water phase with hydrophilic and into an oil phase with the lipophilic active ingredients. For both phases, separated stirring tanks for heating and dissolving exist. The raw materials are weighed in succession and fed via the vacuum.
  - List of the water-soluble ingredients.
- Preparation of the water phase
  Fill the water into the disinfected premix container or emulsifier via a pump or vacuum. From the water-soluble raw materials, the citrate and citric acid are dosed first to adjust the pH, then the other substances follow. The thickening agent (xanthan gum) is finally added with vigorous stirring. While heating and stirring, they dissolve under vacuum at a temperature of 50 °C. Complete dissolution of the polymer can take up to an hour (check the time in the laboratory). After complete dissolution of all components, the pH control is ensued. In exceptional cases, the pH is readjusted to 5.1 ± 0.2.
  - List of the lipid components
- Preparation of the oil phase
  The preparation of the oil phase takes place in the nitrogen-flushed premix container before provision of vacuum. The individual oils and then the lipid substances are sucked in successively with stirring and then heated to a maximum of 60 °C. Immediately after dissolving all components of the oil phase, the metering into the running emulsifier takes place together with the water phase. Depending on the installed capacity and the container size, the emulsification time is set to about 15–60 minutes. After completion of the macroemulsion, the mixture is cooled to 30 °C with stirring.
- Addition of the sensitive substances
  At 30 °C, the sensitive substances are sucked in. Subsequently, the second emulsification is carried out for about 10–45 minutes (standard value: 66% of the emulsification time at 50 °C). Then, the emulsion is ready and transferred to the reservoir of the bottling plant. There, the sampling takes place for carrying out various measurements to release the batch.

All actions before, during, and after production are carried out in accordance with the GMP-guidelines. The dosage of the substances took place as described in the recipe.

according to GMP (Chapter 11), exact details of the batch must be available. This includes the cream name, recipe, PI number, and in particular the batch number and batch size. All information is linked via the batch number. The plant-related PI, marked with the plant number, should accurately describe the steps with places and times during the batch execution as well as the amounts of used raw materials in a target-actual comparison. Packaging information should include the supplier, invoice, warehouse location, types, and sizes as well as the consumed and available quantities. Only pots/bottles/dispensers suitable for cosmetic products are used. The table could look like the one shown in Table 13.7. The points mentioned there should be measured and reported for each batch/cream.

Table 13.7 Analysis report of the produced batch (an example).

Cream type and brand name: Face cream, moisturizer, …; owner	
Batch No:	**type B-number/R-number/date**
Recipe No:	type R-number/date
(Contract) manufacturer:	Name
Production Instruction:	PIXXYYZZdate01
Plant No:	01
Batch size:	3000 kg
Package supplier:	Name
Package:	AD MMNNdate PP-airless dispenser (Type, No.), 50 ml (30 000) and 100 ml (15 000)
Filling line No:	11, 12
Parameter	**Assessment**
Visual ratings – Appearance – Color – Odor	
Physical tests – pH – Centrifugal test – Viscosity – Density	
Microbiological tests – Germ number (<1000 cfu/g respectively <100 cfu/g) – Effect of the preservative (provocation test)	
Package quality – Visual control – Smear test (if necessary)	
Patch test	
Animal tests	Neither the manufacturer of the cream nor the product owner carried out animal experiments.

The preservation is carried out here according to the guidelines of the ABDA-Federal Association of German Pharmacist Associations (20.8.2009): "Recipe Notes: Preservation of water-containing formulations." There, the system potassium sorbate/citric acid in amounts of 0.2/0.1% is recommended for preservation as well as in many other publications [7]. The amounts are clearly exceeded in the proposed formulations with 0.25/0.16 (safe side). The pH is fixed by addition of citrate (buffer system). A great number of tests have shown that this preservation in the described amounts is sufficient to achieve the required reduction in the number of germs by 3 orders of magnitude after 14 days (see Section 5.3.5). There is no counterexample. Therefore, the preservative loading tests in the specified recipes may be desirable but superfluous because the effect was often checked by various official laboratories. Furthermore, alcohol and the salts as well as urea show a preservative effect. The cream is preferably in a closed airless dispenser. The removal takes place without access of air and without contact with the contents. Overall, a wealth of measures ensures safe conservation.

13.4.6 Cream Exposure to the Skin

The intended use of the cream for face, neck, décolleté, eyes, body, arms, hands, feet, and the recommended number of applications per day are indicated on the label. However, it happens that people use the cream differently or additionally. A common case is the use of a facial cream for the hands or a hand cream for the arms. For example, the face, neck, décolleté, and hands must be considered in the exposure calculation for the face cream. Another deviation from the recommendations is that the cream is not used twice, but three times a day ($F = 3$). The applied quantities differ significantly, depending on whether it is a frugal, normal, or abundant application (0.7, 1.0, and 1.3 mg/cm^2). The size of the corresponding areas can be found in Table 13.8, derived from various references. The retention value R deals with the amount, which remains on the skin. For creams, R is 1 (100%), on the other side shows R a value of 0.01 (1%) for shower gels because the gel is rinsed off by the water.

Two cases are assumed for the estimation of dermal exposure. First, the calculation is done with the information on the label that relates to the predetermined application (area and frequency), and second, a different application with a higher cream quantity. The calculation is based on the general equation (Eq. (13.1)). The examples, illustrated in Table 13.9, describe the use of a face cream and body lotion. According to Eq. (13.2), the dermal absorption of raw materials contained in the cream must be estimated as a percentage. The calculation is done preferably

Table 13.8 Typical application amounts of creams for an average person (60 kg; $R = 1$).

Area	Size (cm^2)	Frequency of use (d^{-1})	Specific quantity (mg/cm^2)	Amount (g/d)	Daily exposure (mg/kg/d)
Body lotion (without back)	11 000[a]	1	1	11	183
		2	1	22	367
		2	1.3	29	477
Face cream	570	1	1	0.57	9.5
		2	1	1.14	19
		3	1	1.71	28.5
Neck	320	2	1	0.64	11
Décolleté	800[a]	2	1	1.6	27
Hand cream	860	2	1	1.72	29
Foot cream	1 500[a]	1	1	1.5	25
		2	1.3	3.9	65

a) Estimation

Table 13.9 Dermal exposure of a cream and of a raw material in 5% concentration ($R = 1$) with 10% and 20% permeation.

Cream type	Area	Area size A (cm²)	Frequency per day F (d⁻¹)	Specific amount G (mg/cm²)	Dermal exposure of cream E_{derm} (mg/kg/d)	Percutaneous permeation P (0.1 = 10%)	Systemic expose dose of raw material SED (mg/kg/d) Limit: 10
Face cream	Face (normal use)	570	2	1	19	0.2	0.19
Face cream	Face, neck, décolleté, hands (expanded use)	2 550	2	1 1.3	85 110	0.1 0.2	0.42 1.1
Body lotion	Body – without back	11 000	1	1	183	0.1	0.92
			2	1	367	0.1	1.84
	– with back	16 000	1	1	267	0.2	2.67
						0.5	6.68

for higher concentrated ingredients that means for concentrations of 2–5% and for problematic substances. Even higher concentrations occur only for oils. The dermal absorption rate is usually estimated between 0% and 50%; higher values are likely to be rare. The calculation ensues with 10%, 20%, and 50%, although many ingredients do not penetrate at all or only to a small extent. As shown, the maximum SED of 10 mg/kg/d is not reached even under extreme conditions with 70% dermal absorption rate. Equations and tables of daily dermal exposures as well as guidance on safety assessment can be found in the literature [8–11]. Of these references, the SCCS description contains the most reliable information ([8], with many references and legal regulations).

$$E_{\text{derm}} = (G \times A) \times F \times R/K \tag{13.1}$$

With:

E_{derm} (mg/kg/d)	=	dermal exposure related to the body weight
G (mg/cm²)	=	specific applied amount
A (cm²)	=	creaming area
F (d⁻¹)	=	application frequency per day
R (–)	=	retention ($R = 1$ for creams)
K (kg)	=	body weight (bw set to 60 kg)

In estimating the potential hazards of creams, it is common practice to consider and evaluate each raw material separately. Basis for this represents the CosIng database, in which problematic substances are commented by the SCCS (see Figure 1.4). Their hints and restrictions should be followed. In addition, the annexes of the Cosmetics Regulation provide information about the limited use of some substances. The last version of the Cosmetics Regulation (EC) introduced a further method of assessment [8] in Annex 1, which in some cases provide additional information for the usual ingredients. Because of the large treated areas, which can be more than a power of 10 above the other applications, body lotions strain the body considerably more than other forms of application. Usually, body lotions are less commonly used as facial or hand creams, often only once a week. With more frequent use, such as twice a day or even more with sunscreen, it may lead to limitations in the formulation after the new material evaluation, as shown later. Therefore, the ingredients of body lotions need to be checked very carefully. This statement applies even more to sunscreens. According to the SCCS recommendation, an amount of 18 g/d should be consider here, which is clearly not enough. A value of 30 g/d would correspond to two normal applications per day and should represent the minimum for safety calculations.

In order to assess possible risks, the estimation ensues using the SED value. The required dermal absorption (DA_a) can be determined in the laboratory. This value allows the calculation of the SED in accordance with Eq. (13.3). So far, only few results of the dermal absorption are available for cosmetic ingredients. Therefore, an SED estimate is made using Eq. (13.2), as demonstrated in Table 13.9. The tolerable dose of substances, described by the NOAEL (no observed adverse effect level, Eq. (13.4)), is difficult to find for cosmetics ingredients and sometimes not available as a reading. In addition to the SED, NOAEL is required to calculate the MoS [8], which must be above 100 and should be stated in the safety information if possible. Otherwise, the SED < 10 is sufficient. The procedure for calculating the safety margin when using the cream for babies and children is described in Section 13.4.6.

$$SED = E_{derm} \times C_{rm} \times P \tag{13.2}$$

With:

SED (mg/kg bw/d) = systemic exposure dose; dermal exposure of a raw material related to the body weight

C_{rm} (%/100%) = concentration of the raw material

P (%/100%) = fraction that diffuses into the skin (measured or estimated)

$$SED = DA_a \times A \times F/60\,kg \tag{13.3}$$

With:

SED (mg/kg bw/d)	=	systemic exposure dose; dermal exposure of a raw material related to the body weight
DA_a (mg/cm^2)	=	dermal absorption reported as amount/cm^2, resulting from an assay under in-use mimicking conditions
A (cm^2)	=	skin surface area expected to be treated with the finished cosmetic product
F (d^{-1})	=	frequency of application of the finished product ($F \geq 1$)
60 kg	=	default human body weight

$$\text{MoS} = \text{NOAEL}/\text{SED} \tag{13.4}$$

With:

MoS (–)	=	margin of safety
NOAEL (mg/kg/d)	=	no observed adverse effect level

The procedure for calculating the MoS should be described using the example of urea. It is known as a well-penetrating substance with a NOAEL of 2000 mg/kg/d. Other substances show values between 200 and 10 000, mostly over 1000. For the calculation of the MoS, an absorption rate of 10–50% was assumed (Table 13.10). Probably, in this particular case, the true values are in the range. For many water-soluble ingredients and some lipophilic substances, the values should be less than 10% and not more. In the past, people wrongly assumed that the ingredients do not penetrate the skin. Today, we know that in fact some ingredients do not penetrate, others overcome the barrier in small proportions and few in significant quantities (see Sections 4.4 and 4.5). The harmless vegetable oils formulated in creams may desirably penetrate the skin and promote skin formation. According to the state of knowledge, there is no danger from these substances, which can be used without restrictions according to the Annexes of the Cosmetics Regulation. Also, the MSDS suggest no restrictions (mostly suitable for food).

Problems with substances that show NOAEL values of 200–3000 mg/kg bw/d can only occur when many applications take place daily on the whole body and the absorption rate is high. For example, for the sunscreen with four applications per day, urea may fall below the safety limit of 100. Here, it is advisable to reduce the urea amount from 3% to 2.5%. Sun cream and lotion have to be considered very critically because of the frequent application during the day and the abundant treatment up to 2 mg/cm^2, in particular with regard to the UV-absorbing substances. For other creams (face, foot, and hand), the contaminated areas are much smaller, so no problems occur, even if the NOAEL drops to 200.

The filling of the cream was carried out under nitrogen, so that the cream in the container is largely free of oxygen. In the present substances and under the conditions prevailing in the cream (concentrations and temperatures), a chemical reaction can be ruled out. There are no changes in color and odor and physical stability (emulsion, viscosity), confirmed in comparable formulations even after

Table 13.10 Calculation of the safety margin for 3% **urea** (NOAEL = 2000 mg/kg bw/d) with an expected dermal absorption of 10% and the unlikely case of a 50% absorption.

Cream	Area size A (cm^2)	Frequency of use (d^{-1})	Dermal exposure of cream E_{derm} (mg/kg/d)	Dermal absorption (%/100)	Systemic exposure dose SED (mg/kg/d) Limit: <10	Margin of safety MoS (–) Limit: >100
Face	570	2	19	0.1	0.06	35 000
		2		0.5	0.29	7 000
Face (expanded)	2 550	3	128	0.5	1.91	1 046
Body lotion (without back)	11 000	2	367	0.1	1.10	1 800
		2	367	0.5	5.51	363
		3	550	0.1	1.65	1 212
		3	550	0.5	8.25	242
Sunscreen (whole body[a)])	16 000	2	693	0.1	2.08	962
				0.5	10.40	192
		3	1 040	0.1	3.12	641
				0.5	15.60	128
		4	1 390	0.1	4.16	481
				0.5	20.80	96

a) Abundant treatment, calculated with 1.3 mg/cm^2

years. As many years of experience show, no degradation of sorbic acid due to the stabilization with citric acid/citrate can be observed, so that the cream is well preserved even after three years. For these reasons, the manufacturer can guarantee a shelf life of 30 months at storage temperatures below 22 °C. After opening the container, the following guaranteed shelf lives (recommendation) are valid: 3 months in the pot, 6 months in bottles and dispensers, and 12 months in airless dispensers (see Section 10.8).

Personal Note 3: In recent years, the provisions on safety assessment, including the GMP Directive, have increased considerably without real improvement being achieved in practice. Upon full compliance with the regulations, a 100-page document will be established, in which much information from the MSDS and analysis certificate as well as other literature will be copied. We recommend a shortened version, which contains the most important data and comments, as everyone has access to the MSDS. For the safety assessment of cosmetic care creams, the additional estimations of the SED and NOAEL only make sense for application on the whole body (large areas), especially for babies and when the cream is applied several times a day and the raw material absorbed well. These few substances should be named by the experts with the SED and NOAEL values, so that small and medium-sized enterprises can work properly and quickly. Qualified for this should be the SCCS. Apart from that, the producer must be able to rely on the appendices of the Cosmetics Regulation. If problems are seen, limit values must be specified there (for example, essential oils).

13.4.7 Safety Consideration for Babies and Children

Information about the specifics of baby skin as well as the description of interesting and suitable ingredients of creams for babies is in provided Section 7.4. For the calculation of the exposure load of babies and children, the indication of the surface sizes (face and body) is missing in the extensive SCCS document [8]. These can be estimated by dividing the weight ratio by the size ratio of children and adults (y), and then multiply that value by the area of the adult's body (A; Eq. (13.7)). The y-values are certainly too high but probably approximate the true conditions and should be on the safe side (Table 13.11). According to Eq. (13.8), for one-year-old babies, the specific burden is about double and in infants three times as high as in adults. It follows a twice respectively three times as high SED. Therefore, pay attention to the formulation of body lotions for toddlers and babies because high concentrated, fast penetrating substances can come close to the limit of 10 when used twice a day. The calculation of the SED and MoS values takes place as described for adults, using the factor 2 or 3 if the product is also to be suitable for babies. Many manufacturers set the limit for children over three years. Again, the factor 2 (instead of 1.68) should be taken for a quick estimate.

$$E_{\text{derm,c}} = (G \times A \times y) \times F \times R / K_{\text{c}} \tag{13.5}$$

With:

$E_{\text{derm,c}}$ (mg/kg/d) = dermal exposure per day related to the body weight of children

G (mg/cm^2) = specific applied amount

Table 13.11 Relative dermal exposure dose of a cream in babies and children ($E_{\text{derm,c}}$).

Age	Weight K (kg)	Size S (cm)	Factor y (–)	$E_{\text{derm,c}}/E_{\text{derm}}$
0.25	6	60	0.28	2.83
1	11	83	0.38	2.05
3	16	101	0.45	1.68
6	24	121	0.56	1.40
9	34	138	0.70	1.23
12	48	156	0.87	1.09
>20	60	170[a]	1	1

a) Estimated average size between men and women.

A (cm^2) = creaming area

F (d^{-1}) = application frequency per day

R (–) = retention ($R = 1$ for creams)

K_c (kg) = body weight of children (age-related)

With the area ratio y:

$$y = A_c/A \tag{13.6}$$

$$A_c = A \times y \tag{13.7}$$

$$E_{\text{derm,c}}/E_{\text{derm}} = y \times K/K_c = S/S_c \tag{13.8}$$

With:

S = size

S_c = size of the child

The evaluation of the equations takes place in Table 13.12, in which the values of a standard adult are compared with those calculated for the babies. It is remarkable that the creaming of the baby's head results in relatively low SED values even when used twice a day. Using the discussed ingredients, the daily use of a body lotion is unproblematic for babies. Strong loads arise only in the unlikely double application per day and in particular at high concentration of the considered substance (5%) and a permeation of 10%. If a permeation of 20% is to be expected under these conditions, unacceptably high values can result. Therefore, it is recommended to halve the use concentration of this substance from 5% to 2–3% in order to increase the safety margin in the unlikely case of a double application per day.

Most creams contain perfumes in an amount of 0.1–0.8%, usually around 0.5%. Essential oils, and especially perfume oils, are to be viewed critically because of their potential for allergies and sensitization as well as skin and eye irritation. The SED value, based on a possible penetration of 75% for pure liquids (may be maximum 33% for a thickened phase), indicates no problems, in particular for the adults. The MoS might offer a different statement since most of the corresponding NOAEL values are probably low. Measured values are often in the range of 150–500 mg/kg/d (lavender and rosemary: 300 mg/kg/d), so that babies with one application per day can be already close to the limit, depending on the NOAEL. Therefore, the MoS must be calculated exactly. It is advisable to use a maximum of 0.3%. That is enough to cover the inherent odor and to give a delicate fragrance. The example demonstrates that the SED value alone does not represent the whole reality, but the MoS shows a more extensive meaning (see example in the next chapter).

A concrete example of a P.I.F. will be represented in the next chapter on the basis of a body lotion because of the heavy burden. The product should also be suitable for children from three years.

Table 13.12 Systemic exposure dose for a three-month-old infant and for a one-year-old baby in comparison with an adult; ($G = 1$ mg/cm^2, $R = 1$).

Cream area		Area size	Frequency per day	Amount	Dermal exposure of cream	Percutaneous permeation	Systemic expose dose of raw material
a: adult		A (cm^2) (estimated)	F (d^{-1})	(g/d)	E_{derm} (mg/kg/d)	P (–) (0.1 = 10%, 0.2 = 20%)	SED (mg/kg/d)
b: 3 mo baby							Limit: 10
c: 1 yr baby							
Concentration of the ingredient: $C_{rm} = 0.05$, 5%							
Face:	a	570	2	1.14	19	0.2	0.19
	b	160	2	0.32	53	0.2	0.53
	c	217	2	0.43	39	0.2	0.39
Head:	b	320	2	0.64	107	0.2	1.07
	c	433	2	0.87	79	0.2	0.79
Body:	a	16 000	1	16	267	0.1	1.33
			2	32	533	0.1	2.66
			2	32	533	0.2	5.33
	b	4 480	1	4.48	747	0.1	3.73
			2	8.96	1 493	0.1	7.47
			2	8.96	1 493	0.2	**14.94**
	c	6 080	1	6.08	553	0.1	2.76
			2	12.16	1 105	0.1	5.53
			2	12.16	1 105	0.2	**11.06**
Perfume: $C_{rm} = 0.005$, 0.5%							
Body:	a	16 000	1	16	267	0.33	0.44
			1	16	267	0.75	1.00
	b	4 480	1	4.48	747	0.33	1.23
			1	4.48	747	0.75	2.80
			2	8.96	1 494	0.33	2.47
	c	6 080	1	6.08	553	0.33	0.91
			1	6.08	553	0.75	2.07
			2	12.16	1 105	0.33	1.52

13.5 Example for a P.I.F. (Body Lotion)

Part A: Cosmetic product safety information

0. Information about the manufacturer and product description

Marleen Body lotion/Recipe No. MBL 1201/21.10.2018

Airless dispensers 200 ml

Manufacturer or contract manufacturer

- Carecos Kosmetik GmbH
  D-77694 Kehl
  Germany

Product owner

- GERCOS GmbH
  D-76437 Rastatt, Orchideenstr. 13
  Germany

Contact person

- K. Stemberger
  info@gercos.de
  http://www.gercos.de/impressum.html

Description of the body lotion:

Body Lotion

The lotion with aloe is a nourishing cream for everyday use, suitable for children over the age of 3, adults, and the elderly. After labeling, the application ensues once a day after showering to moisturize the body and restore the skin barrier by refatting. It is a pumpable, easily distributable o/w emulsion based on natural oils such as almond oil, evening primrose oil, and shea butter. As known moisturizing ingredients, glycerin and sorbitol as well as urea and aloe (0.1 powder = 20% gel) are formulated. Vitamins E and C operate for antioxidation and panthenol for soothing the skin. Thanks to a buffer system, the pH in the finished product is at a constant value of 5.1. An airless dispenser (200 ml) supplies the cream without air access. All ingredients are described positively in the literature. Label is made in accordance with the Cosmetics Regulation.

- The cream contains no parabens, mineral oils, waxes, dyes, and perfumes.

  The label comprises the following:

  - Name of the responsible company: *GERCOS, D-76437 Rastatt, Germany;* www.gercos.de
  - Intended use: *Skin nourishing body lotion with refatting and moisturizing effects for children over three years and all adults*
  - Content of the dispenser: *200 ml*
  - Minimum shelf life, shelf life after opening: *30 m; 12 m*
  - Storage condition: *4–22 °C**
  - Application notes: *Use the body lotion once a day after showering*

(Continued)

- Batch number: *on the bottom*
- INCI list: see later
- Animal experiments: *no animal testing*[a)]
- Warnings: *Avoid eye contact and the use on damaged skin.*
- Bar code

* Depending on the country, this information is required

1. Quantitative and qualitative composition of the cosmetic product
The recipe with the INCI names and EC numbers is shown in Table A. Furthermore, the functions of each ingredient in the formulation are given in column V. The abbreviated toxicological profile can be found in columns VI–VIII. As the ingredients are commonly used and known nonhazardous substances, this description suffices, with the exception of essential oil. In assessing the irritation, the high dilution and pH must be considered so that mild irritations are negligible. However, in addition, because cream in the eye is always unpleasant, the warning "*Avoid eye contact*" should be on the label. The last column indicates the supplier of the raw material, who also provided the MSDS and the certificate of analysis. The lot number assigned by the quality laboratory is used for tracing the raw materials.

For the essential oil, the composition/analysis is needed, provided by the supplier, for estimating safety levels. For geranium, a more general but sufficient analysis is presented (Table B). Other analyses demonstrate that the composition strongly depends on the harvest and country [A1]. Therefore, the safety assessment belongs to the purchased product [A2]. Variation of this product [A3] may differ in some values and in the MSDS. Because of their allergic potential, the substances to be declared in accordance with the Cosmetics Regulation must be identified from the analysis, which are then named in the INCI list.

There are no toxicity data on the essential oil itself. Conclusions on safety can only be drawn from the individual constituents, so the NOAEL might be in the range of 250 mg/kg/d.

INCI

Aqua, Prunus Amygdalus Sativa Kernel Oil, Oenothera Biennis Oil, Butyrospermum Parkii Butter, Cetearyl Alcohol, Glycerin, Sorbitol, Panthenol, Niacinamide, Alcohol, Urea, Ceteareth-30, Dimethicone, Tocopherol, Helianthus Annuus, Sodium Citrate, Glyceryl Stearate, Pelargonium Graveolens Stem Leaf Oil, Xanthan Gum, Potassium Sorbate, Citric Acid, Aloe Barbadensis Leaf Juice Powder, Allantoin, Citronellol*, Geraniol*, Linalool*, Citral*, and Limonene*

* Contained in Pelargonium Graveolens (Geranium)

Remarks on production for avoiding by-products:

The manufacturing takes place with GMP compliant quality assurance (Table C). Because of the manufacturing conditions under vacuum, oxidation

Table A Composition of the body lotion with the recipe number MBL R-1201/21.10.2018.

	I	II	III	IV	V	VI	VII	VIII	IX
No.	**Ingredient**	**(%)**	**INCI name**	**EC number**	**F[a]**	**Oral toxicity LD_{50} (mg/kg rat)**	**(a) Irritant (b) NOAEL (mg/kg/d)**	**Limit conc.**	**Supplier/lot no.**
1	Water	66.4; 61.4	Aqua	231-791-2	S		(a) — (b) —		Carecos
2	Sweet almond oil	4	Prunus Amygdalus Sativa Kernel Oil	291-063-5	Emo, Li	—	(a) — (b) —	—	Lamotte, PA 502
3	Evening primrose oil	4	Oenothera Biennis Oil	289-859-2	Emo, Li, Nd	—	(a) — (b) —	—	Lamotte, OB 503
4	Shea butter	3.5	Butyrospermum Parkii Butter	293-515-7	Li, Emo, CA	—	(a) — (b) —	—	Lamotte, BP 504
5	Cetyl stearyl alcohol	3.5	Cetearyl Alcohol	267-008-6	CoE, Emo, CA, RF, FO, IV	2200	(a) Eye, skin, (b) 2 000	—	BTC, CA 505
6	Glycerol	3	Glycerin	200-289-5	Hu, S	>11 500	(a) Eye, (b) 2 000	—	Caesar & Loretz, Gl 506
7	Sorbitol	3	Sorbitol	200-061-5	Hu	24 300 NOAEL 4 500	(a) — (b) 4 500	—	Caesar & Loretz, So 507

Table A (Continued)

	I	II	III	IV	V	VI	VII	VIII	IX
No.	Ingredient	(%)	INCI name	EC number	F[a]	Oral toxicity LD_{50} (mg/kg rat)	(a) Irritant (b) NOAEL (mg/kg/d)	Limit conc.	Supplier/lot no.
8	D-Panthenol	2	Panthenol	201-327-3/ 240-540-6	SC, PV	21 000	(a) — (b) 1 000	—	Caesar & Loretz, Pa 508
9	Magnesium ascorbyl phosphate	1.5	Magnesium Ascorbyl Phosphate	CAS114040-31-2, 113170-55-1	AO, SC, Vit.	>2 000	(a) Eye, skin, (b) 1 000	—	Making Cosmetics MA 509
10	Ethanol	1.5	Alcohol	200-578-6	Ab, LM	6 200–15 000	(a) Eye, (b) 1 730	—	Caesar & Loretz, Al 510
11	Urea	1.5	Urea	200-315-5	ASt, Ker, NMF/Hu	8 200, 8 471	(a) — (b) 2 250	—	Caesar & Loretz, Ur 511
12	Cetyl stearyl ether 30EO	1.5	Ceteareth-30	68439-49-6 CAS	E, Su	>5 000	(a) Eye, (b) 250	—	Caesar & Loretz, Ce 512
13	Silicone oil M500	1.5	Dimethicone	63148-62-9 CAS/9006-65-9 CAS/9016-00-6 CAS	EmoKC, Li	>5 000	(a) – (b) Not toxic	—	Caesar & Loretz, Di 513
14	Vitamin E (natural, D-α; 0.7)/sunflower oil (0.3)	0.85	Tocopherol/ Helianthus Annuus	59-02-9 (D) CAS/232-273-9	AF, AO, AW, UV, Vit. Hl	>5 000	(a) – (b) Not toxic	—	Caesar & Loretz, To 514

15	Trisodium citrate dihydrate	0.64	Sodium Citrate	200-675-3	AHA Hu, Cx, Bu	>8 000	(a) — (b) 2 500	—	Carl Roth, SC 515
16	Glycerol mono-stearates	0.4	Glyceryl Stearate	250-705-4/ 286-490-9	E, Emo IV, RF	>5 000	(a) – (b) 2 500	—	Carl Roth, GS 516
17	Geranium	0.3	Pelargonium Graveolens Stem Leaf Oil	290-140-0	AI, O	>3 000, allergy potential, Table B	(a) Eye, (b) 250	—	Saffire Blue, PG 517
	See Table A								
18	Xanthan	0.3	Xanthan Gum	234-394-2	Est, IV	>5 000	(a) — (b) >1 250	—	Carl Roth, SC 518
19	Potassium sorbate	0.25	Potassium Sorbate	246-376-1	P	8 000	(a) Eye, (b) 3 000	0.6 acid	Caesar & Loretz, PS 519
20	Citric acid monohydrate	0.16	Citric Acid	201-069-1	AHA Hu, Cx, Bu	11 700	(a) Eye, (b) 2 500	—	Caesar & Loretz, CA 520
21	Aloe vera powder (1 : 200)	0.1	Aloe Barbadensis Leaf Juice Powder	287-390-8/ 305-181-2	So, Hu	>3 000 mouse	(a) — (b) >1 000	—	Satcotek AV 521
22	Allantoin	0.1	Allantoin	202-592-8	Hl, KC	>5 000	(a) Eye, skin (b) Not toxic	—	Euro OTC Pharma Al 522

a) Ab, antibacterial; AHA, alpha-hydroxy acid; AI, anti-inflammatory, antifungal, antiseptic; AF, antifoam; AO, antioxidant; ASt, antistatic agents; AW, antiwrinkles; Bu, buffer; CA, consistency agents; CoE, coemulsifier; Cx, complexing agent; E, emulsifier; Emo, emollients, skin softener; Est, emulsion stabilizing; FO, foam stabilizer; Hl, supporting the healing process; Hu, humectant; IB, improvement skin barrier; IV, increase viscosity; Ker, keratolytic; KC, skin caring; Li, lipids; NMF, natural moisturizing factor; Nd, effective against atopic dermatitis; O, odor; P, preservative; PV, provitamin; RF, refatting; S, solvent; SC, skin care agents; So, soothing; Su, surfactant; UV, UV protection.

Table B Composition of a Pelargonium Graveolens Stem Leaf Oil ([A2], Geranium, EC 290-140-0).

Main components	CAS number	EC number (EINECS)	Portion in the oil (%)	LD_{50} oral (mg/kg rat)	LD_{50} dermal rabbit (mg/kg)	Proportion in the cream (ppm)	NOAEL (mg/kg/d) (ECHA[a)][A4])
Citronellol[a)]	106-22-9	203-375-0	<42	3450	2650	<1260	300[a)]
Geraniol[a)]	106-24-1	203-377-1	<25	3600	5000	<750	550[a)]
Linalool[a)]	78-70-6	201-134-4	<12	2790	5610	<360	117[a)]
Menthone	10458-14-7	233-944-9	<5	2180	—	<150	<200[a)]
Citral[a)]	5392-40-5	226-394-6	<4	4960	2550	<120	570[a)]
α-Pinene	80-56-8	201-291-9	<2	2700	—	<60	250
β-Pinene	127-91-3	242-060-2	<2	3300	>5000	<60	222
Limonene[a)]	5989-27-5	227-813-5	<1	4400	5000	<30	215
	Geranium essential oil			3000	3000	3000	NOAEL 250
	(all values estimated)						
Conversion:			$0.1\% = 1\%_0 = 1000\,\text{ppm}$				
			$0.3\% = 3\%_0 = 3000\,\text{ppm}$				
		1% von 0.3%	$0.003\% = 0.03\%_0 = 30\,\text{ppm}$				
			Limit for the declaration = 10 ppm				

a) These substances must be declared as potential allergens in the INCI list.

Table C Production Instruction.

Production instruction for the Marleen Body Lotion

(Recipe No: MBL R-1201/21.10.2018)

The cleanliness of the disinfected production plant is verified.

- First of all, the sensitive substances are separated. These substances require 5% water, which is subtracted from the calculated water amount.
  - Urea, D-Panthenol, Geranium oil.
- Then, the formulation is divided into a water phase with hydrophilic and into an oil phase with the lipophilic active ingredients. For both phases, separated stirring tanks for heating and dissolving exist. The raw materials are weighed in succession and fed via the vacuum.
  - List of the water-soluble components: Aqua, Trisodium citrate dihydrate, Citric acid monohydrate, Glycerol, Sorbitol, Magnesium ascorbyl phosphate, Ethanol, Allantoin, Cetylstearylether 30EO, Potassium sorbate, Aloe vera powder (1 : 200), Xanthan.
- Preparation of the water phase.
  First, add the water into the premix stirred tank 03/01. From the water-soluble raw materials, the citrate and citric acid are dosed first to adjust the pH, followed by the other substances. The thickening agent (xanthan gum) is finally added with vigorous stirring. While heating, they dissolve under vacuum at a temperature of maximum 50 °C during stirring. Complete dissolution of the polymer takes half an hour. After complete dissolution of all components ensues the pH control. Target: pH 5.1. Later, the water phase is metered into the running emulsifier E03 together with the finished oil phase.
  - List of the lipid components: Sweet almond oil, Evening primrose oil, Sheabutter, Cetylstearyl alcohol, Silicone oil M500, Vitamin E (natural, D-α; 0.7)/sunflower oil (0.3), Glycerol monostearates.
- Preparation of the oil phase and the emulsion
  The preparation of the oil phase takes place in the nitrogen-flushed premix stirred tank 03/02 before provision of vacuum. The individual oils and then the lipid substances are sucked in successively with stirring and then heated to a maximum of 60 °C. Immediately after dissolving all lipophilic components, the metering into the running emulsifier E03 ensues together with the water phase. The emulsification time is set to 30 minutes at a temperature of 50 °C. After completion of the macroemulsion, the mixture is cooled to 30 °C with stirring.
- Preparation of the sensitive substances

At 30 °C, urea and panthenol are suspended in 5% water (container 03/12) and sucked in. Geranium oil follows via container 03/11. Subsequently, the second emulsification is carried out for 22 minutes. Then, the emulsion is ready and transferred to the reservoir of the bottling line. There, the sampling undergoes various measurements to release the batch (see Table D).

The production is carried out in accordance with the GMP guidelines in a suitable plant. The dosage of the substances took place as described in the recipe.

reactions are unlikely. In particular, the separation of the sensitive ingredients from the oil and water phase and the limitation of the temperature levels to the lowest possible temperatures prevent side reactions. The immediate adjustment of the pH in the water phase before adding the other substances helps to minimize reactions. The use of airless dispensers and the constant pH value ensure stability against possible reactions during storage.

2. Physical/chemical characteristics and stability of the cosmetic product
Table D show the analysis report of the body lotion after production. The manufacture of the 1000 kg cream took place in batch plant 03 and the bottling in line 11. After testing the physical and microbiological parameters with subsequent release, the product was filled into 5000 airless dispensers each 200 ml.

3. Microbiological quality
See Table D. All criteria are met. There are only a few germs in the lotion and the preservative works as expected.

4. Impurities, traces, information about the packaging material
The quality of the raw materials is suitable for foodstuff and/or cosmetics. The concentrations of impurities are negligible in the cosmetic product. From the annexes of the Cosmetics Regulation and the certificates of analysis, no danger can be derived by secondary components, also there exist no appropriate comment of one or more ingredient by the SCCS.

The primary packing material is an airless dispenser made of opaque PP. It does not contain any harmful substances and is intended for use in high-quality cosmetics.

5. Normal and reasonably foreseeable use
Normally, a body lotion is applied once a week to the complete body. The label recommends a daily use after showering. In exceptional cases, adults might use the cream twice daily, but not small children.

6. Exposure to the cosmetic product
Depending on the source, there are several, sometimes extremely different NOAELs in the literature. In these cases, the NOAEL value for reproductive toxicity was chosen here. However, that is not crucial, as mostly even half of the NOAEL levels do not cause any problems in the safety calculation. From the data, it can be deduced that the NOAEL of the cream should be in the range of 500–2000, without considering the water and the oils. Including them would yield a value that is about four times higher. Here, a NOAEL of 500 would be sufficient to achieve an MoS of over 100 with the calculated SED values (see Table E).

The following information must be available:

- The site of application: entire body and body without back (normal case).
- The surface area of application: 16 000, respectively, 11 000 cm^2 for adults; 7200 cm^2 for three-year-old children
- The amount of product applied per area: 1 mg/cm^2.
- The frequency of use for adults: 2.28 times/d (recommend by the SCCS)
- The frequency of use for children: 1 time/d
- The normal and reasonably foreseeable exposure route(s): dermal
- The targeted (or exposed) population(s): Adults and children over three years
- Retention factor: 1.0

7. Exposure to the substances
In Table E, examples of unproblematic raw materials such as cetearyl alcohol at a concentration of 3.5% with a NOAEL of 2000 mg/kg/d [A5] and panthenol (2%, NOAEL 1000 mg/kg/d) are listed. Rose Geranium oil (0.3%, 250 mg/kg/d)

Table D Analysis report.

Marleen Body Lotion with Aloe Vera (GERCOS)	
Batch No:	**MBL B-30579/R-1201/25.10.2018**
Recipe No:	MBL R-1201/21.10.2018
Contract manufacturer:	Carecos Kosmetik
Production Instruction:	PI R-1201/22.10.2018/P03
Plant No:	P03
Batch size:	1000 kg
Package supplier:	Packari
Package	AP 5003/02.10.2018 PP-airless dispenser (Art. No.: 905040_K05200); 200 ml (5000)
Filling line No:	11

Parameter	**Assessment**
Visual ratings	
– Appearance	– Typical glossy emulsion
– Color	– White
– Odor	– Weak scent of roses
Physical tests	
– pH (4.9–5.5)	– 5.1
– Centrifugal test (40 °C/30 000 g);	– Passed
– Minimum shelf life	– 30 m[a]
– Shelf life after opening	– 12 m[a]
– Viscosity (3–10 Pa s; Haake 10 s^{-1})	– About 4 Pa s (22 °C)
– Density (1.00–1.06 g/cm^3)	– 1.031
– Solids	– Contains no particles
Microbiogical tests	
– Germs (<1000 cfu/g)	– <10 cfu/g
– Preservative stress test	– Passed after 2 wk
Package quality	
– Visual control	– Quality suitable for cosmetic creams, without complaints
– Smear test	– Passed
Patch test	– No sign of allergies (14 subjects)
Animal tests	Neither the manufacturer of the cream nor the product owner induces animal experiments

a) These monthly values result from the centrifuge test and the experience with similar formulations in airless dispensers.

Table E Dermal exposure of the lotion and the systemic exposure dose of some raw materials.

Person	Area size A (cm^2) (estimated)	Frequency per day F (d^{-1})	Amount (g/d)	Dermal exposure of cream E_{derm} (mg/kg/d)	Percutaneous permeation P (–)(0.1 = 10%), Estimated maximum values	Systemic expose dose of raw material SED (mg/kg/d) Limit: 10	Margin of safety (–)
					Cetearyl alcohol 3.5% (= 0.035)		NOAEL = 2 000 mg/kg/d
Adult	16 000	1	16	267	0.3	2.80	713
	11 000	2.28	25	418	0.3	4.39	456
Children age: 3	7 200	1	7.2	450	0.3	4.73	423
					Panthenol 2% (= 0.02)		NOAEL = 1 000 mg/kg/d
Adult	16 000	1	16	267	0.2	1.07	935
	11 000	2.28	25	418	0.2	1.67	599
Children age: 3	7 200	1	7.2	450	0.2	1.80	556
					Magnesium ascorbyl phosphate 1.5% (= 0.015)		NOAEL = 1 000 mg/kg/d
Adult	16 000	1	16	267	0.2	0.80	1 248
	11 000	2.28	25	418	0.2	1.25	797
Children age: 3	7 200	1	7.2	450	0.2	1.39	741
					Ceteareth-30 1.5% (= 0.015)		NOAEL = 250 mg/kg/d
Adult	16 000	1	16	267	0.1	0.40	624
	11 000	2.28	25	418	0.1	0.63	399
	11 000	2.28	25	418	0.2	1.25	199
Children age: 3	7 200	1	7.2	450	0.1	0.68	370

(Continued)

					Potassium sorbate 0.25% (= 0.002 5)		NOAEL = 3 000 mg/kg/d
Adult	16 000	1	16	267	0.1	0.07	45 000
	11 000	2.28	25	418	0.1	0.10	29 000
Children age: 3	7 200	1	7.2	450	0.1	0.11	27 000
					Sorbic acid 0.19% (= 0.001 9)		NOAEL = 300 mg/kg/d
Adult	16 000	1	16	267	0.1	0.05	6 000
	11 000	2.28	25	418	0.1	0.08	3 800
Children age: 3	7 200	1	7.2	450	0.1	0.09	3 500
		1	7.2	450	0.33	0.28	1 063
					Rose Geranium Pelargonium Graveolens 0.3% (= 0.003)		NOAEL = 250 mg/kg/d
Adult	16 000	1	16	267	0.33	0.26	946
	11 000	2.28	25	418	0.33	0.41	604
	11 000	2.28	25	418	0.5	0.63	399
	16 000	2.28	36	608	0.5	0.91	274
Children age:3	7 200	1	7.2	450	0.33	0.45	562
	7 200	1	7.2	450	0.5	0.68	370
	7 200	2	14.4	900	0.5	1.35	185

For all ingredients, the margin of safety is well above 100.

must be considered separately, as displayed in Table B. The calculated SED and MoS values are given in Table E. Others, not listed substances, known from the literature and often used, indicate high NOAELs and are formulated in a lower concentration than cetearyl alcohol, which means safe in application. Therefore, the calculation of all ingredients results in MoS values of well over 100. Of these, no or a very little risk is assumed. The calculation is carried out with unlikely high values, which means the estimated permeation is between 10% and 50%, depending on the substance. Additional security exists as an increasing permeation does not lead to a different assessment, except in unrealistic cases. All SED values are below 10, all MoS over 100. As sorbic acid is formed from potassium sorbate, it was listed in maximum possible concentration and proves safe, as known from the literature.

Ceteareths show high LD_{50} values but low NOAELs. The Cosmetic Ingredients Review (CIR) expert panel [A6] notes that there is enough data to rate Ceteareths as "safe as used" for cosmetics in concentration up to 3% and more also because it is not expected to be systemically available [A7]. For Steareth-20, which is very similar to Cetereth-30, no systemic toxicity could be measured [A7]. There is no concrete NOAEL value for Ceteareth-30, but for the group of ethoxylated fatty alcohols. In this group, the short-chain fatty ethoxylates react more aggressively to the skin. The long-chain C_{16}/C_{18} fatty ethoxylates have proven themselves for decades in cosmetic formulations (Lanette O) after reduction of the by-products. They can be considered as "nontoxic" such as the C_{18}-Steareth and show significantly higher NOAEL values compared to the lowest level in the group (NOAEL > 100). Synthetically produced Ceteareth-25 has a NOAEL of 250 mg/kg bw/d [A8]. Therefore, natural-based Ceteareth-30 should have the same or a higher value. Because of the solubility in water and the molecular weight, a low permeation is expected, which should be below 5%, but the worst scenario might be 10%. The MoS is well above 100, the use causes no problem.

Observe the indication that Ceteareth possibly enhances the absorption of substances. The fact should be considered in the formulation of baby lotions. For adults, the effect can be quite positive. As all polyethylene glycol (PEGs) and PEG derivatives, Ceteareth should not be applied to damaged skin [A6, A7, A9]. Warning: *Avoid use on damaged skin.* Manufacturers of PEG derivatives must continue their efforts to remove impurities and by-products such as ethylene oxide and 1,4-dioxane [A7]. However, the quality offered today is safe for use that means for the ethoxylated alcohol.

- 1,4-Dioxane maximum 10 mg/kg (10 ppm), equals less than 150 ppb in the cream;
- Total ethylene and diethylene glycol maximum 0.3% (by weight), equals less than 45 ppm in the cream;
- Ethylene oxide maximum 1 mg/kg (1 ppm), equals less than 15 ppb in the cream;
- Heavy metals (combined) maximum 10 mg/kg (10 ppm), equals less than 150 ppb in the cream.

References [A10, A11] provide information to glycerol compounds and potassium sorbate.

8. Toxicological profile of the substances

A toxicological profile can be found in Table A, which contains the most important data. Some MSDS and the sources cited below may provide additional information in some cases. The following data bases and sources have been consulted for compiling toxicity data for the ingredients:

- CosIng, http://ec.europa.eu/consumers/cosmetics/cosing
- CIR (Cosmetic Ingredients Review) opinions
- Bibra, www.bibra-information.co.uk
- SCCS (Scientific Committee of Consumer Safety) opinions
- IUCLID (International Uniformed Chemical Information Database) datasets
- ECHA (European Chemicals Agency)
- EFSA (European Food Safety Authority)
- HERA (Human and Environmental Risk Assessment on ingredients of household cleaning products)
- HSDB (Hazardous Substances Databank)
- EPA (United States Environmental Protection Agency)
- FDA (United States Food and Drug Administration)
- IPCS INCHEM (International Program on Chemical Safety).

References for the example:

[A1] Tisserand, R. and Young, R. (2014). *Essential Oil Safety: A Guide for Health Care Professionals*, 2e. Churchill Livingstone Elsevier.

[A2] https://www.saffireblue.ca/shop/docs/MSDS/MSDS-ROSE-GERANIUM-EO.pdf (accessed 23 October 2018).

[A3] https://www.essentialoilsdirect.co.uk/geranium-pelargonium_graveolens-essential_oil.html (accessed 23 October 2018).

[A4] https://echa.europa.eu/registration-dossier/-/registered-dossier/13515/7/6/1 (accessed 25 October 2018).

[A5] http://www.sasoltechdata.com/MSDS/NAFOL1618F.pdf (accessed 26 October 2018).

[A6] Ceteareth. https://online.personalcarecouncil.org/ctfa-static/online/lists/cir-pdfs/pr35.pdf (accessed 28 October 2018).

[A7] Fruijtier-Pölloth, C. (2005). Safety assessment on polyethylene glycols (PEGs) and their derivatives as used in cosmetic products, *Toxicology* 214: 1–38, Elsevier Ireland Ltd. https://www.ncbi.nlm.nih.gov/pubmed/16011869 (accessed 28 October 2018) and https://pdfs.semanticscholar.org/0823/e31852418da589797c46fd5a4dd253325658.pdf (accessed 28 October 2018).

[A8] Ceteareth-30. http://www.sasoltechdata.com/MSDS/EMULDAS25.pdf (accessed 28 October 2018).

[A9] Ceteareth-30. http://journals.sagepub.com/doi/abs/10.1177/109158189901800306 (accessed 28 October 2018).

[A10] NOAEL Gycerol compounds. https://hpvchemicals.oecd.org/ui/handler.axd?id=f29255ef-74da-4be5-8c43-6c2bbe6adf5e (accessed 26 October 2018).

[A11] Potassium sorbate. http://dissemination.echa.europa.eu/Biocides/ActiveSubstances/1350-08/1350-08_Assessment_Report.pdf (accessed 26 October 2018).

9. Undesirable effects and serious undesirable effects

As comparable formulations are already on the market, no undesirable effects are to be expected. On the other hand, customer complaints can be traced via the batch number to get to the bottom of the causes. Traceability and testing are ensured by the GMP compliant file and retention samples.

10. Information on the cosmetic product

See point 0: Information about the manufacturer and product description.

Part B: Cosmetic product safety assessment

1. Assessment conclusion

The formulation and production took place according to the Cosmetics Regulation (EC) No. 1223/2009. Based on the information provided by the manufacturer and the toxicity data compiled for the ingredients, it can be concluded that the Marleen Body Lotion do not cause any side effects under the intended conditions of use. The product application may be designated as unproblematic and safe. After filling in airless dispensers, there is no air above the cream. Because of the oxygen-free content and the constant pH through the buffer, a longer storage is possible.

2. Labeled warnings and instructions of use

- Warnings: *Avoid eye contact and the use on damaged skin.*
- Application notes: *Skin nourishing body lotion with refatting and moisturizing effects for children over three years and all adults. Use the body lotion once a day after showering.*

The effect of the body lotion results from the ingredients that are known for skin moisturizing and refatting.

3. Reasoning

Toxicity: The Marleen Body Lotion is formulated in accordance with the Cosmetics Regulation (EC) and their Annexes. The lotion represents a flowable emulsion that contains well-known ingredients extensively used in different cosmetic products. The toxicity profiles of the raw materials do not indicate any specific worries for any of the ingredients. Reactions in the cream are highly unlikely (oxygen exclusion and temperature levels).

The packaging material (cosmetic quality) does not contain any harmful substances. Interaction between the primary packaging material and the product is not expected. The raw materials are checked and suitable for the body lotion.

Microbiology: The germ count after production was far below the permissible limit. The provocation test was passed after 14 days. Preservative, constant pH, and exclusion of air minimize the likelihood of microbiological change of the lotion.

Irritation: Because of the high dilution in the cream and pH adjustment, Cetearyl Alcohol, Magnesium Ascorbyl Phosphate, Alcohol, Ceteareth-30, Pelargonium Graveolens, Potassium Sorbate, Citric Acid, and Allantoin (see Table A) probably do not cause irritation. These substances are also not known as very irritating. Potential allergic ingredients of the geranium oil present in a concentration above 10 ppm, such as *citronellol, geraniol, linalool, citral,* and *limonene,* must be declared in the INCI list (see Tables B and C). Also, because they are present in an environment with caring ingredients, the patch test with the lotion yielded no allergic reaction.

Margin of Safety: For adults, an application frequency of 2.28/d (SCCS-recommendation) was used instead of the proposed single application per day. The body area estimate yielded 16 000 and for three-year-old children 7200 cm^2. SED and MoS were calculated for a "standard" adult (60 kg) and a three-year-old child. All SEDs are well below the limit value of 10. The ingredients with the lowest NOAEL value are the essential oil and the ethoxylated fatty alcohol (maximum tolerable dose of both: 250 mg/kg/d). Again, the calculation leads to MoS values clearly over the limit of 100 (Table E), even with worse assumptions. In addition, the CIR expert panel rated Ceteareths as "safe as used" for cosmetics. All other substances have NOAELs over 1000 and are far away from safety limits. Overall, the lotion can be considered as safe in the intended application.

4. Assessor's credentials and approval of part B

The Safety Report has been prepared to the best of our knowledge and experience. All data and information as well as the supplements and amendments to the Cosmetics Regulation known to the assessor have been considered (as of 10/2018).

Dr. Wilfried Rähse D-40589 Düsseldorf E-Mail: Raehse1@t-online.de
Chemist (Technical University, TU-Berlin)
Independent Expert
(9 years of industrial experience in cosmetics;
product and process developments, safety assessments).

Düsseldorf, 31.10.2018
Germany

Safety Assessment
(summary)

Marleen Body Lotion
(Recipe No. MBL R-1201/21.10.2018)

Product owner:	GERCOS GmbH, D-76437 Rastatt
Contract Manufacturer:	Carecos Kosmetik GmbH, D-77694 Kehl
Packaging:	200 ml Airless dispenser (PP)
Target group:	Adults, children over 3
Area of application:	Body (approx. 16,000, 7200 cm^2)
Amount of product applied:	
– g product/application:	dermal 16 g, 7.2 g
– mg product/cm^2:	approximately 1.0 mg/cm^2
– frequency/day:	1/d

Conclusion

The Marleen Body Lotion is formulated and produced according to the Cosmetics Regulation (EC) No. 1223/2009 of the EUROPEAN PARLIAMENT and of the COUNCIL of 30 November 2009, considering recent supplements and amendments. The toxicity profiles of the ingredients do not indicate any problems. With the estimated values for the Systemic Exposure Dose (SED) and the available NOAEL values for the tolerated dose, the Margin of Safety (MoS) for all ingredients was checked. For both adults and three-year-old children, all calculated values are well above the limit. The Marleen body lotion ensures safe use.

D-40589 Düsseldorf (Germany), 31 October 2018

Dr. W. Rähse
(Chemist, independent expert)

13.6 Learnings

- ✓ The P.I.F. includes product and manufacturing information with a detailed safety assessment.
- ✓ A table discloses the quantitative composition with the usual and the INCI names as well as the associated EC/CAS numbers and the functions of the individual ingredients.

- ✓ For each raw material, there exists a MSDS. Besides other data, this contains the toxicity profile of each substance.
- ✓ The MSDS include the acute oral toxicity and eye and skin irritation, mostly measured in rats.
- ✓ Ingredients that show mutagenic or carcinogenic behavior as well as reproductive toxicity should not be used in cosmetics.
- ✓ Perfumes and essential oils have to be checked for declarable substances that may cause allergic reactions. These substances must be specified in the INCI list.
- ✓ The area treated with a lotion or cream (face, body, and hands) is used to estimate the cream amount applied to a 60 kg person and the SED using percutaneous permeation.
- ✓ From the NOAEL, which represents the maximum tolerable dose of the tested substance, in combination with the SED, the MoS can be calculated, which must exceed 100.
- ✓ If all ingredients are above the limit, then a safe application can be assumed.

References

1 Compliance with Regulation 1223/2009 on Cosmetic Products, Cosmetics Europe Guidelines on the Product Information File (P.I.F.) Requirement, 2015. file:///D:/3%20B%C3%BCcher,%20Ver%C3%B6ffentl.,%20Vortr%C3%A4ge/Buch%20Effective%20skin%20creams/4%20Literatur/Updated_Cosmetics_Europe_PIF_Guidelines_-_2015_-_Update.pdf (accessed 29 September 2018).

2 Cosmetics Regulation in English. Regulation (EC) No 1223/2009 of the European Parliament and of the Council of 30 November 2009 on cosmetic products. https://eur-lex.europa.eu/LexUriServ/LexUriServ.do?uri=OJ:L:2009:342:0059:0209:en:PDF (accessed 05 May 2019).

3 Carl Roth (2014). Safety data sheet according to 1907/2006/EC, Article 31, Potassium sorbate. https://www.carlroth.com/downloads/sdb/en/N/SDB_NC09_GB_EN.pdf (accessed 3 October 2018).

4 Caelo (2018). Sicherheitsdatenblatt gemäß Verordnung (EG) 1907/2006, Kaliumsorbat. file:///D:/3%20B%C3%BCcher,%20Ver%C3%B6ffentl.,%20Vortr%C3%A4ge/Buch%20Effective%20skin%20creams/4%20Literatur/2350.pdf (accessed 3 October 2018).

5 NIH U.S. National Library of Medicine, National Center for Biotechnology Information, PubChem, Potassium sorbate. https://pubchem.ncbi.nlm.nih.gov/compound/potassium_sorbate#section=Top (accessed 23 December 2018)

6 Safety data sheet, From Wikipedia, the free encyclopedia, 2019. https://en.wikipedia.org/wiki/Safety_data_sheet (accessed 08 May 2019).

7 Seidel, K. (2015). Rezepturen konservieren, Deutsche Apotheker Zeitung DAZ 43/2015, https://www.deutsche-apotheker-zeitung.de/daz-az/2015/daz-43-2015/rezepturen-konservieren (accessed 6 October 2018).

8 SCCS (Scientific Committee on Consumer Safety) (2016). SCCS Notes of Guidance for the Testing of Cosmetic Ingredients and Their Safety Evaluation, 9th revision, 29 September 2015, SCCS/1564/15, revision of 25 April 2016; file:///D:/3%20Bücher,%20Veröffentl.,%20Vorträge/Buch%20Effective%20skin%20creams/4%20Literatur/sccs_o_190.pdf (accessed 7 October 2018).

9 DGK-Arbeitsgruppe "Sicherheitsbewerter" (2005). DGK-Vorschlag zu Kernelementen einer Sicherheitsbewertung. *SÖFW* 131 (8): 41–50. file:///D:/3%20Bücher,%20Veröffentl.,%20Vorträge/Buch%20Effective%20skin%20creams/4%20Literatur/Sicherheit_2005.pdf (accessed 1 October 2018).

10 GÖCH-Arbeitskreis Kosmetik (2006). Empfehlungen zur Sicherheitsbewertung kosmetischer Mittel. file:///D:/3%20Bücher,%20Veröffentl.,%20Vorträge/Buch%20Effective%20skin%20creams/Empfehlung_GOECH-AK_Sicherheitsbewertung_2007Endfassung_02_06_081.pdf (accessed 1 October 2018).

11 Cosmetics Europe (2015). Compliance with regulation 1223/2009 on cosmetic products, Cosmetics Europe Guidelines on the product information file (P.I.F.) requirement. file:///D:/3%20Bücher,%20Veröffentl.,%20Vorträge/Buch%20Effective%20skin%20creams/4%20Literatur/Cosmetics_Europe_P.I.F._Guidelines_-_2015_-_Update%20(2).pdf (accessed 7 October 2018).

Appendix A

Formulations

Table A.1 Formulation of a moisturizer for the night (w/o emulsion) according to the current organic trend without ethoxylated products, silicone, mineral oils, and parabens (not based on the described modules).

No.	Ingredient	Amount (%)	INCI name	EC number
1	Water (cosm.)	47.85	Aqua	231-791-2
2	Sweet almond oil	7	Prunus Amygdalus Sativa Kernel Oil	291-063-5
3	Evening primrose oil	5	Oenothera Biennis Oil	289-859-2
4	Hemp oil	5	Cannabis Sativa Seed Oil	289-644-3
5	Shea butter	5	Butyrospermum Parkii Butter	293-515-7
6	Urea	5	Urea	200-315-5
7	Sorbitan stearate	4.5	Sorbitan Stearate	215-664-9
8	Glycerol	3	Glycerin	200-289-5
9	Sorbitol	3	Sorbitol	200-061-5
10	D-Panthenol	2.5	Panthenol	201-327-3/ 240-540-6
11	Nicotinamide	2	Niacinamide	202-713-4
12	Glycine	2	Glycine	200-272-2
13	Cetyl Palmitate	2	Cetyl Palmitate	208-736-6
14	Ethanol	2	Alcohol	200-578-6
15	Dermofeel® viscolid palm oil free	1	Hydrogenated Rapeseed Oil	283-532-8
16	Vitamin E (natural, D-α; 0.7)/sunflower oil (0.3)	1	Tocopherol/ Helianthus Annuus	59-02-9 (D) CAS/232-273-9

(Continued)

Cosmetic Creams: Development, Manufacture and Marketing of Effective Skin Care Products,
First Edition. Wilfried Rähse.

Table A.1 (Continued)

No.	Ingredient	Amount (%)	INCI name	EC number
17	Trisodium citrate dihydrate	0.64	Sodium Citrate	200-675-3
18	Lavender	0.5	Lavandula Angustifolia Oil	289-995-2
19	Xanthan	0.3	Xanthan Gum	234-394-2
20	Potassium sorbate	0.25	Potassium Sorbate	246-376-1
21	Hyaluronic acid	0.2	Hyaluronic Acid	232-678-0
22	Citric acid monohydrate	0.16	Citric Acid	201-069-1
23	Allantoin	0.1	Allantoin	202-592-8
	Sum (without water)	52.15		

Table A.2 Suggestion for a vitalizing day cream (based on alternative V3, Table 8.3).

No	Ingredient	Amount (%)	INCI name	EC number
1	Water (cosm.)	57.55	Aqua	231-791-2
2	Sweet almond oil	4	Prunus Amygdalus Sativa Kernel Oil	291-063-5
3	Cetyl stearyl alcohol	3.5	Cetearyl Alcohol	267-008-6
4	Evening primrose oil	3	Oenothera Biennis Oil	289-859-2
5	Hemp oil	3	Cannabis Sativa Seed Oil	289-644-3
6	Shea butter	3	Butyrospermum Parkii Butter	293-515-7
7	Glycerol	3	Glycerin	200-289-5
8	Sorbitol	3	Sorbitol	200-061-5
9	Magnesium ascorbyl phosphate	2.5	Magnesium Ascorbyl Phosphate	114040-31-2/ 113170-55-1 CAS
10	D-Panthenol	2.5	Panthenol	201-327-3/ 240-540-6
11	Nicotinamide	2	Niacinamide	202-713-4
12	α-Lipoic acid	2	Thioctic Acid	214-071-2
13	Ethanol	2	Alcohol	200-578-6
14	Sodium lactate	1.6	Sodium Lactate	200-772-0/ 212-762-3

Table A.2 (Continued)

No	Ingredient	Amount (%)	INCI name	EC number
15	Cetyl stearyl ether 30EO	1.5	Ceteareth-30	68439-49-6 CAS
16	Silicone oil M350	1.5	Dimethicone	63148-62-9 CAS/9006-65-9 CAS/9016-00-6 CAS
17	Vitamin E (natural, D-α; 0.7)/sunflower oil (0.3)	1	Tocopherol/ Helianthus Annuus	59-02-9 (D) CAS/232-273-9
18	Lactic acid	0.9	Lactic Acid	200-018-0
19	Trisodium citrate dihydrate	0.64	Sodium Citrate	200-675-3
20	Glycerol monostearates	0.5	Glyceryl Stearate	250-705-4/ 286-490-9
21	Rosemary	0.4	Rosmarinus Officinalis Leaf Oil	283-291-9
22	Xanthan	0.3	Xanthan Gum	234-394-2
23	Potassium sorbate	0.25	Potassium Sorbate	246-376-1
24	Citric acid monohydrate	0.16	Citric Acid	201-069-1
25	Hyaluronic acid	0.1	Hyaluronic acid	232-678-0
26	Allantoin	0.1	Allantoin	202-592-8

Table A.3 Proposal for a potent anti-aging day cream (based on alternative AA3, Table 8.7).

No	Ingredient	Amount (%)	INCI name	EC number
1	Water (cosm.)	54.95	Aqua	231-791-2
2	Kinetin	4.1	Kinetin	208-382-2
3	Sweet almond oil	4	Prunus Amygdalus Sativa Kernel Oil	291-063-5
4	D-Panthenol	3.5	Panthenol	201-327-3/ 240-540-6
5	Cetyl stearyl alcohol	3.5	Cetearyl Alcohol	267-008-6
6	Evening primrose oil	3	Oenothera Biennis Oil	289-859-2

(Continued)

Table A.3 (Continued)

No	Ingredient	Amount (%)	INCI name	EC number
7	Hemp oil	3	Cannabis Sativa Seed Oil	289-644-3
8	Shea butter	3	Butyrospermum Parkii Butter	293-515-7
9	Glycerol	3	Glycerin	200-289-5
10	Sorbitol	3	Sorbitol	200-061-5
11	Magnesium ascorbyl phosphate	2.5	Magnesium Ascorbyl Phosphate	114040-31-2/ 113170-55-1 CAS
12	Nicotinamide	2	Niacinamide	202-713-4
13	Glycine	2	Glycine	200-272-2
14	Ethanol	2	Alcohol	200-578-6
15	Cetyl stearyl ether 30EO	1.5	Ceteareth-30	68439-49-6 CAS
16	Silicone oil M350	1.5	Dimethicone	63148-62-9 CAS/9006-65-9 CAS/9016-00-6 CAS
17	Vitamin E (natural, D-α; 0.7)/sunflower oil (0.3)	1	Tocopherol/ Helianthus Annuus	59-02-9 (D) CAS/232-273-9
18	Trisodium citrate dihydrate	0.64	Sodium Citrate	200-675-3
19	Glycerol monostearates	0.5	Glyceryl Stearate	250-705-4/ 286-490-9
20	Xanthan	0.3	Xanthan Gum	234-394-2
21	Potassium sorbate	0.25	Potassium Sorbate	246-376-1
22	Rose	0.2	Rosa Damascena Flower Oil	290-260-3
23	Retinaldehyde	0.2	Retinal	204-135-8
24	Citric acid monohydrate	0.16	Citric Acid	201-069-1
25	Hyaluronic acid	0.1	Hyaluronic acid	232-678-0
26	Allantoin	0.1	Allantoin	202-592-8

Table A.4 Formulation of an antiacne tonic.

No	Ingredient	Amount (%)	INCI name	EC number
1	Water (cosm.)	43.4	Aqua	231-791-2
2	Ethanol	30	Alcohol	200-578-6
3	Sandalwood	5	Santalum Album Wood Oil	284-111-1
4	Magnesium ascorbyl phosphate	5	Magnesium Ascorbyl Phosphate	114040-31-2/ 113170-55-1 CAS
5	Nicotinamide	4	Niacinamide	202-713-4
6	Glycerol	3	Glycerin	200-289-5
7	Sorbitol	3	Sorbitol	200-061-5
8	Urea	3	Urea	200-315-5
9	Pyridoxine-HCl	2	Pyridoxine-HCl	200-386-2
10	Sodium lactate	0.8	Sodium Lactate	200-772-0/ 212-762-3
11	Lactic acid	0.5	Lactic Acid	200-018-0
12	Xanthan	0.3	Xanthan Gum	234-394-2

Table A.5 Formulation of an after-sun cream (after AS2).

No	Ingredient	Amount (%)	INCI name	EC number
1	Water (cosm.)	54.95	Aqua	231-791-2
2	Shea butter	6	Butyrospermum Parkii Butter	293-515-7
3	D-Panthenol	5	Panthenol	201-327-3/ 240-540-6
4	Sweet almond oil	4	Prunus Amygdalus Sativa Kernel Oil	291-063-5
5	Cetyl stearyl alcohol	3.5	Cetearyl Alcohol	267-008-6
6	Evening primrose oil	3	Oenothera Biennis Oil	289-859-2
7	Hemp oil	3	Cannabis Sativa Seed Oil	289-644-3
8	Glycerol	3	Glycerin	200-289-5
9	Sorbitol	3	Sorbitol	200-061-5
10	Urea	2.9	Urea	200-315-5

(Continued)

Table A.5 (Continued)

No	Ingredient	Amount (%)	INCI name	EC number
11	Nicotinamide	2	Niacinamide	202-713-4
12	Ethanol	2	Alcohol	200-578-6
13	Cetyl stearyl ether 30EO	1.5	Ceteareth-30	68439-49-6 CAS
14	Silicone oil M350	1.5	Dimethicone	63148-62-9 CAS/9006-65-9 CAS/9016-00-6 CAS
15	Magnesium ascorbyl phosphate	1	Magnesium Ascorbyl Phosphate	114040-31-2/ 113170-55-1 CAS
16	Vitamin E (natural, D-α; 0.7)/sunflower oil (0.3)	1	Tocopherol/ Helianthus Annuus	59-02-9 (D) CAS/232-273-9
17	Trisodium citratedihydrate	0.64	Sodium Citrate	200-675-3
18	Glycerol monostearates	0.5	Glyceryl Stearate	250-705-4/286-490-9
19	Chamomile (Roman)	0.35	Anthemis Nobilis Flower Oil	283-467-5
20	Xanthan	0.3	Xanthan Gum	234-394-2
21	Aloe vera powder (1 : 200)	0.25	Aloe Barbadensis Leaf Juice Powder	287-390-8/305-181-2
22	Potassium sorbate	0.25	Potassium Sorbate	246-376-1
23	Citric acid monohydrate	0.16	Citric Acid	201-069-1
24	Hyaluronic acid	0.1	Hyaluronic acid	232-678-0
25	Allantoin	0.1	Allantoin	202-592-8

Table A.6 Formulation of an intensive bedsore cream for prophylaxis (according to PU2; based on the 80% modules mixture).

No	Ingredient	Amount (%)	INCI name	EC number
1	Water (cosm.)	44.95	Aqua	231-791-2
2	Silicone oil (M500)	6	Dimethicone	63148-62-9 CAS/9006-65-9 CAS/9016-00-6 CAS
3	D-Panthenol	4.5	Panthenol	201-327-3/ 240-540-6

Table A.6 (Continued)

No	Ingredient	Amount (%)	INCI name	EC number
4	Jojoba oil	4	Simmondsia Chinensis Seed Oil	289-964-3
5	Sweet almond oil	4	Prunus Amygdalus Sativa Kernel Oil	291-063-5
6	Urea	4	Urea	200-315-5
7	Cetyl stearyl alcohol	3.5	Cetearyl Alcohol	267-008-6
8	Evening primrose oil	3	Oenothera Biennis Oil	289-859-2
9	Hemp oil	3	Cannabis Sativa Seed Oil	289-644-3
10	Shea butter	3	Butyrospermum Parkii Butter	293-515-7
11	Glycerol	3	Glycerin	200-289-5
12	Sorbitol	3	Sorbitol	200-061-5
13	Sea salt	3	Sea Salt	231-598-3
14	Nicotinamide	2	Niacinamide	202-713-4
15	Ethanol	2	Alcohol	200-578-6
16	Cetyl stearyl ether 30EO	1.5	Ceteareth-30	68439-49-6 CAS
17	Ylang-Ylang	1.4	Cananga Odorata Leaf Oil	281-092-1
18	Magnesium ascorbyl phosphate	1	Magnesium Ascorbyl Phosphate	114040-31-2/ 113170-55-1 CAS
19	Vitamin E (natural, D-α; 0.7)/sunflower oil (0.3)	1	Tocopherol/ Helianthus Annuus	59-02-9 (D) CAS/232-273-9
20	Trisodium citrate dihydrate	0.64	Sodium Citrate	200-675-3
21	Glycerol monostearates	0.5	Glyceryl Stearate	250-705-4/ 286-490-9
22	Xanthan	0.3	Xanthan Gum	234-394-2
23	Potassium sorbate	0.25	Potassium Sorbate	246-376-1
24	Hyaluronic acid	0.2	Hyaluronic acid	232-678-0
25	Citric acid monohydrate	0.16	Citric Acid	201-069-1
26	Allantoin	0.1	Allantoin	202-592-8

Table A.7 Formulation of an anti-cellulite cream (based on alternative C2).

No	Ingredient	Amount (%)	INCI name	EC number
1	Water	ad. 100	Aqua	231-791-2
2	Jojoba oil	6	Simmondsia Chinensis Seed Oil	289-964-3
3	Protalgel	4	Algae Extract, Hydrolyzed Algin, Carrageenan	—
4	Lemon	4	Citrus Limon Seed Oil	296-174-2/ 284-515-8/ 285-359-3
5	Fucus extract (or powder)	4	Fucus Vesiculosus Extract (or Powder)	283-633-7
6	Polyglyceryl-3 dicitrate/stearate	4	Polyglyceryl-3 Dicitrate/Stearate	—
7	Myristylmyristat	4	Myristyl Myristat	221-787-9
8	D-Panthenol	3.5	Panthenol	201-327-3/ 240-540-6
9	Kelp extract (or powder)	3	Laminaria Digitata Extract (or Powder)	289-980-0
10	Wheat germ oil	3	Triticum Vulgare Oil	281-689-7
11	Safflower oil	3	Carthamus Tinctorius Seed Oil	232-276-5
12	Borage seed oil	3	Borago officinalis Seed Oil	225234-12-8 CAS
13	Glycerol	3	Glycerin	200-289-5
14	Sorbitol	3	Sorbitol	200-061-5
15	Nicotinamide	2	Niacinamide	202-713-4
16	Ethanol	2	Alcohol	200-578-6
17	Dead Sea salt	1	Sea Salt (Dead Sea Salt)	231-598-3
18	Magnesium ascorbyl phosphate	1	Magnesium Ascorbyl Phosphate	114040-31-2/ 113170-55-1 CAS
19	Vitamin E (natural, D-α; 0.7)/sunflower oil (0.3)	1	Tocopherol/ Helianthus Annuus	59-02-9 (D) CAS/232-273-9
20	Trisodium citrate dihydrate	0.64	Sodium Citrate	200-675-3
21	Lysolecithin, de-oiled (powder)	0.5	Lysolecithin	288-318-8
22	Xanthan	0.3	Xanthan Gum	234-394-2
23	Potassium sorbate	0.25	Potassium Sorbate	246-376-1
24	Citric acid monohydrate	0.16	Citric Acid	201-069-1
25	Hyaluronic acid	0.1	Hyaluronic Acid	232-678-0
26	Allantoin	0.1	Allantoin	202-592-8

Table A.8 Formulation of a depilatory cream.

No	Ingredient	Amount (%)	INCI name	EC number
1	Water	61.85	Aqua	231-791-2
2	Sweet almond oil	6	Prunus Amygdalus Sativa Kernel Oil	291-063-5
3	Urea	6	Urea	200-315-5
4	Cetyl stearyl alcohol	6	Cetearyl Alcohol	267-008-6
5	Potassium thioglycolate	6	Potassium Thioglycolate	200-677-4
6	Cetyl stearyl ether 30EO	6	Ceteareth-30	68439-49-6 CAS
7	Calcium hydroxide	2	Calcium Hydroxide	215-137-3
8	Silicone oil M350	2	Dimethicone	63148-62-9 CAS/9006-65-9 CAS/9016-00-6 CAS
9	Glycerol	1.5	Glycerin	200-289-5
10	Potassium hydroxide	0.9	Potassium Hydroxide	215-181-3
11	Vitamin E (natural, D-α; 0.7)/sunflower oil (0.3)	0.7	Tocopherol/ Helianthus Annuus	59-02-9 (D) CAS/232-273-9
12	Geranium	0.5	Pelargonium Roseum Leaf Oil	290-144-2
13	Sodium salicylate	0.4	Sodium Salicylate	200-198-0
14	Carbopol Ultrez 30	0.15	Acrylates/C10-C30 Alkyl Acrylate Crosspolymer	—

Appendix B

MSDS Niacinamide

Vertellus

Niacinamide
Revision Date: 15 Mar 2017
MSDS0137 (ENG)
page 1 of 9

SAFETY DATA SHEET

SECTION 1: Identification of the substance/mixture and of the company/undertaking

1.1. Product identifier **Niacinamide**

Synonyms: Nicotinamide; Vitamin B3; Niacinamide Feed Grade; Niacinamide Free Flow

Chemical Abstracts Registry No: 98-92-0

1.2. Relevant identified uses of the substance or mixture and uses advised against

Animal & human nutrition, chemical intermediate, personal care

1.3. Details of the supplier of the safety data sheet

Vertellus Specialty Chemicals (Nantong) Co., Ltd.
#9 Shengkai Road NETDZ
Nantong, Jiangsu, China. 226009
Phone: 86-513-83591318

e-mail Address: sds@vertellus.com

1.4. Emergency telephone number **CHEMTREC (USA):** 1-800-424-9300 (collect calls accepted)
CHEMTREC (International): 1-703-527-3887 (collect calls accepted)
NRCC (China): +86 025-85477110

SECTION 2: Hazards identification

2.1. Classification of the substance or mixture

(According to Regulation (EC) No 1272/2008)	(According to Directive 67/548/EEC)
Serious Eye Irritation Category 2	**Symbol:** Xi **Risk Phrases:** R36: Irritating to the eyes. **Safety Phrases:** S26: In case of contact with eyes, rinse immediately with plenty of water and seek medical advice.

2.2. Label elements

Hazard Symbols (Pictogram):

Signal Word: Warning

Hazard Precautions: H319 - Causes serious eye irritation.

Prevention Precautionary Statements: P280 - Wear protective gloves/protective clothing/eye protection/face protection.

First Aid Precautionary Statements: P305+P351+P338 - IF IN EYES: Rinse cautiously with water for several minutes. Remove contact lenses, if present and easy to do. Continue rinsing.
P337+P313 - If eye irritation persists: Get medical advice/attention.

** Indicates a trademark owned by the Vertellus group of companies registered in the United States and elsewhere.*

Cosmetic Creams: Development, Manufacture and Marketing of Effective Skin Care Products,
First Edition. Wilfried Rähse.

SAFETY DATA SHEET

Storage Precautionary Statements: Not required.

Disposal Precautionary Statements: Not required.

2.3. Other hazards

Other Hazards: WARNING! MAY FORM COMBUSTIBLE DUST CONCENTRATIONS IN AIR (DURING PROCESSING).

SECTION 3: Composition/information on ingredients

3.1. Substances or 3.2. Mixtures

Ingredient	CAS Number	Concentration (weight %)	EC Number	CLP Inventory/ Annex VI	EU DSD Classification (67/548/EEC)	EU CLP Classification (1272/2008)
Niacinamide	98-92-0	~ 100	202-713-4	Not listed.	Xi R36	Eye Irrit. 2; H319

NOTE: See Section 8 for exposure limit data for these ingredients. See Section 15 for trade secret information (where applicable). See Section 16 for the full text of the R-phrases above.

SECTION 4: First aid measures

4.1. Description of first aid measures

Skin Contact: Wash exposed area twice with soap and water. The exposed area should be examined by medical personnel if irritation or pain persists after the area has been washed.

Eye Contact: Rinse eyes immediately with large amounts of water for at least 15 minutes, occasionally lifting the eyelids. Seek medical advice if symptoms persist.

Inhalation: Remove from exposure area to fresh air immediately. If breathing has stopped, give artificial respiration. Keep affected person warm and at rest. Seek medical advice if symptoms persist.

Ingestion: If swallowed, contact physician or poison control center immediately. Give oxygen if respiration is shallow. Do not give anything by mouth to an unconscious person.

4.2 Most important symptoms and effects, both acute and delayed

Acute: Niacinamide is an eye irritant, but does not irritate the skin. May cause respiratory irritation upon exposure to dusty conditions. In humans, nausea with or without vomiting was the main effect after acute exposure and was generally seen after doses in excess of 5 grams/day; no effects were persistent.

Delayed Effects: None known.

4.3. Indication of any immediate medical attention and special treatment needed

Note to Physician: No specific indications. Treatment should be based on the judgment of the physician in response to the reactions of the patient.

SAFETY DATA SHEET

SECTION 5: Firefighting measures

5.1. Extinguishing media

Appropriate Extinguishing Media: Water fog, foam, carbon dioxide, or dry chemical

5.2. Special hazards arising from the substance or mixture

Hazardous Products of Combustion: Cyanide and nitrogen oxides may be released during thermal decomposition.

Potential for Dust Explosion: Niacinamide presents a significant dust explosion hazard unless properly handled.

- Maximum Explosion Pressure = 8.0 bar
- Maximum Rate of Pressure Rise = 885 bar/s
- K_{st} = 240 bar•m/s
- Minimum Ignition Energy = 3 - 5 mJ
- Limiting Oxygen Concentration = 13 - 14%
- Minimum Explosible Concentration = 50 - 60 g/m^3.

Refer to NFPA 654, Standard for the Prevention of Fire and Dust Explosions from the Manufacturing, Processing, and Handling of Combustible Particulate Solids, for safe handling. Refer to European standards: EN1127-1, EN14491, EN14797, EN14373, and EN15089 for safe handling of and controlling explosive atmospheres in the workplace.

Special Flammability Hazards: This product is an organic solid. As such, in its finely divided form, this product has the potential to present a dust explosion hazard under certain conditions. Please review the dust explosion data enclosed in this section. Handle this product in a manner that prevents dust generation and accumulation, and refer to National Fire Protection Association (NFPA) Standard 654 for further information on prevention of dust explosions.

5.3. Advice for firefighters

Basic Fire Fighting Guidance: Wear self-contained breathing apparatus and protective clothing. Normal firefighting procedures may be used. Avoid generating dust. Fine dust dispersed in air in sufficient concentrations, and in the presence of an ignition source is a potential dust explosion hazard.

SECTION 6: Accidental release measures

6.1. Personal precautions, protective equipment and emergency procedures

Evacuation Procedures: Isolate the hazard area and deny entry to unnecessary and unprotected personnel.

Special Instructions: See Section 8 for personal protective equipment recommendations. Remove all contaminated clothing to prevent further absorption. Decontaminate affected personnel using the first aid procedures in Section 4. Leather shoes that have been saturated must be discarded.

6.2. Environmental precautions

Prevent releases to soils, drains, sewers and waterways.

6.3. Methods and material for containment and cleaning up

Remove all ignition sources. Ventilate the area of spill or leak. Wear protective equipment during clean-up. Material can then be collected for later disposal. After collection of material, flush area with water. Dispose of the material in accordance with standard practice for disposal of potentially hazardous materials as required by applicable federal, state or local laws. Dust deposits should not be allowed to accumulate on surfaces, as these may form an explosive mixture if they are released into the atmosphere in sufficient concentration.

Avoid dispersal of dust in the air (i.e., clearing dust surfaces with compressed air). Nonsparking tools should be used.

6.4. Reference to other sections

Refer to section 8 for information on selecting personal protective equipment. Refer to section 13 for information on spilled product, absorbent and clean up material disposal instructions.

SECTION 7: Handling and storage

7.1. Precautions for safe handling

Precautions for Unique Hazards:	This material may present a dust explosion hazard in solid form and is sensitive to ignition by electrostatic discharge. Maintain areas below flammable vapor / explosive dust concentrations.
Practices to Minimize Risk:	Wear appropriate protective equipment when performing maintenance on contaminated equipment. Wash hands thoroughly before eating or smoking after handling this material. Do not eat, drink or smoke in work areas. Prevent contact with incompatible materials. Avoid spills and keep away from drains. Handle in a manner to prevent generation of aerosols, vapors or dust clouds. To reduce the risk of dust explosion, the recommendations for facility and process design, control of ignition sources and fugitive dust, fire protection, training and maintenance outlined in "NFPA 654: Standard for the Prevention of Fire and Dust Explosions from the Manufacturing, Processing, and Handling of Combustible Particulate Solids" should be followed. Implementing a housekeeping program to control the accumulation of dust on work surfaces is critical to reducing the risk of catastrophic secondary dust explosions.
Special Handling Equipment:	Not applicable.

7.2. Conditions for safe storage, including any incompatibilities

Storage Precautions & Recommendations:	Protect containers against physical damage. Maintain dry, ventilated conditions for storage. Keep away from strong acids, strong bases and oxidizing agents. Do not store with poisons. Minimize dust generation and accumulation. Routine housekeeping should be instituted to ensure that dusts do not accumulate on surfaces. Dry powders can build static electricity charges when subjected to friction of transfer and mixing operations. Provide adequate precautions, such as electrical grounding and bonding, or inert atmospheres.
Dangerous Incompatibility Reactions:	Avoid strong acids, strong bases, and oxidizing agents.
Incompatibilities with Materials of Construction:	None known

7.3. Specific end use(s)

If a chemical safety assessment has been completed an exposure scenario is attached as an annex to this Safety Data Sheet. Refer to this annex for the specific exposure scenario control parameters for uses identified in subsection 1.2.

SECTION 8: Exposure controls/personal protection

8.1. Control parameters

Country	Occupational Exposure Limit
Latvia	1 mg/m^3
New Zealand	Particulates: 10 mg/mg^3 (inhalable); 3 mg/m^3 (respirable)

SAFETY DATA SHEET

United States (OSHA)	Particulates: 15 mg/m^3 (total dust); 5 mg/m^3 (respirable fraction)
United States (NIOSH)	Particulates: 10 mg/m^3
Belgium, Canada (Quebec), Singapore, South Korea	Particulates: 10 mg/m^3

Air Monitoring Method: Gravimetric analysis for total particulate and respirable fraction (<10 microns).

8.2. Exposure controls

Also see the annex to this SDS (if applicable) for specific exposure scenario controls.

Other Engineering Controls: All operations should be conducted in well-ventilated conditions. Local exhaust ventilation should be provided. It is recommended that all dust control equipment such as local exhaust ventilation and material transport systems involved in handling of this product contain explosion relief vents or an explosion suppression system or an oxygen-deficient environment. Ensure that dust-handling systems (such as exhaust ducts, dust collectors, vessels, and processing equipment) are designed in a manner to prevent the escape of dust into the work area (i.e., there is no leakage from the equipment).

Personal Protective Equipment: Work uniform or impervious clothing. Impervious gloves and boots. Safety glasses or chemical goggles. NIOSH approved dust mask, or negative pressure respirator with dust or HEPA cartridges as necessary.

Respirator Caution: Observe OSHA regulations for respirator use (29 CFR 1910.134). Air-purifying respirators must not be used in oxygen-deficient atmospheres.

Thermal Hazards: Not applicable.

Environmental Exposure Controls: The level of protection and types of controls necessary will vary depending upon potential exposure conditions. Select controls based on a risk assessment of local circumstances. If user operations generate dust, fumes, gas, vapor or mist, use process enclosures, local exhaust ventilation or other engineering controls to keep worker exposure to airborne contaminants below any recommended or statutory limits.

SECTION 9: Physical and chemical properties

9.1. Information on basic physical and chemical properties

Appearance, State & Odor (ambient temperature): White crystalline powder, with essentially no odor.

Molecular Formula:	$C_6H_6N_2O$	**Molecular Weight:**	122.13
Vapor Pressure:	< 1 mm Hg	**Evaporation Rate:**	Not applicable.
Specific Gravity or Density:	1.4 @ 25°C	**Vapor Density (air = 1):**	No data available.
Boiling Point:	150 - 160°C	**Freezing / Melting Point:**	124 - 131°C
Solubility in Water:	500,000 mg/L @ 25°C	**Octanol / Water Coefficient:**	log Kow = -0.37
pH:	pKa = 3.35 @ 20°C	**Odor Threshold:**	No data available.
Viscosity:	Not Applicable	**Autoignition Temperature:**	No data available.
Flash Point and Method:	360°F (182°C) Tag Open Cup	**Flammable Limits:**	No data available.
Flammability (solid, gas):	No data available.	**Decomposition Temperature:**	No data available.
Explosive Properties:	Not explosive.	**Oxidizing Properties:**	Not an oxidizer.

SAFETY DATA SHEET

SECTION 10: Stability and reactivity

10.1. Reactivity	Not classified as dangerously reactive.
10.2. Chemical stability	Stable
10.3. Possibility of hazardous reactions	Will not occur.
10.4. Conditions to avoid	Avoid static discharge and generation of dust. Thermal decomposition begins at 150°C.
10.5. Incompatible materials	Avoid strong acids, strong bases, and oxidizing agents.
10.6. Hazardous decomposition products	Cyanide and nitrogen oxides may be released during thermal decomposition.

SECTION 11: Toxicological information

11.1. Information on toxicological effects

Acute Oral LD_{50}:	> 3500 mg/kg (rat) > 2500 mg/kg (mouse)	Niacinamide
Acute Dermal LD_{50}:	> 2000 mg/kg (rabbit)	Niacinamide
Acute Inhalation LC_{50}:	> 3.8 mg/L (rat, 4 hr)	Niacinamide
Skin Irritation:	Non-irritating to skin.	
Eye Irritation:	Moderately irritating to eyes.	
Skin Sensitization:	Not sensitizing (Weight of evidence)	
Mutagenicity:	This material was tested and found to be non-mutagenic in the Ames assay and Mouse Micronucleus test. Equivocal test results occurred in the Unscheduled DNA Synthesis assay in rat primary hepatocytes.	
Reproductive / Developmental Toxicity:	In a 28-day oral toxicity test in rats, no effects on reproductive organs were observed in either sex. In a developmental toxicity study in rats using niacin, the NOAEL for maternal toxicity was 200 mg/kg/d (body weight changes) and the NOAEL on reproductive toxicity and developmental toxicity was 200 mg/kg/d (decreased placental and male pup body weight). No teratogenic effects were observed.	
Carcinogenicity:	This material is not listed by IARC, NTP or OSHA as a carcinogen. No test data is available that indicates this material is a carcinogen.	
Target Organs:	None Known.	
Primary Route(s) of Exposure:	Skin contact and absorption, eye contact, and inhalation. Ingestion is not likely to be a primary route of exposure.	
Most important symptoms and effects, both acute and delayed	Niacinamide is an eye irritant, but does not irritate the skin. May cause respiratory irritation upon exposure to dusty conditions. In humans, nausea with or without vomiting was the main effect after acute exposure and was generally seen after doses in excess of 5 grams/day; no effects were persistent. Delayed Effects: None known.	
Additive or Synergistic effects:	None known.	

Vertellus

SAFETY DATA SHEET

SECTION 12: Ecological information

12.1. Toxicity	EC_{50} (24h) ***Daphnia magna*** > 1000 mg/L LC_{50} (96h) ***Poecilia reticulata*** (guppy) > 1000 mg/L EC_{50} (72h) ***Scenedesmus subspicatus*** > 1000 mg/L	Niacinamide
12.2. Persistence and degradability	Material is readily biodegradable under aerobic conditions.	
12.3. Bioaccumulative potential	Not expected to bioconcentrate in aquatic species.	
12.4. Mobility in soil	This material is soluble in water. Its adsorption to soil and sediment should not be significant.	
12.5. Results of PBT and vPvB assessment	This substance is not a PBT or vPvB.	
12.6. Other adverse effects	No data available.	

SECTION 13: Disposal considerations

13.1. Waste treatment methods

US EPA Waste Number:	Non-Hazardous
Waste Classification: (per US regulations)	Non-Hazardous
Waste Disposal:	NOTE: Generator is responsible for proper waste characterization. State hazardous waste regulations may differ substantially from federal regulations. Dispose of this material responsibly, and in accordance with standard practice for disposal of potentially hazardous materials as required by applicable international, national, regional, state or local laws, and environmental protection duty of care principles. Do NOT dump into any sewers, on the ground, or into any body of water. For disposal within the EC, the appropriate classification code according to the European Community List of Wastes should be used. Note that disposal regulations may also apply to empty containers and equipment rinsates.

SECTION 14: Transport information

The following information applies to all shipping modes (DOT/IATA/ICAO/IMDG/ADR/RID/ADN), unless otherwise indicated:

14.1. UN number	Not applicable	**14.2. UN proper shipping name**	Chemicals, n.o.s. (Niacinamide)
14.3. Transport hazard class(es)	Not applicable	**14.4. Packing group**	Not applicable
14.5. Environmental hazards	Not applicable		
14.6. Special precautions for user	Cannot be stored or shipped with TOXIC materials		
NA Emergency Guidebook Numbers:	Not applicable	**IMDG EMS:**	Not applicable;
14.7. Transport in bulk according to Annex II of MARPOL73/78 and the IBC Code			Not applicable.

SAFETY DATA SHEET

SECTION 15: Regulatory information

15.1. Safety, health and environmental regulations/legislation specific for the substance or mixture

Chemical Inventory Lists:	Status:		
USA TSCA:	Listed	EINECS:	202-713-4
Canada(DSL/NDSL):	DSL	Japan:	(5)-736
Korea:	KE-29935	Australia:	Listed
China:	Listed	Philippines:	Listed
Taiwan:	Listed	New Zealand:	Listed

WHMIS Classification: Class D, Division 2, Subdivision B: Irritant.

German Water Hazard Classification: ID Number 2244, hazard class 1 - low hazard to waters ***(Nicotinamid)***

SARA 313: Not listed.

Reportable Quantities: Not applicable.

Other Regulatory Listings:

- Included in US Food and Drug Administration's (US FDA) Priority-Based Assessment of Food Additives database.
- "Generally Regarded as Safe" (GRAS) by US Food and Drug Administration (21 CFR 184.1).
- Approved as cosmetic product additive under European Cosmetic Products Directive 76/768/EEC, Section I listing.

HMIS:

NFPA:

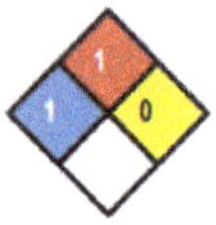

15.2. Chemical safety assessment

A chemical safety assessment is not required as this substance is not classified as hazardous.

SECTION 16: Other information

Full text of R phrases in Section 3: R36: Irritating to the eyes.

Key Data Sources:

- Select Committee on GRAS Substances (SCOGS) (1979). Opinion: Niacinamide (nicotinamide), SCOGS-Report Number: 108
- European Food Safety Authority, 2012. Scientific Opinion on the safety and efficacy of niacin (nicotinic acid and nicotinamide) as a feed additive for all animal species based on a dossier submitted by Vertellus Specialties Belgium BV1, 2. EFSA Panel on Additives and Products or Substances used in Animal Feed (FEEDAP)3, 4
- OECD SIDS, UNEP Publications (2002). 3-Pyridinecarboxaldehyde (Nicotinamide): SIDS Initial Assessment Report for SIAM 15.

Training Advice: Not applicable.

Vertellus

Niacinamide
Revision Date: 15 Mar 2017
MSDS0137 (ENG)
page 9 of 9

SAFETY DATA SHEET

Legend of Abbreviations:

ACGIH = American Conference on Governmental Industrial Hygienists.
CAS = Chemical Abstracts Service.
CFR = Code of Federal Regulations.
DSL/NDSL = Domestic Substances List/Non-Domestic Substances List.
EC = European Community.
EINECS = European Inventory of Existing Commercial Chemical Substances.
ELINCS = European List of Notified Chemical Substances.
EU = European Union.
GHS = Globally Harmonized System.
LC = Lethal Concentration.
LD = Lethal Dose.
NFPA = National Fire Protection Association.
NIOSH = National Institute of Occupational Safety and Health.
NTP = National Toxicology Program.
OSHA = Occupational Safety and Health Administration
PEL = Permissible Exposure Limit.
RQ = Reportable Quantity.
SARA = Superfund Amendments and Reauthorization Act of 1986.
TLV = Threshold Limit Value.
WHMIS = Workplace Hazardous Materials Information System.

Important Note: Please note that the information contained herein is furnished without warranty of any kind. Users should consider these data only as a supplement to other information gathered by them and must make independent determinations of suitability and completeness of information from all sources to assure proper use and disposal of these materials and the safety and health of employees and customers. Recipients are advised to confirm in advance of need that the information is current, applicable, and suitable to their circumstances. The information contained herein may change without prior notice. THIS SAFETY DATA SHEET SUPERSEDES ALL PREVIOUS EDITIONS.

Revision Date:	15 Mar 2017	**Original Date of Issue:**	13 July 1995
Issued by:	Regulatory Management Department	**Email:**	SDS@Vertellus.com
Revision Details:	Revised supplier of SDS and emergency phone number.		

http://www.vertellus.com/Documents/msds/Niacinamide%20(Nicotinamide)%20English.pdf

Index

a

acid mantle 104, 108, 121, 131, 203, 237, 248
acne 17, 49, 126, 198–200, 204, 212, 217, 231, 233–236, 290
Acne comedonica 235
Acne conglobata 235
acne creams
  formulations 287–288
Acne papulo-pustulosa 235
acne vulgaris 235
administrative costs 59, 83, 84, 89
after sun creams/lotions formulations 288–289, 292, 459
age spots 197, 200, 201, 204, 212, 213, 244, 251
agitators 338, 349
airless dispenser 154, 163, 232, 276, 361–364, 399, 402, 422, 428, 429, 433, 437, 443–445, 450, 452
alcoholic facial waters 235
algae 221, 223, 224, 226, 295, 298, 462
allantoin 139, 223, 238, 239, 247, 287, 300, 438, 443, 451
allergies 135, 160, 218, 242, 243, 260, 293, 310, 330, 435
almond oil 114, 139, 207, 208, 210, 245, 287, 293, 306, 422, 437, 443
aloe vera 194, 196, 216, 284, 288, 443aloe vera, 194, 196, 216, 284, 288, 443
α-hydroxy acids (AHA) 191, 193, 247, 277, 288, 300
α-linolenic acid (ALA) 205
α-lipoic acid (ALA) 163, 213, 277, 284, 288
amino acids 123, 167, 170, 189, 191, 192, 194, 196, 213, 216, 224, 225, 241, 277, 288
amphiphilic ingredients 123
amphiphilic liposomes 167–168
analysis certificate 420, 424, 433
analysis report 426–429, 444, 445
ANNEX of the Cosmetics Regulation 160, 161, 167, 179, 260, 417, 420, 422
anti-acne creams, for blemished skin 231
anti-acne tonic formulations 459
anti-aging creams
  aloe vera 216
  α-lipoic acid 213
  amino acid 213–214
  Coenzyme Q10 214
  2-dimethylaminoethanol 214
  enzymes 216
  formulations 284, 457
  grapes and raspberries 215
  green tea 215
  hyaluronic acid 214
  kinetin 215
  phytohormones 214–215
  polyphenols 215–216
  retinol 213
  vitamin C 213
anti-cellulite cream formulations 462
anti-herpes cream 245–246

Cosmetic Creams: Development, Manufacture and Marketing of Effective Skin Care Products, First Edition. Wilfried Rähse.

antioxidants 134, 135, 140, 153, 162, 163–167, 170–171, 182–183, 198, 200–201, 208, 211, 212–216, 217–218, 223, 227, 236, 269, 271, 277, 288, 300
aromatherapy 218, 314, 316, 318, 332
ascorbic acid 189, 225
ascorbyl palmitate 150, 164, 197
ASEAN Cosmetic Directive (ACD) 16–17
athlete's foot 231, 243–244, 300–302
atopic dermatitis 205, 242, 246, 411
Australia 38, 49, 264
Autoxidation 163

b

baby/babies 18, 33, 37, 43, 51, 108, 148, 161, 167, 175, 193, 223, 227, 231, 234, 237–239, 261, 271, 289–294, 306, 431, 433–436, 448
baby creams
  formulations 289
baby's skin creams 237
balance sheet 59, 62–64
Bancroft rule 140, 170
Basal cell carcinoma 253
BASF 84–86, 105, 146, 149, 319
basic modules 134
batch production 345–349
batch turnover number (BTO) 353
bedsore cream formulations 295, 296, 460
bedsores 175, 231, 239, 271, 295, 296, 460
beeswax 1, 179, 239, 293
Beiersdorf 55, 56, 87, 90, 96
benzoyl peroxide 235, 287
Berloque Dermatitis 330
biotin 196, 200
blackhead 231–235
black skin cancer 253
blemished skin, anti-acne creams for 232
body lotion 247
  analysis report 444
  description 437
  exposure to product 444
  exposure to substances 444, 446–448
  formulations 301
  microbiological quality 444
  normal and reasonably foreseeable use 444
  packaging material 444
  production instruction 443
  quantitative and qualitative composition 438–443
  toxicological profile 449
borage oil 211, 282
bottles 75, 154, 249, 332, 342, 344, 361–366, 375, 427, 433
brand names, searching for 9
branded article concept 47, 92, 97
brands 9, 31, 37–38, 47–51, 53–54, 81, 82, 91, 95–98, 309, 313
broad-spectrum filter 254, 255, 260
brick-mortar-model 109–110
buffer substances 163–167

c

callus removal cream formulations 300–301
cellulite 20, 218, 231, 239–240
Carbopol 158
cellulite creams
  formulations 295, 297, 298, 299–300
Center for Food Safety and Applied Nutrition (CFSAN) 15
centrifuge test 154
ceramides 107, 108, 110, 113, 170, 189, 199, 204, 242
  chemical structures 110
chamomile 135, 216, 218, 239, 271, 288, 293, 327
Chemical Abstracts Service 12
China
  sunscreens 264
Chinese cosmetics market 37–38
Chypre 311
citral 319, 321, 438, 451
citric acids 134, 153, 161–163, 187, 193–194, 301, 335, 390, 424, 428, 433, 451

clean-in-place (CIP) 73, 348–349, 374, 383, 392–396
clinical acne 231, 233
coemulsifier 131, 134, 140, 141, 146–151, 155, 157, 170, 247, 423, 441
coenzyme 163, 170, 196, 199, 200, 213, 214, 225, 271, 284
Coenzyme Q10 163, 170, 200, 213, 214, 284
color index (CI) 7, 167
Colorimeter 407, 409
comedones 203, 205, 232, 233, 235, 287, 400
competition-oriented pricing 95–97
complete protection 261, 269
complexing agents 134, 135, 140, 163–164, 166
concrete 19, 251, 318, 435, 448
consumer goods 50, 51, 55, 61, 65, 84, 86, 90–91, 92–93, 309, 328, 383, 387
continuous production plant 350, 351
cornea 108, 120, 121, 126, 189–191, 200, 201, 225, 246, 249
corneocytes 107, 109–112, 121–123, 189, 198, 406
corneodesmosomal pathway 123, 127
Corneofix® 406
Corneometer® 403
cornified envelope 107
CosIng 11–12, 14, 15, 431
cosmeceuticals 21, 182, 231
  anti-herpes cream 245–246
  baby skin creams 237–239
  blemished skin, anti-acne creams for 231–235
  body lotion 247
  cellulite 239–240
  delimitation 232
  eczema 242–243
  foot cream 243–244
  hand cream 244–245
  itchy shins cream 248
  pressure ulcers 239
  reddened skin, after sun creams and lotions for 236
  self-tanning cream 248–249
  sunscreens
    application and warning notices 261–262, 264–265
    radiation-induced damage 252–254
    solar radiation 249–250
    substances 254–255, 259–260
    sun protection measurement 265–268
    UV rays character 250–252
  thick horny layer cream 246–247
  unwanted hair removal 241
Cosmetic and Personal Care Industry 12–13
cosmetic GMP (cGMP) 371–372, 375, 377, 380
Cosmetic Ingredient Dictionary 12, 13
cosmetic product safety report (CPSR)
  assessment 419
  information 417–419
Cosmetic Products Notification Portal (CPNP) 9, 11
cosmetic market
  China 37–38
  Europe 31–35
  Germany 27–30
  Japan 35–37
  size 39–45
  USA 35
  World 38–39
cosmetics
  definition 2–3
  delimitation 18–22
  history 1
  ingredient databases 11–15
  label lettering and trademark 7–9
  mandatory registration 11
  regulations
    in Asia 16–17
    in Europe 4–6
    in United States 15–16
  typical cosmetic products 3–4
cosmetics regulation (EC) directive 9
COSMOS standard 181
cost-intensive processes 65
cost-oriented 85, 92–93

crushing 225, 310, 353
CutiScan 406
Cutometer® 406
cyclodextrins 135, 153, 277, 284

d

Dead Sea 108, 240
demand-oriented pricing 93–95
depilatory cream 241
  formulations 463
dermal exposure 429 -434, 446
dermatics 187–188
dermis 104–106, 114–117, 169, 194, 200, 250, 252–253, 331–332
dexpanthenol 3, 199
dihydroxyacetone (DHA) 248–249
2-dimethylaminoethanol (DMAE) 214, 284
direct costs 62–63, 79, 81–82, 83, 84–85
D-panthenol 139, 199
drugs 15, 19, 20, 22, 91, 112, 121, 124, 126, 168–170, 184, 187, 188, 190, 198, 223, 253, 261
Directive 93/42/EEC 18
dry skin 108, 126, 176, 188, 192, 194, 207, 217, 218, 231, 243, 244, 246, 247, 248, 276, 282–283, 288

e

Eau
  de Cologne (EdC) 311, 314, 330
  de Toilette (EdT) 311, 314
  de Parfum (EdP) 314
EBIT 56, 59, 60, 84, 85, 86, 87, 90, 92
ecolabels 178, 180
economic importance
  Chinese market 37–38
  Estée Lauder Companies 53
  European market 31–35
  Germany
    market and distribution channels 27–30
    shopping on internet 30–31
  global market volume 39–44
  Japan 35
  K.K. Shiseidō / Japan 53–54
  L'Oréal 48
  largest manufacturers worldwide 45–47
  Nivea 55
  Procter & Gamble Company 51
  trends 45
  Unilever 50
  USA 35
  world division in market regions 38–39
eczema 108, 121, 187, 188, 210, 211, 231, 242–243, 283
elastosis 252
electricity consumption 72
electropolishing 384, 387, 388, 390–392
emollients 146, 155, 191, 194, 423, 441
emulsifiers 140
  antioxidants, complexing agents and buffer substances 163–167
  consistency 155, 157–158
  macro emulsion
    boundary conditions 146
    C-chain compounds 142
    chemical structures 143
    ethoxylated fatty alcohol-based surfactants 142
    HLB 141
    nonionic surfactants 147–148
    oil droplets 140–141
    sorbitan and glycerides esters 146
    stability 141
    surfactants 142
    water droplets 140
    water-repellent and protective properties 148
  mini (nano) emulsions 149
  preservation 158–163
  stability 149, 152–154
energy costs 64, 71–72, 75, 89
Enfleurage 318
enhancers 121–122, 132, 170, 203
enzymes 216, 233, 254, 402
epicutaneous test 399
epidermis 2, 104–106, 109, 114–117, 169, 188, 189, 191, 200, 204, 246, 248, 252, 271, 276, 331, 405

Essence absolue 318
essential oils 124, 125, 138–139, 153, 216–219, 221, 240, 245, 252, 277, 287, 288, 293, 295, 310, 311, 312, 314, 315–318, 319, 321, 322, 325, 327–328, 330, 335, 402, 425, 426, 434–435, 438, 449, 451
Estée Lauder Companies 53
ethosomes 126, 167, 170, 177
ethoxylated emulsifiers 146, 177
ethylenediaminetetraacetic acid (EDTA) 164–165
Eudermine 54
Europe
- cosmetics legal regulations 4–6

European Community Number (EC Number) 11, 12, 438
European Cosmetics Regulation 2, 16, 159, 231
European Federation for Cosmetic Ingredients (EFfCI) 372
European Hygienic Engineering and Design Group (EHEDG) 379
- manufacturer of cosmetics 379
- microorganisms 381
- production manager and operator aware 380

European Inventory of Existing Commercial Chemical Substances (EINECS) 12, 418
European List of Notified Chemical Substances (ELINCS) 12, 418
European market
- economic importance 31–35
  - shopping on internet 31–35

European Scientific Committee on Consumer Safety (SCCS) 15, 161, 261, 400, 401, 424, 425, 430–431, 433, 434, 444, 451
European Union Intellectual Property Office (EUIPO) 9
evening primrose oil 79, 138–139, 141, 194, 205, 210–211, 238, 282 , 422, 437, 443
excipients 63, 121, 131–135, 138, 140–167, 170, 182, 187, 275, 293, 332, 420
extracted substances from plants (liquid and dried) 221
extraction 122, 131, 178, 205, 206, 219, 221, 223, 225, 227, 311, 315–319, 332
Extrait de Parfum 314
Extruder 169, 205
eye area with sun protection cream 304

f

Fast Optical In Vivo Topometry of Human Skin (FOITS) 408
fatty acids (FA) 107, 109, 113, 117, 122, 125, 131, 139, 141–142, 155, 162, 168–170, 188, 189, 201, 203–205, 206–207, 209–211, 212, 216, 224, 237, 255
FD&C Act 2, 19, 20
Federal Association of German Industry and Trade (BDIH) 179
filaggrin 107
filling line 65, 68, 71, 73 -75, 77, 336, 358, 361–366, 384, 392
film-forming polymers 191
Firmenich 312, 332
fixed production costs 75
Flavors & Fragrances (F & F) 312
FOITS 408
folic acid 199
Food & Drug Administration (FDA) 2, 13, 15–16, 17, 161, 166, 182, 261–262, 264, 449
foot cream 244
- formulations 300

formulations 275
- acne creams 287–288
- after sun creams/lotions 288–289, 459
- anti-acne tonic 459
- anti-aging creams 284, 457
- baby creams 289, 292–293
- bedsore cream 295
- body lotion 301, 303
- callus removal cream 300–301
- cellulite creams 295
- depilatory cream 463

formulations (*contd.*)
  eye area with sun protection cream 304, 306
  foot cream 300
  hand creams 300
  moisturizers 276–277, 455
  sunscreen cream 307
  vitalizing creams 277, 282, 456
fragrances 27, 33–34, 37, 44, 53, 127, 135, 167, 176, 178–179, 182, 217–218, 245, 275–276, 277, 293, 309–321, 325, 327–332, 435
free-radical-inhibiting antioxidants 164
Frictiometer 405, 407
fungi 104, 208, 242–244, 300

g

γ-linolenic acid (GLA) 205, 242, 243, 282
γ-tocopherol 201
gels 30, 108, 155, 157, 169, 194, 196, 214, 248, 268, 277, 288, 313, 329, 429
generally recognized as safe (GRAS) 161, 166
German Cosmetics Ordinance 4
German Industry Association for Personal Care and Detergents 377
German Medicines Act 21
German Patent and Trademark Office 9
German Pharmacopoeia (DAB and DAC) 187
Germany
  economic importance
    market and distribution channels 27–30
    shopping on internet 30–31
geranium 218, 240, 245, 301, 327, 438, 442 -444, 449, 451
ginkgo leaves 215
Givaudan 312, 332
global company
  profit and loss accounts 86
global company cost
  direct cost 81–82
  Henkel 79
  internal cost accounting 80
  L'Oréal 79
  overhead and performance accounting 82–84
  principle 79
  published annual reports 84
global cosmetics market Size 39–45
glossy layer 107
glycerol 125, 131, 134, 139, 141, 155, 161, 168, 178, 192, 203, 205, 219, 221, 238, 276, 293, 300, 424–425
glycine 167, 170, 191, 215
Good Manufacturing Practice (GMP) regulations 4–6, 9, 16, 17, 73, 343, 349, 371–379, 400, 417, 426, 433, 438, 450
grapes 215
green tea 215
Griffin 141 170

h

hand creams 244
  formulations 300
heat collapse 253
hemp oil 3, 138, 139, 153, 167, 205, 210, 287, 318
Henkel Fragrance Center 313
herpes 231, 245–246
high-density polyethylene (HD-PE) 362–363
high-pressure homogenizer 149, 152, 343, 349, 356 -361
homogenizer 149, 152, 336–345, 347–349, 351–361, 366
horny cells, layer 106, 122, 198
humectants 191, 192, 276, 292, 423, 441
hyaluronic acid 77, 123–125, 134, 139, 170, 190–191, 194, 199, 214, 240, 271, 276–277, 284, 335
hydrolysates 191, 194, 271
hydrophilic/lipophilic balance (HLB)-values 123, 124, 140, 141, 170, 171, 352
hygienic productions
  cleaning-in-place 392–396
  EHEDG-Guidelines 379–383

equipment materials
choice of materials 384–385
metal surfaces smoothing 387–392
problem 383–384
stainless steel 385–387
good manufacturing practice rules
change control 375
complaint and recall 375
deviations 375
documentation 375–376
equipment 373–374
finished products 374
internal audit 375
premises 373
production 374
quality control laboratory 374–375
scope 372
staff 373
starting and packaging materials 374
subcontracting 375
unspecified products treatment 375
waste 375
hyperkeratosis 233, 238, 246, 250

i

itchy shins cream 248
Indentometer 405, 408
India, Middle East and Africa (IMEA) 39, 53
indirect overhead costs 82
influence of brand 97
Information Network of Departments of Dermatology 135
Ingredients from the Sea 223–225
intercellular lipids 107–108, 111
internal cost accounting 80
International Flavors & Fragrances 312
International Nomenclature Cosmetic Ingredient (INCI) 7, 9, 11, 13, 134, 149, 167, 182, 216, 240, 255, 293, 295, 331–332, 418, 420, 422, 425–426, 438, 451
International Organic and Natural Cosmetics Corporation (IONC GmbH) 180
Italian barber Giovanni Paolo Feminis 311

j

Japan
economic importance 35
Japanese Government 17
jars 65, 362
jasmine 3, 167, 218, 288, 293, 318, 327, 328
Johnson & Johnson 46, 90
jojoba oil 207–208, 241, 293, 295

k

K.K. Shiseidō / Japan 53–54
Karité butter 207
keratin 107, 122, 189, 196, 199–200, 241, 248
keratinocyte 107, 109, 112, 251
kinetin 215

l

L'Oréal 48
global share 48
manufacturing cost 79
market share 48, 49
operational divisions 48, 49
product portfolio and global sales regions 48
label 3, 7–10, 13, 16, 20, 21, 73, 167, 176, 177, 179, 181, 182, 241, 261–265, 271, 307, 330, 365, 366, 374, 413, 417–419, 422, 425, 429, 437, 438, 444, 450
labor costs 65–71, 73, 75, 81, 89
lactic acid 166, 192, 225
lanolin 179, 239, 293
large production plants 345
largest cosmetics manufacturers worldwide 45–47
laser diffraction 148, 152, 154, 155, 359
lavender 135, 167, 216 -218, 240, 244, 300, 310, 311, 315, 327, 329, 435
lauric acid 203
lecithin 167–170, 179, 194, 206

lethal dose of 50% rats (LD_{50}) 420
light callosity 250, 252
linoleic acid 109, 110, 204, 211, 243, 295
linolenic acid 169, 194, 203, 205, 210, 242–243, 282, 287
lipid matrix 107, 110
  chemical structure 113
lipids 107, 110–111, 117, 120–126, 131, 132, 135, 139, 145, 147, 157, 167–170, 176, 182, 189–190, 194, 201, 203–205, 207, 209, 233, 242–243, 244, 245, 247, 253, 260, 277, 293, 300, 301
lipophilic substances 109, 110, 114–120, 124, 127, 424, 426, 432
liposomes 126
  effects 168
  lecithin 168
  preparation 169
  small vesicles production 169
L'Oréal 48
  global share 48
  manufacturing cost 79
  market share 48, 49
  operational divisions 48, 49
  product portfolio and global sales regions 48
  profit and loss account 59, 60
  sales and earnings 48, 49

m

macro emulsion
  boundary conditions 146
  C-chain compounds 142
  chemical structures 143
  HLB 141
  nonionic surfactants 147–148
  oil droplets 140–141
  sorbitan and glycerides esters 146
  stability 141
  surfactants 142
  water droplets 140
  water-repellent and protective properties 148
magnesium ascorbyl phosphate 277, 443, 451
Mallorca Acne 253
mangosteen 240, 295
manufacturing costs
  development products 61
  dependency 73, 75–76
  energy costs 71–72
  installation and building 65
  labor 65–66, 68, 71
  market price determination 61–62
  production 62
  raw material 76–77, 79
  rough calculation 59
margin of safety (MoS) 293, 419, 420, 432, 447, 451
market price 47, 61–62, 82, 85, 89, 91–93, 97, 313
Marleen Body Lotion 437, 443, 450, 452
material safety data sheet (MSDS) 420, 424, 432, 433, 438, 449, 465–473
materials for equipment 383-392
Mazeration 318
Medical Device Directive (MDD) 18
medical devices 18–19, 371
Medicinal product 19–21, 162
Medicinal Product Act 184
medicines 21–22, 180, 182, 184, 187–188, 193, 199, 211, 312
melanin 20, 200, 233, 248, 250–252, 267, 409
Mercosur Common Market 264
metal surfaces smoothing 387
methylparaben 160
Mexameter® 407, 409
Mexoplex® 269, 271, 306
Mexoryl® 269
microbiological quality 377, 418, 444
microbiological tests 118
micro emulsions 149
microorganisms 103, 108, 131, 153, 159, 163, 171, 178, 183, 188, 189, 192, 243, 244, 282, 293, 358, 363, 379, 381–384, 392, 400–402
mineral oil 177, 180, 194, 205, 241, 242, 253, 293
mini-emulsion 149, 343, 355–360

minimal erythema dose (MED) 266, 267
MoistureMap device 409
moisturizers formulations 276
moisturizing substances 139
monoterpene hydrocarbons 321
multiple emulsions 140, 149, 154
myristic acid 203
myrrh 218, 316

n

nano-emulsion 342, 355
nanoparticles 9, 168, 170, 254, 255, 260, 268, 269, 425
national brand register 9
National Industrial Chemicals Notification & Assessment Scheme (NICNAS) 264
natural cosmetics 176
natural fragrances
  chemical composition 319, 321, 327
  extraction 315–319
natural moisturizing factor (NMF) 77, 78, 188, 189–190, 191–192, 194, 216, 245, 293
Nestlé Group 48
neurodermatitis 175, 205, 248, 411
New Drug Application (NDA) process 261
niacinamide safety data sheet 465
nicotinamide 170, 199, 301, 420
Nivea 55–56, 96
No-Longer-Polymers-List (NLP-List) 12
no observed adverse effect level (NOAEL) 431, 432
Novartis 86
nuclear magnetic resonance (NMR) 114–115, 117

o

occlusion 117, 122, 123, 155, 190, 191, 194, 205, 275, 403
odor neutralizer 133
ointment 1, 21–22, 94, 187, 188, 193, 246, 271, 310–311, 376
oleic acid 169, 203, 210–211
"one-pot" method 337, 342
operating income 51, 54, 59
operating profit 49, 84
overhead costs 59, 62, 63, 79, 81–83, 93
over-the-counter (OTC) 19–22, 27, 94, 176, 184, 235, 247, 261, 271
oxidations 135, 149, 152, 153, 159, 162, 163, 170, 193, 198, 200, 201, 203, 210, 212, 235, 251–253, 255, 260, 269, 304, 307, 329, 335, 350, 361, 438

p

P.I.F, *see* product information file (P.I.F.)
package, packaging 3, 7–9, 11, 15, 16, 21, 31, 62, 64, 65, 73, 75, 91, 135, 149, 152, 154, 159, 167, 180 -182, 231, 261, 309, 329, 331, 361, 365, 366, 372, 374, 375, 377, 378, 382, 417, 418, 426, 427, 444, 450
palmitic acid 203, 207
papules, pustules 232–235 253
parabens 159–161, 177, 180, 455
Paracelsus 146
patch test 119, 135, 329, 330, 399, 414, 451
peeling 193, 216, 235, 247, 249, 271
penetration
  lipophilic substances 114, 124
    essential oils 124
    polar substances 125
    skin tolerance test 119
    solid products 125
    standardized measurement methods 120
    vegetable oils 124
    skin creams effects 120
  stratum corneum
    brick-mortar-model 109
    corneocytes 122
    enhancer 121
    extraordinary mechanical stability 109
    horny layer 121
    imaginable way 111
    lipid matrix 113

penetration (*contd.*)
  modern concept 111
  non-polar organic substances 108
  occlusion 122–123
  parameters 121
  permeability 121
  possible pathways 110
  skin appendages 111
  skin barrier function 108
  sufficient water soluble 123
  transepidermal water loss 109
peptides 123, 168, 170, 180, 183, 184, 213, 225, 227, 284
perfumes 167
  composition and markets 312–315
  history 310–312
  importance 309–310
  natural fragrances
    chemical composition 319, 321, 327
    extraction 315–319
  of products 314–315
  personal care 329
  product design possibility 327–329
  safety 330–332
permeability 121–123, 125, 170, 362, 363
persistent pigment darkening (PPD) value 261, 267, 307
Persian Ibn Sina 311
pharmaceutical products 91, 393, 425
phase inversion temperature (PIT) emulsion 140
phenoxyethanol 159–160
phosphatidylcholine 126, 168, 169, 199
phospholipids 106, 126, 168–170, 194, 203–204, 206, 211, 216
photoaging 201, 252
pH regulators 134
phytohormones 214–215
phytosterols 203, 207, 211, 240, 269, 293
pickering emulsion 154
pickling 384, 388–391
plant extracts 177, 220, 223, 227, 238, 240
polyacrylic acid 158
polyethylene glycols (PEGs) 122, 142, 308, 448, 449
polyethylene terephthalate (PET) 361
polymers 125, 134, 149, 157, 158, 170, 171, 190, 191, 194, 214, 268, 384, 402, 408, 427, 443
polymorphic light eruption (PALE) 253
polyphenols 215–216, 223–224
pre-acne 231–232
preservatives 5, 16, 17, 23, 131, 132, 134, 135, 140, 153, 154, 159–163, 177, 179, 181, 182, 187, 238, 242, 247, 276, 287, 293, 306, 363, 364, 400, 402, 421, 428, 429, 441, 444, 451
pressing 182, 205, 225, 317, 318
pressure ulcers 239, 295
price-quantity strategy 47, 92, 97
pricing depending on customer 91
pricing methods
  competition-oriented pricing 95–97
  cost-oriented 92–93
  demand-oriented pricing 93–95
  depending on customer 91–92
  influence of brand 97–98
  pre-set profit targets 91
  site-specific prime costs 91
Procter & Gamble Company 31, 45, 47, 51–52, 54–55, 90
product information file (P.I.F.)
  babies and children safety consideration 434–435
  cream analysis report 426–429
  cream composition 422, 424
  cream exposure to skin 429–434
  CSPR
    assessment 419
    information 417–419
  EU Cosmetics Regulation 417
  identification and description 422
  instruction 426
  safety data sheet 420–421
  structure 422–436
  toxcologic profile 424–425
product safety data sheet (PSDS) 420
Product safety report 417–419

production
  batch 345, 347–349
  bottles and filling lines 361–366
  continuous 350
  laboratory equipment and pilot plant 342–345
  methods 335–337
  mini-emulsion 355–360
  scale-up 350–354
  stirring and homogenizing tools 337–339, 341–342
production costs 62, 64–65, 73–75, 79, 81, 85
prophylaxis creams 239, 253, 295, 297, 460
propioni bacteria 235
propionibacterium acnes 233
prostaglandin 243
protection for the whole day 261
protection grades (PA+) 268
provitamin B_5 199, 200
pyridoxine 199, 287
pyrrolidone carboxylic acid (PCA) 191

q
quality
  emulsion stability 399
  microbial checks 400–402
  physical measurements
    advertising claims support 409–410
    firmness and elasticity 406–408
    gloss and color 409
    moisture and sebum 403, 405–406
    probes and devices 403
  safety analysis 399
  tests by customers 410–411, 413–414
quality evaluation 399–400
quasi drugs 17, 21, 184

r
raspberries 215, 332
raw material cost 73, 76–79, 84, 89, 245
raw materials 62–64, 73, 75-81, 84, 89, 98, 163, 177, 179, 180, 205, 223, 225, 245, 311, 319, 335, 342, 348–350, 367, 372–375, 379, 382, 384, 392, 419, 420, 424–427, 430–433, 438, 443, 444, 446, 450, 453
relative operating margin 51
Remedies Advertising Act 8
retinol 196–199, 213, 271
rose 135, 216, 218, 240, 245, 318, 327, 444
rotor/rotor homogenizer 342
rotor/stator homogenizer 340, 366

s
safety assessment 5, 6, 9, 17, 23, 61, 119, 399, 417, 419, 420, 422, 430, 433, 438, 449–452
safety data sheet (SDS) 15, 420, 424
safflower oil 211, 300
salicylic acid 161, 179, 188, 196, 247
sandalwood 1, 217, 218, 316, 327, 425
scale-up 343, 350–354, 366
scenting of rooms 314
Scientific Committee of the Consumer Safety's (SCCS) recommendations 15, 161, 261, 400–401, 424–425, 430–431, 433, 434, 444, 449, 451
Scientific Committee on Cosmetic Products Intended for Consumers (SCCNFP) 331
Sebufix® 406
sebaceous glands/ducts 104, 111, 126, 204, 233 -235, 244, 248, 406
sebum 103, 114, 126, 183, 199, 203, 204, 207, 208, 233–235, 237, 243, 271, 287, 289, 403, 406
Sebumeter® 406, 407
sheabutter 207, 443
Shiseido 87, 89, 90
shunt pathway 112
Single Customs Union 264
skin
  dermis 104–105
  epidermis 105–107

skin (*contd.*)
  gender differences 126
  horny layer 107–108
  stratum basale 106
  stratum corneum 107
  stratum granulosum 107
  stratum lucidum 107
  structure 103–108
  subcutis 104
skin barrier 108, 109, 120, 121, 125, 146, 189, 193, 212, 238, 242, 243, 406, 414, 437, 441
skin care cream
  active ingredients
    from sea 223–225
    origin 225
  additives for color and fragrance 167
  anti-aging cream
    aloe vera 216
    α-Lipoic acid 213
    amino acid 213
    Coenzyme Q10 214
    2-dimethylaminoethanol 214
    enzymes 216
    grapes and raspberries 215
    green tea 215
    hyaluronic acid 214
    ingredients 212
    kinetin 215
    oxidative process 212
    phytohormones 214–215
    polyphenols 215–216
    retinol 213
    vitamin C 213
  cosmeceuticals 182–184
  cosmetic cream 176–182
  emulsifiers 140
    antioxidants, complexing agents and buffer substances 163–167
    consistency 155, 157–158
    macro emulsions 140–142, 146–149
    mini (nano) emulsions 149
    preservation 158–163
    stability 149, 152–154
  essential oil 216–218
  extracts from plants 219, 221, 223
  liposomes 167–170
  medicines 184, 187–188
  modules 132, 134–135, 138–140
  moisture 188
    moisturizing substances 191–194, 196
    natural moisturizing factor 188–191
  natural and actual condition 175
  natural cosmetics 176–182
  structure 131–132
  tasks 175
  vegetable oil 201
    in fats 205–208, 210–212
    natural fatty acids 203–205
  vitalizing substances 196, 198–201
skin care products 29–30, 35, 43–44, 54, 118–119, 131, 142, 148, 149, 153, 182, 208, 241, 359, 407
skin cleansing and care 3, 43
skin eczema 242
Skin Glossymeter 409
Skin-Visiometer® 408
SkinCeutical products 49
skinstratum spinosum 106, 107
smoothing cream 282, 283, 288
sodium citrate 167, 424, 438
sodium fluoride 11
  CosIng database 12
  US-database 13
sodium lactate 166, 192, 194
sodium phosphates 166
SoFiTix broad-spectrum powder 255
solar radiation 249–250, 265
sorbic acid 162
sorbitol 132, 135, 159, 161–163, 179, 192, 244, 246, 282, 293, 300, 401, 433, 448
sorbic acid (check) 132, 135, 159, 161–163, 171, 179, 244, 246, 282, 293, 300, 401, 433, 448
soybean 168, 194
special cosmetic creams 231
special cosmetics 17, 21, 271
specific energy input 341, 354, 356, 360
spinalioma 253
spiny cell layer 107

sprays 2, 44, 196, 248, 255, 268–269, 313, 314, 328, 329, 392, 418
squalene 117, 193–194, 203, 225, 269
squamous cell carcinoma 253
squeezing 225, 311, 361
stainless steel 337–338, 342–343, 348, 377, 379–380, 382, 383–392, 394
State Food and Drug Administration (SFDA) 264
steam distillation 216, 219, 311, 315–318, 332
sterilize-in-place (SIP) systems 374, 392–394
stirrer 338, 339, 342–344, 347–349, 352, 366
stratum basale 106–107
stratum corneum 106–114, 117, 121–123, 126, 188–189, 191, 201, 204–205, 210, 214, 255, 403, 405–406
stratum granulosum 107, 108
stratum lucidum 106, 107
stratum spinosum 107
subcutis 104–105, 114, 253
sun allergy 253
sun blocker 261
sun creams
  for reddened skin 236
sun erythema 252
sunburn 196, 236, 237, 250, 252, 265
sun protection factor (SPF) 6, 238, 261, 307
sunscreen cream
  formulation 307
sunscreens
  application and warning notices 261–262, 264–265
  classification and labeling 265
  radiation-induced damage
    black skin cancer 253
    elastosis 252
    heat collapse 253
    polymorphic light eruption 253
    spinalioma 253
    sunburn 252
  solar radiation 249
  substances 254–255, 259–260
  sun protection measurement 265–268
  UV rays character 250
sunscreens labeling 263, 265
surface area of application 444
surplus 59
sweet almond, oil 139, 207, 208, 293, 443
sweat glands 104, 111, 189, 237, 238
Symrise 312, 313, 332
Systemic Exposure Dose (SED) 420, 430–431, 434–436, 444, 448, 451

t
tanning cream 248–249
terpenes 109, 121, 124, 319, 321
tetrahydrocannabinol, (THC) 317
Tewameter® 405, 409
TEWL 109
thickeners 117, 122, 131, 132, 134, 135, 139, 140, 152, 155, 156, 158, 161, 163, 170, 171, 275, 293, 336
thick horny layers cream 246
thioglycolic acid 14
thioglycol salts 241
tibia 248
tinea pedis 243
Tinosorb® 260, 269
titanium dioxide 179, 238, 254, 255, 259, 261, 264, 267, 361
tocopherol 163, 225, 438
Tocopherolacetat 187
toxicological profile 419, 420, 424–426, 438, 449–450
trademark 9
transepidermal water loss (TEWL) 108–110, 188–190, 192–194, 199, 203, 205, 208, 215, 248, 405–406
transfersomes® 126, 170
triglycerides 115, 117, 125, 131, 141, 201, 203–205, 219, 269, 317
typical cosmetic products 3

u
ubihydroquinone 214
ubiquinol 214, 284
ultrasound 169, 355, 360, 389

ultraviolet (UV) 5, 134, 189, 231, 250
Unilever 45–47, 50–51, 54–55
unit operations 131, 225
United States
  economic importance 35
unwanted hair removal 241
urea 109, 123, 180, 188–193, 242–243, 247, 277, 282, 300, 301, 335, 429, 432–433
US Federal Food, Drug, and Cosmetic Act 15
UVA/UVB-filter 254, 259
UV rays character 250–252

V
vaseline 123, 187, 194, 293
vegetable oil 201
  almond oil 208, 210
  borage oil 211
  evening primrose oil 210
  hemp oil 210
  jojoba oil 207–208
  natural fatty acids 203
  safflower oil 211
  sheabutter 207
  wheat germ oil 211
vesicles 126, 169, 194
Visioline® 408
Visioscan® 406, 408
Viscosity 79, 118, 124, 125, 134, 142, 148, 155, 157, 159, 203, 240, 293, 306, 335, 338, 339, 349, 351, 352, 375, 392, 399, 402, 414, 432, 441
vitalizing 196–201, 247, 274, 277–281, 284, 456
vitalizing creams
  formulations 277, 456
vitalizing substances
  biotin 200
  composition 199
  dexpanthenol 199
  folic acid 199
  nicotinamide 199
  pyridoxine 199
  retinol 196, 198
  vitamin C 200–201
  vitamin E 201
vitamin A 153, 183, 196, 212, 224, 284
vitamin C 153, 163, 170, 198–200, 212, 213, 277
vitamins 125, 132, 139, 152, 163, 168, 176, 182–184, 196–201, 206, 210–211, 212–213, 216, 223, 239, 244, 247, 252, 269, 271, 277, 284, 288, 295, 300, 301, 335, 361, 420

W
warnings 7, 9, 11, 12, 15, 241, 261, 262, 264, 330, 413, 418, 419, 425, 438, 448, 450
water-soluble polymers 157–158, 163
waxes 118, 134, 155, 157, 179, 180, 182, 194, 239, 241, 245, 306, 318, 389
wheat germ oil 211
wheat hydrolyzate 194
whitehead 234
w/o-emulsions 138–139, 140–142, 146, 149, 152, 153, 187, 269, 275, 292, 307, 402, 407, 455
Wrinkles 1, 127, 183, 193, 198–201, 208, 212–214, 236, 252, 271, 408–409

X
xanthan 134, 158, 191, 194, 335

Z
zinc oxide 238, 239, 254, 255, 259, 261, 262, 264, 267, 269

www.ingramcontent.com/pod-product-compliance
Lightning Source LLC
Chambersburg PA
CBHW080733150926
51501CB00005B/162
* 9 7 8 3 5 2 7 3 4 3 9 8 0 *